Differential Equations

Differential Equations

Foundations and Applications

by

Hervé Reinhard

Professor at the Conservatoire National des Arts et Métiers

Translated by

A. Howie

Macmillan Publishing Company
A Division of Macmillan, Inc.
New York

English translation © 1987 North Oxford Academic Publishers Ltd

Original French language edition
(Équations différentielles: Fondements et applications)
© BORDAS 1982

Revised and updated 1987

English edition first published 1987
by North Oxford Academic Publishers Ltd,
a subsidiary of Kogan Page Ltd, 120 Pentonville Road,
London N1 9JN

Macmillan Publishing Company, 866 Third Avenue,
New York, NY10022
Collier Macmillan Canada Inc.

Library of Congress Catalog Card Number
86-62412

ISBN 0-02-948090-6

Printed and bound in Great Britain

Contents

Foreword

The aim of this work is above all educational. Its main purpose is to establish a link between 'major treatises' on the one hand and elementary books and those intended for users on the other. It seems to me, and to many of my colleagues, that the automation engineer, the physicist or the chemist, who would like to have a better understanding of the mathematical methods he uses, or to discover new ones, has not in general had the opportunity of studying the 'classics' intended for mathematicians and written in their language. In particular, it is difficult to recognize, in the form in which they present themselves, the problems in which the practical man is interested. For his part, the mathematician has everything to gain by realizing the practical significance of certain theoretical considerations.

This work is also intended for the student of mathematics. It will bring him into contact with many aspects of the theory and practice of differential equations.

It seems to me essential, in establishing this link, not to shrink from presenting mathematical concepts which truly correspond to the situation; I am sure that it is always profitable to proceed in this way. This is particularly true for differential equations: some of the most important problems which utilize them are related to the qualitative features of solutions (behaviour, stability, sensitivity to perturbations, existence of vibrations).

It seems necessary to show, for example, how to express in mathematical terms a sentence as significant and yet as imprecise as this: 'The solutions of two neighbouring equations resemble each other (or do not resemble each other).' Conversely, I have taken care to show clearly the practical significance of the mathematical concepts which occur in the course of the exposition, and I invite the reader continually to keep in mind the interplay between the theoretical and the practical aspects of the problems studied.

Having reviewed some very elementary results, I have introduced, as the subject demands, the minimum of mathematical concepts which will allow satisfactory interpretation of the topics under discussion. To allow a better understanding of these, I have made use of analogies and have taken advantage of physical or geometrical intuition while taking care to state exactly the limitations of these comparisons and the pitfalls to be avoided. I have attempted, as far as possible, to choose presentations which will enable the reader eventually to approach works of deeper significance without being completely overwhelmed.

For every important idea, I have stated and demonstrated several theorems as examples illustrating the essential point; I have thus been led to make very restrictive assumptions which reduce the practical scope of the conclusions. I have therefore indicated the more commonly used theorems either by outlining the proof or by omitting it altogether; I have indicated, however, whether the difficulty was only of a technical nature or of another kind. For readers more familiar with mathematics I have often indicated the general lines of proofs which are easy to construct.

I have illustrated the exposition with many examples and have included 250 exercises. The extent of the work and the level at which I have set it have imposed restrictions on me and I have had to be selective.

Since my aim was to produce a work which would be useful in many situations, I have added to the questions which are traditionally included in treatises on differential equations some elements which are closely linked to them and which do not as a rule appear in the same books.

In particular, the reader will find some elements of control theory for finite and infinite horizons, an introduction to the theory of stability, and a brief study of linear operators with applications to Green's functions, integral equations, the Sturm–Liouville problem and, through the use of special functions, partial differential equations. One section gives results relating to asymptotic expansions. An important chapter is devoted to periodic problems and particularly to different methods of deriving approximations to periodic solutions. A list of the principal points covered is given in the table of contents.

The chapters are divided into parts and sections. Each chapter and each part starts with an introduction which indicates the concepts with which it deals both from a mathematical and from a practical point of view. Details of these sections also appear in the table of contents.

The first two chapters are indispensable for the understanding of the following chapters; the other chapters, however, may be studied separately, the few references necessary being carefully indicated in the following way: in each chapter the propositions are numbered within the section and part. Thus a reference to proposition 2.1.3 is a reference within the chapter to the third proposition (3) in the first section (1) of the second part (2). If it refers to another chapter, this is explicitly stated.

With regard to the bibliography, I have not given many titles. The major treatises to which I have referred are those of Coddington and Levinson, Hale, and Hartmann, and, to a lesser degree, that of Yosida. Cesari's book, quoted as a reference, covers many aspects of the study of differential equations and ends with a bibliography of more than 2000 entries. On the subject of chapters 3 and 4, the reader may find the books of Lee and Markus, and of Rouche, Habets and Laloy to be useful reference works. Lastly, we must mention the two remarkable works of Arnold. I hope that the present book will enable many more people to read these texts and others which are particularly related to the subject.

I should like to thank my colleagues who have encouraged me to write this book and who have read over certain sections of it. I am particularly indebted to Michel Demazure, Professor at the Polytechnique, for the guidance and advice which he has given me.

I shall be happy to receive any comments and corrections which the reader would care to send me.

Hervé Reinhard
Paris, April 1982

I am pleased to thank the translator of this edition: his work and that of the technical checker have been outstanding. I should also like to acknowledge the quality of the OMEGA Publishing Services team charged with practical realization of the book, and, on a broader front, the publisher.

March 1986

Notation

$\mathbb{R}$ denotes the set of real numbers, $]a, b[= \{x : a < x < b\}$ whereas $[a, b] = \{x : a \leqslant x \leqslant b\}$. $\mathbb{R}^d = \{x = x_1, \ldots, x_d\}$ where $x_i \in \mathbb{R} \;\; \forall i$.

$\mathbb{C}$ denotes the set of complex numbers, and $\text{Re}\, z$ and $\text{Im}\, z$ the real part and the imaginary part of a complex number z.

If E is a normed vector space, $O \in E$ is an open set if $\forall x \in O$ $\exists \varepsilon : \{y, \|x - y\| < \varepsilon\} \subset O$. F is a closed set if every convergent sequence in E, of elements $\{x_n\}_{n \in \mathbb{N}}$ of F, has as limit an element of F.

If M is a matrix, $\tilde{M}$ denotes the transposed matrix. If V is a vector, $\tilde{V}$ is the transposed vector.

Index of notation

Chapter 1

Generalities

Part 1: Introduction

1.1. Definitions

Definition 1.1.1

Let $F: D \leadsto \mathbb{R}$ be a mapping defined over a domain D of the space $\mathbb{R}^{n+2}$ with values in $\mathbb{R}$. We define a **scalar differential equation of order n** as

$$F\left(t, x, \frac{\mathrm{d}x}{\mathrm{d}t}, \ldots, \frac{\mathrm{d}^n x}{\mathrm{d}t^n}\right) = 0 \tag{1}$$

A **solution** of this equation is a function

$$\varphi: \begin{array}{c} I \to \mathbb{R} \\ t \leadsto \varphi(t) \end{array}$$

defined and n-times differentiable over an interval I (bounded or not) of $\mathbb{R}$ and such that

$$\left[t, \varphi(t), \frac{\mathrm{d}\varphi}{\mathrm{d}t}, \ldots, \frac{\mathrm{d}^n\varphi}{\mathrm{d}t^n}\right] \in D \qquad \forall t \in I \tag{2a}$$

$$\forall t \in I \qquad F\left[t, \varphi(t), \frac{\mathrm{d}\varphi}{\mathrm{d}t}, \ldots, \frac{\mathrm{d}^n\varphi}{\mathrm{d}t^n}\right] = 0 \tag{2b}$$

If I is bounded, say $I = [A, B]$, we require that the relation (2b) is satisfied at the point A by right-derivatives and at the point B by left-derivatives.

We say that the equation is in **normal form** (or solved for the derivative of highest order) if it can be written in the form

$$\frac{\mathrm{d}^n x}{\mathrm{d}t^n} = f\left(t, x, \frac{\mathrm{d}x}{\mathrm{d}t}, \ldots, \frac{\mathrm{d}^{n-1}x}{\mathrm{d}t^{n-1}}\right)$$

Definition 1.1.2

Let $F_1, \ldots, F_d$ be d functions $F_i: D_i \to \mathbb{R}$ defined over domains D_i of $\mathbb{R}^{1+(p+1)d}$; we define a **differential system of order p** as a system of the form

$$F_1\left(t; x_1, \ldots, x_d; \frac{dx_1}{dt}, \ldots, \frac{dx_d}{dt}; \ldots; \frac{d^p x_1}{dt^p}; \ldots; \frac{d^p x_1}{dt^p}, \ldots, \frac{d^p x_d}{dt^p}\right) = 0$$

$$F_2\left(t; \qquad\qquad \ldots \qquad\qquad \frac{d^p x_d}{dt^p}\right) = 0 \quad (3)$$

$$\vdots$$

$$F_d\left(t; x_1, \ldots, x_d; \qquad \ldots \qquad \frac{d^p x_d}{dt^p}\right) = 0$$

A solution of (3) is a set of d functions

$$\varphi_1 \ldots \varphi_d: \begin{array}{c} I \to \mathbb{R} \\ t \rightsquigarrow \varphi_i(t) \end{array}$$

defined over an interval I (bounded or not) of $\mathbb{R}$ and such that

$$\forall i \quad \forall t \in I \quad \left[t; \varphi_1(t), \ldots, \varphi_d(t); \ldots, \frac{d^p \varphi_d}{dt^p}(t)\right] \in D_i$$

$$\forall i \quad \forall t \in I \quad F_i\left[t; \varphi_1(t), \ldots, \frac{d^p \varphi_d}{dt^p}(t)\right] = 0$$

If I is bounded, we use the same convention as before.

We note that the family $(F_i)_{i=1,\ldots,d}$ actually defines a mapping F from $\mathbb{R}^{pd+d+1}$ to $\mathbb{R}^d$ and that a solution, in fact, defines over I a vector $\varphi(t)$ of components $\varphi_i(t)_{i=1,\ldots,d}$, so that (3) can be written more simply

$$F\left(t, \varphi, \frac{d\varphi}{dt}, \ldots, \frac{d^p \varphi}{dt^p}\right) = 0 \quad \text{where } \varphi: I \to \mathbb{R}^d \quad (3')$$

Provided that we consider vectors we can thus content ourselves with the study of a single equation; we shall see that it will suffice even to study an equation of the first order.

Proposition 1.1.3

Every scalar differential equation of order n is equivalent to a system of the first order and therefore to a differential equation of the first order with functions having values in $\mathbb{R}^n$.

Proof For the equation

$$F\left(t, x, \frac{dx}{dt}, \ldots, \frac{d^n x}{dt^n}\right) = 0 \quad (4)$$

let

$$x_1 = x, \qquad x_2 = \frac{dx}{dt}, \ldots, \qquad x_n = \frac{d^{n-1} x}{dt^{n-1}}$$

given the system

$$\left.\begin{array}{c} \dfrac{dx_1}{dt} = x_2, \ \ldots, \ \dfrac{dx_{n-1}}{dt} = x_n \\[3mm] F\left(t, \ x_1, \ x_2, \ \ldots, \ x_n, \ \dfrac{dx_n}{dt}\right) = 0 \end{array}\right\} \tag{5}$$

If $\varphi(t)$ is a solution of (4), $\varphi: I \to \mathbb{R}$, φ is differentiable n times and $[t, \varphi(t), \ldots, d^n\varphi(t)/dt^n] \in D \quad \forall t \in I$. If we put $\varphi_1(t) = \varphi(t)$, $\varphi_2(t) = d\varphi/dt, \ldots$, $\varphi_n(t) = d\varphi_{n-1}/dt$, the vector $(\varphi_1, \ldots, \varphi_n)$ is clearly a solution of (5); conversely if $(\psi_1, \ldots, \psi_n)$ is a solution of (5) and if we put $\psi(t) = \psi_1(t)$ then $\psi(t)$ is a solution of (4).

We are interested primarily in equations in normal form and therefore we shall study the equations $dx/dt = F(t, x)$ where $x \in \mathbb{R}^p$.

Let

$$\frac{dx}{dt} = F(t, x) \tag{6}$$

be a differential equation in which $x \in \mathbb{R}$ and F is continuous; if φ is a solution of (6), i.e. $d\varphi/dt = F[t, \varphi(t)]$, then by integrating from t_0 to t we obtain

$$\varphi(t) = \varphi(t_0) + \int_{t_0}^{t} F[u, \varphi(u)] \, du \tag{7}$$

Every continuous solution of (7) is differentiable and is then a solution of (6). The form (7) is called the **integral form** of (6); if x takes values in $\mathbb{R}^d$ we write the same relation by integrating vectors.

From now on we shall be considering equations $dx/dt = F(t, x)$ in which $x \in \mathbb{R}^d$.

Definition 1.1.4

Let $dx/dt = F(t, x)$ be a differential equation in which $x: \mathbb{R} \to \mathbb{R}^d$, $F: \mathbb{R} \times \mathbb{R}^d \to \mathbb{R}^d$, and let φ be a solution defined over an interval I, i.e. a function $I \to \mathbb{R}^d$ such that $d\varphi/dt = F[t, \varphi(t)] \ \forall t \in I$.

We define an **integral curve** as the set of points $[t, \varphi(t)]$ in which t covers I; this is a set of points of $\mathbb{R}^{d+1}$.

An **orbit** is the set of points $[\varphi(t)]$ in which t covers I; this is a set of points of $\mathbb{R}^d$. The space $\mathbb{R}^d$ (in which the solutions take their values) is called the **phase space**.

1.2. Geometrical interpretation

Suppose that in $\mathbb{R}^3$ (as an example) Γ is an integral curve and M is a point of this curve with coordinates $x = \varphi_1(t)$, $y = \varphi_2(t)$, $z = t$ (see p. 22); the 'tangent vector' to Γ at M has components $d\varphi_1/dt$, $d\varphi_2/dt$ and 1, i.e. $F_1[t, \varphi(t)]$, $F_2[t, \varphi(t)]$ and 1 (where we denote by F_1 and F_2 the coordinates of F).

For such an equation the phase space is $\mathbb{R}^2$, an orbit has equation $x = \varphi_1(t)$, $y = \varphi_2(t)$ and the tangent vector at a point a has components $F_1[t, \varphi(t)]$ and $F_2[t, \varphi(t)]$.

Example   Let

$$\frac{d^2x}{dt^2} + x = 0 \quad \text{where } x \in \mathbb{R} \tag{8}$$

Putting $x_1 = x$, $x_2 = dx/dt$ and

$$X = \begin{pmatrix} x_1 \\ x_2 \end{pmatrix}$$

Eqn (8) is equivalent to the equation

$$\left.\begin{array}{l} \dfrac{dx_1}{dt} = x_2 \\[2mm] \dfrac{dx_2}{dt} = -x_1 \end{array}\right\} \quad \text{where} \quad \frac{dX}{dt} = F(X) \quad \text{with} \quad F(X) = \begin{bmatrix} F_1(X) \\ F_2(X) \end{bmatrix} = \begin{pmatrix} x_2 \\ -x_1 \end{pmatrix} \tag{9}$$

We know that the solutions of (8) are $x(t) = A\cos(t + \vartheta)$. The integral curves are curves in $\mathbb{R}^3$ and the orbits are curves in $\mathbb{R}^2$ which are projections of the integral curves (Fig. 1.1). It often happens that we can determine the orbits without being able to determine the integral curves exactly.

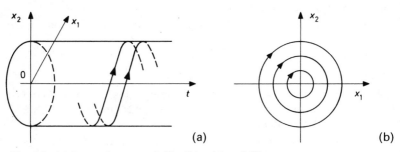

Fig. 1.1. (a) *Integral curves of* (8); (b) *orbits of* (8).

In many situations (but not all) t represents time and the orbits are trajectories which we can determine without knowing the speed at which they are traversed. We are often interested in cases in which F does not depend on t.

Definition 1.2.1

We speak of a **vector field** in a domain D of $\mathbb{R}^d$ when we are given, at every point x of D, a vector $F(x)$.

The equation $dx/dt = F(x)$ is called the **autonomous equation** (or **autonomous system** or again **dynamic system**) associated with this field.

A point a of D such that $F(a) = 0$ is called an **equilibrium point** (or **fixed point** or **critical point**). $\varphi(t) \equiv a \; \forall t$ is then a solution of the autonomous equation. A point a at which $F(a) \neq 0$ is called a **regular point**.

Example Let

$$\frac{d^2 x}{dt^2} + x = 0 \tag{10}$$

This equation is an autonomous equation whose vector field is given at every point

$$x = \begin{pmatrix} x_1 \\ x_2 \end{pmatrix}$$

of $\mathbb{R}^2$ by

$$F(x) = \begin{pmatrix} x_2 \\ -x_1 \end{pmatrix}$$

The point

$$x = \begin{pmatrix} 0 \\ 0 \end{pmatrix} = 0$$

the origin of coordinates in $\mathbb{R}^2$, is an equilibrium point of the system.

By simply drawing the vector field we can often get an idea of the orbits. The vector field associated with Eqn (10) is shown in Fig. 1.2.

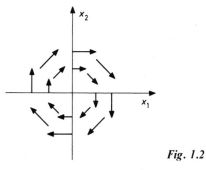

Fig. 1.2

Definition 1.2.2 (Cauchy problem)
Let

$$\frac{dx}{dt} = F(t, x) \tag{11}$$

be a differential equation in $\mathbb{R}^d$. Let $t_0 \in \mathbb{R}$ and $x_0 \in \mathbb{R}^d$; we say that φ is a solution of the Cauchy problem relative to (t_0, x_0) if φ is a solution of (11) such that $\varphi(t_0) = x_0$.

A solution of the Cauchy problem is an integral curve passing through the point (t_0, x_0); its orbit is thus a curve of $\mathbb{R}^d$ passing through x_0.

We shall demonstrate in this chapter that in many cases there is existence and uniqueness of solutions to the Cauchy problem relative to (t_0, x_0), i.e. there exists an interval I such that $t_0 \in I$ (often $I = \mathbb{R}$) and a unique solution such that $\varphi(t_0) = x_0$.

Notation 1.2.3

We shall denote by $\varphi(t, t_0, x_0)$ the solution to the Cauchy problem relative to (t_0, x_0) when the latter is unique.

This involves an abuse of language since this solution ought to be written $t \rightsquigarrow y(t, t_0, x_0)$. However, this abuse does not normally lead to confusion.

Proposition 1.2.4

When there is existence and uniqueness for the Cauchy problem at every point (t_0, x_0) there is one orbit, and one only, passing through each point of the phase space.

It must be noted that the solution relative to (t_0, x_0) is defined for $t \in I$, where t is greater than or less than t_0, so that the case shown in Fig. 1.3 cannot occur if there is uniqueness. The preceding proposition is then evident.

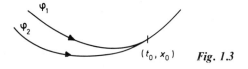

(t_0, x_0) *Fig. 1.3*

We shall return to the very important problem of non-uniqueness for the Cauchy problem, but we already notice that if the equation is not in normal form and if at one and the same point (t_0, x_0) there exist several values v_0 such that $F(t_0, x_0, v_0) = 0$ the equation $F(t, x, \mathrm{d}x/\mathrm{d}t)$ will have several solutions passing through (t_0, x_0).

Example $(\mathrm{d}x/\mathrm{d}t)^2 = 1$, $x \in \mathbb{R}$. For every point (t_0, x_0) there is an infinity of solutions.

See *Exercises 1 to 5* at the end of the chapter.

Part 2: Functions from $\mathbb{R}^p$ into $\mathbb{R}^q$

We have seen that every normal equation leads to an equation of the form $dx/dt = F(t, x)$ where $x \in \mathbb{R}^d$ and $F: \mathbb{R}^{d+1} \to \mathbb{R}^d$.

We denote by $\mathbb{R}^d$ the vector space of points $x = (x_1, \ldots, x_d)$ with the usual operations of addition and multiplication by a scalar. The dimension of this space is d. In order to discuss limits we are led to define the notion of distance; we shall deduce this from the notion of norm.

Before discussing neighbouring differential equations, we need to introduce the notion of proximity for functions F and G from $\mathbb{R}^p$ into $\mathbb{R}^q$.

2.1. Norms

NORMS IN A VECTOR SPACE (CONSTRUCTED OVER $\mathbb{R}$ or $\mathbb{C}$)

Definition 2.1.1

A **norm** over a vector space E is a mapping from E into $\mathbb{R}^+$ such that if we denote by $\|V\|$ the norm of V

$$\|V\| = 0 \Leftrightarrow V = 0 \tag{12a}$$

$$\|V + W\| \leq \|V\| + \|W\| \qquad \forall V, W \in E \text{ (triangular inequality)}* \tag{12b}$$

$$\|\lambda V\| = |\lambda| \, \|V\| \qquad \forall V \in E, \forall \lambda \in \mathbb{R} \text{ or } \mathbb{C} \tag{12c}$$

We define the **distance associated with the norm** $\|\cdot\|$ as the quantity $d(V, W) = \|V - W\|$.

Remark   In the relation (12a) it would be necessary, for complete rigour, to distinguish the zero element of the vector space and the element 0 of $\mathbb{R}^+$.

Example   Let $x = (x_1, \ldots, x_d)$ be an element of $\mathbb{R}^d$; the Euclidean norm of x is defined by

$$\|x\| = \left(\sum_{i=1}^d x_i^2 \right)^{1/2}$$

The associated distance is the ordinary distance.

With every norm we can associate the notion of limit and hence of differentiation and of integration.

---

* It can be deduced in fact from this that

$$|\|U\| - \|V\|| \leq \|U - V\| \leq \|U\| + \|V\|$$

Definition 2.1.2
We say that

$$\lim_{t \to t_0} V(t) = V_0$$

if $\forall \varepsilon > 0$

$$\exists \eta : |t - t_0| \leqslant \eta \Rightarrow \|V(t) - V_0\| \leqslant \varepsilon$$

We say that

$$\lim_{t \to \infty} V(t) = V_0$$

if $\forall \varepsilon > 0$

$$\exists T : t \geqslant T \Rightarrow \|V(t) - V_0\| \leqslant \varepsilon$$

From this we can deduce ideas of continuity.

A priori the notion of convergence which we have just defined depends on the norm chosen; it could be possible that for one choice of norm $\lim V(t) = V_0$ but that this would not be true for another choice. As it happens, this never occurs in $\mathbb{R}^d$.

Definition 2.1.3
Two norms denoted by $\|\cdot\|$ and $\|\|\cdot\|\|$ are **equivalent** over a vector space E if there exist two strictly positive numbers A and B such that

$$\forall V \in E \qquad A\|V\| \leqslant \|\|V\|\| \leqslant B\|V\|$$

It is clear that in this case $\lim V(t) = V_0$ for the norm $\|\cdot\|$ iff $\lim V(t) = V_0$ for the norm $\|\|\cdot\|\|$.

Proposition 2.1.4
All norms over the space $\mathbb{R}^d$ are equivalent.

This proposition will be demonstrated in an exercise (see p. 55).
Since all norms are equivalent, the following property which is true for the Euclidean norm is true for all norms.

Property 2.1.5

$$\lim_{t \to t_0} V(t) = V \Leftrightarrow \forall i \in (1, \ldots, d) \qquad \lim V_i(t) = V_i$$

where $V = (V_1, \ldots, V_d)$.

We know that the $d \times d$ matrices form a vector space of dimension d^2; in the same way, we have

$$\lim_{t \to t_0} A(t) = A \Leftrightarrow \lim_{t \to t_0} A_{ij}(t) = A_{ij} \qquad \forall i, j$$

Thus, whatever the norm, a vector, and hence a matrix, is continuous iff its components are continuous.

Compatible norms of matrices and of vectors

Definition 2.1.6

Let $\|\cdot\|$ be a norm over $\mathbb{R}^d$ and $\|\|\cdot\|\|$ be a norm over the vector space of matrices; then these norms are compatible if $\forall x \in \mathbb{R}^d$ and $\forall\ d \times d$ matrices A

$$\|Ax\| \leqslant \|\|A\|\| \cdot \|x\|$$

and if

$$\|\|A \cdot B\|\| \leqslant \|\|A\|\| \cdot \|\|B\|\| \qquad \forall\ d \times d \text{ matrices } A, B$$

We shall always use compatible norms. We shall verify in Exercise 9 (p. 54) that, if $\|\cdot\|$ is a norm over $\mathbb{R}^d$, the quantity

$$\sup_{x:\ \|x\|=1} \|Ax\|$$

is a matrix norm compatible with the norm chosen over $\mathbb{R}^d$.

See *Exercises 6 to 12* at the end of the chapter.

DIFFERENTIATION AND INTEGRATION OF FUNCTIONS FROM $\mathbb{R}$ INTO $\mathbb{R}^q$

Definition 2.1.7

Let V be a vector whose components V_i are differentiable; we denote by dV/dt the vector of components dV_i/dt.

Similarly if the elements A_{ij} of a matrix are differentiable we denote by dA/dt the matrix of components dA_{ij}/dt.

If the components of V are continuous $\int V(s)\,ds$ denotes the vector of components $\int V_i(s)\,ds$. We define $\int A(s)\,ds$ similarly.

Proposition 2.1.8

For every norm

$$\left\| \int_a^b V(s)\,ds \right\| \leqslant \int_a^b \|V(s)\|\,ds$$

and similarly

$$\left\|\left\| \int_a^b A(u)\,du \right\|\right\| \leqslant \int_a^b \|\|A(u)\|\|\,du$$

2.2. Tangent and differential mappings

We know that a **linear mapping** of $\mathbb{R}^p$ into $\mathbb{R}^q$ is defined as a mapping A such that $\forall\,\lambda,\mu\in\mathbb{R}$ and $\forall\,x, y\in\mathbb{R}^p$ $A(\lambda x + \mu y) = \lambda A(x) + \mu A(y)$. In a coordinate system a linear mapping is represented by a matrix A such that $A(x) = A\cdot x$; a mapping $x\to Ax + b$ where $b\in\mathbb{R}^q$ is called an **affine mapping**.

We shall denote by $\mathcal{L}(\mathbb{R}^p, \mathbb{R}^q)$ the vector space of linear mappings from $\mathbb{R}^p$ into $\mathbb{R}^q$.

DERIVATIVE

Let $x\rightsquigarrow f_1(x)$ and $x\rightsquigarrow f_2(x)$ be two mappings from $\mathbb{R}^p$ into $\mathbb{R}^q$. To measure their distance apart in the neighbourhood of a point a we use the expression

$$m_a(r) = \sup_{x:\,\|x-a\|\leqslant r} \|f_1(x) - f_2(x)\|$$

Definition 2.2.1

We say that f_1 and f_2 are tangential at a if

$$\lim_{r\to 0} \frac{m_a(r)}{r} = 0$$

This naturally implies that $f_1(a) = f_2(a)$, and this does not depend on the norm chosen. We can easily verify that it is an equivalence relation, i.e. that

(a) f_1 is tangential to f_1,
(b) f_1 is tangential to $f_2 \Leftrightarrow f_2$ is tangential to f_1, and
(c) if f_1 is tangential to f_2 and f_2 is tangential to f_3, then f_1 is tangential to f_3.

Example **Tangent to a curve of $\mathbb{R}^2$, $y = f(x)$:** let Γ be the graph of the equation $y = f(x)$ such that $f'(x)$ exists in the neighbourhood of a point a (Fig. 1.4). We denote by Δ a straight line passing through the point A with coordinates $(a, f(a))$.

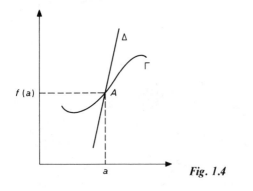

Fig. 1.4

Let us compare Δ and Γ in the neighbourhood of A; in this case

$$m_a(r) = \sup_{|x-a| \leqslant r} |f(x) - f(a) - k(x-a)|$$

Now $f(x) - f(a) = (x-a)[f'(a) + \eta(x)]$ where η is a function such that

$$\lim_{x \to a} \eta(x) = 0$$

Thus

$$m_a(r) = \sup_{|x-a| \leqslant r} [|x-a| \, |f'(a) - k + \eta(x)|]$$

If $f'(a) \neq k$, $m_a(r) \geqslant r[f'(a) - k + \eta(r)]$ and $m_a(r)/r$ cannot have zero for limit: the straight line is not a tangent to the curve.

If $f'(a) = k$

$$m_a(r) = \sup_{|x-a| \leqslant r} [|x-a| \, |\eta(x)|]$$

Now by hypothesis $\forall \varepsilon > 0 \; \exists r_0 : \forall x \leqslant r_0 \; \eta(x) \leqslant \varepsilon$. Thus, if $r \leqslant r_0$, $m_a(r) \leqslant r\varepsilon$, so that $[m_a(r)/r] \leqslant \varepsilon$. It follows that Δ and Γ are tangential.

With the definition that we have adopted, a straight line is tangential to a curve differentiable at a point a iff the slope of the line is equal to the derivative at a.

Proposition 2.2.2

Two affine mappings which are tangential are identical.

Proof Let

$$f_1 : x \rightsquigarrow Ax + b \quad \text{where } b \in \mathbb{R}^q, \; A \in \mathscr{L}(\mathbb{R}^p, \mathbb{R}^q)$$

$$f_2 : x \rightsquigarrow Bx + c \quad \text{where } c \in \mathbb{R}^q, \; B \in \mathscr{L}(\mathbb{R}^p, \mathbb{R}^q)$$

If f_1 and f_2 are tangential at a, $Aa + b = Ba + c$ and

$$m_a(r) = \sup_{\|x-a\| \leqslant r} \|(A - B)(x-a)\|$$

If $A \neq B$, $A - B \neq 0$; i.e. there exists $V \in \mathbb{R}^p$ such that $(A-B)V \neq 0$. With every r let us associate

$$x = a + \frac{rV}{\|V\|}$$

It is clear that $\|x - a\| \leqslant r$ and that

$$(A - B)(x - a) = \frac{r}{\|V\|}(A - B)V$$

Then

$$\forall r \quad \frac{m_a(r)}{r} \geq \frac{1}{\|V\|} \|(A-B)V\| > 0$$

so that $m_a(r)/r$ cannot have zero for limit. Consequently, if f_1 and f_2 are tangential at a, $A = B$ and $f_1 \equiv f_2$.

Corollary 2.2.3
There exists at most one affine mapping tangential at a point to a given mapping.

Definition 2.2.4
Let f be a mapping from $\mathbb{R}^p$ into $\mathbb{R}^q$. We say that f is differentiable if there exists an affine mapping tangential to f at a or equally well if there exists a linear mapping denoted $Df(a)$ such that

$$\lim_{x \to 0} \frac{1}{\|x\|} [f(x+a) - f(a) - Df(a)x] = 0 \qquad Df(a) \in \mathscr{L}(\mathbb{R}^p, \mathbb{R}^q)$$

If we define $\varepsilon(x)$ by the relation $f(x+a) - f(x) - Df(a)x = \|x\|\varepsilon(x)$ the preceding condition states that

$$\lim_{x \to 0} \varepsilon(x) = 0$$

The linear mapping $Df(a)$ is called the **derivative of f at the point a**. If $a \to Df(a)$ is continuous, we say that f is of class C^1; we define similarly the class C^r of mappings r times continuously differentiable.

Definition 2.2.4 also contains the assertion of an equivalence which we shall prove.

Let g be the affine mapping tangential to f at $a: g(y) = Ay + b$. Writing $A = Df(a)$ and taking account of the fact that $g(a) = f(a)$ we have

$$g(y) = Df(a)[y - a + f(a)]$$

and

$$m_a(r) = \sup_{\|y-a\| \leq r} \|f(y) - f(a) - Df(a)(y-a)\|$$

or, putting $y - a = x$,

$$m_a(r) = \sup_{\|x\| \leq r} \|f(x+a) - f(a) - Df(a)x\|$$

By definition f is differentiable and $Df(a)$ is its derivative iff

$$\lim_{r \to 0} \frac{m_a(r)}{r} = 0$$

Suppose this is the case; then

$$\forall \, \varepsilon > 0 \qquad \exists \, r_0 : \forall \, r \leqslant r_0 \qquad \frac{m_a(r)}{r} \leqslant \varepsilon$$

Therefore

$$\forall \, x : \|x\| \leqslant r \leqslant r_0 \qquad \|f(x+a) - f(a) - Df(a)x\| \leqslant \varepsilon r \leqslant \varepsilon r_0$$

In particular if $\|x\| = r$, $\|f(x+a) - f(a) - Df(a)x\| \leqslant \varepsilon \|x\|$, which establishes the relation given in Definition 2.2.4. Let us suppose now that this relation is verified and imagine that $m_a(r)/r$ does not have zero as its limit. Thus we suppose that there exists $\varepsilon : \forall \, \eta \, \exists \, r_0 < \eta$ such that $m_a(r_0) > r_0 \varepsilon$, i.e. that there exists $x : \|x\| \leqslant r_0 < \eta$ such that

$$\|f(x+a) - f(a) - Df(a)x\| > \varepsilon r_0 \geqslant \varepsilon \|x\|$$

This is the negation of the relation appearing in Definition 2.2.4.

The existence of the derivative of F is related to the existence of the partial derivatives of the components of F; moreover DF can be expressed as a function of these partial derivatives. The following theorem is fundamental.

Theorem 2.2.5

Let F be a mapping $\mathbb{R}^p \to \mathbb{R}^q$ with coordinates $F_i(x_1, \ldots, x_p), i = 1, \ldots, q$. If F is differentiable at the point a the partial derivatives $(\partial F_i / \partial x_j)(a)$ exist. Conversely, if $\partial F_i / \partial x_j$ exists and is continuous in the neighbourhood of a, $\forall \, i$ and j, then F is differentiable. DF is then represented by the matrix

$$DF(a) = \left\{ \frac{\partial F_i}{\partial x_j} \right\}_{i,j} (a) \qquad \text{denoted also} \qquad \frac{D(F_1, \ldots, F_q)}{D(x_1, \ldots, x_p)} (a)$$

which is called the **Jacobian matrix**. If the partial derivatives of order r are continuous, f is of class C^r.

Proof Let $e_1, \ldots, e_p$ be a basis of $\mathbb{R}^p$ and let $x_1, \ldots, x_p$ be the corresponding coordinates. Suppose F is differentiable at a and let $A = Df(a)$ so that

$$\lim_{x \to 0} \frac{1}{\|x\|} [F(x+a) - F(a) - Ax] = 0$$

in particular if $x = (0, \ldots, 0, h, \ldots, 0)$ where $x_i = h$. Then

$$\lim_{h \to 0} \frac{1}{h} |F_j(x+a) - F_j(a) - A_{ji}h| = 0 \quad \text{for all } j$$

which shows that $(\partial F_j / \partial x_i)(a)$ exists and is equal to A_{ij}. This establishes the result.

Suppose now that $\partial F_j / \partial x_i$ exists for all i and j in a neighbourhood of the point $a = (a_1, \ldots, a_p)$ and let us calculate $F_j(x+a) - F_j(a)$ varying the coordinates one at a time. We then obtain the formula

$$F_j(x+a) - F_j(a) = F_j(x_1 + a_1, \ldots, x_p + a_p) - F_j(a_1, x_2 + a_2, \ldots, x_p + a_p)$$

$$+ F_j(a_1, x_2 + a_2, \ldots, x_p + a_p) - F_j(a_1, a_2, x_2 + a_3, \ldots, x_p + a_p)$$

$$\vdots$$

$$+ F_j(a_1, a_2, \ldots, a_{p-1}, x_p + a_p) - F_j(a_1, a_2, \ldots, a_p)$$

By definition of the derivative, each of the terms can be written in the form

$$\frac{\partial F_j}{\partial x_k}(a_1, a_2, \ldots, a_k, x_{k+1} + a_{k+1}, \ldots, x_p + a_p)x_k + x_k \varepsilon_k{}^j(x_k)$$

where

$$\lim_{x_k \to 0} \varepsilon_k{}^j(x_k) = 0$$

We have assumed that the partial derivatives are continuous; consequently

$$\frac{\partial F_j}{\partial x_k}(a_1, a_2, \ldots, a_k, x_{k+1} + a_{k+1}, \ldots, x_p + a_p) - \frac{\partial F_j}{\partial x_k}(a_1, \ldots, a_p) = \delta_k{}^j(x)$$

and

$$\lim \delta_k{}^j(x) = 0$$

Then

$$F_j(x+a) - F_j(a) - \sum_{k=1}^{p} \frac{\partial F_j}{\partial x_k}(a) x_k = \sum_{k=1}^{p} [\varepsilon_k{}^j(x) + \delta_k{}^j(x)]x_k$$

Let

$$A = \left(\frac{\partial F_j}{\partial x_k}\right)_{j,k}$$

and let $\|V\| = |V_1| + \ldots + |V_q|$. For this norm

$$\|F(x+a) - F(a) - Ax\| = \sum_{j=1}^{q} \left| F_j(x+a) - F_j(a) - \sum_{k=1}^{p} \frac{\partial F_j}{\partial x_k}(a) x_k \right| = \varepsilon(x)\|x\|$$

where

$$\lim_{x \to 0} \varepsilon(x) = 0$$

which establishes the differentiability of F and identifies $Df(a)$ and A.

PARTICULAR CASE: FUNCTIONS FROM $\mathbb{R}^n$ INTO $\mathbb{R}$ (SURFACES)

In this case $Df \in \mathscr{L}(\mathbb{R}^n, \mathbb{R})$. Such a mapping can in fact be identified with a vector (its representation in a coordinate system is a row matrix) and we can write $Df \cdot V = a_1 V_1 + \ldots + a_n V_n = \langle A, V \rangle$, the scalar product of the vector A with components $a_1, \ldots, a_n$ and the vector V. This allows us to recover the standard formula for scalar functions of several variables differentiable at the point a:

$$F(x_1 + a_1, x_2 + a_2, \ldots, x_n + a_n) - F(a_n, \ldots, a_n)$$

$$= \frac{\partial F}{\partial x_1}(a) x_1 + \ldots + \frac{\partial F}{\partial x_p}(a) x_p + \varepsilon(x)$$

Definition 2.2.6

Let $f \colon \mathbb{R}^n \to \mathbb{R}$ possess continuous partial derivatives of the second order; then the **gradient** of f is the vector with components $\partial f / \partial x_i$, and the **Hessian matrix** is the matrix $H(f)$ with elements $\partial^2 f / \partial x_i \, \partial x_j$. If $V \in \mathbb{R}^n$

$$Df(a) \cdot V = \langle \operatorname{grad} f(a), V \rangle$$
$$D^2 f(a) \cdot V = H(f)(a) \cdot V$$

Finally if $t \in \mathbb{R}$, $V \in \mathbb{R}^p$, we can express $f(a + tV) - f(a)$ in the form

$$f(a + tV) - f(a) = t \langle \operatorname{grad} f(a), V \rangle + \frac{t^2}{2!} \langle H(f)(a) \cdot V, V \rangle + t^2 \varepsilon(t)$$

where

$$\lim_{t \to 0} \varepsilon(t) = 0$$

The quantity $\langle \operatorname{grad} f(a), V \rangle$ is then called the **derivative of f in the direction V** taken at the point a (this is the derivative of $t \rightsquigarrow f(a + tV)$).

Remark The first formula in Definition 2.2.6 is the definition of the gradient. $Df(x)$, being identified with $\operatorname{grad} f(x)$, is a differentiable mapping $\mathbb{R}^n \to \mathbb{R}^n$; since $\partial f / \partial x_i$ is assumed to be differentiable and even, in an indirect way, continuous, its derivative exists and is then an element of $\mathscr{L}(\mathbb{R}^n, \mathbb{R}^n)$ represented by the Hessian matrix: this is the second formula of Definition 2.2.6.

The last formula is a Taylor formula which is obtained in the same way as the standard formula. It gives the variation in the function f when one moves from the point a in the direction of the vector V; in particular it establishes that

$$\lim_{t \to 0} \frac{1}{t} |f(a + tV) - f(a) - t \langle \operatorname{grad} f(a), V \rangle| = 0$$

The Taylor formula, as in the case of functions of real variables, allows us to state conditions for f to have an extremum. Before expressing these conditions we note that the Hessian matrix is symmetric and that the expression $\langle H(f) \cdot x, x \rangle$ where $x = (x_1, \ldots, x_n)$ is a quadratic form. We say that $H(f)$ is positive if this quadratic form is positive, i.e. if $\forall x \ \langle H(f) \cdot x, x \rangle \geqslant 0$, and we write $H(f) \geqslant 0$. Similarly we define $H(f) > 0$ if $\langle H(f) \cdot x, x \rangle$ is greater than zero for every $x \neq 0$.

We can then demonstrate the following proposition.

Proposition 2.2.7

Let $f: \mathbb{R}^n \to \mathbb{R}$ possess a derivative Df and a second derivative $D^2 f$ at every point of an open set D.

If x_0 is a local minimum of f, $Df(x_0) = 0$ and $D^2 f(x_0) \geqslant 0$.
If $Df(x_0) = 0$ and if $D^2 f(x_0) > 0$, x_0 is a strict minimum.

(By reversing the sign in each expression we obtain conditions for a maximum.)

The demonstration follows from Definition 2.2.6.

Example   Let $z = f(x, y)$, where z is a surface, and let

$$p = \frac{\partial f}{\partial x}, \quad q = \frac{\partial f}{\partial y}, \quad r = \frac{\partial^2 f}{\partial x^2}, \quad s = \frac{\partial^2 f}{\partial x \, \partial y}, \quad t = \frac{\partial^2 f}{\partial y^2}$$

$Df(x_0)$ is identified with the vector

$$\begin{pmatrix} p \\ q \end{pmatrix}$$

and $D^2 f$ with the Hessian matrix

$$\begin{pmatrix} r & s \\ s & t \end{pmatrix}$$

The preceding conditions are then $p = q = 0$ and $rt - s^2 \geqslant 0$, $r > 0$; or $rt - s^2 > 0$ for the sufficient condition where $r > 0$.

DERIVATIVE OF A COMPOSITE FUNCTION AND FIRST INTEGRALS

We can easily prove the following theorem, which is the analogue of the theorem for differentiation of composite functions in the case of functions $\mathbb{R} \to \mathbb{R}$.

Theorem 2.2.8

Let F and G be two mappings from $\mathbb{R}^p$ into $\mathbb{R}^q$ and from $\mathbb{R}^q$ into $\mathbb{R}^r$ respectively. We assume that $DF(x)$ exists in the neighbourhood of a point a and $DG(y)$ exists in the neighbourhood of $F(a)$. Then

$$D(G \circ F)(a) = DG[F(a)] \circ DF(a)$$

where $DF(a) \in \mathscr{L}(\mathbb{R}^p, \mathbb{R}^q)$, $DG[F(a)] \in \mathscr{L}(\mathbb{R}^q, \mathbb{R}^r)$ and $D(G \circ F)(a) \in \mathscr{L}(\mathbb{R}^p, \mathbb{R}^r)$.

We can illustrate this by the scheme

$$x \overset{F}{\rightsquigarrow} F(x) \overset{G}{\rightsquigarrow} (G \circ F)(x)$$

$$\mathbb{R}^p \rightarrow \mathbb{R}^q \rightarrow \mathbb{R}^r$$

$$V \overset{DF}{\rightsquigarrow} DF \cdot V \overset{DF}{\rightsquigarrow} DG \cdot DF \cdot V$$

Derivative of a function along integral curves

Theorem 2.2.9

Let

$$\frac{dx}{dt} = F(x) \tag{13}$$

be a differential equation where $x \in \mathbb{R}^p$ and let $\varphi(t)$ be a solution of this equation. Let $U(y)$ be a differentiable function from $\mathbb{R}^p \rightarrow \mathbb{R}$. Then

$$\frac{d}{dt} U[\varphi(t)] = \langle \operatorname{grad} U[\varphi(t)], F[\varphi(t)] \rangle$$

Proof This is a simple application of the preceding theorem since

$$\frac{d\varphi}{dt} = F[\varphi(t)]$$

$$DU[\varphi(t)] \cdot V = \langle \operatorname{grad} U[\varphi(t)], V \rangle$$

and

$$\frac{d}{dt} U[\varphi(t)] = DU[\varphi(t)] \cdot \frac{d\varphi}{dt} = \langle DU[\varphi(t)] \cdot F[\varphi(t)] \rangle$$

Definition 2.2.10

A **first integral** of Eqn (13) is defined as a function U, of class C^1, which is constant on the integral curves; it is thus a function such that

$$\sum \frac{\partial U}{\partial x_i} (x) F_i(x) = 0 \quad (U \text{ differentiable})$$

If we know a first integral we then know that the orbit passing through a point a is supported by the curve $U(x) = U(a)$. We often make use of this property.

See *Exercises 7 to 22* at the end of the chapter.

ELEMENTARY STUDY OF THE VAN DER POL EQUATION

This is the equation

$$\frac{dx}{dt} = y - x^3 + x$$

$$\frac{dy}{dt} = -x$$

where

$$\frac{d^2x}{dt^2} + \frac{dx}{dt}(3x^2 - 1) + x = 0$$

As shown in Fig. 1.5, we denote by Γ the curve $y = x^3 - x$, by Γ^+ the part of this curve for which $x > 0$, by Γ^- the part for which $x < 0$, by Y^+ the part of the axis $x = 0$ for which $y > 0$, and by Y^- the part for which $y < 0$. Also (A), (B), (C) and (D) show the regions determined by these four curves.

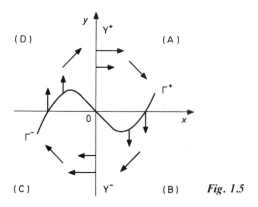

Fig. 1.5

We note at once that, if $(x(t), y(t))$ is a solution, then $(-x(t), -y(t))$ is also a solution.

Figure 1.5 shows the behaviour of the vector field. We suppose that for every $(t, x(0), y(0))$ the Cauchy problem has a unique solution (this will be established in Section 3.4).

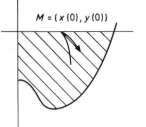

$M = (x(0), y(0))$

Fig. 1.6

Let $M = (x(0), y(0))$ be a point of region (A) (Fig. 1.6); at this point $dy/dt < 0$ and $dx/dt > 0$. Let us assume that we remain in (A); then $x(t) > x(0)$ and

$$y(t) = \int_0^t \frac{dy}{ds}\, ds \leqslant y(0) - x(0)t$$

which is absurd. Thus every solution starting from a point M of (A) leaves that region along the curve Γ^+. We can show in the same way that the solutions run through the four regions (A), (B), (C) and (D) in succession.

To study how the orbits spiral round the point 0, we consider an initial point p situated on Y^+ and we define the sequence of intersections of the orbit with Y^+ and Y^- respectively, as $\sigma_1(p), \sigma_2(p), \ldots, \sigma_n(p), \ldots$ on Y^+ and $\alpha_1(p), \alpha_2(p), \ldots$ on Y^- (Fig. 1.7). Let us examine the orbit starting from the point $-p$; it cuts Y^+ at the point $-\alpha_1(p)$ since the trajectories have the property of symmetry already indicated. Thus an orbit is periodic iff $\alpha_1(p) = -p$. Thus we shall study the function $\delta(p) = |\alpha_1(p)|^2 - |p|^2$.

Consider the point $B = (1, 0)$ (Fig. 1.8); there exists a point p_0 of Y^+ such that the solution starting from p_0 cuts Γ^+ at the point B. Clearly there exists only one such point because of the uniqueness property, and it is easy to see that

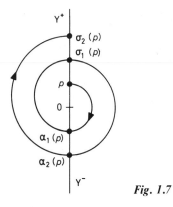

Fig. 1.7

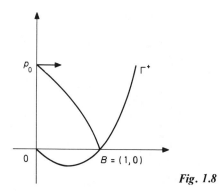

Fig. 1.8

there is a point for the following reason. Let $T(p)$ be the 'time' at which Γ^+ is reached starting from p. It will suffice to study

$$y(T(p)) = p - \int_0^{T(p)} x(s)\,ds$$

to reach our conclusion (see the exercises at the end of this section). For this purpose we shall demonstrate the following lemma.

Lemma
If $0 < |p| < |p_0|$ then $\delta(p) > 0$; if $|p| \geq |p_0|$ δ is monotonic decreasing and $\delta(p) \rightarrow -\infty$ if $|p| \rightarrow \infty$.

If we admit for the moment the truth of this lemma (see Fig. 1.9), it implies that there exists a unique q_0 such that $\delta(q_0) = 0$ and thus a unique periodic solution.

If $p > q_0$, $\alpha_1(p) < -q_0$ and, as $\delta(p) < 0$, $\alpha_1(p) > -p$ and thus $\sigma_1(p) < p$ and sequentially we determine that

$$p > \sigma_1(p) > \sigma_2(p) > \ldots > \sigma_n(p) > \ldots > q_0$$

(see Fig. 1.10). Thus $\sigma_n(p)$ has a limit q_1 but $\sigma(q_1) = q_1$, so that $\sigma_n(p) \rightarrow q_0$. We can show that if $p < q_0$ the behaviour is similar.

Conclusion   There exists one periodic solution. All other solutions tend towards this, which is a limiting cycle.

Proof of the lemma   Let $t_1(p)$ be the instant when the trajectory starting from p arrives at the point $\alpha_1(p)$ and $W(x, y) = \frac{1}{2}(x^2 + y^2)$; then

$$\delta(p) = 2W[\alpha_1(p)] - 2W(p)$$
$$= \int_0^{t_1} \frac{d}{dt} W[x(t), y(t)]\,dt$$
$$= \int_0^{t_1} x^2(t)[1 - x^2(t)]\,dt$$

If $|x(t)| \leq 1$ $\forall t \leq t_1$, then $\delta(p) > 0$; this is the case if $0 < |p| < |p_0|$.

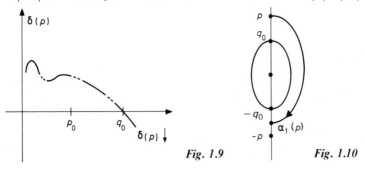

Fig. 1.9   Fig. 1.10

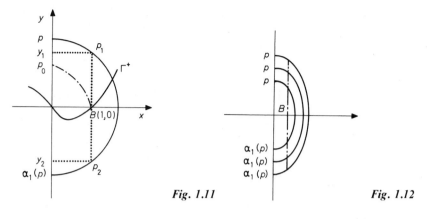

Fig. 1.11   **Fig. 1.12**

Let $p > p_0$, p_1 and p_2 be the points indicated on Fig. 1.11 and u and v be the instants when the solution reaches these points.

$$\delta(p) = \int_0^u x^2(s)[1 - x^2(s)]\,ds + \int_u^v x^2(s)[1 - x^2(s)]\,ds + \int_v^{t_1} x^2(s)[1 - x^2(s)]\,ds$$

$$\int_0^u x^2(s)[1 - x^2(s)]\,ds = \int_0^1 x^2(1 - x^2)\frac{dt}{dx}\,dx = \int_0^1 \frac{x^2(1 - x^2)}{y(x) - x^3 + x}\,dx$$

This term is positive and decreases to zero if $|p| \nearrow \infty$. In the same way

$$\int_v^{t_1} x^2(s)[1 - x^2(s)]\,ds = \int_1^0 \frac{x^2(1 - x^2)}{y(x) - x^3 + x}\,dx$$

is positive and decreases to zero. There remains

$$\int_u^v x^2(s)[1 - x^2(s)]\,ds = \int_{y_1}^{y_2} x^2(y)[1 - x^2(y)]\frac{dt}{dy}\,dy$$

$$= \int_{y_2}^{y_1} x(y)[1 - x^2(y)]\,dy$$

This term is negative and tends to $-\infty$ if $|p| \nearrow \infty$ as one can realize by displacing the arc $(p, \alpha_1(p))$ towards the right (Fig. 1.12) (see the exercise below).

This completes the proof of the lemma.

We can generalize this method by studying α and σ. (As an exercise the reader could complete the proof following the indications given.)

DERIVATIVE: GEOMETRIC INTERPRETATION

Volume element

Taking $\mathbb{R}^3$ as an example, consider three vectors (V_1, V_2, V_3). We know that the absolute value of their determinant $|\det(V_1, V_2, V_3)|$ represents the volume of the parallelepiped determined by these vectors. Let A be a linear transformation and let AV_1, AV_2, AV_3 be the transformed vectors obtained from (V_1, V_2, V_3). The new volume is

$$|\det(AV_1, AV_2, AV_3)| = |\det A \parallel \det(V_1, V_2, V_3)|$$

In a linear transformation $|\det A|$ is the coefficient of variation of volume (it is well known that this does not depend on the coordinate system).

If a transformation F is differentiable it behaves in the neighbourhood of each point like the linear transformation DF and the volume variation is then locally $|\det DF|$. This quantity is called the Jacobian and in a coordinate system where (x_1, x_2, x_3) becomes (u_1, u_2, u_3) it is denoted by

$$\left| \det \frac{D(x_1, x_2, x_3)}{D(u_1, u_2, u_3)} \right| = \begin{vmatrix} \dfrac{\partial x_1}{\partial u_1} & \dfrac{\partial x_1}{\partial u_2} & \dfrac{\partial x_1}{\partial u_3} \\[2mm] \dfrac{\partial x_2}{\partial u_1} & \dfrac{\partial x_2}{\partial u_2} & \dfrac{\partial x_2}{\partial u_3} \\[2mm] \dfrac{\partial x_3}{\partial u_1} & \dfrac{\partial x_3}{\partial u_2} & \dfrac{\partial x_3}{\partial u_3} \end{vmatrix} = J$$

Similarly the volume element $dx_1\, dx_2\, dx_3$ becomes $J\, du_1\, du_2\, du_3$ and it is precisely this which we write when evaluating triple integrals by change of coordinates; if Δ is the image of D

$$\int_D f(x_1, x_2, x_3)\, dx_1\, dx_2\, dx_3$$

$$= \int_\Delta f[x_1(u_1, u_2, u_3), x_2(u_1, u_2, u_3), x_3(u_1, u_2, u_3)]$$

$$\times \left| \det \frac{D(x_1, x_2, x_3)}{D(u_1, u_2, u_3)} \right| du_1\, du_2\, du_3$$

Tangent vector and tangent plane

Let S be a region in $\mathbb{R}^p$, e.g. a surface, and let γ be a curve traced on S, i.e. a mapping

$$[0, T] \to S$$
$$t \rightsquigarrow \varphi(t)$$

The curve γ, the image of $[0, T]$ under φ, will also be denoted by φ.

We assume that φ is differentiable and denote the tangent vector or velocity vector by the vector $d\varphi/dt$. Let a be a point of S and let V be the tangent vector to φ at a (Fig. 1.13). Let F be a transformation from $\mathbb{R}^p$ into $\mathbb{R}^q$ which results in Σ corresponding to S and ψ corresponding to φ, where ψ is defined by

$$\psi(t) = F[\varphi(t)]$$

Suppose that F is differentiable; then so is ψ and the tangent vector W to ψ

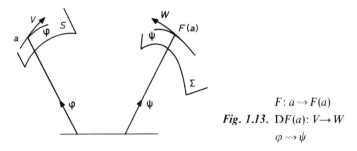

$$F: a \rightsquigarrow F(a)$$

Fig. 1.13. $\mathrm{D}F(a): V \to W$

$$\varphi \rightsquigarrow \psi$$

at $F(a)$ is $(\mathrm{d}/\mathrm{d}t)(F \circ \varphi)(a)$ whose coordinates are

$$\frac{\mathrm{d}}{\mathrm{d}t}(F \circ \varphi)_k(a) = \sum_{i=1}^{p} \frac{\partial F_k}{\partial x_i} \frac{\mathrm{d}\varphi_i}{\mathrm{d}t}$$

Thus

$$W = \mathrm{D}F(a)V$$

Hence, if F transforms a to $F(a)$ and the curve φ to ψ, $\mathrm{D}F(a)$ transforms the tangent vector V to φ into the tangent vector W to ψ.

The tangent linear mapping is a linear mapping which transforms 'tangent spaces' into 'tangent spaces'.

Proposition 2.2.11

Let $F(x, y, z) = 0$ be the equation of a surface S in $\mathbb{R}^3$ and suppose that F is differentiable. Let (a, b, c) be a point of this surface. The equation of the tangent plane to S at (a, b, c) is

$$\frac{\partial F}{\partial x}(a, b, c)(x - a) + \frac{\partial F}{\partial y}(a, b, c)(y - b) + \frac{\partial F}{\partial z}(a, b, c)(z - c) = 0$$

This is by definition the equation of the affine mapping tangential to F at the point (a, b, c). Every tangent vector to a curve traced on S lies in the tangent plane.

Proof   Let $(x, y, z) \rightsquigarrow F(x, y, z)$ be a differentiable mapping. The derivative of F at a point (a, b, c) is the mapping

$$(x, y, z) \rightsquigarrow \mathrm{D}F(a, b, c)\begin{pmatrix} x \\ y \\ z \end{pmatrix} = \frac{\partial F}{\partial x}(a, b, c)\,x + \frac{\partial F}{\partial y}(a, b, c)\,y + \frac{\partial F}{\partial z}(a, b, c)\,z$$

The tangential affine mapping is

$$G = \mathrm{D}F(a, b, c)\begin{pmatrix} x \\ y \\ z \end{pmatrix} + k$$

where k is such that $F(a, b, c) = 0 \Leftrightarrow G(a, b, c) = 0$. Thus

$$\frac{\partial F}{\partial x}(a, b, c)(x - a) + \frac{\partial F}{\partial y}(a, b, c)(y - b) + \frac{\partial F}{\partial z}(a, b, c)(z - c) = 0$$

If γ is a curve traced on S with equation $x(t)$, $y(t)$, $z(t)$, then $F[x(t), y(t), z(t)] = 0 \; \forall t \in [0, T]$ and thus

$$\frac{\partial F}{\partial x}\frac{dx}{dt} + \frac{\partial F}{\partial y}\frac{dy}{dt} + \frac{\partial F}{\partial z}\frac{dz}{dt} = 0$$

which states that the tangent vector to γ lies in the tangent plane.

Change of variables

Let F be a differentiable mapping from $\mathbb{R}^p$ into $\mathbb{R}^p$ which defines a change of variables. We have just seen that F transforms a vector field into another vector field and a tangent curve to these vectors into another tangent curve to the vectors of the transformed field, i.e. it transforms a solution of a differential equation into a solution of the transformed equation. The problem is to move in the opposite direction, i.e. to obtain the solution of the initial equation starting from the transformed equation. This is possible only if DF is invertible. It is convenient then to consider problems in domains in which DF^{-1} exists.

Example The simplest example is that of conversion to **polar coordinates**. In a plane, let $x = r \cos \vartheta$, $y = r \sin \vartheta$ and $J = |r|$. The mapping is then not one-to-one over the entire plane, and we must choose a suitable domain, e.g. $r \geqslant \rho$ and $0 < \vartheta \leqslant 2\pi$ or $0 < r < \infty$ and $0 < \vartheta < 2\pi$. We must remember this when we study the neighbourhood of the origin.

Straightening theorem

The following theorem, which is difficult to demonstrate, is fundamental to understanding how solutions of a differential equation behave in the neighbourhood of a 'regular' point. It states that, in fact, one can always 'rectify' the solutions and arrive locally at one equation $dx_i/dt = 1$ involving one single coordinate.

Theorem 2.2.12

Let $(e_1, e_2, \ldots, e_n)$ be a basis of $\mathbb{R}^n$, a a point of $\mathbb{R}^n$ and $f : \mathbb{R}^n \to \mathbb{R}^n$ a differentiable mapping such that $f(a) \neq 0$. Let

$$\frac{dx}{dt} = f(x) \tag{14}$$

(We say that a is a regular point of (14).)

There exists then a neighbourhood G of a and a differentiable and invertible transformation T in G such that, at any point of G, $DT \cdot f(x) = e_1$. If we put

$y = T(x)$, (14) then becomes

$$\frac{dy}{dt} = e_1 \tag{15}$$

(see Fig. 1.14).

The solution of (15) is obviously

$$y_1 = k_1 + t$$

$$y_2 = k_2$$

$$\vdots$$

$$y_3 = k_n$$

See *Exercises 23 to 25* at the end of the chapter.

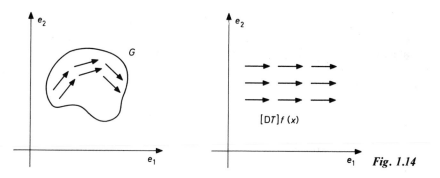

Fig. 1.14

INVERSION THEOREMS

We recall that a mapping $F: \mathbb{R}^n \to \mathbb{R}^n$ is invertible if $\forall\, y \in \mathbb{R}^n$ there exists x such that $F(x) = y$. For a linear mapping represented by a matrix we are concerned with the idea of an invertible matrix. The following theorem states that, if DF is invertible at a, then F is invertible in a neighbourhood of a.

Theorem 2.2.13

Let F be a differentiable mapping from $\mathbb{R}^n$ into $\mathbb{R}^n$. Let a be a point such that $DF^{-1}(a)$ exists. Then there exists a neighbourhood G of a in which F is invertible.

The following theorem is of the same type.

Theorem 2.2.14

Let $V_1, \ldots, V_n$ be n vectors in $\mathbb{R}^n$ and let $M = (V_1, V_2, \ldots, V_n)$ be the matrix whose columns are the V_i. We know that the vectors are independent iff M^{-1} exists; dependence between the vectors is expressed by the fact that one of them is a linear combination of the others (at least).

In a similar way we say that n functions $f_1, \ldots, f_n \colon \mathbb{R}^n \to \mathbb{R}$ are independent if no differentiable function H exists (not identically zero) such that $H(f_1, \ldots, f_n) = 0$.

Theorem 2.2.13 implies that n differentiable functions $f_1, \ldots, f_n$ are independent iff the n vectors

$$\begin{pmatrix} \dfrac{\partial f_1}{\partial x_1} \\[2ex] \vdots \\[2ex] \dfrac{\partial f_1}{\partial x_n} \end{pmatrix} \cdots \begin{pmatrix} \dfrac{\partial f_n}{\partial x_1} \\[2ex] \vdots \\[2ex] \dfrac{\partial f_n}{\partial x_n} \end{pmatrix}$$

are independent, i.e. if

$$\det \frac{D(f_1, \ldots, f_n)}{D(x_1, \ldots, x_n)} \neq 0$$

These two theorems express a further aspect of the proximity of F and DF.

Before we indicate another consequence of Theorem 2.2.13 we must make the notion of partial derivative more precise.

Let $\mathbb{R}^{r_1}$ be the vector space generated by the first r_1 coordinates of $\mathbb{R}^p$, and let $\mathbb{R}^{r_2}$ be the vector space generated by the $r_2 = p - r_1$ other coordinates.

Every point of $\mathbb{R}^p$ can be written in a unique form $x = (\tilde{x}_1, \tilde{x}_2)$ where $\tilde{x}_1 \in \mathbb{R}^{r_1}$ and $\tilde{x}_2 \in \mathbb{R}^{r_2}$. We express this by writing $\mathbb{R}^p = \mathbb{R}^{r_1 + r_2} = \mathbb{R}^{r_1} \times \mathbb{R}^{r_2}$.

Definition 2.2.15

We say that F is differentiable with respect to $\tilde{x}_1$ at the point $a = (a_1, a_2)$ if there exists a linear mapping from $\mathbb{R}^{r_1}$ into $\mathbb{R}^q$, denoted by $(\partial F / \partial \tilde{x}_1)(a)$ and called a partial derivative, such that

$$\lim_{\tilde{x}_1 \to a_1} \frac{1}{\|x_1 - a_1\|} \left[F(\tilde{x}_1, a_2) - F(a_1, a_2) - \frac{\partial F}{\partial \tilde{x}_1}(a)(\tilde{x}_1 - a_1) \right] = 0$$

If $\partial F_j / \partial x_i$ exists and is continuous for $i \leqslant r_1$ and $j = (1, \ldots, q)$, then $\partial F / \partial \tilde{x}_1$ exists and is represented in a coordinate system by the matrix

$$\begin{bmatrix} \dfrac{\partial F_1}{\partial x_1}(a) & \dfrac{\partial F_1}{\partial x_{r_1}}(a) \\[2ex] \vdots & \\[2ex] \dfrac{\partial F_q}{\partial x_1}(a) & \dfrac{\partial F_q}{\partial x_{r_1}}(a) \end{bmatrix}$$

Conversely, if $\partial F / \partial \tilde{x}_1$ exists, the partial derivatives in question exist.

We shall now state a fundamental theorem to indicate the conditions under which a relation $F(x, y) = 0$ allows the definition of y as a function of x.

Theorem 2.2.16 (Implicit functions)
Let

$$F: \begin{array}{c} \mathbb{R}^p \times \mathbb{R}^q \to \mathbb{R}^q \\ (x, y) \rightsquigarrow F(x, y) \end{array}$$

and let (x_0, y_0) be such that $F(x_0, y_0) = 0$. We assume that $(\partial F/\partial y)(x_0, y_0)$ exists and is invertible in a neighbourhood of (x_0, y_0). Then there exists a neighbourhood $W(x_0)$ and a unique function g defined over $W(x_0)$ such that $F[x, g(x)] = 0$, i.e. $F(x, y) = 0 \Leftrightarrow y = g(x)$. We say that g is defined implicitly.

If F is differentiable, so is g, and we find $Dg(x)$ by differentiating the equation $F[x, g(x)] = 0$; thus

$$\frac{\partial F}{\partial x}[x, g(x)] + \frac{\partial F}{\partial y}[x, g(x)] \circ Dg(x) = 0$$

i.e.

$$Dg(x) = -\left\{\frac{\partial F}{\partial y}[x, g(x)]\right\}^{-1} \circ \frac{\partial F}{\partial x}[x, g(x)]$$

In $\mathbb{R}^2$ the theorem is quite obvious; $F(x, y) = 0$ defines a curve in $\mathbb{R}^2$ and it is sufficient to examine Fig. 1.15.

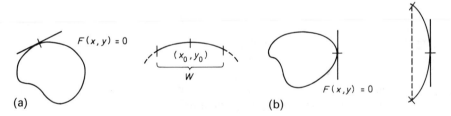

(a)   (b)

Fig. 1.15. (a) $(\partial F/\partial y)^{-1}$ *exists and* $y = g(x)$ *is defined.* (b) $(\partial F/\partial y)^{-1}$ *does not exist; there are two values of* y *for each* x.

Application   Let $F(t, x, dx/dt)$ be a differential equation in $\mathbb{R}^d$. If there exists a neighbourhood of (t_0, x_0, x_0') such that $(\partial F/\partial x')(t_0, x_0, x_0')$ is invertible, there exists in this neighbourhood a function f such that $dx/dt = f(t, x)$.

It is possible that there exists a function $\varphi(t)$ which satisfies $F(t, \varphi(t), d\varphi/dt) = 0$ and

$$\det \frac{\partial F}{\partial x'}\left[t, \varphi(t), \frac{d\varphi}{dt}\right] = 0$$

at the same time, over an interval; such a solution, having a particular character, is called a **singular integral**.

Part 3: Approximate solutions, exact solutions and comparison of solutions of neighbouring equations

Let

$$\frac{dx}{dt} = f(t, x) \tag{16}$$

where $x \in \mathbb{R}^d$; we shall be concerned with the solution of the Cauchy problem, i.e. with the search for solutions φ such that $\varphi(t_0) = x_0$ for given t_0 and x_0. We recall that for a scalar differential equation of order n the initial data are

$$y(0), \frac{dy}{dt}(0), \dots, \frac{d^{n-1}y}{dt^{n-1}}(0)$$

In this part we shall use two fundamental mathematical tools which will be introduced in the two following sections.

3.1. Cauchy sequences—complete spaces

Let $a_1, a_2, \dots, a_n, \dots$ be a sequence of real numbers. We say that

$$\lim_{n \to \infty} a_n = a$$

if $\forall \varepsilon > 0 \; \exists n_0(\varepsilon): n \geqslant n_0 \Rightarrow |a_n - a| \leqslant \varepsilon$. Now let $m \geqslant n_0(\varepsilon): |a_n - a_m| \leqslant 2\varepsilon$. Then, replacing ε by $\varepsilon/2$, when the sequence converges,

$$\forall \varepsilon > 0 \quad \exists n_0(\varepsilon): n, m \geqslant n_0(\varepsilon) \Rightarrow |a_n - a_m| \leqslant \varepsilon \tag{17}$$

A sequence which satisfies condition (17) is called a Cauchy sequence; the condition itself is called the **Cauchy condition**.

Suppose now that a sequence $a_1, \dots, a_n, \dots$ satisfies (17) and let

$$\alpha_p = \inf_{n \geqslant p} a_n \quad \text{and} \quad \beta_p = \sup_{n \geqslant p} a_n$$

The sequence α_p is increasing towards α, the sequence β_p is decreasing towards β and (17) implies that

$$\forall \varepsilon > 0 \quad \exists n_0(\varepsilon): p \geqslant n_0(\varepsilon) \Rightarrow |\beta_p - \alpha_p| < \varepsilon$$

Thus $\beta = \alpha$ so that the sequence $\{a_n\}$ converges to $\alpha = \beta$. Hence in $\mathbb{R}$ a sequence converges iff it is a Cauchy sequence. To verify that it converges it is unnecessary to know the limit and it is the verification which is essential. Moreover, this is a means of defining numbers as limits of convergent sequences; e.g. we define

$$\exp(1) = \lim_{n \to \infty} \left(1 + \frac{1}{1!} + \frac{1}{2!} + \dots + \frac{1}{n!} \right)$$

Definition 3.1.1

Let E be a vector space with a norm; let $\{u_n\}$ be a sequence of elements of E such that

$$\forall \varepsilon > 0 \quad \exists n_0(\varepsilon): n, m \geqslant n_0(\varepsilon) \Rightarrow \|u_n - u_m\| \leqslant \varepsilon$$

The sequence $\{u_n\}$ is called a **Cauchy sequence**. Every convergent sequence is a Cauchy sequence. Conversely, if every Cauchy sequence is convergent we say that E is a **complete space**.

Examples (1) We have just seen that $\mathbb{R}$ is complete. Naturally it is the same for $\mathbb{R}^d$ $\forall d < \infty$.

(2) $]0, \infty]$ is not complete, since the Cauchy sequence $u_n = 1/n$ does not converge in this space.

(3) The set of rational numbers $\mathbb{Q}$ (the vector space constructed from the set of rationals) is not complete.

Definition 3.1.2 (Uniform convergence)

We say that a sequence f_n of functions from $\mathbb{R}^p$ into $\mathbb{R}^q$ converges uniformly to a function f if

$$\lim_{n \to \infty} \sup_{x \in \mathbb{R}^p} \|f_n(x) - f(x)\| = 0$$

If the sequence f_n is defined over a domain D we take the $\sup_{x \in D}$.

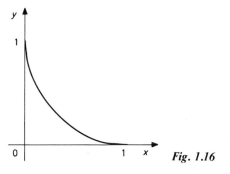

Fig. 1.16

Example The sequence $y_n \colon]0, 1] \to \mathbb{R}$ defined by $y_n(x) = (1 - x)^n$ (Fig. 1.16) converges at every point to zero since $y_n(x) \to 0$ $\forall x \in]0, 1]$ even though

$$\forall n \colon \sup_{x \in]0,1]} |f_n(x) - 0| = 1$$

In contrast, the sequence

$$y_n = \frac{1}{n} \cos x$$

converges uniformly to zero, for

$$\sup_{x \in]0,1]} \| f_n(x) - 0 \| = \frac{1}{n}$$

Definition 3.1.3

We say that a sequence of functions f_n is a Cauchy sequence for uniform convergence if

$$\forall \varepsilon > 0 \quad \exists N(\varepsilon): n, m \geqslant N(\varepsilon) \Rightarrow \sup_{x \in \mathbb{R}^p} \| f_n(x) - f_m(x) \| \leqslant \varepsilon \qquad (18)$$

The quantity

$$\| f \| = \sup_{x \in \mathbb{R}^p} \| f(x) \|$$

is called the norm of f for uniform convergence; condition (18) can also be written

$$\forall \varepsilon > 0 \quad \exists N(\varepsilon): n, m \geqslant N(\varepsilon) \Rightarrow \| f_n - f_m \| \leqslant \varepsilon$$

Theorem 3.1.4

The vector space of continuous functions from $\mathbb{R}^p$ into $\mathbb{R}^q$ is complete for uniform convergence; every Cauchy sequence of continuous functions converges uniformly to a continuous function.

Proof   Let $\{ f_n \}$ be a Cauchy sequence, and let x be fixed. The sequence $\{ f_n(x) \}$ is a Cauchy sequence in $\mathbb{R}^q$; thus it has a limit, say $f(x)$.
By hypothesis

$$\forall \varepsilon > 0 \quad \exists N(\varepsilon): n, m \geqslant N(\varepsilon) \Rightarrow \sup_x \| f_n(x) - f_m(x) \| \leqslant \varepsilon$$

Let us fix n and make m tend to infinity; then, for every x, $f_m(x) \to f(x)$ so that $\| f_n(x) - f_m(x) \| \to \| f_n(x) - f(x) \|$ and consequently

$$\sup_x \| f_n(x) - f(x) \| \leqslant \varepsilon$$

i.e. f_n converges uniformly to f.
It remains to verify that f is continuous. Now for all points x and y of $\mathbb{R}^p$

$$\| f(x) - f(y) \| \leqslant \| f(x) - f_n(x) \| + \| f_n(x) - f_n(y) \| + \| f_n(y) - f(y) \|$$

For given ε

$$\exists N(\varepsilon): n \geqslant N(\varepsilon) \Rightarrow \| f(x) - f_n(x) \| \leqslant \frac{\varepsilon}{3}$$

and

$$\| f(y) - f_n(y) \| \leqslant \frac{\varepsilon}{3} \qquad \forall\, x, y$$

Let us choose such an n. Now, since f_n is a continuous function, we can choose $\| x - y \| \leqslant \delta(\varepsilon)$ so that $\| f_n(x) - f_n(y) \| \leqslant \varepsilon/3$. Thus with ε we can associate a $\delta(\varepsilon)$ such that $\| x - y \| \leqslant \delta(\varepsilon) \Rightarrow \| f(x) - f(y) \| \leqslant \varepsilon$.

Figure 1.17 illustrates the notion of uniform convergence.

Fig. 1.17

3.2. Lipschitzian functions and Gronwall's lemma

Definition 3.2.1
A mapping

$$f: \begin{array}{l} \mathbb{R} \times \mathbb{R}^d \supset U \to \mathbb{R}^d \\ (t, x) \leadsto f(t, x) \end{array}$$

is called **k-Lipschitzian in x** if $\forall (t, x_1), (t, x_2) \in U$

$$\| f(t, x_1) - f(t, x_2) \| \leqslant k \| x_1 - x_2 \| \qquad (k \text{ does not depend on } t)$$

f is called locally Lipschitzian if $\forall (t_0, x_0) \in U$ there exists a neighbourhood $V(t_0, x_0)$ of (t_0, x_0) in which f is $k(t_0, x_0)$-Lipschitzian.

Theorem 3.2.2
Every function of class C^1 (in t and x) is locally Lipschitzian and is therefore Lipschitzian in every closed and bounded set Ω. (Ω is bounded in x and in t.)

Proof   We assume that f is now of class C^1 in a convex domain U, i.e. a domain such that $x, y \in U \Rightarrow \vartheta x + (1 - \vartheta)y \in U \;\; \forall \vartheta \in [0, 1]$ (in other words, the segment $[x, y]$ lies completely in $U \;\; \forall x, y \in U$).

Then let $u = y - x$ and $g(t, s) = f(t, x + su)$. From Theorem 2.2.8

$$\frac{\partial g}{\partial s}(t, s) = \frac{\partial f}{\partial x}(t, x + su)u$$

Now

$$f(t, y) - f(t, x) = g(t, 1) - g(t, 0) = \int_0^1 \frac{\partial f}{\partial x}(t, x + su)u\, ds$$

$\partial f / \partial x$, being by hypothesis a continuous function, is bounded in every bounded set; thus if $(x, y) \in V(t_0, x_0)$ (convex) then $(\partial f / \partial x)(t, x + su)$ is bounded by $k(t_0, x_0)$, so that

$$\| f(t, y) - f(t, x) \| \leqslant k(t_0, x_0) \| y - x \|$$

We shall not establish the second point, which follows from the compactness of Ω; this means that, from every partition of Ω into an arbitrary number of neighbourhoods $V(t_i, x_i)$, we can extract one partition comprising only a finite number of such neighbourhoods, so that it is then sufficient to take the maximum of the $k(t_i, x_i)$ appearing in this finite set (or a multiple of this maximum).

The following lemma is very useful.

Theorem 3.2.3 (Gronwall's lemma)

Let φ be a continuous function over an interval $[0, T]$, and let us assume that there exists $k > 0$ and a function f continuous over $[0, T]$ such that

$$\varphi(t) \leqslant f(t) + k \int_0^t \varphi(u)\, du \qquad \forall\, t \in [0, T]$$

then

$$\varphi(t) \leqslant f(t) + k \int_0^t f(u) \exp[k(t - u)]\, du \qquad \forall\, t \in [0, T]$$

Particular cases

(a) If $0 \leqslant \varphi(t) \leqslant k \int_0^t \varphi(u)\, du$ then $\varphi \equiv 0$

(b) If $\varphi(t) \leqslant a + k \int_0^t \varphi(u)\, du$ then $\varphi(t) \leqslant a \exp(kt)$

(c) If $\varphi(t) \leqslant \alpha t + k \int_0^t \varphi(u)\, du$ then $\varphi(t) \leqslant \frac{\alpha}{k} [\exp(kt) - 1]$

Let

$$\vartheta(t) = \int_0^t \varphi(u)\, du$$

Then

$$\vartheta'(t) = \varphi(t) \leqslant f(t) + k \int_0^t \varphi(u)\, du = f(t) + k\vartheta(t)$$

Let

$$\eta(t) = \exp(-kt)\vartheta(t) \qquad \eta'(t) \leqslant \exp(-kt)f(t)$$

Then

$$\eta(t) = \int_0^t \eta'(u)\,du \leqslant \int_0^t \exp(-ku)f(u)\,du$$

Thus

$$\vartheta(t) \leqslant \int_0^t \exp[k(t-u)]f(u)\,du$$

and as $k > 0$

$$\varphi(t) \leqslant f(t) + k\int_0^t f(u)\exp[k(t-u)]\,du$$

As an exercise we can prove the following generalization. Suppose that there exists $k(u) > 0$ such that

$$\varphi(t) \leqslant f(t) + \int_0^t k(u)\varphi(u)\,du$$

Then

$$\varphi(t) \leqslant f(t) + \int_0^t k(u)f(u)\exp\left[\int_u^t k(v)\,dv\right]du$$

If in addition f is non-decreasing

$$\varphi(t) \leqslant f(t)\exp\int_0^t k(u)\,du$$

3.3. ε-approximate solutions

Let

$$\frac{dx}{dt} = f(t, x) \tag{19}$$

be a differential equation. Then a function φ such that

$$\frac{d\varphi}{dt} - f[t, \varphi(t)] = 0$$

is a solution; e.g. the trajectory of a moving body of mass m subjected to a force F is such that

$$\frac{d^2\varphi}{dt^2} - \frac{1}{m}F = 0$$

If we are concerned with physical magnitudes we have, of course, only approximate equality and we might say that

$$\left|\frac{d^2\varphi}{dt^2} - \frac{1}{m}F\right| \leqslant \varepsilon$$

We say that we are concerned with an ε-approximate solution.

Another way of understanding an ε-approximate solution is the following. The construction of a solution consists of starting from a point along a tangent vector and then of following the vector field by constructing a curve which is always tangential to this field. If, during this construction, we allow a 'difference' of ε between the tangent vector and the field vector, we obtain an ε-approximate solution.

Finally, when we construct—as we shall do later—solutions as sums of series in which there occur small parameters, and when we stop at a certain stage of the construction, we obtain an ε-approximate solution.

Definition 3.3.1
Let

$$\frac{dx}{dt} = f(t, x) \qquad x \in \mathbb{R}^d \tag{20}$$

be a differential equation. By an ε-approximate solution we mean a mapping $\varphi: I \to \mathbb{R}^d$ C^1-piecewise and such that

(a) $(t, \varphi(t))$ belongs to the domain of definition of f $\forall t \in I$

(b) $\left\|\dfrac{d\varphi}{dt} - f[t, \varphi(t)]\right\| \leqslant \varepsilon$ $\forall t \in I$

By definition there can be a finite number of points of I (if I is bounded) or a denumerable number of points of I (if I is infinite) at which φ is not differentiable. In this case left- and right-derivatives exist at the points of discontinuity and it is required that they satisfy relation (b).

Figure 1.18 illustrates the construction in $\mathbb{R}$ of an ε-approximate solution. Starting from t_0 with a given tangent, we continue along the tangent until the difference between the right-gradient and that of the field vector at the point reached is equal to ε. When the difference exceeds ε (say at (x_1, t_1)) we take as a new solution the new field vector and follow it until the difference attained again exceeds ε. Thus we construct a sequence (t_n, x_n), which we wish to study.

Theorem 3.3.2
Let $f: \mathbb{R} \times \mathbb{R}^d \to \mathbb{R}$ be defined and continuous over a domain D of $\mathbb{R} \times \mathbb{R}^d$; let (t_0, x_0) be a point of D, I an open bounded interval containing t_0 and $B(x_0, r) = \{x: \|x - x_0\| \leqslant r\}$ a ball with centre x_0 such that $I \times B(x_0, r) \subset D$.

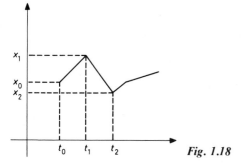

Fig. 1.18

Let M be the maximum of f over $I \times B(x_0, r)$ and

$$J = I \cap \left[t_0 - \frac{r}{M}, t_0 + \frac{r}{M} \right]$$

For every $\varepsilon > 0$, there exists an ε-approximate solution, defined over J, which is piecewise linear and such that $\varphi(t) \in B(x_0, r)$, $\varphi(t_0) = x_0$.

Comments (1) It should be carefully noted that, if $I =]a, b[$, the ε-approximate solution is defined for $a < t \leqslant t_0$ as well as for $t_0 \leqslant t < b$.

(2) If f is defined in the entire space $\mathbb{R}^{d+1}$ we can contemplate finding an ε-approximate solution defined for every $t \in \mathbb{R}$. For this, it is tempting to choose r very large, but if f is not bounded in $\mathbb{R} \times \mathbb{R}^d$ this operation will cause M to increase (so that one ought to denote it in this case by $M(r)$). Thus it is possible that we can define ε-approximate solutions only over an interval which is much smaller than that over which f is defined. This is also the case for exact solutions.

(3) Once r and M are fixed we can choose ε as small as we wish. Solutions are still defined over the whole of J.

Proof of Theorem 3.3.2 Let $t_0 \in I$. We shall construct an ε-approximate solution over the interval

$$I \cap \left[t_0, t_0 + \frac{r}{M} \right] = J_1 \qquad (21)$$

A solution over

$$J_2 = I \cap \left[t_0 - \frac{r}{M}, t_0 \right] \qquad (22)$$

could be constructed similarly, the union of (21) and (22) constituting an ε-approximate solution over J.

We shall let $J_1 = [t_0, T]$; then $\forall\, t \in J_1$

$$|t - t_0| \leqslant (T - t_0) \leqslant \frac{r}{M}$$

Let

$$\varphi_0(t) = x_0 + (t - t_0)f(t_0, x_0) \qquad \varphi_0(t_0) = x_0$$

and

$$\frac{d\varphi_0}{dt} = f(t_0, x_0) \qquad \forall\, t \in J_1$$

Moreover

$$\|\varphi_0(t) - x_0\| \leqslant (t - t_0)M \leqslant \frac{r}{M}M = r$$

Therefore $\varphi_0(t) \in B(x_0, r)$ $\quad \forall t \in J_1$. Thus $f[\varphi_0(t), t]$ is defined $\forall t \in J_1$ and

$$\left\| \frac{d\varphi_0}{dt} - f[t, \varphi_0(t)] \right\| = \| f(t_0, x_0) - f[t, x_0 + (t - t_0)f(t_0, x_0)] \|$$

is a continuous function of t which is zero for $t = t_0$. Therefore there exists t_1 such that if $t \leqslant t_1$

$$\left\| \frac{d\varphi_0}{dt} - f[t, \varphi_0(t)] \right\| \leqslant \varepsilon$$

Then, if $t_1 \geqslant T$, φ_0 is an ε-approximate solution over J_1. If $t_1 < T$ we start again from (t_1, x_1) where $x_1 = \varphi_0(t_1)$, after verifying that $x_1 \in B(x_0, r)$ and that $\|x_1 - x_0\| \leqslant (t_1 - t_0)M$.

For $r_1 = r - \|x_1 - x_0\|$, $B(x_1, r_1) \subset B(x_0, r)$ and f defined over $[t_1, T] \times B(x_1, r_1)$, over which set $\|f(t, x)\| \leqslant M$ (Fig. 1.19), then $\forall t \in [t_1, T]$

$$t - t_1 \leqslant T - t_1$$

$$= T - t_0 - (t_1 - t_0)$$

$$\leqslant \frac{r}{M} - \frac{\|x_1 - x_0\|}{M}$$

$$= \frac{r_1}{M}$$

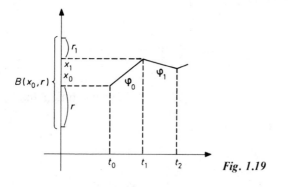

Fig. 1.19

We now put $\varphi_1(t) = x_1 + (t - t_1)f(t_1, x_1)$. There exists a greater real number t_2 such that

$$\forall t \in [t_1, t_2] \qquad \left\| \frac{d\varphi_1}{dt} - f[t, \varphi_1(t)] \right\| \leq \varepsilon$$

If $t_2 \geq T$ the process terminates; otherwise we start again from (x_2, t_2) and so on.

Thus we construct the following:

(a) a sequence $t_0 < t_1 < t_2 < \ldots < t_n < T$ of real numbers;
(b) a sequence $x_0, x_1, \ldots, x_n$ of points of $B(x_0, r)$;
(c) a sequence of linear functions $\varphi_0, \varphi_1, \ldots, \varphi_n$ defined over $[t_0, t_1]$, $[t_1, t_2], \ldots, [t_n, T]$ respectively and such that the function φ (equal to φ_0 over the first interval, φ_1 over the second, ..., and φ_n over $[t_n, t_{n+1}]$) is an ε-approximate solution such that $\varphi(t)$ is within the ball $B(x_0, r)$.

A priori we can only affirm that $\lim t_n \leq T$ even though the theorem affirms that T can be attained. Let

$$t' = \lim_{n \to \infty} t_n$$

Suppose that $t' < T$; since $\|x_{n+1} - x_n\| \leq M(t_{n+1} - t_n)$ the sequence $\{x_n\}$ is a Cauchy sequence and thus

$$\lim_{n \to \infty} x_n = x'$$

Moreover $x' \in B(x_0, r)$ for $\|x_n - x_0\| \leq r \Rightarrow \|\lim x_n - x_0\| \leq r$.

Since t' is taken to be strictly smaller than T, the operation described above never ends, and the sequence t_n is strictly increasing; thus at each stage the solution constructed from t_n to t_{n+1} is no longer suitable for $t \in]t_{n+1}, t']$, and this can arise only from the fact that the derivative diverges too quickly from $f(t_n, x_n)$.

Now this derivative of φ_n is $f[t, \varphi_n(t)]$; since $x_n \to x'$, $t_n \to t'$ and f is continuous in (t', x'). It is then absurd to suppose that the deviation remains large, and therefore inconceivable to suppose that $t' < T'$.

It is easy to interpret the above reasoning correctly.

$$\forall \varepsilon > 0 \quad \exists \eta: \left. \begin{cases} |t' - u| \leq \eta \\ \|x' - x\| \leq \eta \end{cases} \right\} \Rightarrow \|f(t', x') - f(u, x)\| \leq \frac{\varepsilon}{2}$$

Now

$$\|\varphi_n(t) - x'\| \leq \|x' - x_n\| + \|\varphi_n(t) - x_n\| \leq \|x' - x_n\| + M(t - t_n)$$

$$\forall t: t_n \leq t \leq t'$$

$$\exists n_1: n \geq n_1 \Rightarrow t' - t_n \leq \eta \quad \text{and} \quad \|x' - \varphi_n(t)\| \leq \eta$$

and therefore

$$\|f(t', x') - f[t, \varphi_n(t)]\| \leqslant \frac{\varepsilon}{2}$$

$$\exists \, n_2 : n \geqslant n_2 \Rightarrow \|f(t_n, x_n) - f(t', x')\| \leqslant \frac{\varepsilon}{2}$$

since $t_n \rightarrow t'$ and $x_n \rightarrow x'$.

Thus, if $n \geqslant \sup(n_1, n_2)$, $\|f(t_n, x_n) - f[t, \varphi_n(t)]\| \leqslant \varepsilon \quad \forall t \in [t_n, t']$, which expresses the contradiction indicated above.

Example: linear equations   Let $A(t) \in \mathcal{L}(\mathbb{R}^d, \mathbb{R}^d) \ \forall t$ and $B(t) \in \mathbb{R}^d \ t$ (A and B are defined over $\mathbb{R}$).

Theorem 3.3.3
Let

$$\frac{\mathrm{d}x}{\mathrm{d}t} = A(t)x + B(t) \tag{23}$$

be a linear equation in which A and B are continuous over $\mathbb{R}$. For every interval I and for all $\varepsilon > 0$, there exists an ε-approximate solution over I.

Proof   We can apply Theorem 3.3.2 with the same notation. Let I be fixed and bounded such that $\forall t \in I \ \|A(t)\| \leqslant \alpha$ and $\|B(t)\| \leqslant \beta$. Then

$$M(r) \leqslant \sup_{(t,x) \in I \times B(x_0, r)} [\|A(t)x\| + \|B(t)\|] \leqslant \alpha(\|x_0\| + r) + \beta$$

so that

$$\frac{M(r)}{r} \leqslant \frac{\alpha\|x_0\| + \beta}{r} + \alpha$$

and we can, for example, choose r so that $M(r)/r \leqslant 2\alpha$. Thus, starting from (t_0, x_0), we construct an ε-approximate solution in $[t_0 - 1/2\alpha, t_0 + 1/2\alpha]$. Note that α depends only on the interval I. We can start again from $t_0 + 1/2\alpha$ and $t_0 - 1/2\alpha$ respectively; we know that we are starting from an (unknown) point of $B(x_0, r)$ and that by choosing r_1 we can move by a distance $r_1/M(r_1)$ which is always greater than or equal to $1/2\alpha$. It is clear that after a finite number of operations we shall have covered I.

See *Exercises 26 to 28* at the end of the chapter.

3.4. Exact solutions; comparison of solutions

We have noted that if, over an interval I, there exists an ε_0-approximate solution for a given ε_0, then there exist in fact, over the same interval, ε-

approximate solutions for every ε. It is natural to think that by making ε tend towards zero we obtain an exact solution; Gronwall's lemma effectively allows us to state that this is true when f is Lipschitzian. The key result is the following.

Lemma 3.4.1

Let $dx/dt = f(t, x)$ be a differential equation in $\mathbb{R}^d$; we assume f to be k-Lipschitzian over an interval I in which are defined two solutions $\varphi_1 : I \to \mathbb{R}^d$, ε_1-approximate, and $\varphi_2 : I \to \mathbb{R}^d$, ε_2-approximate. Let $t_0 \in I$, $x_1 = \varphi_1(t_0)$ and $x_2 = \varphi_2(t_0)$; then

$$\|\varphi_1(t) - \varphi_2(t)\| \leqslant \|x_1 - x_2\| \exp(k|t - t_0|) + (\varepsilon_1 + \varepsilon_2) \frac{\exp(k|t - t_0|) - 1}{k}$$

Proof To simplify the notation we shall put $t_0 = 0$ and shall assume $t > 0$. By hypothesis

$$\left\| \frac{d\varphi_1}{dt} - f[t_1, \varphi_1(t)] \right\| \leqslant \varepsilon_1$$

and

$$\left\| \frac{d\varphi_2}{dt} - f[t, \varphi_2(t)] \right\| \leqslant \varepsilon_2$$

Then

$$\left\| \frac{d\varphi_1}{dt} - \frac{d\varphi_2}{dt} \right\| \leqslant \varepsilon_1 + \varepsilon_2 + \|f[t, \varphi_1(t)] - f[t, \varphi_2(t)]\|$$

$$\leqslant \varepsilon_1 + \varepsilon_2 + k\|\varphi_1(t) - \varphi_2(t)\|$$

Let

$$\varphi = \varphi_1 - \varphi_2 : \|\varphi(t) - \varphi(0)\| \leqslant \int_0^t [\varepsilon_1 + \varepsilon_2 + k\|\varphi(u)\|] \, du$$

$$\leqslant \int_0^t [\varepsilon_1 + \varepsilon_2 + k\|\varphi(0)\|] \, du$$

$$+ \int_0^t k\|\varphi(u) - \varphi(0)\| \, du$$

and therefore, according to Gronwall's lemma,

$$\|\varphi(t) - \varphi(0)\| \leqslant \frac{\varepsilon_1 + \varepsilon_2 + k\|\varphi(0)\|}{k} [\exp(kt) - 1]$$

Hence

$$\|\varphi_1(t) - \varphi_2(t)\| \leqslant \|\varphi(0)\| + \|\varphi(t) - \varphi(0)\|$$

$$\leqslant \|x_1 - x_2\| \exp(kt) + \frac{\varepsilon_1 + \varepsilon_2}{k} [\exp(kt) - 1]$$

We can now state the theorem concerning existence and uniqueness of solutions, which we shall prove to a large extent.

EXACT SOLUTIONS

Theorem 3.4.2 (Existence and uniqueness theorem)
Let

$$f: \begin{array}{l} \mathbb{R} \times \mathbb{R}^d \to \mathbb{R}^d \\ (t, x) \rightsquigarrow f(t, x) \end{array}$$

Let Ω be an open set of $\mathbb{R} \times \mathbb{R}^d$ in which f is continuous and locally Lipschitzian in x.

If $(t_0, x_0) \in \Omega$ there exists a greatest interval I and a unique solution $\varphi: I \to \mathbb{R}^d$ such that $\varphi(t_0) = x_0$. I is called the maximal interval and φ the maximal solution of

$$\frac{dx}{dt} = f(t, x) \tag{24}$$

Proof (a) According to Theorem 3.3.2 we know that for every $\varepsilon > 0$ there exists an interval J containing t_0 over which we can define an ε-approximate solution such that $\varphi(t_0) = x_0$ and which remains within a ball $B(x_0, r)$ for

$$t \in J = \left[t_0 - \frac{r}{M}, t_0 + \frac{r}{M} \right]$$

Now let $\{\varepsilon_n\}$ be a sequence converging towards zero and $\{\varphi_n\}$ be the corresponding sequence of ε_n-approximate solutions constructed on J. According to Lemma 3.4.1

$$\|\varphi_n(t) - \varphi_m(t)\| \leqslant (\varepsilon_n + \varepsilon_m) \frac{\exp[k(t - t_0)] - 1}{k}$$

(where k is a Lipschitz constant relative to J (cf. Theorem 3.2.2)).

Thus $\{\varphi_n\}$ is a Cauchy sequence for uniform convergence; it therefore converges uniformly on J towards a continuous function φ (cf. Theorem 3.1.4). Since $\varphi_n(t) \in B(x_0, r)$ $\forall n$, $\varphi(t)$ belongs also to $B(x_0, r)$; since $\varphi_n(t_0) = x_0$ $\forall n$, $\varphi(t_0) = x_0$, and to establish that φ is a solution of (24) it remains to verify that

$$\varphi(t) = x_0 + \int_0^t f[u, \varphi(u)] \, du$$

Now

$$\left\| \frac{d\varphi_n}{dt}(u) - f[u, \varphi_n(u)] \right\| \leqslant \varepsilon_n \qquad \forall\, t \in J$$

Therefore

$$\left\| \varphi_n(t) - x_0 - \int_{t_0}^{t} f[u, \varphi_n(u)]\, du \right\| \leqslant \varepsilon_n |t - t_0|$$

Thus

$$\left\| \varphi(t) - x_0 - \int_{t_0}^{t} f[u, \varphi(u)]\, du \right\|$$

$$\leqslant \|\varphi(t) - \varphi_n(t)\| + \left\| \varphi_n(t) - x_0 - \int_{t_0}^{t} f(u, \varphi_n(u)\, du \right\|$$

$$+ \left\| \int_{t_0}^{t} \{ f[u, \varphi(u)] - f[u, \varphi_n(u)] \}\, du \right\|$$

$$\leqslant \|\varphi(t) - \varphi_n(t)\| + \varepsilon_n |t - t_0| + k \sup_{u \in J} \|\varphi(u) - \varphi_n(u)\|\, |t - t_0|$$

each of these terms tending to zero with n uniformly in t.

This establishes the existence of a solution in J; we shall accept that there is a maximal interval I.

(b) We shall now prove uniqueness. Suppose first that f is Lipschitzian over every $I \times \Omega$ and let φ_1 and φ_2 be two exact solutions such that $\varphi_1(t_0) = \varphi_2(t_0) = x_0$. Then, φ_i being ε-approximate for all ε,

$$\|\varphi_1(t) - \varphi_2(t)\| \leqslant \frac{2\varepsilon}{k} \{ \exp[k(t - t_0)] - 1 \}$$

Therefore

$$\varphi_1(t) \equiv \varphi_2(t)$$

If f is only locally Lipschitzian (Ω therefore being not bounded), we complete the argument in the following way (which we can accept).

A set $A \subset I$ is open in I if there exists an open set G of $\mathbb{R}$ such that $I \cap G = A$. We know that a set A can be both open and closed in I iff $A = \varnothing$ or $A = I$.

Then if $A = \{t: \varphi_1(t) = \varphi_2(t)\}$, A is closed since $\varphi_1 - \varphi_2$ is continuous and A is not empty.

Moreover $\forall\, u \in A$ there exists an ε such that $\forall\, t, |t - u| \leqslant \varepsilon$, $\varphi_1(t) \equiv \varphi_2(t)$ and so A is open. Consequently $A = I$.

Examples (1) The preceding theorem applies whenever f is of class C^1, which is the case in the earlier exercises for which we have admitted existence and uniqueness.

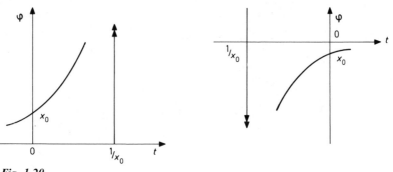

Fig. 1.20

(2) It often happens that the maximal interval is bounded. Let $dx/dt = x^2$, $x \in \mathbb{R}$; $f(t, x)$ is locally Lipschitzian and $\varphi(t) \equiv 0$ is the only solution passing through the origin.

If $x_0 \neq 0$ the solution to the Cauchy problem relative to $(0, x_0)$ is $\varphi(t) = x_0/(1 - x_0 t)$ defined over $]-\infty, 1/x_0[$ if $x_0 > 0$ or $]1/x_0, +\infty[$ if $x_0 < 0$ (Fig. 1.20).

(3) Autonomous equations $dx/dt = f(x)$: let $\varphi : [a, b] \to \mathbb{R}^d$ be a maximal solution and $\psi : [a - c, b - c] \to \mathbb{R}^d$ be defined by $\psi(t) = \varphi(t + c)$. ψ is also a solution which is necessarily maximal, so that if we translate a maximal solution with respect to t we obtain another maximal solution whose projection on the phase space is naturally the same; thus there exists a maximal orbit passing through every point of the phase space.

Theorem 3.4.2 can naturally be applied to scalar equations.

Theorem 3.4.3
Let

$$f : \begin{array}{l} \mathbb{R} \times \mathbb{R}^n \to \mathbb{R} \\ (t, x) \rightsquigarrow f(t, x) \end{array}$$

where $x = (x_1, x_2, \ldots, x_n) \in \mathbb{R}^n$. Suppose that in an open set Ω f is continuous and locally Lipschitzian with respect to the variables $(x_1, \ldots, x_n)$ (we know that it is sufficient that $\forall i \in 1, \ldots, n$, $\partial f/\partial x_i$ is continuous).
Let

$$\frac{d^n x}{dt^n} = f\left(t, x, \frac{dx}{dt}, \ldots, \frac{d^{n-1} x}{dt^{n-1}}\right)$$

be a differential equation of order n. Then, whatever the initial 'point' $[x_1(0), \ldots, x_n(0)]$, there exists a unique maximal solution φ defined over a maximal interval I such that

$$\varphi(t_0) = x_1(0); \quad \frac{d\varphi}{dt}(t_0) = x_2(0); \ldots; \quad \frac{d^{n-1}\varphi}{dt^{n-1}}(t_0) = x_n(0)$$

Comments Firstly, like ε-approximate solutions, exact solutions relative to (t_0, x_0) are defined for $t \leqslant t_0$ as well as for $t \geqslant t_0$ and the uniqueness implies that two solutions cannot cross.

A second comment concerns the comparison of ε-approximate solutions and exact solutions.

Comparison of ε-approximate solutions and exact solutions

According to Lemma 3.4.1 if φ_ε is an ε-approximate solution in an interval I and φ is the exact solution (supposed unique) for the same initial value at t_0

$$\|\varphi_\varepsilon(t) - \varphi(t)\| \leqslant \frac{\varepsilon}{k} [\exp(k|t - t_0|) - 1]$$

Over a bounded interval I, given η, we can then choose ε so that, $\forall t \in I$, $\|\varphi_\varepsilon(t) - \varphi(t)\| \leqslant \eta$ (this only states again that, over every bounded interval, φ_{ε_n} converges uniformly towards φ).

Suppose that it is known that φ is **periodic** with unknown period T; also suppose that we have found φ_ε which is periodic with period T_ε. Then ε can be chosen so that the distance between the orbits is less than η. Suppose in fact that (t_0, x_0) is the initial point for $|t - t_0| \leqslant \sup(T_\varepsilon, T)$; we can choose ε so that $\|\varphi_\varepsilon(t) - \varphi(t)\| \leqslant \eta$ (Fig. 1.21). However, we have then completely covered the two orbits, so that we can only return to the same regions; naturally, since $T_\varepsilon \neq T$, there is in general a difference of phase and the distance between $\varphi_\varepsilon(t)$ and $\varphi(t)$ passes periodically through a maximum, yet the orbit of φ_ε is an approximation of the orbit of φ which is useful in practice.

See **Exercises 29 to 36** at the end of the chapter.

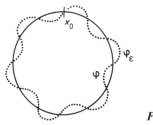

Fig. 1.21

COMPARISON OF THE SOLUTIONS OF TWO NEIGHBOURING EQUATIONS

Theorem 3.4.4
Let

$$\frac{dx}{dt} = f(t, x) \tag{25}$$

$$\frac{dx}{dt} = g(t, x) \tag{26}$$

be two differential equations in $\mathbb{R}^d$. We assume that f and g are continuous and that f is k-Lipschitzian in x. Let φ be a solution of (25) defined over an interval I and such that $\varphi(t_0) = x_0$, and let ψ be a solution of (26) defined over the same interval I and such that $\psi(t_0) = x_0$; finally suppose that, over I, ψ belongs to the domain of definition of f and that

$$\| f[t, \psi(t)] - g[t, \psi(t)] \| \leqslant \varepsilon \qquad \forall\, t \in I$$

Then

$$\| \varphi(t) - \psi(t) \| \leqslant \frac{\varepsilon}{k} [\exp(k|t - t_0|) - 1] \qquad \forall\, t \in I$$

Proof

$$\| \varphi(t) - \psi(t) \| = \left\| \int_{t_0}^t f[u, \varphi(u)] - g[u, \psi(u)]\, du \right\|$$

$$\leqslant \int_{t_0}^t \| f[u, \psi(u)] - g[u, \psi(u)] \|\, du$$

$$+ \int_{t_0}^t \| f[u, \varphi(u)] - f[u, \psi(u)] \|\, du$$

$$\leqslant \varepsilon|t - t_0| + k \int_{t_0}^t \| \varphi(u) - \psi(u) \|\, du$$

and we complete the argument with the help of Gronwall's lemma.

A theorem which is comparable and frequently applied is the following; its proof is more complicated, although analogous, but we shall not give it.

Theorem 3.4.5
Let

$$\frac{dx}{dt} = f(t, x) \tag{27}$$

and

$$\frac{dy}{dt} = g(t, y) \tag{28}$$

be two differential equations where f and g are continuous and f is k-Lipschitzian in x. Suppose that $\varphi(t)$ is a common solution of (27) and (28), such that $\varphi(t_0) = x_0$, which satisfies

$$\| g[t, \varphi(t) + u] - f[t, \varphi(t) + u] \| \leqslant \lambda \| u \|^\alpha$$

(where $\alpha \geqslant 1$) when $\| u \| \leqslant R$ and $|t - t_0| \leqslant r$.

Let $M = \lambda R^\alpha + kR$ and $h = \inf(r, R/2M)$. Let ψ_1 be a solution of (27) and ψ_2 be a solution of (28) such that $\psi_1(t_0) = y_0 = \psi_2(t_0)$.

Then if $|t - t_0| \leqslant h$ and $\| x_0 - y_0 \| \leqslant Mh$

$$\| \psi_1(t) - \psi_2(t) \| \leqslant \| y_0 - x_0 \|^\alpha \frac{\lambda}{k} \exp\left(\frac{\alpha}{2}\right) [\exp(kh) - 1]$$

Application: linearization of small displacements

Let

$$\frac{dx}{dt} = f(t, x) \qquad x \in \mathbb{R}^n \tag{29}$$

be a system describing the evolution of a system as a function of time. Suppose that $f(t, 0) \equiv 0$ so that $\varphi(t) \equiv 0$ is a solution of (29).

Suppose that $\partial^2 f_k / \partial x_i \, \partial x_j$ exists for all i, j and k and is bounded over $[0, T] \times B(0, R)$ so that there exists an A such that

$$\| f(t, u) - Df(u)(t, 0)u \| \leqslant A \| u \|^2 \qquad \forall t \in [0, T], \| u \| \leqslant R$$

Let

$$\frac{dx}{dt} = Df(x)(t, 0)x \tag{30}$$

$\varphi(t) \equiv 0$ is also a solution of (30) and we can apply the preceding theorem.

Let ψ_1 be a solution of (29) and ψ_2 be a solution of (30) such that $\psi_1(t_0) = x_0 = \psi_2(t_0)$. Then there exist constants h, M and C such that

$$\| x_0 \| \leqslant Mh$$

and

$$|t| \leqslant h \Rightarrow \| \psi_1(t) - \psi_2(t) \| \leqslant C \| x_0 \|^2$$

We shall see later that if $\varphi(t) \equiv 0$ is 'stable' we can then have an upper bound valid for all t.

See *Exercises 37 to 41* at the end of the chapter.

SUCCESSIVE APPROXIMATIONS: PICARD'S METHOD

Theorem 3.4.6

Let

$$f: \begin{array}{l} \mathbb{R} \times \mathbb{R}^d \to \mathbb{R}^d \\ (t, x) \rightsquigarrow f(t, x) \end{array}$$

and

$$\frac{dx}{dt} = f(t, x) \tag{31}$$

Let $(t_0, x_0) \in \mathbb{R} \times \mathbb{R}^d$ and J be an interval of $\mathbb{R}$ containing t_0. Suppose that f is continuous and k-Lipschitzian in x over $J \times B(x_0, r)$; let M be an upper bound of $\| f(t, x)\|$ over this set and $I = J \cap (t_0 - r/M, t_0 + r/M)$. Let

$$x_0(t) \equiv x_0 \qquad x_1(t) = x_0 + \int_{t_0}^{t} f[u, x_0(u)]\, du$$

$$\vdots$$

$$x_n(t) = x_0 + \int_{t_0}^{t} f[u, x_{n-1}(u)]\, du$$

The sequence x_n is well defined and converges uniformly over I to the solution $x(t)$ of the Cauchy problem of (31) relative to (t_0, x_0); moreover

$$\sup_{t \in I} \| x_n(t) - x(t)\| \leqslant \frac{M}{k} \sum_{m=n+1}^{\infty} \frac{k|t - t_0|^m}{m!}$$

Proof   In order that x_n be well defined we must make sure that

$$x_{n-1}(u) \in B(x_0, r) \qquad \forall u \in I$$

Let us verify this by recurrence. Naturally $x_0(u) \in B(x_0, r)$. Now

$$\| x_1(t) - x_0\| \leqslant \int_{t_0}^{t} M\, du = M|t - t_0| \leqslant M \frac{r}{M} = r \quad \text{for } x_1(t) \in B(x_0, r)$$

Similarly

$$\| x_2(t) - x_0\| \leqslant \int_{t_0}^{t} M\, du$$

for $x_1(t) \in B(x_0, r)$, and then $x_2(t) \in B(x_0, r)$, etc.

We shall now verify that the sequence $\{x_n\}$ is a Cauchy sequence for uniform convergence over I. In fact

$$\| x_2(t) - x_1(t)\| \leqslant \int_{t_0}^{t} k\|x_1(u) - x_0(u)\|\, du$$

$$\leqslant Mk \int_{t_0}^{t} |u - t_0|\, du$$

$$= \frac{M}{k} \frac{(k|t - t_0|)^2}{2!}$$

and by recurrence

$$\| x_n(t) - x_{n-1}(t)\| \leqslant \frac{M}{k} \frac{(k|t - t_0|)^n}{n!}$$

Then

$$\| x_m(t) - x_n(t)\| \leqslant \sum_{p=n+1}^{m} \|x_p - x_{p-1}\| \leqslant \frac{M}{k} \sum_{p=n+1}^{\infty} \frac{(k|t - t_0|)^p}{p!}$$

Thus we establish that $\{x_n\}$ is a Cauchy sequence and by making m alone tend to infinity we obtain the upper bound of $\| x_n(t) - x(t)\|$.

Finally we must establish that the limit $x(t)$ is a solution of Eqn (31), i.e. that

$$x(t) - x_0 - \int_{t_0}^t f[s, x(s)] \, ds = 0$$

and for that we repeat exactly the proof on page 40.

In practice we often combine this method with the results of Theorems 3.4.4 or 3.4.5.

Example Let

$$\frac{dX}{dt} = \begin{pmatrix} 1 & -1 \\ t & 0 \end{pmatrix} X \tag{32}$$

be an equation in $\mathbb{R}^2$, i.e. taking coordinates x and y, the system

$$\left. \begin{array}{l} \dfrac{dx}{dt} = x - y \\[2mm] \dfrac{dy}{dt} = tx \end{array} \right\} \tag{33}$$

Let us apply Picard's method for $(t_0 = 0, X_0 = (1, 0))$; we construct

$$X_n(t) = \begin{pmatrix} 1 \\ 0 \end{pmatrix} + \int_0^t \begin{pmatrix} 1 & -1 \\ u & 0 \end{pmatrix} X_{n-1}(u) \, du$$

a relation which can be rewritten as

$$x_n(t) = 1 + \int_0^t [x_{n-1}(u) - y_{n-1}(u)] \, du$$

$$y_n(t) = \int_0^t u x_{n-1}(u) \, du$$

We verify then that

$$x_0(t) \equiv 1 \qquad x_1(t) = 1 + t \qquad x_2(t) = 1 + t + \frac{t^2}{2!} - \frac{t^3}{3!}$$

$$y_0(t) \equiv 0 \qquad y_1(t) = \frac{t^2}{2!} \qquad y_2(t) = \frac{t^2}{2!} + \frac{t^3}{3!}$$

We consider the approximation

$$\tilde{x}(t) = 1 + t + \frac{t^2}{2}$$

$$\tilde{y}(t) = \frac{t^2}{2}$$

Thus

$$\tilde{X}(t) = \begin{pmatrix} \tilde{x}(t) \\ \tilde{y}(t) \end{pmatrix}$$

$\tilde{X}(t)$ is a solution of the equation

$$\frac{d\tilde{X}}{dt} = \begin{pmatrix} 1 & -1 \\ t & 0 \end{pmatrix} \tilde{X} + \begin{pmatrix} 0 \\ -t^2 - \dfrac{t^3}{2} \end{pmatrix} \tag{34}$$

Now let us compare solutions of (33) and (34) for the same initial value. Writing the first equation in the form $dX/dt = f(t, X)$ we establish that

$$f(t, X_1) - f(t, X_2) = \begin{pmatrix} x_1 - x_2 + y_2 - y_1 \\ t(x_1 - x_2) \end{pmatrix}$$

if

$$X_1 = \begin{pmatrix} x_1 \\ y_1 \end{pmatrix} \qquad X_2 = \begin{pmatrix} x_2 \\ y_2 \end{pmatrix}$$

Taking as the norm $\|V\| = |V_1| + |V_2|$ we obtain

$$\|f(t, X_1) - f(t, X_2)\| = (1 + t)|x_1 - x_2| + |y_1 - y_2|$$

$$\leqslant (1 + t)\|X_1 - X_2\|$$

$$\leqslant (1 + T)\|X_1 - X_2\|$$

if $t \in [0, T]$; over this set f is Lipschitzian with constant $1 + T$.
Now let the second equation be written $d\tilde{X}/dt = g(t, \tilde{X})$:

$$\|g(t, \tilde{X}) - f(t, \tilde{X})\| = t^2 + \frac{t^3}{2} \leqslant T^2 + \frac{T^3}{2}$$

Thus

$$\|X(t) - \tilde{X}(t)\| \leqslant \frac{T^2(2 + T)}{2(1 + T)} \{\exp[(1 + T)t] - 1\}$$

$$\leqslant \frac{T^2(2 + T)}{1 + T} \{\exp[(1 + T)T] - 1\}$$

If $T = 0.1$, we obtain 1.1×10^{-3} approximately. If $T = 1$, we obtain 4.8 approximately.

Comments   This method is a theoretical method which allows us to study a phenomenon through a simplified equation. It is not a method for numerical calculation; practical methods depend on a partitioning of the interval $[0, T]$ such that $0 < t_1 < t_2 < \ldots < T$ and on successive calculation of the $x(t_i)$

through

$$x(t_n) = x(t_{n-1}) + \int_{t_{n-1}}^{t_n} f(x, u)\, du$$

(cf. the Runge–Kutta method).

Important remark It is not through mathematical pedantry that we have introduced the notion of k-Lipschitzian functions rather than restricting ourselves to functions of class C^1; the various preceding formulae show in fact the *practical* importance of the value of the constant k.

See *Exercise 42* at the end of the chapter.

Part 4: Regularity of solutions; linear variational equations

Consider the differential equation $dx/dt = f(t, x, \lambda)$ where $x \in \mathbb{R}^d$ and $\lambda \in \mathbb{R}^p$ is a parameter. We assume that f is of class C^1 with respect to the three variables (t, x, λ). For every point (t_0, x_0) of the domain in which f is of class C^1 there exists for every λ a unique maximal solution such that $x(t_0) = x_0$; we shall denote this by $\varphi(t, t_0, x_0, \lambda)$. The object of Part 4 is the study of the dependence of φ on t_0, x_0 and λ.

Definition 4.1.1
Let

$$\frac{dx}{dt} = f(t, x, \lambda) \tag{35}$$

be a differential equation in which f is of class C^1. Let $\varphi(t, t_0, x_0, \lambda)$ be the solution such that $x(t_0) = x_0$. The equation

$$\frac{dz}{dt} = \frac{\partial f}{\partial x}[t, \varphi(t, t_0, x_0, \lambda)]z$$

is called the **linear variational equation relative to** φ.

In this equation $\partial f/\partial x \in \mathscr{L}(\mathbb{R}^d, \mathbb{R}^d)$ and $z \in \mathscr{L}(\mathbb{R}^q, \mathbb{R}^d)$. (If $q = 1$, it is a vector.)

Study of Eqn (35) in the neighbourhood of $\varphi(t, t_0, x_0, \lambda)$
Let x be a solution of (35) and $y(t) = x(t) - \varphi(t, t_0, x_0)$.

$$\frac{dy}{dt} = f[t, y(t) + \varphi(t, t_0, x_0)] - f[t, \varphi(t, t_0, x_0)]$$

$$= \frac{\partial f}{\partial x}[t, \varphi(t, t_0, x_0)]y(t) + y(t)\varepsilon[t, y(t)] \tag{35'}$$

where

$$\lim_{\|u\| \to 0} \varepsilon(t, u) = 0$$

Then let

$$\frac{dy}{dt} = \frac{\partial f}{\partial x}[t, \varphi(t, t_0, x_0)]y \tag{36}$$

If we have the conditions for application of Theorem 3.4.4 (p. 43) or of an analogous theorem allowing us to compare the solutions of (35′) and (36), the linear variational equation in a neighbourhood of φ thus allows the study of $x - \varphi$.

The equation occurs in the study of partial derivatives of $\varphi(t, t_0, x_0, \lambda)$.

Theorem 4.1.2

Let $dx/dt = f(t, x, \lambda)$.

(1) Suppose that f is k-Lipschitzian in x (k independent of λ); then $\varphi(t, t_0, x_0, \lambda)$ is continuous in all four variables.

(2) If f is of class C^m in (t, x, λ), φ is itself of class C^m in (t, x, λ); in particular $\partial\varphi/\partial t_0$ is the solution of

$$\frac{dz}{dt} = \frac{\partial f}{\partial x}[t, \varphi(t, t_0, x_0, \lambda), \lambda]z$$

such that $z(t_0) = -f(t_0, x_0, \lambda) \in \mathbb{R}^d$; $\partial\varphi/\partial x_0$ is the solution of

$$\frac{dz}{dt} = \frac{\partial f}{\partial x}[t, \varphi(t, t_0, x_0, \lambda), \lambda]z$$

such that $z(t_0)$ is the identity element in $\mathscr{L}(\mathbb{R}^d, \mathbb{R}^d)$; $\partial\varphi/\partial\lambda$ is the solution of

$$\frac{dz}{dt} = \frac{\partial f}{\partial x}(t, \varphi, \lambda)z + \frac{\partial f}{\partial\lambda}(t, \varphi, \lambda)$$

such that $z(t_0) = 0 \in \mathscr{L}(\mathbb{R}^p, \mathbb{R}^d)$.

Comments   (1) It must be understood that the continuity of φ, with respect to the pair (t, x_0) for example, entails the proximity of $\varphi(t, t_0, y_0)$ and $\varphi(t, t_0, x_0)$ only for $|y_0 - x_0|$ small and $|t - t_0|$ small.

For example for $dx/dt = x^2$, $\varphi(t, 0, 0) \equiv 0$ and $\varphi(t, 0, a) = a/(1 - at)$, so that if $t \to 1/a$, $\varphi(t, 0, a) \to \infty$ even if a is arbitrarily small.

To obtain conditions for the proximity of $\varphi(t, t_0, x_0)$ and $\varphi(t, t_0, y_0)$ for all t we introduce the notion of **stability**. Thus we shall say that $\varphi(t, t_0, x_0)$ is stable if

$$\forall \varepsilon > 0 \quad \exists \eta: \|x_1 - x_0\| \leqslant \eta \Rightarrow \|\varphi(t, t_0, x_0) - \varphi(t, t_0, x_1)\| \leqslant \varepsilon \quad \forall t \geqslant t_0$$

(2) To understand the stated equations fully, we recall that $\partial\varphi/\partial t_0$ is a vector in $\mathbb{R}^d$ whereas $\partial\varphi/\partial x_0 \in \mathscr{L}(\mathbb{R}^d, \mathbb{R}^d)$, the theorem indicating that its initial value is the identity element; again, $\partial\varphi/\partial\lambda \in \mathscr{L}(\mathbb{R}^p, \mathbb{R}^d)$. One could for example

interpret the second equation by writing that $\forall V \in \mathbb{R}^d$

$$\frac{\partial \varphi}{\partial x_0}(t, t_0, x_0)V = V + \int_{t_0}^t \frac{\partial f}{\partial x}[s, \varphi(s, t_0, x_0)]\frac{\partial \varphi}{\partial x_0}(s, t_0, x_0)V \, ds$$

Let us examine as an example this relation in $\mathbb{R}^2$; we write

$$x = (x_1, x_2) \quad f(t, x) = [f_1(t, x_1, x_2), f_2(t, x_1, x_2)]$$

$$x_0 = (\alpha, \beta) \quad \varphi(t, t_0, x_0) = [\varphi_1(t, t_0, x_0), \varphi_2(t, t_0, x_0)]$$

and

$$V = \begin{pmatrix} V_1 \\ V_2 \end{pmatrix}$$

Thus

$$\begin{pmatrix} \dfrac{\partial \varphi_1}{\partial \alpha}[t, \varphi(t)] & \dfrac{\partial \varphi_1}{\partial \beta}[t, \varphi(t)] \\[2ex] \dfrac{\partial \varphi_2}{\partial \alpha}[t, \varphi(t)] & \dfrac{\partial \varphi_2}{\partial \beta}[t, \varphi(t)] \end{pmatrix}\begin{pmatrix} V_1 \\ V_2 \end{pmatrix}$$

$$= \begin{pmatrix} V_1 \\ V_2 \end{pmatrix} + \int_{t_0}^t \begin{pmatrix} \dfrac{\partial f_1}{\partial x_1}[s, \varphi(s)] & \dfrac{\partial f_1}{\partial x_2}[s, \varphi(s)] \\[2ex] \dfrac{\partial f_2}{\partial x_1}[s, \varphi(s)] & \dfrac{\partial f_2}{\partial x_2}[s, \varphi(s)] \end{pmatrix}$$

$$\times \begin{pmatrix} \dfrac{\partial \varphi_1}{\partial \alpha}[s, \varphi(s)] & \dfrac{\partial \varphi_1}{\partial \beta}[s, \varphi(s)] \\[2ex] \dfrac{\partial \varphi_2}{\partial \alpha}[s, \varphi(s)] & \dfrac{\partial \varphi_2}{\partial \beta}[s, \varphi(s)] \end{pmatrix}\begin{pmatrix} V_1 \\ V_2 \end{pmatrix} ds$$

If we wished to separate the effects of α and β we could choose $V = (1, 0)$ or $V = (0, 1)$.

Indications of the proof of Theorem 4.1.2

We shall establish the first point for an equation in which λ does not appear and we shall show only that $\varphi(t, t_0, x_0)$ is a continuous function of (t, x_0). To simplify the writing we shall omit t_0 in the following.

$$\|\varphi(t_1, x_1) - \varphi(t, x)\| \leqslant \|\varphi(t_1, x_1) - \varphi(t_1, x)\| + \|\varphi(t_1, x) - \varphi(t, x)\|$$

$$\varphi(t_1, x_1) - \varphi(t_1, x) = x_1 - x + \int_{t_0}^{t_1} \{f[s, \varphi(s, x_1)] - f[s, \varphi(s, x)]\} \, ds$$

and, when f is k-Lipschitzian,

$$\|\varphi(t_1, x_1) - \varphi(t_1, x)\| \leqslant \|x_1 - x\| + k\int_{t_0}^{t_1} \|\varphi(s, x_1) - \varphi(s, x)\| \, ds$$

$$\leqslant \|x_1 - x\| \exp(k|t_1 - t_0|)$$

according to Gronwall. Moreover

$$\|\varphi(t_1, x) - \varphi(t, x)\| \leqslant \int_{t_0}^{t_1} \|f[s, \varphi(s, s)]\| \, ds$$

For t and t_1 in the interval $[t_0 - T, t_0 + T]$, $\|f[s, \varphi(s, x)]\| \leqslant M$ and therefore

$$\|\varphi(t_1, x_1) - \varphi(t, x)\| \leqslant \|x_1 - x\| \exp(k|t_1 - t_0|) + M|t_1 - t|$$

If we choose

$$|t_1 - t| \leqslant \frac{\varepsilon}{2M} \quad \text{and} \quad \|x_1 - x\| \leqslant \frac{\varepsilon}{2} \exp(-kT)$$

this quantity is less than or equal to ε.

As far as the second point in Theorem 4.1.2 is concerned, we shall examine $\partial\varphi/\partial t_0$ only for a scalar equation not depending on λ. Let

$$\Delta\varphi(t) = \varphi(t, t_0 + h, x_0) - \varphi(t, t_0, x_0)$$

$$= \int_{t_0+h}^{t} f[s, \varphi(s, t_0 + h, x_0)] \, ds - \int_{t_0}^{t} f[s, \varphi(s, t_0, x_0)] \, ds$$

$$= \int_{t_0+h}^{t} \{f[s, \varphi(s, t_0 + h, x_0)] - f[s, \varphi(s, t_0, x_0)]\} \, ds$$

$$- \int_{t_0}^{t_0+h} \{f(t_0, x_0) + f[s, \varphi(s, t_0, x_0)] - f(t_0, x_0)\} \, ds$$

By applying Taylor's formula to the first term, we obtain

$$\int_{t_0+h}^{t} \frac{\partial f}{\partial x}[s, \varphi(s, t_0, x_0)] \Delta\varphi(s) \, ds + \int_{t_0+h}^{t} \Delta\varphi(s)\varepsilon[\Delta\varphi(s)] \, ds$$

where

$$\lim_{u \to 0} \varepsilon(u) = 0$$

so that

$$\frac{\Delta\varphi(t)}{h} = \int_{t_0+h}^{t} \frac{\partial f}{\partial x}[s, \varphi(s, t_0, x_0)] \frac{\Delta\varphi(s)}{h} - f(t_0, x_0) + \eta(h)$$

We can then prove (a fact which is not evident) that

$$\lim_{h \to 0} \eta(h) = 0$$

and we justify the tendency to the limit when h tends to zero, to obtain

$$\frac{\partial\varphi}{\partial t_0}(t) = \int_{t_0}^{t} \frac{\partial f}{\partial x}[s, \varphi(s, t_0, x_0)] \frac{\partial\varphi}{\partial t_0}(s) \, ds - f(t_0, x_0)$$

and we see where the initial value of $\partial\varphi/\partial t_0 = -f(t_0, x_0)$ comes from.

In $\mathbb{R}^d$ and for $\partial\varphi/\partial\lambda$ or $\partial\varphi/\partial x_0$ the method is similar; the proof is long and intricate and thus we shall not give it.

See *Exercises 43 to 44* at the end of the chapter.

Exercises

1. Construct the orbits for the equation

$$\frac{d^2x}{dt^2} + x = -h \operatorname{sgn}\left(\frac{dx}{dt}\right) \qquad h > 0$$

(This equation can represent the motion of a pendulum subject to a constant frictional force whose direction changes with the sign of the velocity under certain normal restrictions.)

2. Study the equation $(dx/dt)^2 = 4x$.

3. Trace the orbits of the system

$$\frac{dx}{dt} = y \qquad \frac{dy}{dt} = -x + x^3$$

(without solving the equation).

4.
$$\underset{a_{2n-1}}{\overset{f<0}{\rule{2cm}{0.4pt}}} \quad \underset{a_{2n}}{\overset{f>0}{\rule{2cm}{0.4pt}}} \quad \underset{a_{2n+1}}{}$$

Let $\{a_n\}$ be a sequence of isolated points of $\mathbb{R}$ and let f be a real function such that $f(a_n) = 0 \quad \forall n$:

$$f(x) > 0 \quad \text{if } a_{2n} < x < a_{2n+1}$$

$$f(x) < 0 \quad \text{if } a_{2n-1} < x < a_{2n}$$

Let $dx/dt = f(x)$ be a differential equation.

Suppose f is such that at every point (t_0, x_0) the Cauchy problem has a unique solution; verify that if $a_{2n} < x_0 < a_{2n+1}$

$$\lim_{t \to \infty} \varphi(t, t_0, x_0) = a_{2n+1}$$

whereas

$$\lim_{t \to -\infty} \varphi(t, t_0, x_0) = a_{2n}$$

Prove the analogous results in the other intervals. If the sequence is written $a_0 < a_1 < \ldots < a_N$, study what happens in $]-\infty, a_0]$ and $[a_N, +\infty[$.

5. *Prey–predator problem*: The following equation is often encountered when we wish to study the growth of two populations which influence each other:

$$\frac{dx}{dt} = x(1 - y) \qquad \frac{dy}{dt} = y(x - 1)$$

(a) Construct the vector field.

(b) Let $\{x(0), y(0)\}$ be a point such that $x(0)y(0) > 0$. Show that every solution passing through this point has an orbit such that $xy > 0$ for all t. (The existence and uniqueness of solutions for every point (t, x, y) may be assumed.)

(c) The regions of the quadrant $xy > 0$ in Fig. 1.22 are denoted by I, II, III and IV (boundaries not included). Let $M = (x(0), y(0)) \in (\text{I})$.

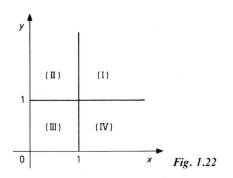

Fig. 1.22

Establish that every solution φ such that $\varphi(0) = M$ moves out of region (I) within a finite time. Deduce from this that the solutions traverse the regions (I), (II), (III), (IV), etc., in succession.

6. Verify that over $\mathbb{R}$ all norms are proportional to the absolute value (which is a norm).

7. For every element $x = (x_1, \ldots, x_d)$ of $\mathbb{R}^d$ consider the quantities $\|x\|_{\max} = \max|x_i|$, $\|x\|_s = \Sigma |x_i|$.

(a) Verify that these are two norms.

(b) If $B(0, \rho) = \{x : \|x\| \leqslant \rho\}$ is the ball of radius ρ centred at the origin, sketch in $\mathbb{R}^2$ the balls corresponding to the above norms as well as the Euclidean norm. Verify the equivalence of all these norms.

8. Verify that, if $\lim V(t) = V$, then $\lim \|V(t)\| = \|V\|$. Is the converse true?

9. Let $\|\cdot\|$ be a norm in $\mathbb{R}^d$ and

$$\||A\|| = \sup_{\{\|x\| = 1\}} \|Ax\|$$

defined for $A \in \mathscr{L}(\mathbb{R}^d, \mathbb{R}^d)$. Verify that

$$\||A\|| = \sup_{\{\|x\| \leqslant 1\}} \|Ax\| = \sup_{x \neq 0} \frac{\|Ax\|}{\|x\|}$$

Verify that this norm over $\mathscr{L}(\mathbb{R}^d, \mathbb{R}^d)$ is compatible with the initial norm. Find the norms constructed in this way starting from $\|\cdot\|_{\max}$ and from $\|\cdot\|_s$.

10. Verify that for all norms $\|\cdot\|$ over $\mathbb{R}^d$ and $\|\|\cdot\|\|$ over the $d \times d$ matrices, there exist constants K_1 and K_2, both strictly positive, such that

$$\|Ax\| \leqslant K_1 \|\|A\|\| \cdot \|x\|$$

$$\|\|A \cdot B\|\| \leqslant K_2 \|\|A\|\| \cdot \|\|B\|\|$$

11. If $x \in \mathbb{R}^d$ we denote by $|x|_e$ the Euclidean norm of x. Let $\mathscr{N}(x)$ be an arbitrary norm.

(a) Verify that $d^{-1/2}|x|_e \leqslant \|x\|_{\max} \leqslant |x|_e$, and hence deduce that there exists $M : \mathscr{N}(x) \leqslant Md|x|_e$.

(b) Verify that if

$$\lim_{n \to \infty} |x_n - x|_e = 0$$

then

$$\lim_{n \to \infty} \mathscr{N}(x_n) = \mathscr{N}(x)$$

for every sequence x_n.

If $B(0, 1) = \{x : |x|_e = 1\}$, verify that $\mathscr{N}(x)$ attains a minimum A and a maximum B over $B(0, T)$ and that $A|x|_e \leqslant (x) \leqslant B|x|_e$. Hence deduce that over $\mathbb{R}^d$ all norms are equivalent.

12. In a vector space of infinite dimension all norms are not in general equivalent.

Let E be the vector space of functions defined over $[0, 1]$ which are continuous except at a finite number of points. Let

$$\|f\|_1 = \int_0^1 |f(x)| \, dx$$

and

$$\|f\|_2^2 = \int_0^1 |f(x)|^2 \, dx$$

Verify that $\|\cdot\|_1$ and $\|\cdot\|_2$ are norms over E.

Show that if $\|f_n\|_2 \to 0$ then $\|f_n\|_1 \to 0$. Find a counter-example showing that the converse is false.

13. Verify that for every norm

$$\lim_{h \to 0} \frac{1}{h} \left\| V(t + h) - V(t) - h \frac{dV}{dt} \right\| = 0$$

14. Verify that

$$\int_a^b V(s)\,ds = \lim \sum_{i=0}^{i=n-1} (t_{i+1} - t_i)V(t_i)$$

when the subdivisions $a \leqslant t_0 < t_1 < \ldots < t_n \leqslant b$ are 'refined' (made smaller).

15. Use the preceding result to prove Proposition 2.1.8.

16. If f is a vector and M and N are matrices calculate

$$\frac{d}{dt}[M(t)f(t)], \quad \frac{d}{dt}[M(t)N(t)] \quad \text{and} \quad \frac{d}{dt}[M^{-1}(t)]$$

in the case where M is invertible.

17. If g is a differentiable function of two variables such that $g(x, y) \in \mathbb{R}$, calculate $(d/dt)g(t, t)$.

18. Study the following two equations:

$$\frac{d^2x}{dt^2} + \omega_0^2(1 + \beta x^2)x = 0 \quad \text{and} \quad \frac{d^2x}{dt^2} - \omega_0^2(1 + \beta x^2)x = 0$$

The first represents (for example) the motion of a unit mass subjected to a positive restoring force proportional to the displacement; the second represents the same phenomenon with a negative restoring force. According as β is negative or positive we speak of a soft or a hard restoring force.

Study the fixed points of these equations.

Verify that the quantities

$$\frac{1}{2}\left(\frac{dx}{dt}\right)^2 + \omega_0^2\left(\frac{x^2}{2} + \beta\frac{x^4}{4}\right)$$

and

$$\frac{1}{2}\left(\frac{dx}{dt}\right)^2 - \omega_0^2\left(\frac{x^2}{2} + \beta\frac{x^4}{4}\right)$$

are first integrals (they correspond to the 'energy'); hence deduce an indication of the form of the orbits.

19. *Prey–predator problem* (see Exercise 5).

(a) Verify that the function $H(x, y) = x - \log x + y - \log y$ is a first integral for $x > 0$ and $y > 0$.

(b) We define a limit cycle as a closed orbit Γ such that there exists a neighbourhood of Γ possessing the following property: every orbit passing through a point of V tends towards Γ (Fig. 1.23).

Prove that, if a limit cycle existed, H would be constant in a neighbourhood of the cycle; hence deduce that no limit cycle exists.

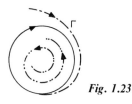

Fig. 1.23

(c) Let α_0 be a point of region I in Fig. 1.24. It may be taken that the solution passing through α_0 at the instant t_0 is defined $\forall\, t \in \mathbb{R}$. We then denote by $t_1, t_2, \ldots, t_n, \ldots$ the instants subsequent to t_0 at which the orbit passing through x_0 cuts the axis $x = 1$ above the point A. We denote by $a_1, a_2, \ldots, a_n, \ldots$ the corresponding points.

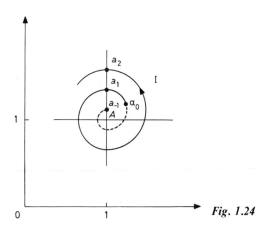

Fig. 1.24

In the same way we define the sequence $t_{-1}, t_{-2}, \ldots$ of instants prior to t_0 at which the orbit has the same properties, and we denote by $a_{-1}, \ldots, a_{-n}, \ldots$ the corresponding points.

Prove that, if the orbit passing through x_0 is not periodic, $H(a_n)$ tends to $H(1,1)$ if $n \to +\infty$ or if $n \to -\infty$.

(d) Prove that $H(1, 1)$ is a strict minimum of $H(x, y)$. Hence deduce that every orbit of the 'prey–predator' equation is periodic.

20. Consider the system

$$\left.\begin{aligned}
\frac{dy}{dt} &= -2x(x - 1)(2x - 1) \\[2mm]
\frac{dx}{dt} &= -2y
\end{aligned}\right\} \tag{37}$$

Verify that the function $x^2(x - 1)^2 + y^2$ is constant on the trajectories of (37); hence deduce the behaviour of the orbits.

21. Consider the **pendulum** equation

$$\frac{d^2\vartheta}{dt^2} + k^2 \sin\vartheta = 0$$

equivalent to the system

$$\frac{dx}{dt} = y \qquad \frac{dy}{dt} = -k^2 \sin x$$

Let

$$H(x, y) = \frac{y^2}{2} + k^2(1 - \cos x)$$

Verify that H is a first integral. Hence deduce the form of the orbits resulting from the value $H(x, y) = a$. If $0 < a < 2k^2$ there exist periodic solutions. If x_m is the maximum amplitude of one of these solutions verify that the period T is

$$T = \frac{2\sqrt{2}}{k} \int_0^{x_m} \frac{du}{(\cos u - \cos b)^{1/2}}$$

Give an expression for T if $b \ll 1$.

22. Establish the Taylor expansion appearing in Definition 2.2.6.

23. Solve the following equations by transforming to polar coordinates:

$$\frac{dx_1}{dt} = x_2 + x_1(1 - x_1{}^2 - x_2{}^2)$$

$$\frac{dx_2}{dt} = -x_1 + x_2(1 - x_1{}^2 - x_2{}^2)$$

24. Use Theorem 2.2.11 to prove the existence of first integrals in the neighbourhood of every regular point of $dx/dt = f(x)$.

25. *Linearized equation*: let $dX/dt = F(X)$ be an equation in $\mathbb{R}^2$; we define the linearized equation as $dY/dt = DF(a)Y$ at a point a at which $F \equiv 0$.
    Write the linearized equation of

$$\frac{d^2x}{dt^2} + f(x)\frac{dx}{dt} + g(x) = 0$$

where f and g are differentiable.
    Example: $d^2\vartheta/dt^2 + \sin\vartheta = 0$ in the neighbourhood of $\vartheta = \pi$.

26. Construct an ε-approximate solution of $dx/dt = 2x + 1$ such that $x(0) = 1$.

27. Verify that

$$x_0 \cos\omega t - \mu\frac{x_0{}^3}{32\omega^2}(\cos\omega t - \cos 3\omega t)$$

(μ is 'small') where $\omega^2 = \omega_0{}^2 + \frac{3}{4}x_0{}^2\mu$ is an ε-approximate solution of

$$\frac{d^2x}{dt^2} + \omega_0{}^2x + \mu x^3 = 0$$

28. Construct an ε-approximate solution of the system in $\mathbb{R}^2$

$$\frac{dx}{dt} = -x + y + 1 \qquad \frac{dy}{dt} = 2x + y$$

such that

$$X(0) = \begin{pmatrix} x(0) \\ y(0) \end{pmatrix} = \begin{pmatrix} 1 \\ -1 \end{pmatrix}$$

29. Given the equation $dx/dt = x^3$ where $x \in \mathbb{R}$, examine the Cauchy problem for $(0, x_0)$. If $0 < \alpha < 1$ examine the same problem for the equation $dx/dt = x^{1+\alpha}$.

30. Let $dx/dt = f(t, x)$ be a differential equation where f is locally Lipschitzian over $\mathbb{R} \times \mathbb{R}$. Let $\varphi(t, t_0, x_0)$ be the maximal solution such that $x(t_0) = x_0$.

(a) Assuming that the maximal interval relative to φ is bounded, let $[S, T]$ be this interval; prove that

$$\lim_{t \to T} \varphi(t, t_0, x_0) = \infty$$

(b) Suppose now that $|f(t, x)| \leqslant g(|x|)$ where g is a function of $\mathbb{R}^+ \to \mathbb{R}^{+*}$ $(g(u) - 0 \;\; \forall u > 0)$ and such that

$$\int_0^\infty \frac{du}{g(u)} = \infty$$

Prove that the maximal interval relative to φ is then the entire space $\mathbb{R}$. (Example: $f(t, x) = x^\beta$.)

31. Let $dx/dt = 1 + x^2$ where $x \in \mathbb{R}$; examine the Cauchy problem for $(0, x_0)$.

32. Let $dx/dt = x^{2/3}$ be a differential equation over $\mathbb{R}^+$.

(a) Verify that $\forall \, T > 0$ there exists an infinity of solutions over $[0, T]$ such that $x(0) = 0$.

(b) If $T \leqslant 3$, verify that there exists only one solution over $[0, T]$ such that $x(0) = -1$; if, on the contrary, $T > 3$ verify that there exists an infinity of solutions over $[0, T]$ such that $x(0) = -1$.

33. Verify that the equation $dx/dt = x^{1/3}(\sin 2t)$ has at least two solutions such that $x(0) = 0$.

34. Verify that the equation

$$\frac{dx}{dt} = \begin{cases} x^{1/2} & \text{for } x \geqslant 0 \\ -(-x)^{1/2} & \text{for } x < 0 \end{cases}$$

has an infinity of solutions such that $x(0) = 0$.

35. Examine the Cauchy problem for the equations

$$\frac{dx}{dt} = \frac{1}{x} \qquad \frac{dx}{dt} = \frac{x}{t} \qquad \frac{dx}{dt} = \frac{x}{t} + 1$$

(involving the value of $x(0)$).

Examine the uniqueness of the Cauchy problem for $dx/dt = x^{4/3}$ according as $x(0) = 0$ or $x(0) = 1$.

36. For the equation $dx/dt = 1/2x$ examine the Cauchy problems for $x(0) = 0$ and $x(1) = 1$.

37. Calculate the constants h, M and C of the result of the application of Theorem 3.4.5, as functions of the coordinates f_i of f.

38. Compare the solutions of $dx/dt = \sin xt$ and $dx/dt = xt$ over the interval $[0, T]$ as functions of x_0.

39. Apply Theorem 3.4.5 to the pendulum equation.

40. Compare the solutions of $dx/dt = t + \sin x$ over the interval $[0, 1]$ for the initial values $x(0) = 0$ and $x(0) = 0.01$.

41. Apply the method of linearization for small displacements to the system

$$\frac{dx}{dt} = tx + y^2 \qquad \frac{dy}{dt} = -y + 2x^2$$

42. Apply the method of Picard to the equations

(a) $\quad \dfrac{dx}{dt} = x + 1 \qquad\qquad x(0) = 1$

(b) $\quad \dfrac{dx}{dt} = 2x + 1 \qquad\qquad x(0) = 1$

(c) $\quad \dfrac{dx}{dt} = t + x \qquad\qquad x(0) = a$

(d) $\quad \dfrac{dx}{dt} = \sin t \cos t \qquad x(0) = 0$

(e) $\quad \dfrac{dx}{dt} = y$

$\qquad\qquad\qquad\qquad\qquad\qquad\qquad\left.\begin{array}{l}\\[1.5em]\\[1.5em]\end{array}\right\}\ x(0) = x_0;\ y(0) = y_0$

$\quad \dfrac{dy}{dt} = -\alpha x - \beta x^3 - \varepsilon y + A \cos \omega t$

(f) $\quad \dfrac{dx}{dt} = x^2 - tx \qquad\qquad x(0) = 1;\ t \in [0, 1]$

43. Let

$$\frac{dx}{dt} = x(1-x) + y \qquad \frac{dy}{dt} = y(1-y)$$

Calculate, in the neighbourhood of the solution $x \equiv 0 \equiv y$, the quantities

$$\frac{\partial x}{\partial x_0}, \qquad \frac{\partial x}{\partial y_0}, \qquad \frac{\partial y}{\partial x_0}, \qquad \frac{\partial y}{\partial y_0}$$

44. Let $dx/dt = \alpha x + \beta$ be a differential equation in $\mathbb{R}$; $x \equiv -\beta/\alpha$ is a solution of this equation. In the neighbourhood of $\alpha = 0$ this quantity is not a continuous function of α. How can this be reconciled with Theorem 4.1.2?

Chapter 2

Linear equations

Definition

We say that the equation $dx/dt = f(t, x)$ is a linear equation in $\mathbb{R}^d$ if it is of the form

$$\frac{dx}{dt} = A(t)x + b(t)$$

where $A(t) \in \mathscr{L}(\mathbb{R}^d, \mathbb{R}^d)$ is a linear mapping, **continuous** in t, and $b(t) \in \mathbb{R}^d$ is a **continuous vector** in $\mathbb{R}^d$.

We shall assume that A and b are defined over the entire space $\mathbb{R}$ so that for every t_0 and every x_0 there exists a unique solution $\varphi(t, t_0, x_0)$ to the Cauchy problem relative to (t_0, x_0) defined over all $\mathbb{R}$.

Except where specified otherwise, we shall use in $\mathscr{L}(\mathbb{R}^d, \mathbb{R}^d)$ the norm

$$\|A\| = \sup_{(x:\, \|x\| = 1)} \|Ax\|$$

Part 1: Resolvent

1.1. Homogeneous equations

VECTOR SPACE OF SOLUTIONS; NOTION OF RESOLVENT

Definition 1.1.1

We define a homogeneous linear equation, or a linear equation with the term $b(t)$ equal to zero, by an equation of the form

$$\frac{dx}{dt} = A(t)x \tag{1}$$

where $x \in \mathbb{R}^d$, $t \rightsquigarrow A(t)$ is continuous and $A \in \mathscr{L}(\mathbb{R}^d, \mathbb{R}^d)$.

Definition 1.1.2

Let $\varphi_1, \varphi_2, \varphi_3, \ldots, \varphi_d$ be a set of d continuous solutions from $\mathbb{R}$ to $\mathbb{R}^d$; we say that these solutions are linearly independent (in the vector space of continuous functions) if the relation

$$a_1\varphi_1(t) + a_2\varphi_2(t) + \ldots + a_d\varphi_d(t) \equiv 0$$

implies that $a_1 = a_2 = \ldots = a_d = 0$.

Theorem 1.1.3

The set of solutions of (1) forms a vector space of dimension d, i.e. every solution can be expressed as a linear function of a system of d **linearly independent** solutions. For every solution φ, there exist $a_1, \ldots, a_d$ such that

$$\varphi(t) = a_1\varphi_1(t) + \ldots + a_d\varphi_d(t)$$

Naturally the basis $\varphi_1, \ldots, \varphi_d$ of the vector space of solutions is not unique. Such a basis is called a **fundamental system of solutions**.

Proof The fundamental fact, related to the homogeneity of the equation, is the following: if φ_1 and φ_2 are two solutions of (1), every linear combination $\alpha_1\varphi_1 + \alpha_2\varphi_2$ is also a solution.

According as we consider real-valued or complex-valued solutions, we shall take the vector space of solutions to be constructed over $\mathbb{R}$ (i.e. $\alpha_i \in \mathbb{R}$) or over $\mathbb{C}$ (i.e. $\alpha_i \in \mathbb{C}$).

We note in particular that $t \rightsquigarrow \varphi(t) \equiv 0$ is the unique solution of (1) such that there exists a t_0 such that $\varphi(t_0) = 0$: this is the neutral element of the vector space of solutions. To establish that this space is of dimension d we choose t_0 and a basis $x_1(t_0), \ldots, x_d(t_0)$ of $\mathbb{R}^d$ constructed from d vectors; then if $\varphi_1, \ldots, \varphi_d$ are the solutions such that $\varphi_i(t_0) = x_i(t_0)$ for $i = 1, 2, \ldots, d$, we shall verify that for all t $\varphi_1(t), \ldots, \varphi_d(t)$ is again a basis of $\mathbb{R}^d$. For, if this were not so, there would exist a real u and numbers $\alpha_1, \ldots, \alpha_d$, not all zero, such that $\alpha_1\varphi_1(u) + \ldots + \alpha_d\varphi_d(u) = 0$, i.e. the solution $\alpha_1\varphi_1 + \ldots + \alpha_d\varphi_d$ would be zero at the instant u, and hence zero at every instant, which is absurd.

If then φ is an arbitrary solution of (1) and ϑ an arbitrary point of $\mathbb{R}$, $\varphi(\vartheta) \in \mathbb{R}^d$ and there exist $\lambda_1, \ldots, \lambda_d$ such that

$$\varphi(\vartheta) = \lambda_1\varphi_1(\vartheta) + \lambda_2\varphi_2(\vartheta) + \ldots + \lambda_d\varphi_d(\vartheta)$$

Consider the solution $\varphi - \lambda_1\varphi_1 - \lambda_2\varphi_2 - \ldots - \lambda_d\varphi_d$; this solution vanishes at ϑ, so that it is identically zero:

$$\varphi = \lambda_1\varphi_1 + \ldots + \lambda_d\varphi_d$$

Notation We shall denote by I the **identity mapping** from $\mathbb{R}^d$ into $\mathbb{R}^d$

defined by $I(x) = x$: in every coordinate system, its matrix is of the form

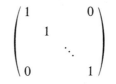

$$\begin{pmatrix} 1 & & & 0 \\ & 1 & & \\ & & \ddots & \\ 0 & & & 1 \end{pmatrix}$$

Resolvent

We have just seen that to obtain solutions it is sufficient to obtain a basis of the vector space which they constitute. We can then study all their properties by studying the properties of bases. There exists naturally an infinity of bases in each coordinate system and these bases depend on the system: the resolvent is a unique linear mapping which summarizes all the properties of the bases and hence of the solutions. We can only obtain it if we can solve the equation, but we can get its properties from the equation.

Instead of stating a proposition affirming that if, in a basis, there exist d independent solutions having a certain property then the solutions will have some other property, we shall state a property relating to the resolvent.

Definition 1.1.4

Let $\varphi_1, \ldots, \varphi_d$ be a basis of solutions of Eqn (1) in a given coordinate system, and let $\hat{\Phi}(t)$ be the matrix whose columns are the coordinates of the vector φ_i:

$$\begin{pmatrix} \varphi_{1,1}(t) & \cdots & \varphi_{d,1}(t) \\ \varphi_{1,2}(t) & & \vdots \\ \vdots & & \\ \varphi_{1,d}(t) & \cdots & \varphi_{d,d}(t) \end{pmatrix}$$

Such a matrix is called a **fundamental matrix**.

The matrix $\hat{\Phi}$ has the following properties: $\hat{\Phi}(t)^{-1}$ exists for all t,

$$\frac{d}{dt}[\hat{\Phi}(t)] = A(t)\hat{\Phi}(t) \quad \text{and} \quad \frac{d}{ds}[\hat{\Phi}(s)]^{-1} = -\hat{\Phi}(s)^{-1}A(s)$$

Theorem 1.1.5 (Resolvent)

Let

$$\frac{dx}{dt} = A(t)x \qquad (2)$$

where $x \in \mathbb{R}^d$; there exists a **linear mapping**

$$\Phi: \begin{array}{l} \mathbb{R} \times \mathbb{R} \to \mathscr{L}(\mathbb{R}^d, \mathbb{R}^d) \\ (t, s) \rightsquigarrow \Phi(t, s) \end{array}$$

called the **resolvent** of (2) or **resolvent of** A, which is the only linear mapping possessing the following properties:

(a) $\Phi(t, t) = I$ $\quad \forall t \quad$ $\Phi(t, s) = \Phi(t, u)\Phi(u, s)$ $\quad \forall s, t, u \in \mathbb{R}$

In particular

$$\Phi(t, u)^{-1} = \Phi(u, t)$$

(b) $\dfrac{\partial}{\partial t}\Phi(t, s) = A(t)\Phi(t, s)$ $\quad$ and $\quad$ $\dfrac{\partial}{\partial s}\Phi(t, s) = -\Phi(t, s)A(s)$

Moreover, the solution $\varphi(t, t_0, x_0)$ such that $\varphi(t_0) = x_0$ is $\Phi(t, t_0)x_0$.

Proof Let $\hat{\Phi}$ be a fundamental matrix and in a coordinate system let $\Phi(t, s) = \hat{\Phi}(t)\hat{\Phi}(s)^{-1}$.

It is clear that Φ satisfies the properties (a) and (b). Let P be the matrix corresponding to a change of coordinates; the linear mapping represented in the initial system by $\Phi(t, s)$ is represented (in the new system) by $P^{-1}\Phi(t, s)P = \Psi(t, s)$, and the mapping A by $P^{-1}AP = B(t)$. Ψ and B naturally satisfy the relations (a) and (b).

If $\varphi(t_0) = x_0$ in a given coordinate system it is clear that

$$\frac{\mathrm{d}}{\mathrm{d}t}\Phi(t, t_0)x_0 = A(t)\Phi(t, t_0)x_0$$

so that in this system $\varphi(t, t_0, x_0) = \Phi(t, t_0)x_0$. If we make a change of coordinates with matrix P (which is formed by taking as columns the vectors of the new basis expressed in the old one) the coordinates of x_0 become $P^{-1}x_0$, $\Phi(t, t_0)$ becomes $P^{-1}\Phi P$ and one can easily verify that $\Phi(t, t_0)x_0$ becomes $P^{-1}[\Phi(t, t_0)x_0]$.

Finally, let $\tilde{\Phi}$ be another linear mapping having the properties (a) and (b): $\tilde{\Phi}(t, t_0)x_0 = \Phi(t, t_0)x_0$ in every coordinate system, so that $\tilde{\Phi} \equiv \Phi$.

Comments (1) $\Phi(t, t_0)$ is the unique solution of $\mathrm{d}M/\mathrm{d}t = A(t)M$ such that $M(t_0) = I$. Hence

$$\Phi(t, t_0) = I + \int_{t_0}^{t} A(s)\Phi(s, t_0)\,\mathrm{d}s$$

(2) Let $s < u < t$ and $V \in \mathbb{R}^d$: $\Phi(t, s)V = \varphi(t, s, V)$ while $\Phi(u, s)V = \varphi(u, s, V)$. The relation $\Phi(t, s) = \Phi(t, u)\Phi(u, s)$ then states that the solution starting from V at s, evaluated at the instant t, is the same as the solution starting from $\Phi(u, s)V$, the point reached at the instant u, and evaluated at the instant t. This is actually obvious, given the uniqueness (Fig. 2.1).

$\Phi(u, s)V = W$

V

$\Phi(t, u)W = \Phi(t, u)\Phi(u, s)V$ $\quad$ *Fig. 2.1*

(3) If one investigates how a fundamental matrix is transformed, it is discovered that it is not transformed like a linear mapping.

Practical method of calculation of the resolvent

Here we show how to calculate the resolvent when we can solve the equation; it will be sufficient to state the following relation:

$$\varphi(t, t_0, x_0) = \Phi(t, t_0)(x_0)$$

An example of the calculation is as follows. Consider

$$\frac{dX}{dt} = \begin{pmatrix} 0 & 1 \\ -\omega^2 & 0 \end{pmatrix} X \quad \text{or} \quad \frac{d^2 x}{dt^2} + \omega^2 x = 0$$

whose solution is

$$x(t) = x_0 \cos \omega(t - t_0) + \frac{x'_0}{\omega} \sin \omega(t - t_0)$$

i.e.

$$X(t) = \left\{ \begin{array}{l} x_0 \cos \omega(t - t_0) + \dfrac{x'_0}{\omega} \sin \omega(t - t_0) \\[2mm] -\omega x_0 \sin \omega(t - t_0) + x'_0 \cos \omega(t - t_0) \end{array} \right\} = \Phi(t - t_0) \begin{pmatrix} x_0 \\ x'_0 \end{pmatrix}$$

We find that

$$\Phi(t - t_0) = \begin{pmatrix} \cos \omega(t - t_0) & \dfrac{1}{\omega} \sin \omega(t - t_0) \\[2mm] -\omega \sin \omega(t - t_0) & \cos \omega(t - t_0) \end{pmatrix}$$

See *Exercises 1 and 2* at the end of the chapter.

COMPARISON, OVER A FINITE INTERVAL, OF NEIGHBOURING LINEAR EQUATIONS

We shall now discover the convenience and simplicity of the notion of resolvent. The following result is very elementary but is nevertheless interesting.

Proposition 1.1.6
Let Φ be the resolvent of $dx/dt = A(t)x$. Then

$$\forall s \leqslant t \qquad \|\Phi(t, s)\| \leqslant \exp \int_s^t \|A(u)\| \, du$$

Let J be a bounded interval on which $\|A(u)\| \leqslant K$; then

$$\|\Phi(t, t_0)\| \leqslant \exp(K|t - t_0|)$$

so that

$$\|\varphi(t, t_0, x_0)\| \leqslant \exp(K|t - t_0|)\|x_0\| \qquad \forall\, t, t_0 \in J$$

Proof $\Phi(t, s)$ is the solution of $\mathrm{d}M/\mathrm{d}t = A(t)M$ such that $M(s) = I$; hence

$$\Phi(t, s) = I + \int_s^t A(u)\Phi(u, s)\,\mathrm{d}u$$

and consequently

$$\|\Phi(t, s)\| \leqslant 1 + \int_s^t \|A(u)\|\,\|\Phi(u, s)\|\,\mathrm{d}u$$

By applying Gronwall's lemma in the general form suggested in the exercise on page 33

$$\left(\text{if } f(t) \leqslant a + \int_0^t k(u)f(u)\,\mathrm{d}u \quad \text{then} \quad f(t) \leqslant \exp\left[\int_0^t k(u)\,\mathrm{d}u\right] \right)$$

one obtains the final result of the proposition. The remainder is an obvious consequence.

The following theorem allows us to compare the solutions of neighbouring equations.

Theorem 1.1.7
Let $A(t)$ and $B(t)$ be two continuous elements of $\mathscr{L}(\mathbb{R}^d, \mathbb{R}^d)$. Let Φ_A and Φ_B be the corresponding resolvents. We consider the equations

$$\frac{\mathrm{d}x}{\mathrm{d}t} = A(t)x \tag{3}$$

$$\frac{\mathrm{d}y}{\mathrm{d}t} = B(t)y \tag{4}$$

Given t_0 and T, there exists K such that $\|A(t)\| \leqslant K \;\; \forall t \in (t_0, t_0 + T)$.
(a) If

$$\|A(t) - B(t)\| \leqslant \eta \qquad \forall\, t \in [t_0, t_0 + T]$$

then

$$\|\Phi_A(t, s) - \Phi_B(t, s)\| \leqslant \exp(KT)[\exp(\eta T) - 1] \qquad \forall\, s, t \in [t_0, t_0 + T]$$

(b) Let $\varphi(t)$ be the solution of (3) such that $x(t_0) = x_0$ and $\psi(t)$ that of (4) such that $y(t_0) = y_0$; then

$$\|\varphi(t) - \psi(t)\| \leqslant \exp[K(t - s)]\{\exp[\eta(t - s)] - 1\}\|x_0\|$$
$$+ \exp[(K + \eta)(t - s)]\|x_0 - y_0\|$$
$$\forall\, t_0 \leqslant s \leqslant t \leqslant T$$

Proof   (a) We know that

$$\frac{d}{dt}[\Phi_A(t,s) - \Phi_B(t,s)] = A(t)\Phi_A(t,s) - B(t)\Phi_B(t,s)$$

$$= A(t)[\Phi_A(t,s) - \Phi_B(t,s)] + [A(t) - B(t)]\Phi_B(t,s)$$

Since $\Phi_A(s,s) = \Phi_B(s,s)$

$$\|\Phi_A(t,s) - \Phi_B(t,s)\|$$

$$\leqslant \int_s^t \|A(u)\| \, \|\Phi_A(u,s) - \Phi_B(u,s)\| \, du + \int_s^t \|A(u) - B(u)\| \, \|\Phi_B(u,s)\| \, du$$

$$\leqslant K \int_s^t \|\Phi_A(u,s) - \Phi_B(u,s)\| \, du + \eta \int_s^t \exp[(K+\eta)(u-s)] \, du$$

Let $f(t) = \|\Phi_A(t,s) - \Phi_B(t,s)\|$.

$$f(t) \leqslant \frac{\eta}{K+\eta} \{\exp[(K+\eta)(t-s)] - 1\} + K \int_s^t f(u) \, du$$

and it suffices to apply Gronwall's lemma:

$$f(t) \leqslant \frac{\eta}{K+\eta} \{\exp[(K+\eta)(t-s)] - 1\}$$

$$+ K \int_s^t \frac{\eta}{K+\eta} \{\exp[(K+\eta)(u-s)] - 1\} \exp[K(t-u)] \, du$$

$$= \exp[K(t-s)] \{\exp[\eta(t-s)] - 1\}$$

$$\leqslant \exp(KT) [\exp(\eta T) - 1]$$

(b) To establish B we note that

$$\varphi(t) - \psi(t) = \Phi_A(t,s)x_0 - \Phi_B(t,s)y_0$$

$$= [\Phi_A(t,s) - \Phi_B(t,s)]x_0 + \Phi_B(t,s)(x_0 - y_0)$$

which implies the relation indicated.

Comments   (1) We note that the first part of the statement says that for all $\varepsilon > 0$ we can find a corresponding $\eta(\varepsilon)$ such that

$$\sup_{t \in [t_0, t_0 + T]} \|A(t) - B(t)\| \leqslant \eta(\varepsilon) \Rightarrow \sup_{s,t \in [t_0, t_0 + T]} \|\Phi_A(t,s) - \Phi_B(t,s)\| \leqslant \varepsilon$$

i.e. the mapping $A \rightsquigarrow \Phi_A$ is continuous for uniform convergence.

(2) The result above can be compared with the results of Propositions 3.4.4 and 3.4.5; this is the object of Exercises 3 and 4.

See ***Exercises 3 to 6*** at the end of the chapter.

LIOUVILLE'S THEOREM

Liouville's theorem studies the variation of volume in the phase space.

Proposition 1.1.8

If $\Phi(t, s)$ is the resolvent of $A(t)$ then

$$\det[\Phi(t, s)] = \exp \int_s^t \text{tr}[A(u)]\, du$$

(This concerns quantities relating to the linear mappings Φ and A.)

If we fix a system of coordinates, for every fundamental matrix $\hat{\Phi}$

$$\det[\hat{\Phi}(t)] = \det[\hat{\Phi}(s)] \exp\left\{ \int_s^t \text{tr}[A(u)] \right\} du$$

Proof To establish the first formula it is sufficient to choose a coordinate system in which the mapping $\Phi(t, s)$ can be written

$$\Phi(t, s) = \begin{pmatrix} \Phi_{1,1}(t, s) & \cdots & \Phi_{1,d}(t, s) \\ \Phi_{i,1}(t, s) & \cdots & \Phi_{i,d}(t, s) \\ \Phi_{d,1}(t, s) & \cdots & \Phi_{d,d}(t, s) \end{pmatrix}$$

so that

$$\frac{\partial}{\partial t} \det[\Phi(t, s)] = \sum_{i=1}^{d} \det \begin{pmatrix} \Phi_{1,1}(t, s) & \cdots & \Phi_{1,d}(t, s) \\ \dfrac{\partial}{\partial t}\Phi_{i,1}(t, s) & \cdots & \dfrac{\partial}{\partial t}\Phi_{i,d}(t, s) \\ \Phi_{d,1}(t, s) & \cdots & \Phi_{d,d}(t, s) \end{pmatrix}$$

$$= \sum_{i,k=1}^{d} \det \begin{pmatrix} \Phi_{1,1}(t, s) & \cdots & \Phi_{1,d}(t, s) \\ A_{i,k}\Phi_{k,1}(t, s) & \cdots & A_{i,k}\Phi_{k,d}(t, s) \\ \Phi_{d,1}(t, s) & \cdots & \Phi_{d,d}(t, s) \end{pmatrix}$$

$$= \sum_{i,k=1}^{d} A_{ik}(t) \det \begin{pmatrix} \Phi_{1,1} & \cdots & \Phi_{1,d} \\ \Phi_{k,1} & \cdots & \Phi_{k,d} \\ \Phi_{d,1} & \cdots & \Phi_{d,d} \end{pmatrix}$$

The only non-zero terms are those for which $k = i$ and the determinants are all equal to $\det[\Phi(t, s)]$, so that

$$\frac{\partial}{\partial t} \det[\Phi(t, s)] = \text{tr}[A(t)] \det[\Phi(t, s)]$$

This implies, by integration, the formula indicated, since $\Phi(s, s) = I$.

The second formula can be deduced from the relation $\hat{\Phi}(t, s) = \hat{\Phi}(t)\hat{\Phi}(s)^{-1}$.

Interpretation: variation in volume

Let $V_1, V_2, \ldots, V_d$ be d independent vectors of $\mathbb{R}^d$ and let $\varphi_i(t)$ be the solution such that $\varphi_i(t_0) = V_i$, for $i = 1, 2, \ldots, d$. If $W(t)$ is the volume of the parallelepiped determined by $\varphi_1(t), \ldots, \varphi_d(t)$ the preceding formula indicates that

$$W(t) = W(t_0) \exp\left\{ \int_{t_0}^{t} \operatorname{tr}[A(u)] \right\} du$$

Thus, if $\operatorname{tr}[A(u)] \equiv 0$ the volume does not change; however, if $\operatorname{tr}[A(u)] < 0$ $\forall u \in (t_0, t)$ the volume decreases, while if $\operatorname{tr}[A(u)] > 0$ $\forall u \in (t_0, t)$ it increases. (All these ideas are naturally independent of the system of coordinates.)

Example In mechanics we often consider equations of the form

$$\frac{\mathrm{d}^2 q}{\mathrm{d}t^2} = -\operatorname{grad} U(q) \quad \text{where } q \in \mathbb{R}^d, \ U: \mathbb{R}^d \to \mathbb{R} \tag{5}$$

We transform (5) by introducing the coordinates q_i of q as well as their derivatives $p_i = \mathrm{d}q_i/\mathrm{d}t$. Putting

$$T(p) = \sum_{1}^{d} \frac{p_i^2}{2}$$

and

$$H(p, q) = T(p) + U(q)$$

(5) becomes

$$\left. \begin{array}{ll} \dfrac{\mathrm{d}q_i}{\mathrm{d}t} = \dfrac{\partial H}{\partial p_i} & i = 1, \ldots, d \\[3mm] \dfrac{\mathrm{d}p_i}{\mathrm{d}t} = -\dfrac{\partial H}{\partial q_i} & i = 1, \ldots, d \end{array} \right\} \tag{6}$$

The phase space of this system is $\mathbb{R}^{2d}$. H is called the Hamiltonian of the system. In general we suppose that U is of class C^2 so that the second member of (6) is of class C^1.

Theorem 1.1.9 (Liouville's theorem)

For a Hamiltonian system the volume is constant in the phase space.

Proof Let $W(t)$ be the volume defined by the solutions $\varphi_1, \ldots, \varphi_{2d}$ with initial values $V_1, V_2, \ldots, V_{2d}$:

$$\frac{dW}{dt} = \frac{d}{dt} \left| \det[\varphi_1(t), \ldots, \varphi_{2d}(t)] \right|$$

if $\varphi_i(t)$ denotes for each i the solution with initial value V_i.

Since the second member of (6) is of class C^1, the variation in volume in the neighbourhood of a point depends only on the tangential linear mapping and hence, according to the preceding proposition, on the trace of that mapping. Now this trace is the quantity

$$\frac{\partial^2 H}{\partial p_1 \partial q_1} + \frac{\partial^2 H}{\partial p_2 \partial q_2} + \ldots + \frac{\partial^2 H}{\partial p_d \partial q_d} - \frac{\partial^2 H}{\partial q_1 \partial p_1} - \frac{\partial^2 H}{\partial q_2 \partial p_2} - \ldots - \frac{\partial^2 H}{\partial p_d \partial q_d} = 0$$

SCALAR EQUATION OF ORDER n: WRONSKIAN

Let

$$\frac{d^n x}{dt^n} = a_0(t)x + a_1(t)\frac{dx}{dt} + \ldots + a_{n-1}(t)\frac{d^{n-1}x}{dt^{n-1}} \qquad (7)$$

We know that this equation is equivalent to $dX/dt = A(t)X$ where

$$A(t) = \begin{pmatrix} 0 & 1 & 0 & & \ldots & 0 \\ 0 & 0 & 1 & 0 & \ldots & 0 \\ \vdots & & & & & \vdots \\ 0 & 0 & & & \ldots & 1 \\ a_0(t) & & & & \ldots & a_{n-1}(t) \end{pmatrix}$$

Taking account of the particular form of this matrix we shall rewrite the results at the beginning of Section 1.1.

Definition 1.1.10

Let $f_1, \ldots, f_n$ be n functions from $\mathbb{R}$ into $\mathbb{R}$, $f_i : t \rightsquigarrow f_i(t)$, $n-1$ times differentiable. We denote by $W(f_1, \ldots, f_n)$ the function

$$W(f_1 \ldots f_n)(t) = \det \begin{pmatrix} f_1(t) & \ldots & f_n(t) \\ \dfrac{df_1}{dt}(t) & \ldots & \dfrac{df_n}{dt}(t) \\ \vdots & & \vdots \\ \dfrac{d^{n-1}f_1}{dt^{n-1}}(t) & \ldots & \dfrac{d^{n-1}f_n}{dt^{n-1}}(t) \end{pmatrix}$$

$W(f_1, \ldots, f_n)$ is called the **Wronskian** of the functions $f_1, \ldots, f_n$.

If $f_1, \ldots, f_n$ are linearly dependent (see Definition 1.1.2) $W(f_1, \ldots, f_n)$ is zero. The converse is false.

Example   Over the interval $[-1, +1]$ $f_1 = t^2$ and $f_2 = t|t|$ are independent, whereas $W(f_1, f_2) = 0$.

Theorem 1.1.11

The set of solutions of (7) forms a vector space of dimension n. Every system of n solutions of (7) whose Wronskian is non-zero at a point constitutes a basis of this space. Moreover, for every set of n solutions

$$W(t) = W(s) \exp\left[\int_s^t a_{n-1}(u)\, du\right]$$

Proof   We know that (7) is equivalent to

$$\frac{dX}{dt} = A(t)X \tag{8}$$

where X is the vector with coordinates $X_1(t) = x(t)$, $X_2(t) = dx/dt, \ldots, X_n(t) = d^{n-1}x/dt^{n-1}$. To every basis $(\psi_1, \ldots, \psi_n)$ of the vector space of solutions of (8) there corresponds a basis $(\varphi_1, \ldots, \varphi_n)$ of the vector space of solutions of (7) where φ_i is the first coordinate of the vector ψ_i for every i. The converse is also true. We know that a system $(\psi_1, \ldots, \psi_n)$ of solutions of (8) is a basis if $\det(\psi_1, \ldots, \psi_n) \neq 0$ and it suffices for this that it is non-zero at one point. The result stated then follows from the fact that this determinant is the Wronskian of $\varphi_1, \ldots, \varphi_n$.

The last relation expresses the result 1.1.8 when account is taken of the fact that $\mathrm{tr}[A(u)] = a_{n-1}(u)$.

Comment   For the solutions of (7) the relation $W(\varphi_1, \ldots, \varphi_n) \neq 0$ is a necessary and sufficient condition for independence and this implies that n independent functions whose Wronskian vanishes cannot all be solutions of one and the same linear equation of order n.

We shall now derive the solution of the Cauchy problem. The particular form of $A(t)$ implies a particular form for $\Phi(t, t_0)$. The equation $\partial\Phi(t, t_0)/\partial t = A(t)\Phi(t, t_0)$ in fact implies that

$$\Phi(t, t_0) = \begin{pmatrix} \varphi_0(t, t_0) & \cdots & \varphi_{n-1}(t, t_0) \\ \dfrac{\partial}{\partial t}\varphi_0(t, t_0) & \cdots & \dfrac{\partial}{\partial t}\varphi_{n-1}(t, t_0) \\ \vdots & & \vdots \\ \dfrac{\partial^{n-1}}{\partial t^{n-1}}\varphi_0(t, t_0) & \cdots & \dfrac{\partial^{n-1}}{\partial t^{n-1}}\varphi_{n-1}(t, t_0) \end{pmatrix}$$

Moreover $\Phi(t_0, t_0) = I$.

Theorem 1.1.12

Let $t_0 \in \mathbb{R}$ and $(x_0, x_1, \ldots, x_{n-1})$ be a point of $\mathbb{R}^n$. The unique solution of

$$\frac{d^n x}{dt^n} = a_0(t)x + \ldots + a_{n-1} \frac{d^{n-1}x}{dt^{n-1}}$$

such that

$$x(t_0) = x_0 \qquad \frac{dx}{dt}(t_0) = x_1 \qquad \cdots \qquad \frac{d^{n-1}x}{dt^{n-1}}(t_0) = x_{n-1}$$

is

$$x(t) = \Phi_0(t, t_0)x_0 + \Phi_1(t, t_0)x_1 + \ldots + \Phi_{n-1}(t, t_0)x_{n-1}$$

where $\Phi_i(t, t_0)$ is the solution such that

$$\frac{\partial^i \Phi_i}{\partial t^i}(t_0, t_0) = 1$$

while

$$\frac{\partial^j \Phi_i}{\partial t^j}(t_0, t_0) = 0$$

for all $j \neq i$.

(In particular $\Phi_{n-1}(t, t_0)$ is the solution for which all derivatives up to order $n-2$ are zero at t_0, that of order $n-1$ being equal to 1.)

Second-order equations: expression for the resolvent

Proposition 1.1.13

Let

$$x''(t) + a(t)x' + b(t)x = 0 \tag{9}$$

and let f_1, f_2 be two independent solutions of (9).

$$\Phi(t, s) = \frac{1}{W(f_1, f_2)(s)} \begin{pmatrix} f_1(t) & f_2(t) \\ f_1'(t) & f_2'(t) \end{pmatrix} \begin{pmatrix} f_2'(s) & -f_2(s) \\ -f_1'(s) & f_1(s) \end{pmatrix}$$

Proof It is sufficient to note that

$$\begin{pmatrix} f_1 & f_2 \\ f_1' & f_2' \end{pmatrix}(t) = \hat{\Phi}(t)$$

is a fundamental matrix, and to calculate $\hat{\Phi}(t, s) = \hat{\Phi}(t)\hat{\Phi}(s)^{-1}$.

See *Exercises 7 and 8* at the end of the chapter.

SECOND-ORDER EQUATIONS: SEPARATION AND COMPARISON THEOREMS

We shall see how the use of the Wronskian enables us to obtain remarkable results for the equation

$$a_0(t)x'' + a_1(t)x' + a_2(t)x = 0 \tag{10}$$

Theorem 1.1.14 (Sturm)

Let f_1 and f_2 be two independent solutions of (10); between two consecutive zeros of f_1 there is exactly one zero of f_2.

Proof Let u and v be two consecutive zeros of f_1 such that $f_1(u) = 0$, $f_1(v) = 0$ and $f_1(t) \neq 0$ for $u < t < v$.

$$W(f_1, f_2)(t) = f_1(t)f_2'(t) - f_2(t)f_1'(t)$$

and therefore $f_2(u)$ and $f_2(v)$ are non-zero.

Suppose that $f_2(t) \neq 0$ for all $t \in [u, v]$; f_1/f_2 is then well defined and is zero at u and at v, so that its derivative vanishes at a point c of $]u, v[$, but

$$\frac{d}{dt}\left(\frac{f_1}{f_2}\right) = -\frac{W(f_1, f_2)}{f_2{}^2} \neq 0$$

Thus there is at least one zero of f_2 between u and v; if there were two, say t_1 and t_2, there would be, by symmetry, a zero of f_1 between t_1 and t_2, which is impossible since u and v are by hypothesis two consecutive zeros.

Equivalent forms of second-order equations

To study an equation $a_0(t)x'' + a_1(t)x' + a_2(t)x = 0$, we shall naturally suppose that we confine ourselves to segments of $\mathbb{R}$ in which $a_0(t) \neq 0$.

(a) If we put

$$x(t) = u(t) \exp\left[-\frac{1}{2}\int_0^t \frac{a_1(s)}{a_0(s)}\,ds\right]$$

Eqn (10) becomes

$$\frac{d^2u}{dt^2} + q(t)u = 0$$

The zeros of x and of u are obviously the same.

(b) Equation (10) is also equivalent to the equation

$$\frac{d}{dt}\left[p(t)\frac{dx}{dt}\right] + q(t)x = 0$$

This form is called the **self-adjoint** form of (10).

We prove (a) by verifying that

$$u(t) \exp\left[-\frac{1}{2}\int_0^t \frac{a_1(s)}{a_0(s)}\,ds\right]$$

is a solution of $a_0(t)x'' + a_1(t)x' + a_2(t)x = 0$.

To prove (b) we note that it is equivalent to

$$\exp\left[\int_0^t \frac{a_1(s)}{a_0(s)}\,ds\,x''\right] + \frac{a_1}{a_0}\exp\left[\int_{t_0}^t \frac{a_1(s)}{a_0(s)}\,ds\,x'\right] + \frac{a_2}{a_0}\exp\left[\int_{t_0}^t \frac{a_1(s)}{a_0(s)}\,ds\,x\right] = 0$$

Now

$$\frac{d}{dt}\left\{\exp\left[\int_{t_0}^t \frac{a_1(s)}{a_0(s)}\,ds\right]\right\} = \frac{a_1(t)}{a_0(t)}\exp\left[\int_{t_0}^t \frac{a_1(s)}{a_0(s)}\,ds\right]$$

and if we put

$$p(t) = \exp\left[\int_{t_0}^t \frac{a_1(s)}{a_0(s)}\,ds\right]$$

the first two terms are

$$\frac{d}{dt}\left[p(t)\frac{dx}{dt}\right]$$

Theorem 1.1.15 (Sturm)
Let

$$x'' + q(t)x = 0 \quad \text{and} \quad q(t) \leqslant 0 \tag{11}$$

in an interval (t_1, t_2). Every solution of (11), not identically zero, has then at most one zero in the interval $[t_1, t_2]$.

Proof Let $\vartheta \in [t_1, t_2]$ be a zero of a solution φ. Since φ is not identically zero $\varphi'(\vartheta) \neq 0$. If we take, for example, $\varphi'(\vartheta) > 0$ then there exists $u: \varphi(t) > 0$ for all $t \in]\vartheta, u]$, but then $\varphi''(t) = -q(t)\varphi(t) \geqslant 0$; therefore $\varphi'(t) > 0$ for $t \in]\vartheta, u]$ and so $\varphi(t)$ is strictly increasing in $[\vartheta, u]$. Thus since $\varphi'(u)$ is greater than zero, $\varphi(t)$ is in fact strictly increasing—and therefore non-zero—in $[\vartheta, t_2]$. In the same way if $\lambda < \vartheta$ is another zero of φ, $\varphi(t)$ must be either strictly increasing, and therefore positive for all $t > \lambda$ and in particular for λ, which is impossible, or strictly decreasing, and therefore negative for all $t > \lambda$, which is also impossible.

Theorem 1.1.16 (Sturm)
Let

$$x'' + f(t)x = 0 \tag{12}$$

$$x'' + g(t)x = 0 \tag{13}$$

be two equations and I an interval in which $f(t) > g(t)$. Let φ be a solution of (12) and ψ a solution of (13); then between two consecutive zeros of ψ in I there is at least one zero of φ.

Proof Let u and v be two consecutive zeros of ψ and let $\psi(t) > 0$, for example, in $]u, v[$. Suppose that φ does not vanish and that $\varphi(t) > 0$, for example. $W(\varphi, \psi)(u) = \varphi(u)\psi'(u) \geqslant 0$ whereas $W(\varphi, \psi)(v) = \varphi(v)\psi'(v) \leqslant 0$ (for $\psi'(u) \geqslant 0$ and $\psi'(v) \leqslant 0$).

Now

$$\frac{dW}{dt}(\varphi, \psi)(t) = \varphi(t)\psi''(t) - \psi(t)\varphi''(t) = \varphi(t)\psi(t)[f(t) - g(t)] > 0$$

in $]u, v[$, so that W must be increasing.

If we had supposed that $\varphi(t) < 0$, we should have been obliged to have $W(u) \le 0$ and $W(v) \ge 0$ on the one hand and $dW/dt < 0$ on the other hand.

Examples (1) Let

$$x'' + t^2 x = 0 \tag{14}$$

and

$$x'' + x = 0 \tag{15}$$

The solutions of (15) are $x(t) = A \sin(t + \varphi)$, so that in every interval of length π in $[1, \infty[$ every solution of (14) has at least one zero.

(2) Bessel's equation:

$$t^2 x'' + tx' + (t^2 - v^2)x = 0 \tag{16}$$

Putting $x = u/t^{1/2}$ for $t > 0$ we obtain

$$u'' + \left(1 + \frac{1 - 4v^2}{4t^2}\right)u = 0 \tag{17}$$

In $]0, \infty[$ the zeros of x and u are the same.

Now let

$$u'' + u = 0 \tag{18}$$

A solution of this is $u = \sin(t - \vartheta)$.

(a) If $0 \le v < \frac{1}{2}$, then

$$1 + \frac{1 - 4v^2}{4t^2} > 1 \quad \forall t$$

and there exists at least one zero of each solution of (16) in every interval of length π in $]0, \infty[$.

(b) If $v > \frac{1}{2}$, then

$$1 + \frac{1 - 4v^2}{4t^2} < 1$$

The (possible) zeros of the solutions of Bessel's equation are then separated by more than π.

(c) If $v = \frac{1}{2}$

$$x(t) = \frac{k}{t^{1/2}} \sin(t - \vartheta)$$

and these solutions have zeros exactly a distance π apart.

It can be shown in fact that all Bessel functions have an infinite number of zeros.

Finally, we give the following second comparison theorem.

Theorem 1.1.17 (Sturm)
Let

$$\frac{d}{dt}\left[p_1(t)\frac{dx}{dt}\right] + q_1(t)x = 0 \tag{19}$$

and let φ_1 be a solution of (19). Let

$$\frac{d}{dt}\left[p_2(t)\frac{dx}{dt}\right] + q_2(t)x = 0 \tag{20}$$

and let φ_2 be a solution of (20). Suppose that $\forall\, t \in [a, b]$, $0 < p_2(t) \leqslant p_1(t)$ and $q_2(t) \geqslant q_1(t)$. Then φ_2 has one zero between two consecutive zeros of φ_1 in $[a, b]$.

See *Exercises 9 to 12* at the end of the chapter.

1.2. Non-homogeneous equations

Definition 1.2.1
Consider equations of the form

$$\frac{dx}{dt} = A(t)x + b(t) \qquad \text{where } b(t) \neq 0$$

$$A(t) \in \mathscr{L}(\mathbb{R}^d, \mathbb{R}^d), \; b(t) \in \mathbb{R}^d \tag{21}$$

We note that the set of solutions does *not* constitute a vector space and in particular that the sum of two solutions is not a solution; by contrast, the difference of two solutions is a solution of the associated homogeneous equation

$$\frac{dx}{dt} = A(t)x \tag{22}$$

or, which comes to the same thing, every solution is the sum of a solution of Eqn (21) and a solution of the homogeneous equation (22).

Theorem 1.2.2
For every $(t_0, x_0) \in \mathbb{R} \times \mathbb{R}^d$, the solution of (21) such that $x(t_0) = x_0$ is

$$\varphi(t) = \Phi(t, t_0)x_0 + \int_{t_0}^{t} \Phi(t, s)b(s)\,ds$$

where Φ is the resolvent of A.

Proof (a) As $\Phi(t, t_0)x_0$ is the solution of (22) such that $x(t_0) = x_0$, it is sufficient to verify that

$$\int_{t_0}^{t} \Phi(t, s)b(s)\,\mathrm{d}s$$

is a solution of (21). Now

$$\frac{\partial}{\partial t}\Phi(t, s)b(s) = A(t)\Phi(t, s)b(s)$$

and the norm of this is dominated over (t_0, T) by a constant; thus we know that $\forall u \in (t_0, T)$

$$\frac{\partial}{\partial t}\int_{t_0}^{u}\Phi(t, s)b(s)\,\mathrm{d}s = \int_{t_0}^{u}A(t)\Phi(t, s)b(s)\,\mathrm{d}s$$

so that

$$\frac{\mathrm{d}}{\mathrm{d}t}\int_{t_0}^{t}\Phi(t, s)b(s)\,\mathrm{d}s = \int_{t_0}^{t}A(t)\Phi(t, s)b(s)\,\mathrm{d}s + \Phi(t, t)b(t)$$

(b) The formula giving φ is often called **the formula for variation of constants**; we shall find it in a way similar to the case $d = 1$.

Let $\varphi(t)$ be the solution of (21) such that $\varphi(t_0) = x_0$. Let us define $\psi(t)$ by $\varphi(t) = \Phi(t, t_0)\psi(t)$ ('variation of constants'). Then we have in addition

$$\psi(t) = \Phi(t_0, t)\varphi(t)$$

$$\frac{\mathrm{d}\psi}{\mathrm{d}t} = -\Phi(t_0, t)A(t)\varphi(t) + \Phi(t_0, t)[A(t)\varphi(t) + b(t)] = \Phi(t_0, t)b(t)$$

and by integrating

$$\psi(t) = \psi(t_0) + \int_{t_0}^{t}\Phi(t_0, s)b(s)\,\mathrm{d}s$$

and since $\varphi(t_0) = \psi(t_0)$ we obtain the formula indicated.

We must note that Theorem 1.2.2 enables us to have an upper bound for φ when we have one for $\Phi(t, t_0)$ and for b, which happens for example if

$$\sup_{t}\|A(t) - A\| \leqslant \eta$$

We shall see numerous applications in the following chapter. Moreover, we shall use a slightly more general formula which we shall now prove.

Theorem 1.2.3 (Generalized formula for variation of constants)
Let x be a solution of

$$\frac{\mathrm{d}x}{\mathrm{d}t} = A(t)x + f(t, x) \tag{23}$$

Then

$$x(t) = \Phi(t, t_0)x(t_0) + \int_{t_0}^t \Phi(t, s)f[s, x(s)] \, ds$$

Proof Let

$$y(t) = \Phi(t, t_0)x(t_0) + \int_{t_0}^t \Phi(t, s)f[s, x(s)] \, ds$$

It is clear that

$$y(t_0) = x(t_0) \quad \text{and} \quad \frac{dy}{dt} = A(t)y(t) + f[t, x(t)]$$

(naturally supposing f to be continuous); if $z(t) = y(t) - x(t)$, $z(t_0) = 0$, $dz/dt = A(t)z$ and so $z(t) \equiv 0$ and $y(t) = x(t)$.

We shall now apply Theorem 1.2.2 to the case of scalar equations of order n; we shall always denote by $\varphi_0, \varphi_1, \ldots, \varphi_{n-1}$ the first row of the resolvent (cf. p. 72) of the associated homogeneous equation.

Theorem 1.2.4
Let

$$\frac{d^n x}{dt^n} = a_0(t)x + a_1(t)\frac{dx}{dt} + \ldots + a_{n-1}(t)\frac{d^{n-1}x}{dt^{n-1}} + b(t) \tag{24}$$

and let $[t_0; (x_0, x_1, \ldots, x_{n-1})] \in \mathbb{R} \times \mathbb{R}^d$. The solution of (24) such that

$$x(t_0) = x_0 \qquad \frac{dx}{dt}(t_0) = x_1 \quad \cdots \quad \frac{d^{n-1}x}{dt^{n-1}}(t_0) = x_{n-1}$$

is

$$x(t) = \Phi_0(t, t_0)x_0 + \Phi_1(t, t_0)x_1 + \ldots + \Phi_{n-1}(t, t_0)x_{n-1} + \int_{t_0}^t \Phi_{n-1}(t, s)b(s) \, ds$$

This is, naturally, the sum of the solutions, with the same initial value, of the homogeneous equation and of the non-homogeneous equation when $t = t_0$.

Note that if we know the solution of the homogeneous equation it is sufficient to be able to calculate $\Phi_{n-1}(t, s)$.

Examples (1) *nth primitive:* this is the solution of $d^n x/dt^n = b(t)$ such that

$$x(t_0) = \frac{dx}{dt}(t_0) = \ldots = \frac{d^{n-1}x}{dt^{n-1}}(t_0) = 0$$

The solution of the associated homogeneous equation satisfying these conditions is $x(t) \equiv 0$.

The matrix A is

$$A = \begin{pmatrix} 0 & 1 & & & 0 \\ 0 & 0 & 1 & & 0 \\ & & & \ddots & \\ 0 & 0 & 0 & & 1 \\ 0 & 0 & 0 & & 0 \end{pmatrix}$$

and $\Phi_{n-1}(t, s)$ then satisfies

$$\frac{d^n \Phi_{n-1}(t, s)}{dt^n} = 0$$

$$\frac{d^{n-1} \Phi_{n-1}(s, s)}{dt^{n-1}} = 1$$

and

$$\frac{d^p \Phi_{n-1}(s, s)}{dt^p} = 0$$

if $p < n - 1$; thus we obtain, step by step,

$$\frac{d^{n-1} \Phi_{n-1}(t, s)}{dt^{n-1}} = 1$$

$$\frac{d^{n-2} \Phi_{n-1}(t, s)}{dt^{n-2}} = t - s$$

$$\vdots$$

$$\Phi_{n-1}(t, s) = \frac{(t - s)^{n-1}}{(n - 1)!}$$

The primitive of order n of b is

$$\int_{t_0}^{t} \frac{(t - s)^{n-1}}{(n - 1)!} b(s) \, ds$$

and it is obtained by a single integration.

(2) *Second-order equations:* as on page 73, let

$$x''(t) + a(t)x'(t) + b(t)x(t) = g(t) \tag{25}$$

and let f_1 and f_2 be two independent solutions of the homogeneous equation. The coefficient $\Phi_{n-1}(t, s)$ is

$$\frac{1}{W(f_1, f_2)(s)} [f_1(s)f_2(t) - f_1(t)f_2(s)]$$

and the solution of (25) such that both it and its derivative vanish for $t = s$ is

$$x(t) = \int_s^t \frac{f_1(s)f_2(t) - f_1(t)f_2(s)}{W(f_1, f_2)(s)} g(s)\,ds$$

See *Exercises 13 to 17* at the end of the chapter.

Part 2: Equations with constant coefficients

We are concerned with equations of the form $dx/dt = Ax + b(t)$ where b is continuous from $\mathbb{R}$ into $\mathbb{R}^d$ and $A \in \mathscr{L}(\mathbb{R}^d, \mathbb{R}^d)$ or more generally, as we shall see, $\mathscr{L}(\mathbb{C}^d, \mathbb{C}^d)$.

Let us first examine the following very simple situation: in $\mathbb{R}^2$ let $\vec{i}$ and $\vec{j}$ be two unit vectors and x and y be the coordinates of a point X in this system. Let A be diagonal:

$$A = \begin{pmatrix} a & 0 \\ 0 & b \end{pmatrix}$$

The equation $dX/dt = AX$ can in fact be split up into two equations $dx/dt = ax$ and $dy/dt = by$ whose solutions are $x(t) = x_0 \exp(at)$ and $y(t) = y_0 \exp(bt)$.

We can state that for an initial value $X_0 = x_0 \vec{i}$ the solution $X(t) = x_0 \exp(at)\vec{i}$ is a multiple of the vector $\vec{i}$ which is an eigenvector of A associated with the eigenvalue a. We can also state that the behaviour of this solution at infinity depends only on the sign of the real part of a.

We observe a comparable phenomenon if $X_0 = y_0\vec{j}$ and $X(t) = y_0 \exp(bt)\vec{j}$.

It is well known that a matrix cannot always be diagonalized but we shall show that for an arbitrary matrix there is a similar situation if we consider, instead of eigenvectors, what are called generalized proper subspaces. We shall see that for this purpose it is necessary to consider the problem as one in the d-dimensional complex space $\mathbb{C}^d$, and that the latter may be decomposed into a sum of subspaces associated with each of the different eigenvalues and independent of each other. These subspaces are invariant in the sense that a solution of which one point belongs to a subspace is entirely contained in that subspace. The behaviour at infinity of these solutions depends only on Re λ if λ is the corresponding eigenvalue.

2.1. Results in linear algebra

Let v be an element of $\mathbb{R}^d$; it is represented in a given coordinate system by V. Let $\mathscr{A} \in \mathscr{L}(\mathbb{R}^d, \mathbb{R}^d)$ be a linear mapping which is represented in the same system by a matrix A. If we change the system of coordinates, and if P is the transformation matrix, the new representation of v is $P^{-1}V$ while that of $\mathscr{A}$ is the matrix $P^{-1}AP$.

CHARACTERISTIC EQUATION; EIGENVALUES

The equation (in z) $\det(A - zI) = 0$ is called the **characteristic equation** of the matrix A and of the linear transformation $\mathscr{A}$ of which A is a representation; in fact this equation does not depend on the system of coordinates.

The roots of the characteristic equation of $\mathscr{A}$ are called the **eigenvalues of** $\mathscr{A}$; if $\lambda_1, \ldots, \lambda_r$ are the distinct roots, the characteristic equation can be written as

$$\prod_{i=1}^{r} (z - \lambda_i)^{m_i} = 0 \qquad (\Sigma\, m_i = d)$$

The number m_i is called the **multiplicity** of the root λ_i. Naturally the roots λ_i of the characteristic equation are not necessarily all real.

First let λ be a real root (if there is one) of the characteristic equation. By hypothesis $\det(A - \lambda I) = 0$, which means that $A - \lambda I$ is not invertible and consequently there exists a non-zero vector V such that $(A - \lambda I)V = 0$; such a vector is called an eigenvector associated with λ.

If λ is a complex number we can adopt this line of reasoning only if we are prepared to consider complex vectors, which, however, presents no difficulty.

Let $\mathbb{C}$ be the set of complex numbers. A vector space is constructed over $\mathbb{C}$, i.e. we define on it the operation of multiplication by an element $\lambda \in \mathbb{C}$ (and this operation has suitable properties). We shall denote by $\mathbb{C}^d$ the space of sequences $(z_1, \ldots, z_d)$ of complex numbers.

If we choose a system of coordinates, every element of $\mathbb{C}^d$ is of the form $V + iW$ where V and $W \in \mathbb{R}^d$.

Let $\mathscr{A} \in \mathscr{L}(\mathbb{R}^d, \mathbb{R}^d)$ be represented by a matrix A; it is easy to consider it as an element of $\mathscr{L}(\mathbb{C}^d, \mathbb{C}^d)$ by putting

$$A(V + iW) = AV + iAW$$

$$A[\lambda(V + iW)] = \lambda A(V + iW) \qquad \forall\, \lambda \in \mathbb{C}$$

If $\lambda = \alpha + i\beta$

$$\lambda(V + iW) = (\alpha V - \beta W) + i(\beta V + \alpha W)$$

Taking these conventions into account, we can introduce complex eigenvectors.

Definition 2.1.1

If λ is an eigenvalue (real or not) of A, there exist vectors (real or not) such that

$$(A - \lambda I)V = 0$$

Such vectors are called **eigenvectors** of A (or even eigenvectors of the mapping $\mathscr{A}$).

If $AV = \lambda V$, then $A\bar{V} = \bar{\lambda}\bar{V}$. $\bar{\lambda}$ is also an eigenvalue and $\bar{V}$ is an eigenvector associated with $\bar{\lambda}$.

Example Let

$$A = \begin{pmatrix} 1 & 1 \\ -1 & 1 \end{pmatrix}$$

with characteristic equation $(1 - \lambda)^2 + 1 = 0$; its eigenvalues are therefore $\lambda_1 = 1 + i$, $\lambda_2 = 1 - i$. Thus an eigenvector associated with $1 + i$ is such that

$$\left[\begin{pmatrix} 1 & 1 \\ -1 & 1 \end{pmatrix} - (1 + i) \begin{pmatrix} 1 & 0 \\ 0 & 1 \end{pmatrix} \right] (V + iW) = 0$$

If

$$V + iW = \begin{pmatrix} u \\ v \end{pmatrix}$$

where u and v are complex, we must have

$$\begin{pmatrix} -i & 1 \\ -1 & -i \end{pmatrix} \begin{pmatrix} u \\ v \end{pmatrix} = 0$$

which is satisfied by $u = 1$ and $v = i$ for example. Thus

$$V + iW = \begin{pmatrix} 1 \\ 0 \end{pmatrix} + i \begin{pmatrix} 0 \\ 1 \end{pmatrix}$$

which we write

$$\begin{pmatrix} 1 \\ i \end{pmatrix}$$

$\begin{pmatrix} 1 \\ i \end{pmatrix}$ is an eigenvector associated with $1 + i$

$\begin{pmatrix} 1 \\ -i \end{pmatrix}$ is an eigenvector associated with $1 - i$

PROPER GENERALIZED SPACES

We know that in certain very simple situations, for example if A is symmetric and has distinct eigenvalues, we can find in $\mathbb{R}^d$ a basis of eigenvectors. We know also that, if there are multiple eigenvalues, the situation is more complicated. By admitting complex eigenvectors we can clarify the situation in all cases.

Definition 2.1.2

Let $\mathcal{A}$ be a linear mapping represented in a coordinate system by a matrix A. We define the **kernel** of $\mathcal{A}$—written $\mathcal{K}(\mathcal{A})$—as the vector space of vectors $\mathcal{V}$

such that $\mathscr{A} \cdot \mathscr{V} = 0$, which are thus represented by V such that $AV = 0$:

$$\mathscr{K}(\mathscr{A}) = \{\mathscr{V} : AV = 0\}$$

It is clear that A is invertible iff $\mathscr{K}(\mathscr{A}) = \{0\}$.

So far we have carefully distinguished $\mathscr{A}$ from A and $\mathscr{V}$ from V; from now on we shall drop this distinction.

In our previous notation, the fact that λ is an eigenvalue of A is expressed by $\mathscr{K}(A - \lambda I) \neq \{0\}$; we shall now consider the dimension of this subspace.

Definition 2.1.3

Let λ be an eigenvalue of A and let $m(\lambda)$ be the multiplicity of λ. There exists a smallest integer $v(\lambda)$, called the **index of the eigenvalue** λ, such that

$$\dim \mathscr{K}[(A - \lambda I)^{v(\lambda)}] = m(\lambda)$$

We can verify that $v(\lambda) \leqslant m(\lambda)$.

If $m < v(\lambda)$ $\quad \dim \mathscr{K}[(A - \lambda I)^{m+1}] > \dim \mathscr{K}[(A - \lambda I)^m]$

If $m \geqslant v(\lambda)$ $\quad \dim \mathscr{K}[(A - \lambda I)^{m+1}] = \dim[\mathscr{K}(A - \lambda I)^m]$

The space $E_\lambda = \mathscr{K}[(A - \lambda I)^{v(\lambda)}]$ is called the **proper generalized space** of A associated with the eigenvalue λ so that

$$V \in E_\lambda \Leftrightarrow \forall n \geqslant v(\lambda) : (A - \lambda I)^n V = 0$$

Thus E_λ is composed of vectors such that $(A - \lambda I)V = 0$ or $(A - \lambda I)^2 V = 0$ up to $(A - \lambda I)^{v(\lambda)} V = 0$. The dimension of $\mathscr{K}[(A - \lambda I)^p]$ does not increase further once we have reached $p = v(\lambda)$.

Proposition 2.1.4

If $V \in E_\lambda$ then $AV \in E_\lambda$; we say that E_λ is an **invariant space** under the transformation A.

Proof The proof is immediate. In fact $(A - \lambda I)^n AV = A(A - \lambda I)^n V$ since A commutes with A^p for every integer p; thus if, $\forall n \geqslant v(\lambda)$, $(A - \lambda I)^n V = 0$, under the same conditions $(A - \lambda I)^n AV = 0$, i.e. $AV \in E_\lambda$.

Definition 2.1.5

We say that λ is an **oscillatory eigenvalue** if $v(\lambda) > 1$.

Naturally if λ is a simple eigenvalue, λ is not oscillatory.

Examples (1)

$$A = \begin{pmatrix} 1 & 0 \\ 0 & 1 \end{pmatrix}$$

1 is a double eigenvalue and $\mathscr{K}(I - A) = \mathbb{R}^2$; thus $v(1) = 1$.

(2) Let A be diagonal:

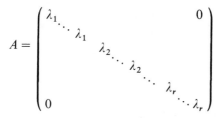

If λ_k is an eigenvalue of order $m(k)$, $\mathscr{K}(\lambda_k I - A)$ is a vector space of dimension $m(k)$: $v(\lambda_k) = 1$.

(3) Since every symmetric matrix is diagonalizable, its eigenvalues are not oscillatory.

(4) If

$$A = \begin{pmatrix} 0 & 1 \\ 0 & 0 \end{pmatrix}$$

0 is an eigenvalue of order 2 and $\mathscr{K}(0 \cdot I - A) = \mathbb{R}^1$ whereas $\mathscr{K}[(0 \cdot I - A)^2] = \mathbb{R}^2$. Thus 0 is an oscillatory eigenvalue.

The fundamental result is that $\mathbb{C}^d$ decomposes into the spaces E_λ.

Definition 2.1.6 (Direct sum of vector spaces)

If F_1 and F_2 are two vector subspaces of $\mathbb{C}^d$, we say that they are **linearly independent** if $\forall x_1 \in F_1$ and $\forall x_2 \in F_2$, x_1 and x_2 are linearly independent $(a_1 x_1 + a_2 x_2 = 0 \Leftrightarrow a_1 = a_2 = 0)$.

We define the direct sum of two independent subspaces F_1 and F_2 as the vector subspace of $\mathbb{C}^d$ of vectors of the form $x_1 + x_2$ where $x_1 \in F_1$ and $x_2 \in F_2$; we denote it $F_1 \oplus F_2$.

$$F_1 \oplus F_2 = \{x \in \mathbb{C}^d; \ \exists\, x_1 \in F_1 \ \exists\, x_2 \in F_2 : x = x_1 + x_2\}$$

This notion generalizes, in a way, that of a basis of coordinates formed from spaces in one dimension; the following theorem states that we can find in $\mathbb{C}^d$ a 'basis' of proper spaces.

Theorem 2.1.7

Let $A \in \mathscr{L}(\mathbb{R}^d, \mathbb{R}^d)$ and let $E_1, \ldots, E_r$ be the proper generalized spaces of A (which are invariant with reference to A):

$$\mathbb{C}^d = E_1 \oplus E_2 \oplus \ldots \oplus E_r$$

Moreover if $i \neq j$ the restriction of $A - \lambda_j I$ to E_i is invertible.

Proof The first part of the statement is a fundamental theorem of linear algebra which we shall not prove.

The second part is easy to verify: if $V \in E_i$

$$W = \frac{V}{\lambda_i - \lambda_j} \in E_i \quad \text{and} \quad (A - \lambda_j I)W = V$$

Theorem 2.1.7 can be used to prove the following result.

Theorem 2.1.8 (Cayley–Hamilton)
Let A be a matrix with characteristic equation

$$\prod_{i=1}^{r} (z - \lambda_i)^{m_i} = 0$$

Then

$$\prod_{i=1}^{r} (A - \lambda_i I)^{m_i} = 0$$

i.e. A is a solution of its own characteristic equation.

Proof   Let

$$U \in \mathbb{C}^d: U = U_1 + \ldots + U_r \quad \text{where } U_i \in E_i$$

$$(A - \lambda_1 I)^{m_1} U = 0 + (A - \lambda_1 I)^{m_1} U_2 + \ldots + (A - \lambda_1 I)^{m_1} U_r$$

$$(A - \lambda_2 I)^{m_2}(A - \lambda_1 I)^{m_1} U = 0 + 0 + (A - \lambda_1 I)^{m_1}(A - \lambda_2 I)^{m_2} U_3 + \ldots + (\ )(\ )U_r$$

and step by step

$$\left(\prod_{i=1}^{r} (A - \lambda_i I)^{m_i} \right) U = 0$$

This being true for every vector of $\mathbb{C}^d$, we obtain the result.

Example   If

$$A = \begin{pmatrix} 1 & 1 \\ -1 & 1 \end{pmatrix} \qquad A^2 - 2A + 2I = 0$$

2.2. Solution of homogeneous equations

SOLUTION

We shall establish that Theorem 2.1.8 allows us to split the problem into simpler problems in which the initial value belongs to a fixed proper subspace. Let us examine this case.

Proposition 2.2.1
Let

$$\frac{dx}{dt} = Ax \quad \text{where } x \in \mathbb{R}^d \tag{26}$$

Let λ be an eigenvalue with index v and $V \in E_\lambda$. The solution of (26) such that $V(0) = V$ is

$$V(t) = \exp(\lambda t) V + (A - \lambda I)t \exp(\lambda t) V + \ldots + (A - \lambda I)^{v-1} \frac{t^{v-1}}{(v-1)!} \exp(\lambda t) V$$

If $V(0) \in E_\lambda$ then $V(t) \in E_\lambda \ \forall \, t$; finally, if $V_1(0) \in E_{\lambda_1}$ and $V_2(0) \in E_{\lambda_2}$, $V_1(t)$ and $V_2(t)$ are independent ($\lambda_1 \neq \lambda_2$).

Proof It is clear that $V(0) = V$ and that $V(t) \in E_\lambda$ since the latter is invariant with reference to A; if $\lambda_1 \neq \lambda_2$, E_{λ_1} and E_{λ_2} are independent so that $V_1(t)$ and $V_2(t)$ are also independent. We calculate

$$\frac{dV}{dt} = \lambda V(t) + (A - \lambda I) \exp(\lambda t) V + \ldots + (A - \lambda I)^{v-1} \frac{t^{v-2}}{(v-2)!} \exp(\lambda t) V$$

$$= \lambda V(t) + (A - \lambda I)V(t) - (A - \lambda I)^v \frac{t^{v-1}}{(v-1)!} \exp(\lambda t) V = AV(t)$$

since $(A - \lambda I)^v V = 0$.

The preceding proposition indicates that $V(t) \in E_\lambda$ when $V(t_0) \in E_\lambda$ for one t_0 and shows the behaviour of $V(t)$ in relation to Re λ.

Definition 2.2.2
We define an invariant subspace of the equation

$$\frac{dx}{dt} = Ax \tag{27}$$

as every subspace E such that if there exists t_0 such that $V(t_0) \in E$ then $V(t) \in E$ for every solution $V(t)$.

Theorem 2.2.3
Let

$$\frac{dx}{dt} = Ax \tag{28}$$

Then every proper generalized subspace of A is an invariant space of (28).
If Re $\lambda < 0$, $\forall \, V \in E_\lambda$

$$\lim_{t \to \infty} \|V(t)\| = 0$$

We say that E_λ is a **stable subspace**.

If $\mathrm{Re}\,\lambda > 0$, $\forall V \in E_\lambda$

$$\lim_{t \to \infty} \|V(t)\| = \infty$$

We say that E_λ is an **unstable subspace**.

To find the solution $\varphi(t, t_0, x_0)$ starting from x_0 at t_0 it is now sufficient to decompose x_0 according to the spaces $E_1, \ldots, E_r$; if $x_0 = x_0^{(1)} + \ldots + x_0^{(r)}$, $\varphi(t, t_0, x_0) = \varphi^{(1)}(t) + \ldots + \varphi^{(r)}(t)$ where $\varphi^{(r)}(t)$ is given by Proposition 2.2.1. We obtain the following result.

Theorem 2.2.4
Let

$$\frac{dx}{dt} = Ax \tag{29}$$

where $x \in \mathbb{R}^d$. The coordinates $x_i(t)$ of every solution $x(t)$ are of the form

$$x_i(t) = \sum_{k=1}^{r} p_{i,k}(t) \exp[\lambda_k(t - t_0)]$$

where $\lambda_1, \ldots, \lambda_r$ are the distinct eigenvalues of A with respective indices $v_1, \ldots, v_r$ and where $p_{i,k}(t)$ is a polynomial of degree $v_k - 1$.
Let

$$\frac{d^n x}{dt^n} = a_0 x + a_1 \frac{dx}{dt} + \ldots + a_{n-1} \frac{d^{n-1} x}{dt^{n-1}} \quad \text{where } x \in \mathbb{R} \tag{30}$$

Every solution of (30) is of the form

$$x(t) = \sum_{k=1}^{r} p_k(t) \exp[\lambda_k(t - t_0)]$$

where $\lambda_1, \ldots, \lambda_r$ are the distinct roots, of multiplicity m_i, of the so-called characteristic equation $\lambda^n = a_0 + a_1 \lambda + \ldots + a_{n-1} \lambda^{n-1}$ and p_k is a polynomial of degree less than or equal to $m_k - 1$.

The second part is simply another form of the corresponding result for the equivalent system with matrix

$$\begin{pmatrix} 0 & 1 & \ldots & 0 \\ 0 & & \ldots & 1 \\ a_0 & a_1 & \ldots & a_{n-1} \end{pmatrix}$$

Real solutions
If λ is not real the above method introduces solutions which are not real; however, we know that $\bar{\lambda}$ is then also an eigenvalue and that V and $\bar{V}$ belong to

E_λ and $E_{\bar\lambda}$ respectively. If $V(t)$ and $\bar V(t)$ are independent Re $V(t)$ and Im $V(t)$ are also independent, since

$$\text{Re } V(t) = \tfrac{1}{2}(V + \bar V)(t) \quad \text{and} \quad \text{Im } V(t) = \frac{1}{2i}(V - \bar V)(t)$$

so that these vectors can be obtained by an invertible transformation, of matrix

$$\begin{pmatrix} \dfrac{1}{2} & \dfrac{1}{2} \\[2mm] \dfrac{1}{2i} & -\dfrac{1}{2i} \end{pmatrix}$$

starting from $V(t)$ and $\bar V(t)$.

Re $V(t)$ and Im $V(t)$ provide two real solutions which are independent of each other and independent of solutions related to the other eigenvalues.

PRACTICAL METHOD

Let

$$\frac{dx}{dt} = Ax \tag{31}$$

with $\lambda_1, \ldots, \lambda_r$ the real eigenvalues of A and $\alpha_1 + i\beta_1, \ldots, \alpha_j + i\beta_j$ the complex eigenvalues. We shall seek to identify solutions of (31) in the form

$$x(t) = \sum_{k=1}^{r} p_k(t) \exp(\lambda_k t) + \sum_{l=1}^{j} \exp(a_l t) \left[q_l(t) \cos(\beta_l t) + r_l(t) \sin(\beta_l t) \right] \tag{32}$$

In this notation the polynomials p, q and r are of degree less than or equal to $m_k - 1$ or $m_l - 1$ respectively and their coefficients are vectors of $\mathbb{R}^d$.

If we seek, for each coordinate, a solution in this form we shall then not be able to choose the polymomials arbitrarily.

Let

$$\frac{d^n x}{dt^n} = a_0 x + a_1 \frac{dx}{dt} + \ldots + a_{n-1} \frac{dx^{n-1}}{dt^{n-1}} \quad \text{where } x \in \mathbb{R} \tag{33}$$

Every solution is of the form of Eqn (32) where the real polynomials p, q, r of degree less than or equal to $m_k - 1$ or $m_l - 1$ respectively are arbitrary.

Example

$$\frac{dx}{dt} = z + v$$

$$\frac{dy}{dt} = x + z + v$$

$$A = \begin{pmatrix} 0 & 0 & 1 & 1 \\ 1 & 0 & 1 & 1 \\ 0 & 0 & 0 & -1 \\ 0 & 0 & 1 & 0 \end{pmatrix} \qquad \det(A - \lambda I) = \lambda^2(\lambda^2 + 1)$$

$$\frac{dz}{dt} = -v$$

$$\frac{dv}{dt} = z$$

The eigenvalues are $\lambda = i$, $\lambda = -i$ and $\lambda = 0$ which is a double eigenvalue, $\mathcal{K}(A)$ being of dimension 1; 0 is an oscillatory eigenvalue. We seek for each coordinate a solution in the form $a \cos t + b \sin t + ct + d$ and find

$$x(t) = x(0) + z(0)(1 - \cos t + \sin t) + v(0)(\cos t + \sin t - 1)$$

$$y(t) = y(0) + 2z(0)(1 - \cos t) + 2v(0) \sin t + [x(0) + z(0) - v(0)]t$$

$$z(t) = z(0) \cos t - v(0) \sin t$$

$$v(t) = v(0) \cos t + z(0) \sin t$$

We recall that, when we can solve the homogeneous equation, we can find d independent solutions and thus find the resolvent and solve the associated non-homogeneous equation.

We shall return later—to remark on certain particular cases—to the non-homogeneous equation, but from now on we know how to integrate it.

PARTICULAR CASES: EULER'S EQUATION AND REDUCTION OF ORDER

Definition 2.2.5
The following equation, for $t > 0$, is called Euler's equation:

$$t^n \frac{d^n x}{dt^n} + a_1 t^{n-1} \frac{d^{n-1} x}{dt^{n-1}} + \ldots + a_{n-1} t \frac{dx}{dt} + a_n x = 0 \qquad (34)$$

Putting $t = \exp(s)$ transforms (34) into an equation with constant coefficients whose solution therefore involves the terms $\exp(\lambda_i s) p_i(s)$ or $t^{\lambda_i} p_i(\log t)$.

We seek directly solutions of this form by substituting in (34).

Example $2t^2 y'' + 3ty' - y = 0$ has independent solutions $t^{1/2}$ and $1/t$ $(t > 0)$.

Reduction of order

Let $dx/dt = A(t)x$ where $x \in \mathbb{R}^d$. Since the set of solutions form a vector space of dimension d, if k independent solutions are known, these can be used to find $d - k$ solutions which are independent among themselves and independent of the preceding ones. A transformation of functions is usually used to find these.

In particular if f is a solution of

$$\frac{d^n x}{dt^n} = a_0 x + \ldots + a_{n-1} \frac{d^{n-1} x}{dt^{n-1}}$$

we put $x = uf$. u is a solution of an equation of order n of which $u \equiv 1$ is one solution. This is thus an equation of order $n - 1$ in u'.

If two solutions f and g are known, we can put $x = uf$, and then we know two solutions $u \equiv 1$ and $u = g/f$ of the equation in u and again a solution $(g/f)'$ of the equation in u'. By putting $v = u(g/f)'$ we then find an equation of order $n - 2$ in v'.

See *Exercises 18 to 20* at the end of the chapter.

2.3. Dynamic systems in $\mathbb{R}^2$

We consider equations of the form $dx/dt = Ax$ where $x \in \mathbb{R}^2$. Scalar linear differential equations of second order with constant coefficients lead to the study of such systems.

We shall suppose that $\det A \neq 0$ so that 0 is the only equilibrium point.

MATRIX A HAS TWO DISTINCT EIGENVALUES λ_1 AND λ_2

Let v_1 and v_2 be the eigenvectors corresponding to λ_1 and λ_2; every solution is of the form

$$x(t) = a_1 \exp(\lambda_1 t)\, v_1 + a_2 \exp(\lambda_2 t)\, v_2$$

Each of v_1 and v_2 generates an invariant subspace.

Case 1: λ_1 and λ_2 real

(a) $\lambda_2 < \lambda_1 < 0$ (Fig. 2.2(a))
Since all solutions tend to zero if $t \to +\infty$, each of v_1 and v_2 generates a stable subspace. The equilibrium point 0 is called a **stable node** or an **attractive node**.

(b) $0 < \lambda_1 < \lambda_2$ (Fig. 2.2(b))
All solutions tend to ∞ if $t \to +\infty$; the invariant subspaces generated by v_1 and v_2 are unstable and 0 is called an **unstable node** or a **repulsive node**.

(c) $\lambda_2 < 0 < \lambda_1$ (Fig. 2.2(c))
The subspace generated by v_1 is unstable. The subspace generated by v_2 is

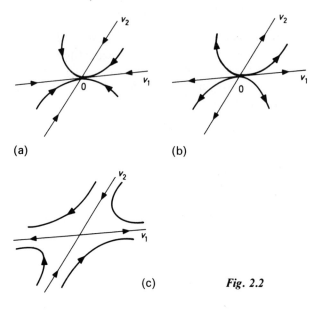

(a)   (b)

(c)   **Fig. 2.2**

stable. Only solutions whose initial value is collinear with v_2 remain bounded. 0 is called a **saddle point**.

Case 2: $\lambda_1 = \alpha + i\beta$, $\lambda_2 = \alpha - i\beta$

The eigenvectors are of the form $v_1 = u + iv$, $v_2 = u - iv$. If we are looking for real solutions we can choose, as a basis of the vector space of solutions, $\mathrm{Re}\left[\exp(\lambda_1 t) v_1\right]$ and $\mathrm{Im}\left[\exp(\lambda_1 t) v_1\right]$, i.e.

$$\exp(\alpha t)(u\cos\beta t - v\sin\beta t) \quad \text{and} \quad \exp(\alpha t)(u\sin\beta t + v\cos\beta t)$$

Every solution is then of the form

$$x(t) = \exp(\alpha t)(a\cos\beta t + b\sin\beta t)u + \exp(\alpha t)(b\cos\beta t - a\sin\beta t)v$$

which can also be written

$$x(t) = A\exp(\alpha t)\left[\cos(\beta t + \varphi)\,u - \sin(\beta t + \varphi)\,v\right]$$

(a) $\alpha = 0$ (Fig. 2.3(a))
The solutions are all periodic. 0 is called a **centre**.

(b) $\alpha < 0$ (Fig. 2.3(b))
All solutions have 0 as limit if $t \to +\infty$. 0 is called a **stable focus** or **stable focal point** or an **attractive focus**.

(c) $\alpha > 0$ (Fig. 2.3(c))
The solutions tend to ∞ in numerical value if $t \to +\infty$. 0 is called an **unstable focus** or **unstable focal point** or a **repulsive focus**.

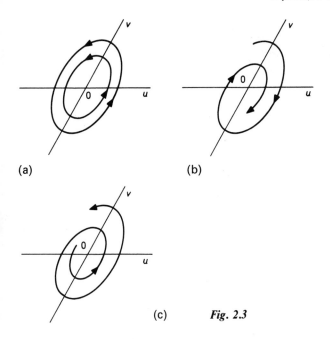

(a)

(b)

(c)   **Fig. 2.3**

We note that the invariant subspaces generated by the eigenvectors v_1 and v_2 which are imaginary are constructed from imaginary vectors so that they no longer appear when we pass over to real and imaginary parts. Nevertheless they still exist.

MATRIX A HAS A DOUBLE EIGENVALUE

This eigenvalue λ is necessarily real.

Case 1: λ *not oscillatory*
(We can verify then that $A = \lambda I$.) There exist two independent eigenvectors v_1 and v_2 associated with the eigenvalue λ and $x(t) = (a_1 v_1 + a_2 v_2) \exp(\lambda t)$.

(a) If $\lambda < 0$, all solutions have 0 as limit if $t \to +\infty$. 0 is a stable node, often called a **sink** (Fig. 2.4(a)).

(b) If $\lambda > 0$, all solutions tend to infinity if $t \to +\infty$. 0 is called a **source** (Fig. 2.4(b)).

A sink is an attractive point, a source a repulsive point.

Case 2: λ *oscillatory*
The dimension of $\mathcal{K}(A - \lambda I)$ is 1; let $u \neq 0$ generate $\mathcal{K}(A - \lambda I)$. The dimension of $\mathcal{K}[(A - \lambda I)^2]$ is 2. Let $w \in \mathcal{K}[(A - \lambda I)^2]$ form with u a basis of $\mathbb{R}^2$; by hypothesis $(A - \lambda I)w \neq 0$, say $(A - \lambda I)w = \alpha u + \beta w$. Thus $0 =$

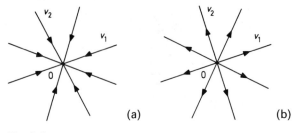

Fig. 2.4

$(A - \lambda I)^2 w = \beta (A - \lambda I) w$ and $\beta = 0$. Therefore

$$(A - \lambda I) w = \alpha u \quad \text{or} \quad (A - \lambda I) \frac{w}{\alpha} = u$$

Let

$$x(0) = au + b \frac{w}{\alpha} = au + bv$$

be an initial value.

$$x(t) = \exp(\lambda t)(au + bv) + (A - \lambda I)t \exp(\lambda t)(au + bv)$$
$$= (a + bt) \exp(\lambda t) u + b \exp(\lambda t) v$$

(a) If $\lambda < 0$ all solutions tend to zero when $t \to +\infty$ and are tangents to the vector u. 0 is an **improper stable node** (attractive) (Fig. 2.5(a)).

(b) If $\lambda > 0$ all solutions tend to infinity if $t \to +\infty$ and tend to zero, tangentially to u, if $t \to -\infty$. 0 is an **improper unstable node** (repulsive) (Fig. 2.5(b)).

Since the signs of det A and tr A determine the nature of the eigenvalues, we can show on a diagram (Fig. 2.6), whose coordinates are det A and tr A, the set of the preceding results. The only element which does not appear on the diagram concerns the character (oscillatory or not) of a double eigenvalue: on

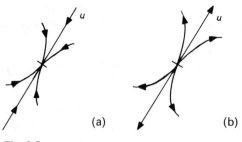

Fig. 2.5

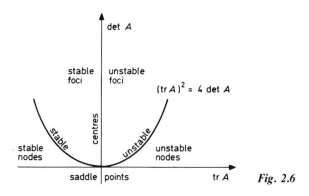

Fig. 2.6

the parabola there can be a sink or an improper stable node on the one hand, or a source or an improper unstable node on the other.

See **Exercises 21 and 22** at the end of the chapter.

Comparison of two dynamic systems in $\mathbb{R}^d$

If we are concerned with two linear systems $dx/dt = Ax$ and $dy/dt = By$ where A and B have the same number of negative eigenvalues, and therefore of positive eigenvalues, the sets of trajectories have the same behaviour. If A and B have the same eigenvalues and they are simple, the solutions of one system can be deduced from those of the other by a linear transformation.

We shall see in a later section that in the neighbourhood of an equilibrium point 0 the solutions of the system $dx/dt = f(x)$ and of the system $dx/dt = Df(0)x$ are of the same nature if the point 0 is a saddle point for the linearized system; for a centre they can be quite different.

2.4. Exponentials of linear mappings

In Section 2.2 we solved equations with constant coefficients without using the general ideas introduced in Part 1.

In $\mathbb{R}$ the solutions of $dx/dt = ax$ are $x(t) = x(t_0) \exp[a(t - t_0)]$ so that the resolvent is the 1×1 matrix $(t, s) \rightsquigarrow \exp[a(t - s)]$.

We shall generalize this result by writing the form $\Phi(t, s) = \exp[A(t - s)]$ where A is a fixed linear mapping.

Definition 2.4.1

Let $\mathscr{A} \in \mathscr{L}(\mathbb{R}^d, \mathbb{R}^d)$ and

$$\frac{dx}{dt} = \mathscr{A}x \tag{35}$$

We call the exponential of the linear mapping $\mathscr{A}$, the resolvent of (35).

If A is a matrix representing $\mathscr{A}$, we denote a matrix representing $\Phi(t, s)$ by $\exp[A(t - s)]$.

Theorem 2.4.2

Let $\mathscr{A} \in \mathscr{L}(\mathbb{R}^d, \mathbb{R}^d)$ be represented in a coordinate system by a matrix A; then

$$\exp(1.0) = I \qquad \exp[A(t+s)] = \exp(At)\exp(As)$$

$$[\exp(At)]^{-1} = \exp(-At) \tag{36}$$

$$\frac{d}{dt}\exp(At) = A\exp(At) = \exp(At)\,A \tag{37}$$

$$\exp(At) = \lim_{n \to \infty} \left(I + At + A^2\frac{t^2}{2!} + \ldots + A^n\frac{t^n}{n!} \right) \tag{38}$$

Proof Let $\Phi(t, s)$ be the resolvent of (35); it is convenient firstly to assure ourselves that $\Phi(t, s)$ depends only on $t - s$. If $\mathscr{K}(t) = \Phi(t+h, s+h)$, $\mathscr{K}(s) = I$ and $d\mathscr{K}/dt = A\mathscr{K}(t)$; the only solution of this equation satisfying $\mathscr{K}(s) = I$ being $\Phi(t, s)$, we have $\Phi(t+h, s+h) = \Phi(t, s)$. Relations (36) and (37) simply express the properties of the resolvent.

To demonstrate (38) assume that we are working in a bounded interval $[0, T]$. If

$$M_n(t) = I + At + \ldots + A^n\frac{t^n}{n!}$$

we shall verify that M_n is a Cauchy sequence.

If $m \geqslant n$,

$$\sup_{t \in [0, T]} \|M_m(t) - M_n(t)\| \leqslant \sum_{n+1}^{\infty} \frac{[\|A\|T]^n}{n!}$$

which is the series remainder of $\exp(\|A\|T)$. Thus $M_n(t)$ converges uniformly over $[0, T]$ to a matrix $M(t)$. Moreover

$$M_1(t) = I + At = I + \int_0^t AM_0(s)\,ds \qquad A\int_0^t M_1(s)\,ds = At + A^2\frac{t^2}{2!}$$

Then

$$M_2(t) = I + \int_0^t AM_1(s)\,ds$$

and by recurrence

$$M_n(t) = I + \int_0^t AM_{n-1}(s)\,ds$$

at the limit, and by uniform convergence over $[0, T]$

$$M(t) = I + \int_0^t AM(s)\,ds$$

which proves that $M(t) = \Phi(t, 0) = \exp(At)$.

Theorem 2.4.2 sometimes allows direct calculation of $\Phi(t, 0)$ either by the obvious method of calculating A^n or by using one or other of the following properties.

Proposition 2.4.3
(a) If P is invertible $P^{-1} \exp(At) P = \exp[(P^{-1}AP)t]$.
(b) If $AB = BA$ then $\exp[(A + B)t] = \exp(At) \exp(Bt) = \exp(Bt) \exp(At)$ and $B \exp(At) = \exp(At) B$.
Conversely, if one of these relations is satisfied, $AB = BA$.

Proof (a) The two members of the first relation represent the same linear mapping $\Phi(t, 0) = \exp(At)$ in one coordinate system and $P^{-1} \exp(At) P$ in another. In the second system, however, A becomes $P^{-1}AP$ and hence $\Phi(t, 0) = \exp[(P^{-1}AP)t]$.
(b) Let $V(t)$ be the solution of $dx/dt = Ax$ such that $V(0) = V$. $BV(t)$ is the solution of $dx/dt = Ax$ such that $BV(0) = BV$ since

$$\frac{d}{dt} BV(t) = BAV(t) = ABV(t)$$

Thus $BV(t) = B \exp(At) V = \exp(At) BV$ for all V and $B \exp(At) = \exp(At) B$.
Let us consider

$$W(t) = \exp(At) \exp(Bt) V \qquad W(0) = V$$

$$\frac{dW}{dt} = A \exp(At) \exp(Bt) V + \exp(At) B \exp(Bt) V = (A + B)W(t)$$

Then $W(t) = \exp(At) \exp(Bt) V = \exp[(A + B)t] V$ which is therefore also equal to $\exp(Bt) \exp(At) V$.
If $B \exp(At) = \exp(At) B$, by differentiation $BA \exp(At) = A \exp(At) B = \exp(At) AB$ so that, for all t, $AB = BA$.
If $\exp[(A + B)t] = \exp(At) \exp(Bt)$, by differentiation

$$(A + B) \exp(At) \exp(Bt) = A \exp(At) \exp(Bt) + \exp(At) B \exp(Bt)$$

Therefore

$$A \exp(At) \exp(Bt) + B \exp(At) \exp(Bt)$$
$$= A \exp(At) \exp(Bt) + \exp(At) B \exp(Bt)$$

and

$$B \exp(At) = \exp(At) B$$

The first relation enables us to calculate $\exp(At)$ for a diagonalizable matrix. If $\lambda_1, \ldots, \lambda_d$ are the eigenvalues (distinct or not) of A,

$$\exists P: \quad P^{-1}AP = \begin{pmatrix} \lambda_1 & & 0 \\ & \ddots & \\ 0 & & \lambda_d \end{pmatrix}$$

$$\exp[(P^{-1}AP)t] = \begin{pmatrix} \exp(\lambda_1 t) & & 0 \\ & \ddots & \\ 0 & & \exp(\lambda_d t) \end{pmatrix}$$

since this is the resolvent of the system $dy_i/dt = \lambda_i y_i$; we then calculate

$$\exp(At) = P \exp[(P^{-1}AP)t] \, P^{-1}$$

The second relation enables us to calculate $\exp(At)$ if we can split A into the sum of two matrices which commute; e.g.

$$A = \begin{pmatrix} a & b \\ -b & a \end{pmatrix} = aI + b\begin{pmatrix} 0 & 1 \\ -1 & 0 \end{pmatrix}$$

$$\exp(At) = \exp(at)\begin{pmatrix} \cos bt & \dfrac{\sin bt}{b} \\ -b\sin bt & \cos bt \end{pmatrix}$$

Comparison with the method of Section 2.2

For given x_0, $\exp(At)x_0$ is the solution such that $x(0) = x_0$. Also let

$$\exp(At) x_0 = \exp[(A - \lambda I)t]\exp(\lambda t) x_0 = \sum_0^\infty \left[(A - \lambda I)^n \frac{t^n}{n!}\right]\exp(\lambda t) x_0$$

If $x_0 \in E_\lambda$, $(A - \lambda I)^n x_0 = 0 \quad \forall n \geqslant v(\lambda)$, so that the above series reduces to

$$\sum_0^{v(\lambda)-1} (A - \lambda I)^n \frac{t^n}{n!} \exp(\lambda t) x_0$$

For such an x_0 this is again Proposition 2.2.1.

Complement

Let us denote by P_{λ_i} the projection operator from $\mathbb{C}^d$ onto E_{λ_i}. The preceding formula can equally well be written

$$\exp(At) = \sum_{i=1}^r \sum_0^{v(\lambda_i)-1} (A - \lambda_i I)^n \frac{t^n}{n!} \exp(\lambda_i t)\, P_{\lambda_i}$$

and, using the notation $f: x \rightsquigarrow \exp(xt)$ and $A \rightsquigarrow \exp(At)$,

$$f(A) = \sum_{i=1}^r \sum_0^{v(\lambda_i)-1} \frac{(A - \lambda_i I)^n}{n!} \frac{d^n f}{dx^n}(\lambda_i) P_{\lambda_i}$$

This formula extends to every holomorphic function in an open set containing the eigenvalues of A; if A is invertible we can thus define $\log A$.

Note   A priori, $\log A$ is a complex matrix even if A is real. Thus for any real invertible A there exists a complex B commuting with A such that $A = \exp(B)$, and there exists a real C, commuting with A, such that $A = \exp(C)$ iff $A = D^2$ (to be verified as an exercise).

See **Exercises 23 and 24** at the end of the chapter.

2.5. The special case when $b(t) = \exp(rt)q(t)$: resonance

If $A \in \mathscr{L}(\mathbb{R}^d, \mathbb{R}^d)$, the solutions of $dx/dt = Ax$ are of the form

$$\sum_{i=1}^{r} p_i(t) \exp(\lambda_i t)$$

where the eigenvalues of A are $\lambda_1, \ldots, \lambda_r$.

If the second term on the right-hand side of $dx/dt = Ax + b(t)$ is of the form $\exp(rt) q(t)$ we can find a particular solution directly; the form of this depends on whether r is or is not equal to one of the eigenvalues.

Theorem 2.5.1

(a) Let

$$\frac{dx}{dt} = Ax + \exp(rt) q(t) \tag{39}$$

where $A \in \mathscr{L}(\mathbb{R}^d, \mathbb{R}^d)$ and $q(t) \in \mathbb{R}^d$. Suppose that q is a polynomial of degree k with coefficients in $\mathbb{R}^d$, i.e. its coordinates are scalar polynomials of degree k. There exists a solution of (39) $y(t) = \exp(rt) p(t)$, where $p(t)$ is a polynomial of degree k if r is not an eigenvalue of A but of degree $k + v$ if r is an eigenvalue of index v.

(b) Let

$$\frac{d^n x}{dt^n} = a_0 x + a_1 \frac{dx}{dt} + \ldots + a_{n-1} \frac{d^{n-1}x}{dt^{n-1}} + \exp(rt) q(t) \tag{40}$$

be a scalar differential equation where q is a polynomial of degree k. There exists a solution of (40) $z(t) = \exp(rt) p(t)$, where p is a polynomial of degree k if r is not a root of the characteristic equation but of degree less than or equal to $k + m$ if r is a root of order m.

Proof   It is naturally sufficient to demonstrate (a). The solution of (39) such that $x(0) = 0$ is

$$\int_0^t \exp[A(t - s)] \exp(rs) q(s) \, ds$$

Let $q(t) = q_0 + q_1 t + \ldots + q_k t^k$ where $q_i \in \mathbb{R}^d$ $\forall i$.

(i) If $r \neq \lambda_i$ is an eigenvalue, $\forall i$, then $(A - rI)^{-1}$ exists. We shall show that we can determine $p(t)$ as a solution of

$$\frac{d}{dt}[\exp(rt) p(t)] = [A(t)p(t) + q(t)]\exp(rt)$$

if

$$rp(t) + \frac{d}{dt} p(t) = Ap(t) + q(t)$$

or again if

$$\frac{d}{dt} p(t) = (A - rI)p(t) + q(t)$$

If $B = A - rI$ and $p(t) = p_0 + p_1 t + \ldots + p_k t^k$ ($p_i \in \mathbb{R}^d$ $\forall i$), we obtain

$$0 = Bp_k + q_k \qquad \text{(coefficient of } t^k)$$

$$kp_k = Bp_{k-1} + q_{k-1} \qquad \text{(coefficient of } t^{k-1})$$

$$\vdots$$

$$2p_2 = Bp_1 + q_1 \qquad \text{(coefficient of } t)$$

$$p_1 = Bp_0 + q_0 \qquad \text{(constant term)}$$

Since B is invertible we can determine step by step the coefficients $p_k = -B^{-1}q_k$, $p_{k-1} = B^{-1}(kp_k - q_{k-1})\ldots$ and finally $p_0 = B^{-1}(p_1 - q_0)$.

(ii) Now suppose r is an eigenvalue of index v. Let E be the proper generalized space associated with r and let $\mathbb{C}^d = E \oplus F$ so that every x is written in the form $x = x_1 + x_2$, $x_1 \in E$, $x_2 \in F$. In particular $q(t) = q_E(t) + q_F(t)$ and $p(t) = p_E(t) + p_F(t)$. $p(t) \exp(rt)$ is a solution of (39) iff

$$\frac{d}{dt} p_E(t) = (A - rI)p_E(t) + q_E(t)$$

$$\frac{d}{dt} p_F(t) = (A - rI)p_F(t) + q_F(t)$$

Over F the restriction of $A - rI$ is invertible (cf. p. 85). We can then find p_F of the same degree as q_F satisfying the relation imposed. Over E the required solution is

$$\int_0^t \exp[A(t-s)] \exp(rs) q_E(s) \, ds = \int_0^t \exp(rs) \sum_{i=0}^{v-1} \frac{(t-s)^i}{i!} A^i q_E(s) \, ds$$

The integrand is $\exp(rs) \pi(s)$ where π is a polynomial of degree $k + v - 1$ whose integral is of degree $k + v$. (To be strictly correct, we would have had to suppose q_E of degree k to obtain this result.)

Case where $r = i\omega$ If $b(t) = \exp(i\omega t) q(t)$ we are looking for solutions of the form $\exp(i\omega t) p(t)$, but we can also take their real part.

If $b(t) = q(t) \cos \omega t$ there exists a solution $p(t) \cos(\omega t + \varphi)$ where the degree of p is equal to that of q if $i\omega$ is not an eigenvalue, and otherwise is greater.

EULER EQUATIONS

Let

$$t^n \frac{d^n x}{dt^n} + \ldots + a_{n-1} t \frac{dx}{dt} + a_n x = 0$$

This equation leads to an equation with constant coefficients if we put $t = \exp(s)$ $(t > 0)$. If, consequently, we consider a right-hand side of the form $t^\alpha p(\log t)$ where p is a polynomial, this right-hand side will be transformed into $\exp(\alpha s) p(s)$ and there will exist a solution $\exp(\alpha s) q(s)$ (where the degree of q depends on the nature of α) and then, after another change of variable, a solution of the form $t^\alpha q(\log t)$.

RESONANCE PHENOMENA

Let us go back to the case where $r = i\omega$ as on page 100; if we consider the right-hand side of an equation as the 'input' of a linear system and the left-hand side as the 'output', in other words the right-hand side as a stimulus and the left-hand side as the response to this stimulus, we can establish that for a periodic and bounded input $\cos \omega t$ we have either an output of the same nature, if ω is not one of the 'natural frequencies' of the system (i.e. an eigenvalue which corresponds to a 'free' oscillation), or by contrast an output $p(t) \cos \omega t$ whose amplitude tends to infinity.

By the **phenomenon of resonance** we mean that equality—or at least proximity—of the excitation frequency and the free (or natural) frequency of the excited system causes a considerable increase in the amplitude of the response.

We shall make all this more precise in the case of second-order equations. We know that they occur in mechanics and in electricity among other applications.

A particle of mass m moving in a straight line with friction proportional to the velocity and a restoring force proportional to the distance has, under the action of a force $G(t)$, a position x such that

$$m \frac{d^2 x}{dt^2} + C \frac{dx}{dt} + \alpha^2 x = G(t) \quad \text{where } C \geqslant 0$$

An R, L, C circuit subjected to an electromotive force $E(t)$ is such that the charge Q on the capacitor satisfies

$$L \frac{d^2 Q}{dt^2} + R \frac{dQ}{dt} + \frac{Q}{C} = E(t)$$

Thus we shall examine the equation

$$\frac{d^2 x}{dt^2} + k \frac{dx}{dt} + \omega^2 x = F(t) \qquad \omega \neq 0, k \geqslant 0 \tag{41}$$

Free system: $\dfrac{d^2x}{dt^2} + k\dfrac{dx}{dt} + \omega^2 x = 0$

Let

$$A = \begin{pmatrix} 0 & 1 \\ -\omega^2 & -k \end{pmatrix}$$

$x \equiv 0$ is a solution and the characteristic equation is $\lambda^2 + k\lambda + \omega^2 = 0$.

(a) $k^2 > 4\omega^2$: *damping.* The two eigenvalues are negative. All solutions $A\exp(\lambda_1 t) + B\exp(\lambda_2 t)$ tend to zero as $t \to +\infty$.

(b) $k = 0$: *free oscillations without damping.* $x(t) = A\cos(\omega t + \varphi)$.

(c) $0 < k^2 < 4\omega^2$: *free oscillations with damping.* $x(t) = A\exp[(-\tfrac{1}{2}k)t] \times \cos(\beta t + \varphi)$ where $\beta = \tfrac{1}{2}(4\omega^2 - k^2)^{1/2}$. The period $2\pi/\beta$ is longer than in the preceding case.

(d) $k^2 = 4\omega^2$: *critical damping.* $x(t) = (At + B)\exp(-\omega t)$ (infinite period).

If $k \neq 0$, all solutions tend to zero: the friction absorbs the 'energy'. In this case $\lambda = \omega$ is oscillatory.

Excited system: $\dfrac{d^2x}{dt^2} + k\dfrac{dx}{dt} + \omega^2 x = F_0 \cos \omega_0 t$

(a) $k \neq 0$. In this case $i\omega_0$ is not an eigenvalue of A and there is no resonance. Let y be a solution of

$$\frac{d^2x}{dt^2} + k\frac{dx}{dt} + \omega^2 x = 0$$

Let x be a solution of the excited system

$$x(t) = y(t) + \frac{F_0}{[(\omega^2 - \omega_0^2)^2 + k^2\omega_0^2]^{1/2}} \cos(\omega_0 t - \vartheta)$$

where

$$\tan \vartheta = \frac{k\omega_0}{\omega^2 - \omega_0^2}$$

Since $k \neq 0$

$$\lim_{t \to \infty} y(t) = 0$$

and after a certain time there remains only the second term which we call the **forced oscillation** (or **sustained oscillation**); its period $2\pi/\omega_0$ is imposed by the excitation force.

If ω_0 is very small $F_0 \cos \omega_0 t \approx F_0 \ \forall t$ and we find as a solution the (forced) constant $x(t) = F_0/\omega^2$; if ω_0 is very large we find $(F_0/\omega_0)\cos \omega_0 t$; the friction no longer has any perceptible effect.

There is no particular relationship between the old and the new amplitudes. If $\omega_0 = \omega$ the second term is $(F_0/k\omega)\sin \omega t$; there is a phase shift of $\pi/2$.

(b) $k = 0$, $\omega_0 \neq \omega$. We obtain the phenomenon of 'beats'. The solution y corresponding to the free system is not damped and

$$x(t) = A\cos(\omega t + \varphi) + \frac{F_0}{\omega^2 - \omega_0^2}\cos\omega_0 t$$

If $\omega = \omega_0 + \varepsilon\,(\varepsilon \ll 1)$ x varies like a sine function of period $2\pi/\omega_0$ modulated by a sine function of period $2\pi/\varepsilon$ and the amplitude becomes very large. There is imperfect resonance, and the amplitude remains **periodic**, although large (Fig. 2.7).

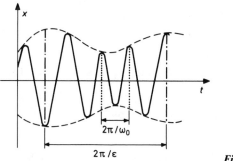

$$2\pi/\omega_0$$

$$2\pi/\varepsilon$$

Fig. 2.7

(c) $k = 0$, $\omega_0 = \omega$, *pure resonance.*

$$x(t) = A\cos(\omega t + \varphi) + \frac{F_0 t}{2\omega}\sin\omega t$$

The amplitude of the second term tends to infinity with t.

Part 3: Linearization in the neighbourhood of an equilibrium point; notion of structural stability

Let

$$\frac{\mathrm{d}x}{\mathrm{d}t} = f(x) \tag{41}$$

be a dynamic system, f a function of class C^1 and a a point of $\mathbb{R}^d$ such that $f(a) = 0$, i.e. an equilibrium point of (41). If $Df(a)$ is the linear mapping tangential to f at the point a, we wish to compare the solutions of

$$\frac{\mathrm{d}x}{\mathrm{d}t} = f(x) \tag{41}$$

$$\frac{\mathrm{d}x}{\mathrm{d}t} = Df(a)(x - a) \tag{42}$$

These equations have a common solution $x \equiv a$. We know that, in the neighbourhood of this equilibrium and for $|t - t_0|$ sufficiently small, the solutions of (41) and (42) are close to each other and we consider the following two questions.

(1) Study solutions whose initial values are close to a, and in particular examine whether they remain close to a for all $t > t_0$.

(2) Compare the behaviour of solutions of (41) and (42) in the neighbourhood of a. For example, for a linear system in $\mathbb{R}^2$, we have described all possible forms of equilibrium; it is natural to enquire whether we find similar forms for neighbouring cases of linear systems.

The first problem is one of stability which will be the subject of Section 3.1; the second is one of structural stability which we shall study in Section 3.2.

3.1. Stability of an equilibrium point and stability of a linear system

We recall that for a dynamic system solutions starting from one and the same point x_0 for different values of t_0 have the same orbit. If f is k-Lipschitzian

$$\varphi(t + t_0, t_0, x_0) - \varphi(t, 0, x_0) = \int_0^t f[\varphi(s + t_0, t_0, x_0)] - f[\varphi(s, 0, x_0)] \, \mathrm{d}s$$

Thus

$$\|\varphi(t + t_0, t_0, x_0) - \varphi(t, 0, x_0)\| \leqslant k \int_0^t \|\varphi(s + t_0, t_0, x_0) - \varphi(s, 0, x_0)\| \, \mathrm{d}s$$

and from Gronwall's lemma

$$\varphi(t + t_0, t_0, x_0) = \varphi(t, 0, x_0)$$

The solution $x(t) \equiv a$ obviously does not depend on t_0. Thus in this section we shall take $t_0 = 0$ and we shall use the notation $\varphi(t, x_0) = \varphi(t, 0, x_0)$.

Definition 3.1.1

Let

$$\frac{\mathrm{d}x}{\mathrm{d}t} = f(x) \tag{43}$$

and $a \in \mathbb{R}^d$: $f(a) = 0$.

(a) We say that a is a **uniformly stable equilibrium** if

$$\forall \varepsilon > 0 \quad \exists \eta(\varepsilon): \|x_0 - a\| \leqslant \eta(\varepsilon) \Rightarrow \|\varphi(t, x_0) - a\| \leqslant \varepsilon \qquad \forall t \geqslant 0$$

(b) We say that a is an **asymptotically uniformly stable equilibrium** if a is uniformly stable and if in addition there exists a neighbourhood of a in which $\varphi(t, x_0)$ has a as its limit, i.e. there exists

$$\rho > 0: \|x_0 - a\| \leqslant \rho \Rightarrow \lim_{t \to \infty} \varphi(t, x_0) = a$$

(c) An equilibrium which is not uniformly stable is called unstable.

Remark In a chapter devoted to the study of problems of stability, we shall study the stability of solutions $\varphi(t, t_0, x_0)$ and this stability will involve t_0; the term 'uniformly' introduced in 3.1.1 corresponds to the fact that η does not depend on t_0 in this case, whereas in general it does depend on it. We shall verify that we can confine ourselves to the case $a = 0$.

Let g be defined by $g(y) = f(y + a)$. It is clear that a is an equilibrium of $dx/dt = f(x)$ iff 0 is an equilibrium of $dy/dt = g(y)$.

Now let

$$\frac{dx}{dt} = f(x) \tag{44}$$

and

$$\frac{dy}{dt} = g(y) \tag{44'}$$

and let $\varphi(t, x_0)$ be the solution of (44) such that $x(0) = x_0$, and $\psi(t, y_0)$ be the solution of (44') such that $y(0) = y_0$; it is clear that $\varphi(t, x_0) - a = \varphi(t, x_0 - a)$.

The origin is stable for (44') iff

$$\forall \varepsilon > 0 \quad \exists \eta(\varepsilon): \|y_0\| \leqslant \eta(\varepsilon) \Rightarrow \|\psi(t, y_0)\| \leqslant \varepsilon$$

i.e.

$$\forall \varepsilon > 0 \quad \exists \eta(\varepsilon): \|x_0 - a\| \leqslant \eta(\varepsilon) \Rightarrow \|\varphi(t, x_0) - a\| \leqslant \varepsilon$$

which is true iff a is a stable equilibrium for (44).

We shall now suppose that 0 is an equilibrium of f and we shall compare the solutions of

$$\frac{dx}{dt} = f(x) \qquad \text{where } f(0) = 0 \tag{45}$$

$$\frac{dx}{dt} = Df(0)x \tag{46}$$

We know that the nature of the equilibrium 0 of (46) depends on the eigenvalues of $Df(0)$; we shall now state this precisely.

Theorem 3.1.2

Let $dx/dt = Ax$ be a linear system. Let $\lambda_1, \ldots, \lambda_r$ be the distinct eigenvalues of A.

(a) 0 is a uniformly stable equilibrium iff $\forall i \ \text{Re} \ \lambda_i \leqslant 0$ and if in addition $\text{Re} \ \lambda_k < 0$ for the oscillatory eigenvalues.

(b) 0 is an asymptotically uniformly stable equilibrium if $\forall i \ \text{Re} \ \lambda_i < 0$.

(c) If there exists an eigenvalue λ such that $\text{Re} \ \lambda > 0$, 0 is unstable.

Proof Let $\lambda_p = \alpha_p + i\beta_p$ (where $\beta_p = 0$ if λ_p is real).

$$\varphi(t, x_0) = \sum_{p=1}^{r} \sum_{n=0}^{v(\lambda_p)-1} (A - \lambda_p I)^n \frac{t^n}{n!} \exp(\alpha_p t)\,(\cos \beta_p t + i \sin \beta_p t) x_0^{(p)}$$

where, as on page 88, $x_0^{(p)}$ denotes the projection of x_0 on E_{λ_p}, the proper generalized space relative to the eigenvalue λ_p.

We obtain the result by comparing the growth of the polynomials and of the exponentials in question; we can state, moreover, that the growth or decay of the solutions is exponential.

The fundamental comparison theorem is the following.

Theorem 3.1.3

Let

$$\frac{dx}{dt} = f(x) \qquad \text{where } f(0) = 0 \tag{47}$$

Let $\lambda_1, \ldots, \lambda_r$ be the eigenvalues of $Df(0)$.

(a) If $\forall i \ \ \mathrm{Re}\,\lambda_i < -\sigma < 0$, $x \equiv 0$ is an **asymptotically** uniformly stable equilibrium of (47).

(b) If $\exists i_0 : \mathrm{Re}\,\lambda_{i_0} > 0$, $x \equiv 0$ is an **unstable** equilibrium of (47).

(c) If $\forall i \ \ \mathrm{Re}\,\lambda_i \leqslant 0$ and $\exists i_0 : \mathrm{Re}\,\lambda_{i_0} = 0$ no conclusion can be drawn.

Proof We shall give the proof of (b) in Chapter 4 (p. 253). Under the hypothesis of (a) there exists $K > 1$ such that $\|\exp(At)\| \leqslant K \exp(-\sigma t) \ \forall t \geqslant 0$. For $A = Df(0)$, (47) can be written

$$\frac{dx}{dt} = Ax + f(x) - Ax$$

so that

$$\varphi(t, x_0) = \exp(At) x_0 + \int_0^t \exp[A(t - s)]\{f[\varphi(s, x_0)] - A\varphi(s, x_0)\}\, ds$$

and

$$\|\varphi(t, x_0)\| \leqslant K \exp(-\sigma t)\|x_0\| + \int_0^t K \exp[-\sigma(t - s)]\|f[\varphi(s, x_0)] -$$

$$A\varphi(s, x_0)\|\, ds$$

By the definition of $Df(0)$, $\forall \alpha > 0 \ \exists \beta(\alpha): \|x\| \leqslant \beta(\alpha) \Rightarrow \|f(x) - Ax\| \leqslant \alpha\|x\|$. Let ε be given and let a fixed α be such that $\alpha K < \sigma$; also let $\beta(\varepsilon) = \inf[\beta(\alpha), \varepsilon]$. Then $\forall x \leqslant \beta(\varepsilon)$, $\|f(x) - Ax\| \leqslant \alpha\|x\|$. Finally let $\eta(\varepsilon) = \beta(\varepsilon)/K$. We shall verify that

$$\|x_0\| \leqslant \eta(\varepsilon) \Rightarrow \|\varphi(t, x_0)\| \leqslant \beta(\varepsilon) \qquad \forall t > 0 \tag{48}$$

which establishes the result since $\beta(\varepsilon) \leqslant \varepsilon$.

Suppose that (48) is not satisfied and let $\vartheta = \inf[s, \|\varphi(s, x_0)\| \geqslant \beta(\varepsilon)]$; then $0 < \vartheta < \infty$ and by continuity

$$\varphi(\vartheta, x_0) = \beta(\varepsilon) \tag{49}$$

If $t < \vartheta$

$$\forall s \leqslant t \qquad \|f[\varphi(s, x_0)] - A\varphi(s, x_0)\| \leqslant \alpha\|\varphi(s, x_0)\|$$

and

$$t < \vartheta \Rightarrow \exp(\sigma t)\|\varphi(t, x_0)\| \leqslant K\|x_0\| + \alpha K \int_0^t \exp(\sigma s)\|\varphi(s, x_0)\| \, ds$$

Then

$$\exp(\sigma t)\|\varphi(t, x_0)\| \leqslant K\|x_0\| \int_0^t \exp[\alpha K(t - s)] \, ds$$

and

$$\|\varphi(t, x_0)\| \leqslant K\|x_0\| \exp[(\alpha K - \sigma)t] < K\|x_0\|$$

Thus

$$\|\varphi(\vartheta, x_0)\| < K\|x_0\| \leqslant \beta(\varepsilon)$$

which contradicts relation (49); it is absurd to suppose that $\vartheta < \infty$. This completes the proof. It is clear then that

$$\lim_{t \to +\infty} \|\varphi(t, x_0)\| = 0$$

Example (a) Let

$$\frac{d^2u}{dt^2} + 4\frac{du}{dt} + u^2 + u = 0 \tag{50}$$

$u \equiv 0$ is a solution, as is $u \equiv -1$. If we put $u = x$ and $du/dt = y$, (50) is equivalent to $dx/dt = y$, $dy/dt = -x - 4y - x^2$. For

$$DF(0, 0) = \begin{pmatrix} 0 & 1 \\ -1 & -4 \end{pmatrix}$$

the eigenvalues $-2 \pm \sqrt{3}$ are negative. $u \equiv 0$ is an asymptotically uniformly stable equilibrium. For

$$DF(-1, 0) = \begin{pmatrix} 0 & 1 \\ 1 & -4 \end{pmatrix}$$

the eigenvalues are $-2 \pm \sqrt{5}$. $u \equiv -1$ is an unstable equilibrium.

For the linear system we have a saddle point. We shall see in the following section that this implies for (50) an analogous form for the set of solutions in the neighbourhood of the equilibrium.

(b) Let

$$\frac{dx}{dt} = y + \varepsilon(x^3 + 2xy^2) \\ \frac{dy}{dt} = -x + \varepsilon y^3 \Biggr\}$$ (51)

$x \equiv y \equiv 0$ is an equilibrium and $\forall \varepsilon$

$$Df(0, 0) = \begin{pmatrix} 0 & 1 \\ -1 & 0 \end{pmatrix}$$

with eigenvalues $\pm i$.
 • If $\varepsilon = 0$ the system is linear, $dx/dt = y$, $dy/dt = -x$, and $(0, 0)$ is stable (not asymptotically). Put $\varepsilon \neq 0$ and $V(x, y) = x^2 + y^2$; let

$$0 < x_0{}^2 + y_0{}^2 = \rho^2 < \frac{1}{\varepsilon} \qquad \rho^2 = (X_0)^2$$

$$\frac{d}{dt} V[x(t), y(t)] = 2\varepsilon(x^2 + y^2)^2$$

$$= 2\varepsilon[V(t)]^2$$

$$V(t) = \frac{\rho^2}{1 - 2\varepsilon\rho^2 t}$$

 • If $\varepsilon > 0$ $V(t)$ is increasing and $V(t) \to \infty$ if $t \to 1/2\varepsilon\rho^2$; however small ρ may be the solution $\varphi(t, X_0)$ tends to infinity: $(0, 0)$ is unstable.
 • If $\varepsilon < 0$ $V(t)$ is decreasing, $\lim V(t) = 0$ and $(0, 0)$ is asymptotically uniformly stable.
 Thus by perturbing the initial system by as small an amount as we wish we obtain either a stable or an unstable equilibrium. Nevertheless all these systems have the same linearized system, but $\text{Re}(\lambda) = 0$.
 It must be clearly understood that $Df(0)x$ and $f(x)$ are in fact close to each other only if x is extremely close to 0.
 See *Exercise 27* at the end of the chapter.

3.2. Structural stability

If $dx/dt = f(x)$ is an autonomous system in $\mathbb{R}^d$, we shall be interested in the following problem: if g is 'close' to f, can we say that the solutions of $dx/dt = g(x)$ 'resemble' those of $dx/dt = f(x)$?

EQUIVALENT DYNAMIC SYSTEMS

We shall suppose f and g to be of class C^1 and we shall characterize their proximity in the following way.

Definition 3.2.1

If G is a set in $\mathbb{R}^d$, we shall say that in G the distance between f and g is smaller than ε in the C^1 sense if

$$\forall x \in G \qquad \| f(x) - g(x) \| \leqslant \varepsilon$$

and

$$\| Df(x) - Dg(x) \| \leqslant \varepsilon$$

and we shall denote this by $\| f - g \|_{1,G} \leqslant \varepsilon$.

To compare the family of orbits of

$$\frac{dx}{dt} = f(x) \tag{52}$$

and

$$\frac{dx}{dt} = g(x) \tag{53}$$

we shall introduce a correspondence between these curves.

We shall denote by $\varphi(t, x)$ the solution of (52) such that $x(0) = x$ and by $\psi(t, x)$ that of (53) such that $x(0) = x$. $t \rightsquigarrow \varphi(t, x)$ and $t \rightsquigarrow \psi(t, x)$ are the parametric forms of the orbits of (52) and (53) passing through a point x.

A mapping $T: \mathbb{R}^d \rightsquigarrow \mathbb{R}^d$ establishes a correspondence between the orbits of (52) and (53) if $\forall x$

$$T[\varphi(t, x)] = \psi[t, T(x)] \qquad \forall t$$

(see Fig. 2.8).

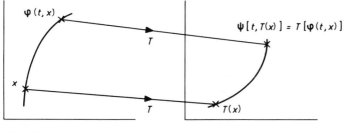

Fig. 2.8

Remark T establishes in fact a correspondence between the orbits provided with their parametric forms (we say that it is a correspondence between the 'flows').

According to the nature of T, we shall encounter several types of equivalence.

Definition 3.2.2

Let

$$\frac{dx}{dt} = f(x) \tag{54}$$

and

$$\frac{dx}{dt} = g(x) \tag{55}$$

be two dynamic systems in $\mathbb{R}^d$.

(a) We say that (54) and (55) are **linearly equivalent** if there exists a linear and invertible function T, from $\mathbb{R}^d$ into $\mathbb{R}^d$, which establishes a correspondence between the orbits of (54) and (55).

(b) We say that (54) and (55) are **differentiably equivalent** if there exists a differentiable function T, with differentiable inverse, which establishes a correspondence between the orbits of (54) and (55); (54) and (55) are C^r **equivalent** if T and T^{-1} are of class C^r.

(c) We say that (54) and (55) are **topologically equivalent** if there exists a continuous invertible function T with continuous inverse, from $\mathbb{R}^d$ into $\mathbb{R}^d$, which establishes a correspondence between the orbits of (54) and (55).

If in one of the three cases T is defined only over a set G we say that the systems are **equivalent in** G in one or the other sense.

It is easy to see that there are connections between the types of equivalence and that linear equivalence implies differentiable equivalence which itself implies topological equivalence.

EQUIVALENT LINEAR SYSTEMS

Theorem 3.2.3

Let

$$\frac{dx}{dt} = Ax \tag{56}$$

and

$$\frac{dx}{dt} = Bx \tag{57}$$

be two linear systems in $\mathbb{R}^d$.

(a) If (56) and (57) are linearly equivalent, A and B have the same eigenvalues.

(b) If A and B have simple eigenvalues, (56) and (57) are linearly equivalent iff the eigenvalues are identical.

Proof With the preceding notation $\varphi(t, x) = \exp(At) x$, $\psi[t, h(x)] =$

$\exp(Bt) h(x)$. If the systems are linearly equivalent, there exists a linear and invertible function T such that $T \exp(At) x = \exp(Bt) Tx \ \forall x \in \mathbb{R}^d$; hence $T \exp(At) T^{-1} = \exp(Bt)$, and $B = T^{-1} A T$ has the same eigenvalues as A.

Suppose now that A and B have the same simple (and therefore real) eigenvalues $\lambda_1, \ldots, \lambda_d$. Let $E_1, \ldots, E_d$ be the corresponding proper spaces for A and $F_1, \ldots, F_d$ the corresponding spaces for B; they are of dimension one and $\mathbb{R}^d = E_1 \oplus \ldots \oplus E_d = F_1 \oplus \ldots \oplus F_d$; thus there exists a basis of eigenvectors of B where $e_i \in E_i$ and $f_i \in F_i \ \forall i$.

Let us define H by $H(e_i) = f_i$. If $V_i \in E_i$ then $HV_i \in F_i$, $\varphi(t, V_i) = \exp(\lambda_i t) V_i$ while $\psi(t, HV_i) = \exp(\lambda_i t) HV_i = H[\exp(\lambda_i t) V_i]$. If $V \in \mathbb{R}^d$

$$V = V_1 + \ldots + V_d$$

and

$$\varphi(t, V) = \sum_{i=1}^{d} \exp(\lambda_i t) V_i$$

$$= \sum_{i=1}^{d} H[\psi(t, HV_i)]$$

$$= H\psi(t, HV)$$

The fundamental theorem follows; we shall not prove it.

Theorem 3.2.4

Let

$$\frac{dx}{dt} = Ax \tag{58}$$

and

$$\frac{dx}{dt} = Bx \tag{59}$$

We suppose $\operatorname{Re} \lambda \neq 0$ for all eigenvalues λ of A or of B. We denote by $m_+(A)$ and $m_-(A)$ the numbers of eigenvalues of A with positive and negative real part respectively, and by $m_+(B)$ and $m_-(B)$ the corresponding numbers for B. Then (58) and (59) are topologically equivalent (see Fig. 2.9) iff $m_+(A) = m_+(B)$.

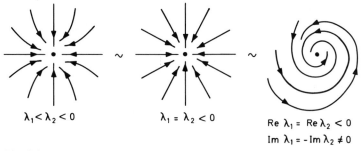

$$\lambda_1 < \lambda_2 < 0 \qquad\qquad \lambda_1 = \lambda_2 < 0$$

$$\operatorname{Re} \lambda_1 = \operatorname{Re} \lambda_2 < 0$$
$$\operatorname{Im} \lambda_1 = -\operatorname{Im} \lambda_2 \neq 0$$

Fig. 2.9. *Examples of topologically equivalent systems.*

Example   Let

$$A = \begin{pmatrix} 1 & 0 \\ 0 & 1 \end{pmatrix} \qquad B = \begin{pmatrix} 1 & 1 \\ 0 & 1 \end{pmatrix}$$

The eigenvalues of A and B are identical, and so the systems are topologically equivalent. However, the eigenvalues are double, so that the systems are not necessarily linearly equivalent. (As an exercise, it can be verified that in fact they are not linearly equivalent.)

If there exists an eigenvalue of zero real part no further conclusion can be drawn.

Example

$$A = \begin{pmatrix} 0 & 1 \\ 0 & 0 \end{pmatrix} \qquad B = \begin{pmatrix} 0 & 1 \\ -1 & 0 \end{pmatrix} \qquad C = \begin{pmatrix} 0 & 1 \\ 1 & 0 \end{pmatrix}$$

The systems have the orbits shown in Fig. 2.10.

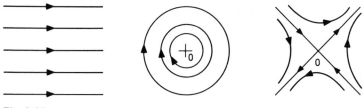

Fig. 2.10

See **Exercises 28 to 31** at the end of the chapter.

We can now study the linear systems near to a linear system.

Theorem 3.2.5

Let $A \in \mathcal{L}(\mathbb{R}^d, \mathbb{R}^d)$; we suppose that for every eigenvalue λ, Re $\lambda \neq 0$. Then there exists $\varepsilon_0 > 0$ such that for every linear mapping P with $\|P\| = 1$ and for every $\varepsilon \leqslant \varepsilon_0$ the systems

$$\frac{dx}{dt} = Ax \tag{60}$$

and

$$\frac{dx}{dt} = (A + \varepsilon P)x \tag{61}$$

are topologically equivalent.

This means that all linear systems sufficiently close to (60) are topologically equivalent to it.

Proof It is easy to see that the condition is sufficient. If we denote by $\lambda_i(\varepsilon)$ the eigenvalues of $A + \varepsilon P$, these are continuous functions of ε so that $\exists \varepsilon_0 : \forall \varepsilon \leqslant \varepsilon_0,\ m_+(A) = m_+(A + \varepsilon P)$.

To show that the condition is necessary suppose that there exists λ_{i_0} such that Re $\lambda_{i_0} = 0$. It is then easy to verify that we can find P such that $\|P\| = 1$ and such that Re $\lambda_{i_0}(\varepsilon) < 0$ while Re $\lambda_{i_0}(-\varepsilon) > 0$, whereas Re λ_j, $j \neq i_0$, does not change sign, i.e. $m_+(A - \varepsilon P) \neq m_+(A + \varepsilon P)$, so that the systems (60) and

$$\frac{dx}{dt} = (A + \varepsilon P)x \qquad \frac{dx}{dt} = (A - \varepsilon P)x$$

cannot be equivalent.

We can see this for example in $\mathbb{R}^2$; let

$$A = \begin{pmatrix} a & b \\ c & d \end{pmatrix}$$

and det $A = 0$, the eigenvalues being 0 and $a + d$. Let

$$P = \begin{pmatrix} 1 & 0 \\ 0 & 0 \end{pmatrix} \qquad A + \varepsilon P = \begin{pmatrix} a + \varepsilon & b \\ c & d \end{pmatrix}$$

with eigenvalues

$$a + d + \frac{\varepsilon a}{a + d} + O(\varepsilon)$$

and

$$\frac{\varepsilon d}{a + d} + O(\varepsilon)$$

The first has the sign of $a + d$ in the neighbourhood of $\varepsilon = 0$, the second that of ε.

STRUCTURAL STABILITY

We shall now compare neighbouring systems which are not necessarily linear, and first of all a system and its linearized form.

Theorem 3.2.6
Let $f: \mathbb{R}^d \to \mathbb{R}^d$ and let a be such that $f(a) = 0$; suppose that f is of class C^1 and that the eigenvalues of $Df(a)$ have a non-zero real part. Then there exists a neighbourhood G of a such that

$$\frac{dx}{dt} = f(x) \tag{62}$$

and

$$\frac{\mathrm{d}x}{\mathrm{d}t} = Df(a)(x-a) \tag{63}$$

are topologically equivalent in G.

This means that, in the neighbourhood of an equilibrium point a satisfying $\mathrm{Re}\,\lambda \neq 0$ for all eigenvalues λ of $Df(a)$, the orbits of the system associated with f and of its linearized form can be deduced from each other by a transformation which is continuous and has a continuous inverse, but as we have seen (p. 112) this implies only a relatively weak resemblance between the orbits.

We can in fact prove a more precise theorem (Samovol's theorem) for which we suppose f to be of class C^∞; in the case of $\mathbb{R}^2$ it takes the following form.

Theorem 3.2.7 (Hartmann)

Let f be of class C^∞ (we could weaken this assumption) and let $Df(a)$ be invertible.

(a) If (63) has a focus at a, the orbits of (62) encircle a and tend towards a (if $t \to +\infty$ or $t \to -\infty$); a is a focus for (62).

(b) If (63) has a saddle point at a, (62) has a saddle point at a. There exist two orbits which are invariant subsets; one of them (stable) is tangential to the stable invariant subspace of (63), whilst the other (unstable) is tangential to the unstable invariant subspace of (63).

(c) If (63) has a node, (62) has a node of the same kind as (63). In the case of an improper node, all the orbits are tangential to those of (63) at the point a; in the case of a proper node they are all tangential to the orbits except for two which are tangential to the separatrices of (63).

(d) If (63) has a source, (62) has a source; if (63) has a sink, (62) has a sink.

(e) If (63) has a centre, (62) has either a centre or a focus.

Figure 2.11 illustrates this theorem.

Definition 3.2.8 (Structural stability)

We say that the system $\mathrm{d}x/\mathrm{d}t = f(x)$ is structurally stable in a set $G \subset \mathbb{R}^n$ if $\exists \varepsilon_0 > 0$ such that, for all g satisfying $\|g - f\|_{1,G} \leqslant \varepsilon_0$ (cf. p. 109), $\mathrm{d}x/\mathrm{d}t = f(x)$ and $\mathrm{d}x/\mathrm{d}t = g(x)$ are topologically equivalent in G.

The following theorem is a consequence of Theorem 3.2.6.

Theorem 3.2.9

Let $f: \mathbb{R}^d \to \mathbb{R}^d$ be such that $f(a) = 0$ and such that the eigenvalues of $Df(a)$ all have a non-zero real part. Then there exists a neighbourhood G of a in which the system $\mathrm{d}x/\mathrm{d}t = f(x)$ is structurally stable.

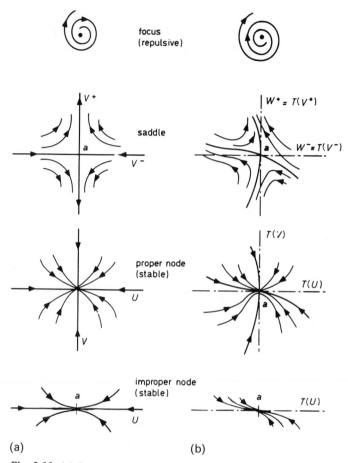

Fig. 2.11. (a) *Linearized system* (63); (b) *original system* (62).

Before giving the proof we shall first state the following result, which we shall not prove but whose statement is significant.

Theorem 3.2.10

Under the same conditions as in Theorem 3.2.9 there exists a neighbourhood G of a with $\varepsilon_0 > 0$ such that, $\forall g: \|g - f\|_{1,G} \leqslant \varepsilon_0$, the system $\mathrm{d}y/\mathrm{d}t = g(y)$ has an equilibrium b, unique in G and such that

$$m_+[Df(a)] = m_+[Dg(b)]$$

and

$$m_-[Df(a)] = m_-[Dg(b)]$$

Proof Even though we restrict the neighbourhood G of Theorem 3.2.10 we can suppose that in this neighbourhood the systems associated with f and $Df(a)$ are topologically equivalent as well as those associated with $Df(a)$ and $Dg(b)$ (since $m_+[Df(a)] = m_+[Dg(b)]$ and $m_-[Df(a)] = m_-[Dg(b)]$); however, $Dg(b)$ then has no eigenvalue with zero real part and all these systems are still topologically equivalent in G to $dx/dt = g(x)$; f is thus structurally stable.

The two preceding theorems have a local character, i.e. they give information for a neighbourhood G of the equilibrium a, and this neighbourhood may be very small.

The following result has a more global character.

Theorem 3.2.11
Let
$$\frac{dx}{dt} = f(x) \tag{64}$$

be an autonomous system in $\mathbb{R}^d$. Let $B = \{x: \|x\| \leqslant 1\}$ and $\partial B = \{x: \|x\| = 1\}$. We suppose f to be of class C^1 and such that $\langle x, f(x) \rangle < 0 \;\; \forall x \in \partial B$ (this is the usual scalar product, and we say that f is re-entrant). We suppose that 0 is the only equilibrium of (64) in B and that $\mathrm{Re}\, \lambda < 0$ for every eigenvalue λ of $Df(0)$.
    Then if
$$\lim_{t \to \infty} \varphi(t, x) = 0 \qquad \forall x \in B$$

the system (64) is structurally stable.

If $d = 2$ we can state an analogous theorem after defining a closed attractive or repulsive orbit: in the first case the neighbouring trajectories follow a spiral path approaching each other; in the second they follow a spiral path moving apart, as in Fig. 2.12.

Theorem 3.2.12
Let
$$\frac{dx}{dt} = f(x) \tag{65}$$

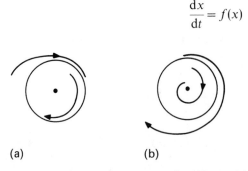

(a)   (b)

Fig. 2.12. (a) *Closed attractive orbit;* (b) *repulsive orbit.*

be an autonomous system in $\mathbb{R}^2$. If $B = \{x: \|x\| \leqslant 1\}$, suppose that '$f$ is re-entrant', that the equilibria of the system are such that Re $\lambda \neq 0$ for all the eigenvalues of the tangential linear mappings, that the closed orbits are either repulsive or attractive and that there is no orbit in B joining two saddle points; then (65) is structurally stable (see *Exercise 36* at the end of the chapter).

Example   The van der Pol equation is structurally stable.

Structural stability and stability of an equilibrium

Theorem 3.2.13
Under the conditions of Theorems 3.2.9 and 3.2.10 every system $dx/dt = g(x)$ where $\|g - f\|_{1,G} \leqslant \varepsilon_0$ has an equilibrium b in G.
If a is stable, b is stable; if a is asymptotically stable, so is b.

In other words, if a structurally stable system has a stable equilibrium every sufficiently close system also has a stable equilibrium; if the first is asymptotically stable, so is the second.

Proof   Let $V \subset T(G)$ and $V = \{y: \|y - b\| \leqslant \varepsilon\}$. To establish that b is stable we must show that we can choose $y (= T(x))$ such that $\psi(t, y) \in V$ $\forall t > 0$ (see Fig. 2.13).

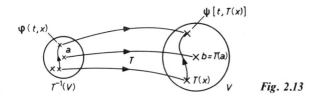

Fig. 2.13

In the same way we can choose x such that $\varphi(t, x) \in T^{-1}(V)$ $\forall t > 0$, which is possible since a is stable. Likewise if

$$\lim_{t \to \infty} \varphi(t, x) = a$$

then

$$\lim_{t \to \infty} \psi[t, T(x)] = b$$

Naturally two systems close to each other but not equivalent can have stable equilibria. Let

$$\frac{dx}{dt} = \begin{pmatrix} 0 & \varepsilon \\ -\varepsilon & 0 \end{pmatrix} x \quad \text{and} \quad \frac{dy}{dt} = \begin{pmatrix} -\varepsilon & 0 \\ 0 & -\varepsilon \end{pmatrix} y$$

In every $\mathbb{R}^n$ $\|f - g\|_{1,\mathbb{R}^n} = \varepsilon$ while 0 is a sink for the second equation and a centre for the first (by changing ε into $-\varepsilon$ we could equally well have a source). This arises from the fact that

$$\frac{dx}{dt} = \begin{pmatrix} 0 & \varepsilon \\ -\varepsilon & 0 \end{pmatrix} x$$

is not structurally stable.

See **Exercises 31 to 35** at the end of the chapter.

Exercises

1. Find the resolvents of the equations

(a)
$$\frac{dX}{dt} = \begin{pmatrix} 1 & 0 \\ 0 & 1 \end{pmatrix} X$$

(b)
$$\frac{dX}{dt} = \begin{pmatrix} 1 & 1 \\ 0 & 1 \end{pmatrix} X$$

2. Consider the equation

$$\frac{dX}{dt} = \begin{pmatrix} 1 + \cos 2t & -1 + \sin 2t \\ 1 + \sin 2t & 1 - \cos 2t \end{pmatrix} X$$

which we denote

$$\frac{dX}{dt} = A(t)X \tag{66}$$

Let

$$S(t) = (\cos t - i \sin t)\begin{pmatrix} \cos t & -\sin t \\ \sin t & \cos t \end{pmatrix}$$

and

$$B = \begin{pmatrix} 2+i & 0 \\ 0 & i \end{pmatrix}$$

(a)   Verify that $dS/dt = A(t)S(t) - S(t)B$; verify that $S^{-1}(t)$ exists.

(b)   If $Y(t) = S^{-1}(t)X(t)$, prove that $dY/dt = BY$.

(c)   Hence deduce an expression for $\Phi(t, 0)$ as well as all the solutions of (66).

3. Apply Theorem 3.4.4 (Chapter 1, p. 43) to the hypotheses of Theorem 1.1.7 and compare the results.

4. Make the same comparison with Theorem 3.4.5 (Chapter 1, p. 44). (Consider the length of the segment $(t_0, t_0 + T)$ over which a good approximation can be obtained.)

5. Let

$$\frac{d^2 x}{dt^2} + [1 + \varepsilon f(t)]x = 0 \tag{67}$$

be an equation in which $|f(t)| \leq M \quad \forall t$; let $(x_0, x_0{}')$ be an initial point. Compare the solutions of (67) and those of

$$\frac{d^2 x}{dt^2} + x = 0$$

over an interval of length T.

6. Compare, over the segment $(1, T + 1)$, the solutions of the two equations

$$t^2 \frac{d^2 x}{dt^2} - 3t \frac{dx}{dt} + 4x = 0$$

$$\frac{d^2 x}{dt^2} - 3\left(\sin \frac{\pi}{2t}\right)\frac{dx}{dt} + 4 \cos\left(\frac{1}{t} - 1\right)x = 0$$

7. Find a necessary and sufficient condition that the equation

$$x'' + p(t)x' + q(t)x = 0 \tag{68}$$

should admit two linearly independent solutions x_1 and x_2 such that $x_1 x_2 = 1$. Supposing that $p(t) = -1/t$, integrate the equation.

8. Find a condition on p and q so that (68) should allow two non-zero solutions x_1 and x_2 such that $x_2 = tx_1$. Integrate the equation in this case.

9. Prove that every solution of $x'' + \sinh tx = 0$ has at most one zero in $]-\infty, 0[$ and an infinity of zeros in $]0, \infty[$.

10. Let $q(t)$ be greater than zero and continuous in $]0, \infty[$; show that every solution of $x'' + q(t)x = 0$ has an infinity of zeros in $]0, \infty[$.

11. Prove that every solution of $x'' + \exp(t)\,x = 0$ has an infinity of zeros in $]0, \infty[$ whereas every solution of $x'' - \exp(t)\,x = 0$ has at most one zero in $]0, \infty[$.

12. Let

$$tx'' + (1 - t)x' + nx = 0 \quad \text{(Laguerre's equation)} \tag{69}$$

Give the self-adjoint form of (69) and study the zeros of the solutions over $]1, \infty[$.

Exercises 13 and 14 may be solved by referring back to Exercises 1 and 2.

13. Solve

$$\frac{dX}{dt} = \begin{pmatrix} 1 & 0 \\ 0 & 1 \end{pmatrix} X + \begin{pmatrix} \sin t \\ \cos t \end{pmatrix} \qquad \text{for given } X(0)$$

$$\frac{dX}{dt} = \begin{pmatrix} 1 & 1 \\ 0 & 1 \end{pmatrix} X + \begin{pmatrix} \exp(t) \\ \exp(t) \end{pmatrix} \qquad \text{for given } X(0)$$

14. Solve

$$\frac{dX}{dt} = \begin{pmatrix} 1 + \cos 2t & -1 + \sin 2t \\ 1 + \sin 2t & 1 - \cos 2t \end{pmatrix} X + \begin{pmatrix} \sin t \\ \cos t \end{pmatrix} \qquad \text{for given } X(t_0)$$

(Cf. Exercise 2.)

15. Find the solution of

$$\frac{d^2 x}{dt^2} + x = b(t)$$

which vanishes, as does its derivative, for $t = t_0$.

16. Find all the solutions of

$$\frac{d^2 x}{dt^2} + x = |t|$$

and of

$$\frac{d^2 x}{dt^2} - x = |t|$$

17. Solve

$$\frac{d^2 x}{dt^2} + 2\frac{dx}{dt} + x = 2 + \sin t$$

Solve the following equations (Exercises 18 to 20).

18. $x'' - 3x' + 2x = 0$ $x'' + x' + x = 0$ $x'' - 6x' + 9x = 0$

19. $x'' + x = \dfrac{1}{\tan t}$ $t^2 x'' - 2tx'' + 2x = 6t^4$ $x'' - x' = 3t + 4\exp(t)$

20.
$$\begin{cases} \dfrac{dx}{dt} = x + 2y + \exp(t) \\[2mm] \dfrac{dy}{dt} = x + y - 2\exp(t) \end{cases} \qquad \begin{cases} \dfrac{dx}{dt} = x + y + 2t \\[2mm] \dfrac{dy}{dt} = 4x + y - \sin t \end{cases}$$

21. Under the conditions of Section 2.3, study the case where det $A = 0$.

22. Study a dynamic system in $\mathbb{R}^3$ in the same way.

23. Consider the equation in $\mathbb{R}^d$

$$\frac{dx}{dt} = A(t)x + b(t) \tag{70}$$

Let φ be the resolvent of $A(t)$.

(a) Let $T > 0$ be given and suppose that $I - \Phi(T, 0)$ is invertible; show that there exists at most one solution of (70) of period T. (Try to find the value of this solution for $t = 0$.)

(b) Supposing that $A(t) \equiv A$, state the form which the preceding condition takes. Suppose that b has period T. Verify that there then exists, under the condition of (a), exactly one solution of period T.

What happens if 1 is an eigenvalue of $\exp(AT)$?

(c) Let

$$\frac{dX}{dt} = \begin{pmatrix} 2 & 0 \\ 1 & 1 \end{pmatrix} X + \begin{pmatrix} f(t) \\ g(t) \end{pmatrix}$$

where f and g have period T ($T > 0$). Show that this equation has one periodic solution and one only. Find this solution.

24. In $\mathbb{R}^d$ consider the equation

$$\frac{dx}{dt} = (\lambda I + P_d)x \tag{71}$$

where

$$P_d = \begin{pmatrix} 0 & 1 & 0 & \cdots & 0 \\ 0 & 0 & 1 & \cdots & 0 \\ \vdots & \vdots & 0 & 1 & \vdots \\ & & & & \ddots & 1 \\ 0 & 0 & \cdots & 0 & 0 \end{pmatrix}$$

(a) Verify that $(P_2)^2 = 0$, $(P_3)^3 = 0$, ..., $(P_d)^d = 0$.

(b) Hence deduce the resolvent of (71) and an expression for the solutions.

25. Solve

$$\frac{d^2x}{dt^2} - x = \exp(t)$$

26. Solve

$$\frac{d^2x}{dt^2} - \frac{2x}{t^2} = t \qquad t > 0$$

$$\frac{d^2x}{dt^2} - \frac{2x}{t^2} = t \log t \qquad t > 0$$

$$\frac{d^2x}{dt^2} - \frac{2x}{t^2} = \log t \qquad t > 0$$

27. Discuss the stability of the equilibria of the following equations:

(a) $\dfrac{d^2x}{dt^2} + \omega^2 x = 0$
(b) $\dfrac{d^2x}{dt^2} + 2\lambda\dfrac{dx}{dt} + x = 0$

(c) $\dfrac{dx}{dt} = -x(1-x)$
(d) $\dfrac{dx}{dt} = (1-x^2)$

(e) $\dfrac{dx}{dt} = -2x(x-1)(2x-1)$
$\dfrac{dy}{dt} = -2y$

28. Write down explicitly the transformations T which enable us to establish the equivalence of the three systems whose orbits are represented after Theorem 3.2.4 (p. 111).

29. Let

$$A = \begin{pmatrix} 1 & 0 \\ 0 & 1 \end{pmatrix} \qquad B = \begin{pmatrix} 1 & 1 \\ 0 & 1 \end{pmatrix}$$

Verify that the corresponding systems are not linearly equivalent.

These systems are topologically equivalent according to Theorem 3.2.4. Let T be the corresponding transformation

$$(x, y) \rightsquigarrow T(x, y) = [X(x, y), Y(x, y)]$$

Supposing that X and Y are differentiable, verify that

$$x\frac{\partial Y}{\partial x}(x, y) + y\frac{\partial Y}{\partial y}(x, y) = Y(x, y)$$

and

$$x\frac{\partial X}{\partial x}(x, y) + y\frac{\partial Y}{\partial y}(x, y) = X(x, y) + Y(x, y)$$

Verify that then $X(x, y) = (ax + by)\log|ax + by|$ and $Y(x, y) = ax + by$. Hence deduce that T is not differentiable.

30. Compare the systems

$$A = \begin{pmatrix} 0 & 1 \\ 0 & 0 \end{pmatrix} \qquad B = \begin{pmatrix} 0 & 1 \\ -1 & 0 \end{pmatrix} \qquad C = \begin{pmatrix} 0 & 1 \\ 1 & 0 \end{pmatrix}$$

31. Let $dx/dt = Ax, dy/dt = By$ be two linear differentiably equivalent systems; let T be the corresponding mapping (we suppose that $T(0) = 0$).

Show that DT establishes a linear equivalence between the two systems.

32. Study the orbits of $dx/dt = x, dy/dt = -y + x^3$. Compare them with those of the linearized system.

33. In $\mathbb{R}^2$ let h be the function defined for $-u \leqslant v \leqslant 0$ by

$$h(u, v) = \left(\frac{v}{u} + 1\right)\left[u^5\left(\frac{v}{u}\right)^3 + (u^5 - 2)\left(\frac{v}{u}\right)^2 - \frac{v}{u} + 1\right]$$

Let φ be defined by

$$\varphi(x, y) = 0 \qquad \text{if } y \leqslant 0$$

$$\varphi(x, y) = 1 \qquad \text{if } y \geqslant x^2 \exp\left(-\frac{x^2}{2}\right) > 0$$

$$\varphi(x, y) = h\left[x^2 \exp\left(-\frac{x^2}{2}\right), \ y - x^2 \exp\left(-\frac{x^2}{2}\right)\right] \quad \text{if } x^2 \exp\left(-\frac{x^2}{2}\right) \geqslant y > 0$$

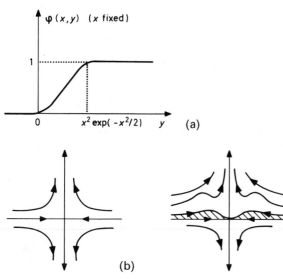

Fig. 2.14. (a) Definition of φ. (b) $\dfrac{dX}{dt} = \begin{pmatrix} -1 & 0 \\ 0 & 1 \end{pmatrix} X.$ (c) $\dfrac{dX}{dt} = \begin{pmatrix} -1 & 0 \\ 0 & 1 \end{pmatrix} X + f(X).$

as shown in Fig. 2.14(a). Let

$$\frac{dx}{dt} = -x$$

$$\frac{dy}{dt} = y + \varphi(x, y)x^2(x^2 - 3)\exp\left(-\frac{x^2}{2}\right)$$

and let

$$X = \begin{pmatrix} x \\ y \end{pmatrix}$$

Denoting

$$\frac{dX}{dt} = \begin{pmatrix} -1 & 0 \\ 0 & 1 \end{pmatrix} X + f(X)$$

verify that the right-hand side is not differentiable and that the orbits of

$$\frac{dx}{dt} = \begin{pmatrix} -1 & 0 \\ 0 & 1 \end{pmatrix} x$$

are transformed as indicated in Figs 2.14(b) and 2.14(c).

34. Let

$$\left. \begin{array}{c} \dfrac{dx}{dt} = 2x + y^2 \\[2mm] \dfrac{dy}{dt} = y \end{array} \right\} \tag{72}$$

Prove that there does not exist a transformation

$$T(x, y) = \begin{cases} f(x, y) \\ g(x, y) \end{cases}$$

of class C^2 establishing a correspondence between the orbits of (72) and those of its linearized form. (Write the defining relation and differentiate f twice with respect to y.)

35. Consider

$$\frac{dx}{dt} = -y \qquad\qquad \frac{dx}{dt} = -y + \varepsilon x(1 - x^2 - y^2)$$

$$\text{and}$$

$$\frac{dy}{dt} = x \qquad\qquad \frac{dy}{dt} = x + \varepsilon y(1 - x^2 - y^2)$$

Compare the orbits of the two systems which are arbitrarily close in $(x^2 + y^2 \leqslant 1)$ by choosing ε close to zero.

36. Let

$$S_\varepsilon : \begin{cases} \dfrac{dx}{dt} = x - x^2 \\[3mm] \dfrac{dy}{dt} = (2x - 1)y + \varepsilon x \end{cases}$$

Find the solution $\varphi_\varepsilon(t)$ such that $\varphi_\varepsilon(0) = (\tfrac{1}{2}, 0)$. What happens if $\varepsilon = 0$? Is S_0 structurally stable? (Cf. Theorem 3.2.11.)

Chapter 3

Analysis of linear systems

Many 'systems' can be described by considering that their state at the instant t, say $s(t)$, is a function of state $e(t)$ of the external world, the dependence often being expressed through a differential equation. Figure 3.1 represents such a system Σ symbolically.

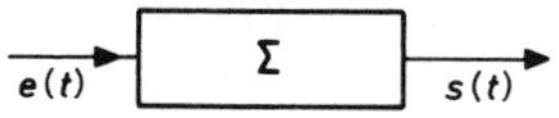

Fig. 3.1. $\dfrac{ds}{dt} = f(s, e)$.

If the corresponding equation is linear we call the system linear. Such systems will be studied in this chapter. In this situation $e(t)$ is called the input of the system and $s(t)$ the output or the response of the system to the input e. For example, the position of a moving body of mass m subjected to an external force e is such that

$$m \frac{d^2 s}{dt^2} = e(t)$$

and the charge on a capacitor in a series (R, L, C) circuit, kept on charge by an electromotive force (e.m.f.) $E(t)$, satisfies

$$L \frac{d^2 Q}{dt^2} + R \frac{dQ}{dt} + \frac{R}{C} = E(t)$$

The examination of a system involves essentially the following three aspects.

(1) First, we naturally investigate under what conditions the outputs from the system are bounded, at least for certain inputs. Then we analyse the responses to certain simple inputs which are frequently encountered: constant or periodic inputs, which are sometimes called stationary inputs. We examine in this case whether the corresponding outputs are of the same type, in particular whether the response to an input of period T is itself of period T; if this is the case we say that the system is functioning in a forced or steady state regime.

If at a fixed instant we change from a stationary input, there appears in general a response which is not steady state; it can happen that this response becomes stabilized as a new steady state regime. Thus the response splits into the sum of two terms: the new steady state regime and a transient regime.

The study of all these problems can be roughly included under the heading **study of the stability of the system**. It necessitates, in particular, the study of the response to certain discontinuous inputs (the conduct of a circuit breaker for example).

(2) It often happens that we do not know the parameters of a system exactly. In the case of linear systems it turns out that it suffices to know the responses to certain inputs to be able to deduce the set of responses to all possible inputs, a fact which in a certain way characterizes the system. If we succeed thus in determining certain parameters of the system, we are said to carry out its **identification**.

(3) In general we wish to obtain certain features in the performance of the output from the system, e.g. we may wish the cost of utilization to be a minimum or the speed of reaction to be as large as possible, or more simply we may require that the output should not depart too far from a function given in advance.

For this purpose we often introduce supplementary elements $u(t)$ whose role is to control the system. This control can be achieved by choosing the element $u(t)$ once and for all, as in Fig. 3.2(a) (we say that the system is operating in open loop), or by using the output, especially by comparing it with its desired value, as in Fig. 3.2(b) (we say then that the system is functioning in closed loop). When it is necessary that the output follows a given function we say that we have a tracking problem.

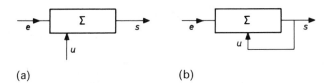

(a)   (b)

Fig. 3.2. (*a*) *Open loop;* (*b*) *closed loop.*

In Part 1 of this chapter we shall study a particular tool which enables us to find, for example, responses to discontinuous inputs and also to define the 'transfer function' of the system, knowledge of which is equivalent to that of the system.

Part 2 will be devoted to the study of the stability of a system and to the behaviour of the solutions at infinity.

In Part 3 we shall present the elements of control theory for linear systems.

Part 1: Laplace transform

We shall be interested in systems with **constant coefficients** and above all in scalar equations of order n in the form

$$a_n \frac{d^n s}{dt^n} + a_{n-1} \frac{d^{n-1}}{dt^{n-1}} + \ldots + a_0 s(t) = e(t) \tag{1}$$

If A is the associated matrix and $\Phi_0(t - s)$, $\Phi_1(t - s), \ldots, \Phi_{n-1}(t - s)$ is the first row of the resolvent of A, then

$$s(t) = \exp(At) s(0) + \int_0^t \Phi_{n-1}(u)e(u)\,du \tag{2}$$

The second term of Eqn (2) does not depend on the initial value. It represents the steady state regime. The first term represents the transient regime when Re $\lambda_i < 0$ for all the eigenvalues of A.

1.1. Laplace transform

Since there must be a definite instant from which we may suppose that the system is subjected to an external action, we shall choose this instant as the origin and we shall suppose all functions to be zero for $t \leq 0$.

DEFINITIONS AND ELEMENTARY PROPERTIES

Definition 1.1.1
If $f: t \leadsto f(t)$ is a function defined for $t > 0$, we say that f is piecewise continuous if
  (a) in every bounded interval $]0, T]$ there exists at most a finite number of points where f is not continuous and
  (b) at every point a where f is not continuous it possesses finite left and right limits which we shall denote

$$\lim_{\substack{t \to a \\ t < a}} f(t) = f(a^-) \quad \text{and} \quad \lim_{\substack{t \to a \\ t > a}} f(t) = f(a^+)$$

which implies that f is bounded over every interval $[T_1, T_2]$ where $0 < T_1 < T_2 < \infty$.

Definition 1.1.2
If $f: t \leadsto f(t)$ is a function defined for $t > 0$, we say that f has **exponential growth** if there exist constants M and c such that $|f(t)| \leq M \exp)ct)$ $\forall t > 0$.
  If f is a piecewise continuous function and if $a_1, \ldots, a_n$ are the points of discontinuity of f in $]0, T]$, we can define

$$\int_0^T f(u)\,du$$

as

$$\int_0^{a_1} f(u)\,du + \int_{a_1}^{a_2} f(u)\,du + \ldots + \int_{a_n}^{T} f(u)\,du$$

Definition 1.1.3

If $f: t \rightsquigarrow f(t)$, defined for $t > 0$, is a piecewise continuous function, we say that f has a Laplace transform at p_0 if

$$\lim_{T \to \infty} \int_0^T \exp(-p_0 u) f(u)\,du$$

exists for the value p_0, and we write

$$\mathscr{L}[f](p_0) = \int_0^\infty \exp(-p_0 u) f(u)\,du \qquad (p_0 \in \mathbb{R})$$

We can verify that, if $\mathscr{L}[f](p_0)$ exists, $\mathscr{L}[f](p)$ exists $\forall p > p_0$. We define the abscissa of convergence as the lower bound of the set $\{p : \mathscr{L}[f](p) \text{ exists}\}$.

The following theorem is known as the **uniqueness theorem**.

Theorem 1.1.4

Let f_1 and f_2 be two piecewise continuous functions with Laplace transforms $\mathscr{L}[f_1]$ and $\mathscr{L}[f_2]$ defined over one and the same domain; if there exists p_0 such that

$$\forall p \geqslant p_0 \qquad \mathscr{L}[f_1](p) = \mathscr{L}[f_2](p)$$

then f_1 and f_2 are identical over their domain of definition except perhaps at the points of discontinuity of one or the other function (finite in number over the whole bounded set). Therefore there exists at most one continuous function having a given Laplace transform.

If $F = \mathscr{L}[f]$ we shall write $f = \mathscr{L}^{-1}[F]$. There are explicit formulae giving f as a function of F (Mellin–Fourier) but we shall not use them. More generally we can define the Laplace transform for p complex, but we shall not use it.

Proposition 1.1.5

Let f be piecewise continuous over $]0, \infty[$ and have exponential growth; then there exist M and c such that

$$|f(t)| \leqslant M \exp(ct) \quad \forall t > 0$$

and $\mathscr{L}[f](p)$ exists over $]c, \infty[$.

Proof Let

$$p > c: \int_0^T |\exp(-pu)f(u)|\,du \leqslant M \int_0^T \exp[-(p-c)u]\,du$$

$$\leqslant \frac{M}{p-c} \quad \forall T$$

The function

$$T \to \int_0^T |\exp(-pu)f(u)|\,du$$

is then increasing and bounded and it has a finite limit.

If $f \geqslant 0$ $f = |f|$ and $\mathscr{L}[f](p)$ exists when $p > c$.

If $f \leqslant 0$ $f = -|f|$ and $\mathscr{L}[f](p)$ also exists.

If f is of variable sign we put $f = f^+ - f^-$ where

$$f^+(t) = \begin{cases} f(t) & \text{if } f(t) > 0 \\ 0 & \text{if } f(t) \leqslant 0 \end{cases}$$

and

$$f^-(t) = \begin{cases} -f(t) & \text{if } f(t) < 0 \\ 0 & \text{otherwise} \end{cases}$$

Then f^+ and f^- are positive, $|f| = f^+ + f^-$ and therefore $\mathscr{L}[f^+](p)$ and $\mathscr{L}[f^-](p)$ exist over $]c, \infty[$, so that $\mathscr{L}[f] = \mathscr{L}[f^+] - \mathscr{L}[f^-]$ exists.

Examples

(a) $\quad f(t) = 1 \qquad \forall t > 0 \quad \mathscr{L}[f](p) = \dfrac{1}{p} \qquad \forall p > 0$

(b) $\quad f(t) = \exp(at) \quad \forall t > 0 \quad \mathscr{L}[f](p) = \dfrac{1}{p-a} \qquad \forall p > a$

(c) $\quad f_1(t) = \cos \omega t \quad \forall t > 0 \quad \mathscr{L}[f_1](p) = \dfrac{p}{p^2 + \omega^2} \qquad \forall p > 0$

$\qquad f_2(t) = \sin \omega t \quad \forall t > 0 \quad \mathscr{L}[f_2](p) = \dfrac{\omega}{p^2 + \omega^2} \qquad \forall p > 0$

We note that if $\mathscr{L}[f](0)$ exists

$$\int_0^\infty f(u)\,du = \lim_{T \to \infty} \int_0^T f(u)\,du$$

exists. It can happen that $\mathscr{L}[f](p)$ exists $\forall p > 0$ even if $\mathscr{L}[f](0)$ does not exist. This is the case for $f = \cos \omega t$ for example.

We have the following two theorems.

Theorem 1.1.6 (Initial value theorem)
If $f(0^+)$ exists

$$\lim_{p \to \infty} p\mathscr{L}[f](p) = f(0^+)$$

Theorem 1.1.7 (Final value theorem)
If

$$\lim_{t \to \infty} f(t) = f(\infty)$$

exists and if $\mathscr{L}[f](p)$ exists $\forall p > -\sigma$ where $\sigma > 0$

$$\lim_{p \to 0} p\mathscr{L}[f](p) = f(\infty)$$

Notation 1.1.8
We denote by $\Upsilon_a(x)$ the function defined by

$$\Upsilon_a(x) = \begin{cases} 0 & \text{if } x \leqslant a \\ 1 & \text{if } x > a \end{cases}$$

Υ_a is called the unit step function at the point a. If f is zero for $x \leqslant 0$ the function $\Upsilon_a(x)f(x-a)$ is called a translation of f to a and it is equal to 0 if $x < a$ and to $f(x-a)$ if $x > a$.

Example: periodic functions (Fig. 3.3)   Let $f: \mathbb{R}_*^+ \rightsquigarrow \mathbb{R}$ be of period $T < \infty$ and

$$f_0: f_0(u) = \begin{cases} f(u) & 0 < u \leqslant T \\ 0 & \text{otherwise} \end{cases}$$

and let $f_n(u) = \Upsilon_{nT}(u)f_0(u - nT)$; then

$$f(t) = \sum_{n \geqslant 0} f_n(t)$$

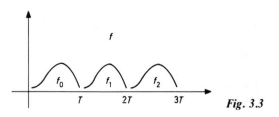

Fig. 3.3

Theorem 1.1.9 (Elementary properties)

(a) If $\mathscr{L}[f_1](p)$ and $\mathscr{L}[f_2](p)$ exist $\forall p \geqslant p_0$, $\mathscr{L}[f_1 + f_2](p)$ exists $\forall p \geqslant p_0$ and $\mathscr{L}[f_1] + \mathscr{L}[f_2] = \mathscr{L}[f_1 + f_2]$.

Likewise $\forall \lambda$ real, $\lambda \mathscr{L}[f_1] = \mathscr{L}[\lambda f_1]$ over $[p_0, \infty[$.

(b) Let

$$f(x) = \sum_{n \geqslant 0} f_n(x)$$

be a function defined by a convergent series over $]0, \infty[$. If p is such that

$$\mathscr{L}\left[\sum_{n \geqslant 0} |f_n|\right](p)$$

exists, then $\mathscr{L}[f](p)$ exists as well as $\mathscr{L}[f_n](p)$ for every n and

$$\mathscr{L}[f](p) = \sum_{n \geqslant 0} \mathscr{L}[f_n](p)$$

(c) If f is such that $\mathscr{L}[f](p)$ exists $\forall p < \sigma$, then $\mathscr{L}[\exp(ax) f](p)$ exists for $p - a > \sigma$ and

$$\mathscr{L}[\exp(ax) f](p) = \mathscr{L}[f](p - a)$$

(d) If p is such that $\mathscr{L}[f](p)$ exists, then the translation of f to a has a Laplace transform defined at p and such that

$$\mathscr{L}[\Upsilon_a(x) f(x - a)](p) = \exp(-pa) \mathscr{L}[f](p)$$

Thus, if f has period T,

$$f(t) = \sum_{n \geqslant 0} \Upsilon_{nT}(t) f_0(t - nT)$$

and

$$\mathscr{L}[f](p) = \frac{\mathscr{L}[f_0](p)}{1 - \exp(-pT)} \qquad p > 0$$

(e) If $\mathscr{L}[f](p/k)$ exists, let $h(x) = f(kx)$; then

$$\mathscr{L}[h](p) = \frac{1}{k} \mathscr{L}[f]\left(\frac{p}{k}\right)$$

The properties (a), (c), (d) and (e) are very easy to verify; property (b) results from an elementary theorem in the theory of integration which affirms that under the conditions of the statement one can integrate term by term the relation

$$\exp(-px) f(x) = \sum_{n \geqslant 0} \exp(-px) f_n(x)$$

Convolution

Definition 1.1.10
If f and g are piecewise continuous over $]0, \infty[$, the function h defined by

$$h(x) = \int_0^x f(t)g(x - t)\,dt$$

is called the convolution of f and g. We often write $h = f * g$.

Theorem 1.1.11
If $\mathscr{L}[f]$ and $\mathscr{L}[g]$ exist $\forall p \geqslant p_0$, then $\mathscr{L}[f * g]$ exists $\forall p \geqslant p_0$ and

$$\mathscr{L}[f * g](p) = \mathscr{L}[f](p)\mathscr{L}[g](p)$$

Proof We shall confine ourselves to the case where f and g have exponential growth so that there exist M, N and a such that $|f(t)| \leqslant M \exp(at)$ and $|g(t)| \leqslant N \exp(at) \, \forall t > 0$. If then $p > a$, we know that $\mathscr{L}[f](p)$ and $\mathscr{L}[g](p)$ exist.

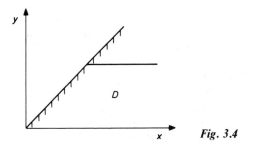

Fig. 3.4

Let D be the domain $0 < x < y$ (Fig. 3.4):

$$\int_D \exp(-px)|f(y)|\,|g(x-y)|\,dx\,dy = \int_0^\infty |f(y)|\left[\int_y^\infty \exp(-px)|g(x-y)|\,dx\right]dy$$

$$\leqslant MN \int_0^\infty \left\{\int_y^\infty \exp[-(p-a)x]\,dx\right\}dy$$

$$= \frac{MN}{(p-a)^2}$$

$$< \infty$$

We know then that the integral

$$\int_D \exp(-px)\,f(y)g(x-y)\,dx\,dy$$

can be calculated by integrating successively with respect to y and then x, or with respect to x and then y:

$$\mathscr{L}[f * g](p) = \int_0^\infty \exp(-px) \left[\int_0^x f(y) g(x-y) \, dy \right] dx$$

$$= \int_0^\infty f(y) \left[\int_y^\infty \exp(-px) g(x-y) \, dx \right] dy$$

$$= \mathscr{L}[f](p) \mathscr{L}[g](p)$$

See **Exercises 1 and 2** at the end of the chapter.

DIFFERENTIATION PROPERTIES

We shall concern ourselves with two types of properties: differentiation or integration of the Laplace transform on the one hand, and the Laplace transform of the derivative, if it exists, on the other.

Theorem 1.1.12

Let f be a piecewise continuous function for $t > 0$, possessing a Laplace transform $\mathscr{L}[f]$ defined for $p > p_0$.

For every integer $n \geqslant 0$, $u^n f(u)$ has a Laplace transform defined for $p > p_0$ by

$$\frac{d^n}{dp^n} \mathscr{L}[f] = (-1)^n \mathscr{L}[u^n f(u)]$$

A Laplace transform is thus always infinitely differentiable.

Proof Let $p > p_0$ and ε be such that $p > p - \varepsilon > p_0$. There exists U such that $u \geqslant U > 0 \Rightarrow u^n \leqslant \exp(\varepsilon u)$ and therefore such that

$$\int_U^\infty u^n |f(u)| \exp(-pu) \, du \leqslant \int_U^\infty |f(u)| \exp[-(p-\varepsilon)u] \, du < \infty$$

$\mathscr{L}[u^n f(u)](p)$ then exists.

We can verify that

$$\frac{d}{dp} \mathscr{L}[f] = -\mathscr{L}[uf(u)]$$

which establishes the theorem. We put $F(p) = \mathscr{L}[f](p)$ and it remains to evaluate the expression

$$\left| \frac{F(p+h) - F(p)}{h} + \int_0^\infty uf(u) \exp(-pu) \, du \right|$$

$$\leqslant \int_0^\infty \frac{|\exp(-hu) - 1 + hu|}{h} |\exp(-pu) f(u)| \, du$$

Since

$$|\exp(-hu) - 1 + hu| \leqslant \frac{h^2 u^2}{2}$$

the preceding expression is smaller than

$$\frac{h}{2} \int_0^\infty u^2 \exp(-pu) |f(u)| \, du$$

Since this integral is finite, the expression tends to zero when h tends to zero.

The following theorem is also useful.

Theorem 1.1.13
If p is such that $\mathcal{L}[f](p)$ exists then

$$\mathcal{L}\left[\int_0^t f(u) \, du\right] = \frac{1}{p} \mathcal{L}[f](p)$$

Proof

$$\int_0^T \exp(-pt) \left[\int_0^t f(u) \, du\right] dt = \int_0^T f(u) \left[\int_u^T \exp(-pt) \, dt\right] du$$

$$= \frac{1}{p} \int_0^T \exp(-pu) f(u) \, du - \frac{1}{p} \exp(-pT) \int_0^T f(u) \, du$$

Since $\mathcal{L}[f](p)$ exists

$$\forall \varepsilon > 0 \quad \exists S: \int_S^\infty \exp(-pu) |f(u)| \, du < \frac{\varepsilon}{2}$$

and for $T > S$

$$\exp(-pT) \int_0^T |f(u)| \, du = \exp(-pT) \int_0^S |f(u)| \, du$$

$$+ \int_S^T \exp[-p(T-u)] \exp(-pu) |f(u)| \, du$$

As each term has zero as limit if $T \to \infty$,

$$\mathcal{L}\left[\int_0^t f(u) \, du\right](p)$$

exists and has the value

$$\frac{1}{p} \mathcal{L}[f](p)$$

Even though the integral of a function having a Laplace transform has itself a Laplace transform, the derivative of a function having a Laplace transform does not always have a Laplace transform.

Foe example

$$\mathcal{L}\left[\frac{1}{x^{1/2}}\right](p) = \frac{\pi^{1/2}}{p^{1/2}}$$

whereas $1/xx^{1/2}$ does not have a Laplace transform.

Theorem 1.1.14
Let f be defined for $t > 0$ and have a finite limit $f(0^+)$ as t decreases to zero. Suppose that f' exists and is continuous and that $\mathcal{L}[f](p)$ exists. Then $\mathcal{L}[f'](p)$ exists and $\mathcal{L}[f'](p) = p\mathcal{L}[f](p) - f(0^+)$.

Proof Let p be such that $\mathcal{L}[f](p)$ exists; then

$$\int_0^T f'(t)\exp(-pt)\,dt = [f(t)\exp(-pt)]_0^T + p\int_0^T f(t)\exp(-pt)\,dt$$

whose limit is $-f(0^+) + p\mathcal{L}[f](p)$.

Generalization
If f is continuous except at a sequence of points $\{a_n\}_{n\in\mathbb{N}}$ and is differentiable at every point of $]a_n, a_{n+1}[$ we can define $\mathcal{L}[f'](p)$ under the following conditions. Suppose that $f(0^+)$, $f(a_n^+)$ and $f(a_n^-)$ exist for all n. Let $\Delta f_n = f(a_n^+) - f(a_n^-)$ and $\Sigma \Delta f_n < \infty$. If $\mathcal{L}[f](p)$ exists

$$\mathcal{L}[f'](p) = p\mathcal{L}[f](p) - \sum_0^\infty \exp(-pa_i)\Delta f_i$$

(The proof is the same as before.)

In the same way, if the derivatives of f exist up to order n and if each has a limit $f^{(r)}(0^+)$ when $t \to 0$ for all $r \leqslant n - 1$

$$\mathcal{L}\left[\frac{d^n f}{dt^n}\right](p) = p^n\mathcal{L}[f](p) - p^{n-1}f(0^+) - p^{n-2}f'(0^+) - \ldots$$

$$- pf^{(n-2)}(0^+) - f^{(n-1)}(0^+)$$

The properties established in the last two subsections (pp. 127 and 133) enable us to calculate many Laplace transforms. Table 3.1 shows a number of transforms.

TABLE OF LAPLACE TRANSFORMS

Some of the results given in Table 3.1 are also valid for complex values of the parameter; they will be stated without comment. The formulae can be established as an exercise.

See *Exercises 3 and 4* at the end of the chapter.

Table 3.1

$\mathscr{L}[f]$	f	$\mathscr{L}[f]$	f
$\dfrac{1}{p}$ $(p>0)$	$1\ (t>0)$	$\dfrac{p}{p^2+\omega^2}$ $(p>0)$	$\cos\omega t$
$\dfrac{1}{p^2}$ $(p>0)$	t	$\dfrac{\omega}{p^2+\omega^2}$ $(p>0)$	$\sin\omega t$
$\dfrac{n!}{p^{n+1}}$ $(p>0)$	$t^n\ (n\ \text{integer}>0)$	$\dfrac{p+a}{(p+a)^2+\omega^2}$	$\exp(-at)\sin\omega t$ (a real or complex)
$\left(\dfrac{n}{p}\right)^{1/2}$	$\dfrac{1}{t^{1/2}}$	$(p>\mathrm{Re}(-a))$	
$\dfrac{1}{p+a}$	$\exp(-at)$ (a real or complex)	$\dfrac{\omega}{(p+a)^2+\omega^2}$	$\exp(-at)\sin\omega t$ (a real or complex)
$(p>\mathrm{Re}(-a))$		(a real, $p>-a$)	
$\dfrac{n!}{(p+a)^{n+1}}$	$t^n\exp(-at)$ (a real or complex)	$\dfrac{\alpha p+\beta}{(p+a)^2+\omega^2}$	$A\exp(-at)\cos(\omega t+\Phi)$ where $$A=\frac{1}{\omega}[\alpha^2\omega^2+(\beta-a\alpha)^2]^{1/2}$$ and $$\Phi=-\arctan\frac{\beta-\alpha a}{\alpha\omega}$$

1.2. Solution of differential equations

We shall establish that, if the input e to a system has a Laplace transform $\mathscr{L}[e]$, the output s also has a Laplace transform $\mathscr{L}[s]$ and that $\mathscr{L}[s]$ is determined by $\mathscr{L}[e]$, a fact which determines s.

Theorem 1.2.1
Let

$$a_n\frac{\mathrm{d}^n s}{\mathrm{d}t^n}+a_{n-1}\frac{\mathrm{d}^{n-1}s}{\mathrm{d}t^{n-1}}+\ldots+a_1\frac{\mathrm{d}s}{\mathrm{d}t}+a_0 s(t)=e(t) \tag{3}$$

We assume that for $p\geqslant p_0$ $\mathscr{L}[e](p)$ exists; we know that p_1 exists such that all the terms of $\exp(At)$ are maximized by $\exp(p_1 t)$; then $\mathscr{L}[s](p)$ exists for $p>p_0$ and p_1, as well as

$$\mathscr{L}\left[\frac{\mathrm{d}^m s}{\mathrm{d}t^m}\right](p) \qquad \forall m\leqslant n$$

and

$$(a_n p^n+a_{n-1}p^{n-1}+\ldots+a_1 p+a_0)\mathscr{L}[s](p)=\mathscr{L}[e](p)+P(p)$$

where P is a polynomial of degree $n - 1$ which vanishes if $s(0^+) = s'(0^+) = \ldots = s^{(n-1)}(0^+) = 0$.

Proof Let

$$S(t) = \begin{pmatrix} s(t) \\ \dfrac{ds}{dt} \\ \vdots \\ \dfrac{d^{n-1}s}{dt^{n-1}} \end{pmatrix} \quad \text{and} \quad E(t) = \begin{pmatrix} 0 \\ \vdots \\ 0 \\ e(t) \end{pmatrix}$$

Let A be the matrix associated with (3). We know that

$$S(t) = \exp(At)\,S(0) + \int_0^t \exp[A(t-u)]\,E(u)\,du$$

Therefore

$$\int_0^T \exp(-pt)\,S(t)\,dt = \int_0^T \exp(-pt)\exp(At)\,dt\,S(0)$$

$$+ \int_0^T \exp(-pt)\left\{\int_0^t \exp[A(t-u)]\,E(u)\,du\right\}dt$$

$$= \int_0^T \exp(-pt)\exp(At)\,dt\,S(0)$$

$$+ \int_0^T \exp(-Au)\left\{\int_u^T \exp[(A-pI)t]\,dt\right\}E(u)\,du$$

If $p > p_1$, $A - pI$ is invertible and the second integral is

$$(A-pI)^{-1}\left\{\exp[(A-pI)T]\int_0^T \exp(-Au)\,E(u)\,du - \int_0^T \exp(-pu)\,E(u)\,du\right\}$$

All the terms have a limit if $T \to \infty$ when $p > p_0$ and $p > p_1$.

We can use this result to solve a differential equation at least for $t > 0$; if we wish to find a solution over the whole line $\mathbb{R}$ we can start by seeking one over $\mathbb{R}^+$ and then examine whether it can be extended.

Examples

(1) $y'' - 3y' + 2y = \exp(3t)$. Find φ such that $\varphi(0) = 0 = \varphi'(0)$.

$$\mathcal{L}[\varphi'' - 3\varphi' + 2\varphi] = p^2 \mathcal{L}[\varphi](p) - 3p\mathcal{L}[\varphi](p) + 2\mathcal{L}[\varphi](p)$$
$$= \mathcal{L}[\exp(3t)](p)$$
$$= \frac{1}{p-3}$$

$$\mathcal{L}[\varphi](p) = \frac{1}{(p-1)(p-2)(p-3)}$$
$$= \frac{1}{2(p-1)} - \frac{1}{p-2} + \frac{1}{2(p-3)}$$

$$\varphi(t) = \frac{1}{2}\exp(t) - \exp(2t) + \frac{1}{2}\exp(3t)$$

(2) $y'' + 2y' + y = 0$. Find φ such that $\varphi(0) = 0$ and $\varphi(1) = 2$. We shall seek φ such that $\varphi(0) = 0$ and $\varphi'(0) = k$ and then determine k. $p^2\mathcal{L}[\varphi](p) - k + 2p\mathcal{L}[\varphi](p) + \mathcal{L}[\varphi](p) = 0$, so $\mathcal{L}[\varphi](p) = k/(p+1)^2$. Now

$$\frac{1}{(p+1)^2} = -\frac{d}{dp}\left(\frac{1}{p+1}\right) = -\frac{d}{dp}\mathcal{L}[\exp(-t)]$$

Therefore $\varphi(t) = kt\exp(-t)$ and the solution sought is $2et\exp(-t)$.

(3) $y'' - 3y' + 2y = f(t)$ where f has a Laplace transform (e.g. $f \leqslant M\exp(dt)$). Find φ such that $\varphi(0) = 0 = \varphi'(0)$.

$$\mathcal{L}[\varphi](p) = \frac{\mathcal{L}[f](p)}{p^2 - 3p + 2}$$
$$= \mathcal{L}[f](p)\left(\frac{1}{p-2} - \frac{1}{p-1}\right)$$
$$= \mathcal{L}[f](p)\mathcal{L}[\exp(2t) - \exp(t)]$$
$$= \mathcal{L}\left\{\int_0^t f(u)[\exp(2t - 2u) - \exp(t - u)]\,du\right\}(p)$$

(cf. Theorem 1.1.11). If $f \leqslant M\exp(at)$, $\mathcal{L}[\varphi]$ is defined for $p > 2$ and $p > a$ (otherwise for $p > p_0$) and

$$\varphi(t) = \int_0^t f(u)\{\exp[2(t - u)] - \exp(t - u)\}\,du$$

(4) *Volterra's equation.* We seek y satisfying the integral equation

$$y(t) = t + \int_0^t \cos(t - u)y(u)\,du$$

Since

$$|y(t)| \leqslant t + \int_0^t |y(u)|\, du \qquad |y(t)| \leqslant \exp(t) - 1$$

from Gronwall's lemma, and y has a Laplace transform defined at least for $p > 1$,

$$\mathscr{L}[y](p) = \frac{1}{p^2} + \mathscr{L}[\cos](p)\,\mathscr{L}[y](p)$$

$$= \frac{1}{p^2} + \mathscr{L}[y](p)\frac{p}{1+p^2}$$

$$\mathscr{L}[y](p) = \frac{1}{p^2}\frac{1+p^2}{1+p^2-p}$$

$$= \frac{1}{p^2} + \frac{1}{p} - \left(p - \frac{1}{2}\right)\left[\left(p - \frac{1}{2}\right)^2 + \frac{3}{4}\right]^{-1} + \frac{1}{2}\left[\left(p - \frac{1}{2}\right)^2 + \frac{3}{4}\right]^{-1}$$

$$= \mathscr{L}[t] + \mathscr{L}[1] - \mathscr{L}\left[\exp\left(\frac{t}{2}\right)\cos\frac{\sqrt{3}}{2}t + \frac{1}{\sqrt{3}}\exp\left(\frac{t}{2}\right)\sin\frac{\sqrt{3}}{2}t\right]$$

Thus

$$y(t) = 1 + t - \exp\left(\frac{t}{2}\right)\left(\cos\frac{\sqrt{3}}{2}t - \frac{1}{\sqrt{3}}\sin\frac{\sqrt{3}}{2}t\right)$$

Equations in $\mathbb{R}^d$

There is no difficulty in generalizing the preceding method to $\mathbb{R}^d$. First we define the Laplace transform of a vector φ defined for $t > 0$; if the coordinates φ_i of φ have a Laplace transform at a point p we put

$$\mathscr{L}[\varphi](p) = \begin{pmatrix} \mathscr{L}[\varphi_1](p) \\ \vdots \\ \mathscr{L}[\varphi_d](p) \end{pmatrix} \in \mathbb{R}^d$$

If A is a matrix, $\mathscr{L}[A\varphi] = A\mathscr{L}[\varphi]$ and we solve the equation $dX/dt = AX + b(t)$ by taking the Laplace transforms of the two parts:

$$\begin{pmatrix} p\mathscr{L}[\varphi_1](p) - \varphi_1(0) \\ \vdots \\ p\mathscr{L}[\varphi_d](p) - \varphi_d(0) \end{pmatrix} = p\mathscr{L}[\varphi](p) - \varphi(0) = A\mathscr{L}[\varphi](p) + \mathscr{L}[b](p)$$

(at a point where $\mathscr{L}[b]$ exists, which implies that $\mathscr{L}[\varphi]$ exists for the same reason as in one dimension). Thus

$$(pI - A)\mathscr{L}[\varphi](p) = \mathscr{L}[b](p) + \varphi(0) \qquad (4)$$

We have to solve (4) to find $\mathscr{L}[\varphi]$; this can always be done when p is greater than all the eigenvalues of A so that $(pI - A)^{-1}$ exists. (This is what we used in Theorem 1.2.1.) (We naturally have this result in one dimension and $\mathscr{L}[\varphi]$ is defined for p greater than all the eigenvalues of the associated matrix, i.e. all the roots of $a_0 p^n + a_1 p^{n-1} + \ldots + a_{n-1} p + a_n = 0$.)

See *Exercises 5 to 7* at the end of the chapter.

1.3. Impulses: Dirac delta, distributions

It often happens that a system is subjected to a stress of large amplitude and of very short duration. Such a stress is called an **impulse**. Its form in general is not precisely stated (even not precisely determinable); an approximate representation of it is given by every curve f_n such that

$$f_n(t) = 0 \qquad \text{if } |t - t_0| \geqslant \frac{1}{n}$$

and otherwise

$$f_n(t) \geqslant 0 \qquad \text{and} \qquad \int_{-\infty}^{+\infty} f_n(u)\,du = 1$$

(see Fig. 3.5(a)). If

$$F_n(t) = \int_{-\infty}^{t} f_n(u)\,du$$

F_n behaves as indicated in Fig. 3.5(b).

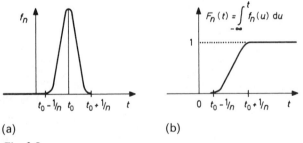

(a)   (b)

Fig. 3.5

An **ideal impulse** at the instant t_0 is the 'limit' of f_n when $n \to \infty$, only this limit does not exist in the ordinary sense whereas that of F_n does. The limit of F_n is

$$\Upsilon_{\{t_0\}}(t) = \begin{cases} 1 & t > t_0 \\ 0 & \text{otherwise} \end{cases}$$

(cf. p. 130).

The following two facts express the same phenomenon. The limiting function of the F_n is not differentiable at t_0 and the functions f_n do not have a limit in the ordinary sense.

We shall generalize the notion of function and that of differentiation so that f_n will have a limit representing a unit impulse and this limit will be the derivative of $\Upsilon_{\{t_0\}}$.

DISTRIBUTIONS, DIRAC DELTA

Definition 1.3.1

We shall denote by $\mathscr{D}$ the vector space of functions which are infinitely differentiable and zero outside a compact set (e.g. $[-A, +A]$).

If $\{\varphi_n\}_{n\in\mathbb{N}}$ is a sequence of elements of $\mathscr{D}$, we say that φ_n tends to zero in $\mathscr{D}$ if all the functions φ_n are zero outside a fixed compact set and if φ_n and all its derivatives tend to zero uniformly when n tends to infinity.

A mapping T from $\mathscr{D}$ into $\mathbb{R}$ is called continuous over $\mathscr{D}$ if

$$\lim_{n\to\infty} T(\varphi_n) = 0$$

for every sequence $\{\varphi_n\}_{n\in\mathbb{N}}$ of $\mathscr{D}$ such that

$$\varphi_n \xrightarrow[n\to\infty]{\mathscr{D}} 0$$

We shall associate with every function f continuous over $\mathbb{R}$ a continuous linear mapping T_f from $\mathscr{D}$ into $\mathbb{R}$ which completely characterizes f, i.e. which is such that $T_{f_1} = T_{f_2}$ iff $f_1 = f_2$. For this purpose we shall put

$$T_f(\varphi) = \int_{-\infty}^{\infty} f(u)\varphi(u)\,du$$

which exists.

It is clear that $\varphi \to T_f(\varphi)$ is linear and that, if $\varphi_n(u) = 0$ $\forall n$ for $|u| \geq A$ and

$$|\varphi_n(u)| \leq \varepsilon/\sup_{|u|\leq A} f$$

$\forall |u| < A$ and $n \geq N_0$, $|T_f(\varphi_n)| \leq \varepsilon$. This means that we are concerned with a continuous linear form.

If we suppose that $T_{f_1} = T_{f_2}$ for $f_1 \neq f_2$, this would imply that in an interval $[u_0 - \eta, u_0 + \eta]$ we would have, for example, $f_1(u) < f_2(u)$; thus for every positive function φ that is zero outside the interval $[u_0 - \eta, u_0 + \eta]$, this would imply $T_{f_1}(\varphi) < T_{f_2}(\varphi)$, which is absurd.

We shall call T_f the distribution associated with f and give the following definition.

Definition 1.3.2
We call every continuous linear mapping from $\mathscr{D}$ into $\mathbb{R}$ a distribution.

Definition 1.3.3
We denote by $\mathscr{D}'$ the vector space of distributions and we say that a sequence T_n tends to T in $\mathscr{D}'$ if $\forall \varphi \in \mathscr{D}$

$$\lim_{n \to \infty} T_n(\varphi) = T(\varphi)$$

and we denote it by $\lim_{n \to \infty} T_n = T$ in $\mathscr{D}'$.

Proposition 1.3.4
If $\{f_n\}_{n \in \mathbb{N}}$ is the sequence of functions introduced on page 140

$$\lim_{n \to \infty} T_{f_n}$$

exists. We denote this by $\delta_{\{t_0\}}$ and we also write

$$\lim_{n \to \infty} (\mathscr{D}') f_n = \delta_{\{t_0\}}$$

This limit distribution is called the **Dirac delta** at t_0. It is not associated with any function, and $\delta_{\{t_0\}}(\varphi) = \varphi(t_0) \ \forall \varphi \in \mathscr{D}$.

It should be noted that the form of the f_n appears only through the fact that $f_n(t) = 0$ if $|t - t_0| > 1/n$, $f_n(t) \geqslant 0$ and

$$\int_{-\infty}^{+\infty} f_n(u) \, du = 1$$

This is the minimal characterization of an approximate impulse.

Proof

$$T_{f_n}(\varphi) = \int_{t_0 - 1/n}^{t_0 + 1/n} \varphi(u) f_n(u) \, du$$

as φ is continuous.

$$\forall \varepsilon > 0 \quad \exists N : n \geqslant N \Rightarrow |\varphi(u) - \varphi(t_0)| \leqslant \varepsilon \quad \forall u \in \left[t_0 - \frac{1}{n}, t_0 + \frac{1}{n} \right]$$

so that

$$n \geqslant N \Rightarrow |T_{f_n}(\varphi) - \varphi(t_0)| \leqslant \varepsilon \quad \text{for } T_{f_n}[\varphi(t_0)] = \varphi(t_0)$$

If we define $\delta_{\{t_0\}}(\varphi)$ by $\varphi(t_0)$ we actually obtain a distribution and there is no function g such that

$$\forall \varphi \in \mathcal{D} \quad \varphi(t_0) = \int_{-\infty}^{+\infty} g(u)\varphi(u)\,du$$

This 'explains' the fact that the functions f_n do not have limits.

We shall now give a meaning to the phrase '$\delta_{\{t_0\}}$ is the "derivative" of $\Upsilon_{\{t_0\}}$', which generalizes the fact that f_n is the derivative of F_n.

If F is a function with continuous derivative f, we shall express this in the following way:

$$\forall \varphi \in \mathcal{D} \quad T_f(\varphi) = -T_F(\varphi')$$

(we note that $\varphi' \in \mathcal{D}$). In fact

$$\int_{-\infty}^{+\infty} F(u)\varphi'(u)\,du = \int_{-A}^{A} F(u)\varphi'(u)\,du \quad \text{if } \varphi(u) = 0 \quad \forall |u| \geqslant A$$

This expression is equal to

$$\left[F(u)\varphi(u)\right]_{-A}^{+A} - \int_{-A}^{+A} f(u)\varphi(u)\,du$$

Now

$$F(A)\varphi(A) = F(-A)\varphi(-A) = 0$$

and

$$\int_{-A}^{A} f(u)\varphi(u)\,du = \int_{-\infty}^{\infty} f(u)\varphi(u)\,du$$

We are led to the following definition.

Definition 1.3.5

We say that a distribution T' is the derivative in the distribution sense of the distribution T if

$$\forall \varphi \in \mathcal{D} \quad T'(\varphi) = -T(\varphi')$$

This definition generalizes the formula for integration by parts and, if f has f' as derivative, $T_{f'} = T_f'$ (the derivative of T_f).

We can combine all the information concerning impulses in a single statement.

Proposition 1.3.6

Let $\{f_n\}$ be a sequence of 'approximate' impulses, i.e. of functions $f_n \geqslant 0$, which are zero for $|t - t_0| \geqslant 1/n$ and such that

$$\int_{-\infty}^{\infty} f_n(u)\, du = 1$$

Then the sequence f_n has as its limit $\delta_{\{t_0\}}$ in the sense of a distribution, and this distribution is the derivative in the distribution sense of $\Upsilon_{\{t_0\}}$ which is the limit of the F_n. Thus

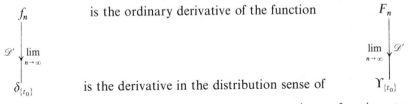

f_n is the ordinary derivative of the function F_n

$\delta_{\{t_0\}}$ is the derivative in the distribution sense of $\Upsilon_{\{t_0\}}$

Dirac delta at t_0 unit step function at t_0

OTHER EXAMPLES OF DISTRIBUTIONS

(a) As the sum of two continuous linear forms is a continuous linear form, the sum of two distributions is a distribution, as also is the product of a distribution by a constant.

(b) If f is a function which is continuous except at certain points, finite in number over every bounded set, the linear form

$$\varphi \rightsquigarrow \int_{-\infty}^{\infty} f(u)\varphi(u)\, du$$

is well defined and we can again associate the distribution T_f with f.

If $\quad T_1 = T_f \quad$ and $\quad T_2 = T_g \quad\quad T_1 + T_2 = T_{f+g}$

If $T = T_f \quad\quad T_{\lambda f} = \lambda T_f$

(c) Since the derivative of a distribution is always defined by the formula

$$T'(\varphi) = -T(\varphi')$$

T' in its turn is differentiable, etc., and we thus obtain new distributions. For example

$$\delta'(\varphi) = -\delta(\varphi') = -\varphi'(0)$$

$$\delta''(\varphi) = -\delta'(\varphi') = \varphi''(0)\ldots \delta^{(k)}(\varphi) = (-1)^k \frac{d^k\varphi}{dt^k}(0)$$

If f is a differentiable function T_f has $T_{df/dt}$ as its derivative, but if f is differentiable except at one point a we cannot speak of its derivative over the whole of $\mathbb{R}$, even though the associated distribution has a derivative. We shall denote by df/dt the derivative of the function f wherever it exists and by $[f']$ the derivative of f in the distribution sense, i.e. the derivative of T_f.

We shall compare the distributions $[f']$ derived from T_f and $T_{df/dt}$ associated with df/dt, which we assume to be continuous at every point except a.

Proposition 1.3.7

Let f be a function of class C^1 except at a point a; let

$$f(a^+) = \lim_{t \downarrow a} f(t) \quad \text{and} \quad f(a^-) = \lim_{t \uparrow a} f(t)$$

Then

$$(T_f)' = [f'] = T_{df/dt} + [f(a^+) - f(a^-)] \delta_{\{a\}}$$

Proof

$$(T_f)'(\varphi) = -\int_{-\infty}^{+\infty} \varphi'(u) f(u) \, du$$

$$= -\int_{-\infty}^{a} \varphi'(u) f(u) \, du - \int_{a}^{+\infty} \varphi'(u) f(u) \, du$$

$$= -f(a^-)\varphi'(a) + \int_{-\infty}^{a} \varphi(u) \frac{df}{du}(u) \, du + f(a^+)\varphi'(a) + \int_{a}^{\infty} \varphi(u) \frac{df}{du}(u) \, du$$

$$= T_{df/dt} + \{[f(a^+) - f(a^-)] \delta_{\{a\}}\}(\varphi)$$

(d) *Series of impulses:* let t_n be a sequence tending to infinity,

$$\lim_{n \to \infty} t_n = \infty$$

and let $\delta_{\{t_n\}}$ be the Dirac delta at the point t_n such that

$$\sum_{n=1}^{N} \delta_{\{t_n\}}(\varphi) = \sum_{1}^{N} \varphi(t_n)$$

defines a sequence of distributions. Since

$$\sum_{1}^{N} \varphi(t_n)$$

has a limit when $n \to \infty$, for all functions $\varphi \in \mathscr{D}$

$$\sum_{1}^{\infty} \delta_{\{t_n\}}(\varphi)$$

is defined and equals

$$\sum_{1}^{\infty} \varphi(t_n)$$

Now

$$\varphi \rightsquigarrow \sum_{1}^{\infty} \varphi(t_n)$$

is a continuous linear form over $\mathscr{D}$, since if

$$\lim_{p \to \infty} \varphi_p = 0$$

in $\mathscr{D}$ then

$$\lim_{p \to \infty} \sum_{n=1}^{\infty} \varphi_p(t_n) = 0$$

$$\mathscr{D} \ni \varphi \rightsquigarrow \sum_{1}^{\infty} \varphi(t_n)$$

defines the distribution

$$\sum_{n=1}^{\infty} \delta(t_n)$$

i.e. the sum of the series $\delta_{\{t_n\}}$.

(e) *Convolution:* let us consider a function constructed in the following way. Let t_0 be fixed and $t \rightsquigarrow f(t - t_0)$. We can consider this function as the action of the distribution $\delta_{\{t_0\}}$ applied to the function $u \rightsquigarrow f(t - u)$, which we can write

$$f(t - t_0) = \int_{-\infty}^{+\infty} f(t - u)\delta_{\{t_0\}} \, du$$

and which we call the **convolution** of f and $\delta_{\{t_0\}}$; we say that $\delta_{\{0\}}$ is the identity element of the convolution.

We sometimes write this expression in the form

$$\int_{-\infty}^{+\infty} f(t - u)\delta(t_0 - u) \, du$$

as if δ were a function; this incorrect notation is a source of confusion and often leads to error. It is better to avoid it.

LAPLACE TRANSFORMS OF DISTRIBUTIONS

If f is a function defined over $\mathbb{R}^+$ we can always arrange that it is defined over $\mathbb{R}$ by putting $f(t) = 0$ if $t < 0$. We can then associate with it a distribution T_f by

$$T_f(\varphi) = \int_{-\infty}^{\infty} \varphi(u)f(u) \, du$$

By definition T_f is a continuous linear form over $\mathscr{D}$ but it often happens that we can extend its definition to a function space larger than $\mathscr{D}$.

Let φ_p be the function defined by

$$\varphi_p(t) = 0 \qquad \text{if } t < 0$$

$$\varphi_p(t) = \exp(-pt) \qquad \text{if } t \geqslant 0$$

To say that $\mathscr{L}[f](p)$ exists is to say that we can extend the definition of T_f to φ_p; we shall say that $p \rightsquigarrow T_f(\varphi_p)$ is the Laplace transform of the distribution T_f.

Definition 1.3.8

If T is a distribution, we say that T has a Laplace transform $\mathscr{L}[T]$ if we can define $T(\varphi_p)$, and we put $\mathscr{L}[T](p) = T(\varphi_p)$.

Theorem 1.3.9

If T has a Laplace transform, all its derivatives have a Laplace transform and $\mathscr{L}[T'](p) = p\mathscr{L}[T](p)$.

Proof By definition the derivative T' of T is defined by $T'(\varphi) = -T(\varphi') \quad \forall \varphi \in \mathscr{D}$. If we assume that we can define $T(\varphi_p)$, we can then define $T(p\varphi_p)$. Now

$$\frac{\mathrm{d}}{\mathrm{d}t} \varphi_p(t) = -p\varphi_p$$

We can then define $T'(\varphi_p) = pT(\varphi_p)$.

Examples

$$\mathscr{L}[\delta_{\{t_0\}}](p) = \exp(-pt_0) \qquad \mathscr{L}[\delta'_{\{t_0\}}] = p\exp(-pt_0)$$

In particular, $\mathscr{L}[\delta] = 1$, $\mathscr{L}[\delta'] = p, \ldots, \mathscr{L}[\delta^{(k)}] = p^k$, and we find again that $\mathscr{L}[\delta_{\{t_0\}}](p) = p\mathscr{L}[\Upsilon_{\{t_0\}}(p)]$ since $\delta_{\{t_0\}}$ is the derivative in the distribution sense of $\Upsilon_{\{t_0\}}$.

Functions of class C^1 piecewise

Suppose that we have the conditions of Proposition 1.3.7. We always let $[f'] = (T_f)'$ and $\mathrm{d}f/\mathrm{d}t$ be the derivative of f wherever it exists, i.e. at every point different from a.

Proposition 1.3.10

If we suppose that $f(0^+)$ exists then $\mathscr{L}[T_{\mathrm{d}f/\mathrm{d}t}]$ and $\mathscr{L}[f']$ exist and $\mathscr{L}[f'](p) = p\mathscr{L}[f](p)$.

Proof

$$\mathscr{L}\left[\frac{df}{dt}\right](p) = p\mathscr{L}[f](p) - f(0^+) - [f(a^+) - f(a^-)]\exp(-pa)$$

$$= \mathscr{L}[T_{df/dt}]$$

Now

$$[f'] = T_{df/dt} + f(0^+)\delta_{\{0\}} + [f(a^+) - f(a^-)]\delta_{\{a\}}$$

$$\mathscr{L}[f'](p) = \mathscr{L}\left[\frac{df}{dt}\right](p) + f(0^+) + [f(a^+) - f(a^-)]\exp(-pa)$$

$$= p\mathscr{L}[f](p)$$

As in Proposition 1.3.7, the latter extends naturally to the case where f is continuous except at a sequence of points a_n, of which there is only a finite number in every bounded set.

Laplace transform of a series of impulses

Let $\{t_n\}$ be a sequence tending to infinity in such a way that the series

$$\sum_{n \in \mathbb{N}} \exp(-pt_n)$$

is convergent; the distribution

$$\sum_{n \in \mathbb{N}} \delta_{\{t_n\}}$$

is defined and has for Laplace transform

$$\mathscr{L}\left[\sum_{n \in \mathbb{N}} \delta_{\{t_n\}}\right](p) = \sum_{n \in \mathbb{N}} \exp(-pt_n)$$

Thus we have the following.

Proposition 1.3.11

$$\mathscr{L}\left[\sum_{n=0}^{\infty} \delta_{\{n\}}\right] = \frac{1}{1 - \exp(-p)}$$

1.4. Application: transfer function, response of a system to a distribution

We should like to generalize the method of Section 1.2 to the case where $e(t)$ is a distribution. This very simple generalization is called **symbolic calculus**; its basis is the following elementary proposition.

Proposition 1.4.1

Let

$$a_n \frac{d^n s}{dt^n} + \ldots + a_0 s = T \tag{5}$$

be a differential equation whose right-hand side T is a distribution and whose left-hand side is the sum of the distribution associated with $a_0 s$, etc., where the derivatives are in the distribution sense (this imposes initial values on s, ds/dt, etc.). If $\mathscr{L}[T](p)$ exists, each term of the left-hand side has a Laplace transform and

$$\mathscr{L}[a_n p^n + a_{n-1} p^{n-1} + \ldots + a_1 p + a_0] \mathscr{L}[T_s](p) = \mathscr{L}[T](p) \tag{6}$$

We have the same result as in Theorem 1.2.1 but, as we are considering distributions, the initial values of s and its derivatives do not appear in Eqn (6) because $\forall T \, \mathscr{L}[T'](p) = p\mathscr{L}[T](p)$.

However, the initial values appear in the following way: if $s(0^+) \neq 0$, $s(0^+)\delta_{\{0\}}$ appears in the distribution associated with ds/dt, $s(0^+)\delta'_{\{0\}}$ in that associated with $d^2 s/dt^2$ and $a_n s(0^+)\delta_{\{0\}}^{(n-1)}$ appears on the left-hand side and must therefore appear on the right.

For example let

$$\frac{d^2 x}{dt^2} + a \frac{dx}{dt} + bx = \lambda\delta_{\{0\}} + \mu\delta'_{\{0\}} + f$$

where f is a continuous function; then x is a continuous function of class C^2:

$$[x'] = x(0^+)\delta_{\{0\}} + \frac{dx}{dt}$$

$$[x''] = x(0^+)\delta'_{\{0\}} + x'(0^+)\delta_{\{0\}} + \frac{d^2 x}{dt^2}$$

Thus

$$x(0^+) = \mu \quad \text{and} \quad x'(0^+) = \lambda$$

Examples

(a)
$$\frac{d^2 x}{dt^2} + x = \delta_{\{0\}}$$

where x is defined for $t \geqslant 0$. A solution φ is such that $\varphi(0^+) = 0$ and $\varphi'(0^+) = 1$:

$$\mathscr{L}[\varphi] = \frac{1}{1 + p^2}$$

and $\varphi(t) = \sin t$ for $t \geqslant 0$.

We can extend φ by putting $\varphi(t) = \sin t \, \Upsilon_{\{0\}}(t)$.

We can verify that $\varphi'(t) = \cos t \, \Upsilon_{\{0\}}(t)$ and that $d^2\varphi/dt^2 = \delta_{\{0\}} - \sin t \, \Upsilon_{\{0\}}(t)$.

(b)
$$\frac{d^2 x}{dt^2} + 2\frac{dx}{dt} + 5x = \delta_{\{1\}}$$

Necessarily $\varphi(0^+) = \varphi'(0^+) = 0$.

$$\mathscr{L}[\varphi] = \frac{\exp(-p)}{p^2 + 2p + 5} = \frac{\exp(-p)}{(p+1)^2 + 4}$$

and

$$\varphi(t) = \begin{cases} 0 & t \leqslant 1 \\ \frac{1}{2}\sin 2(t-1)\exp[-(t-1)] & t > 1 \end{cases}$$

We can verify (but it is not necessary) that on the one hand

$$\varphi'(t) = \begin{cases} \cos 2(t-1) - \frac{1}{2}\sin 2(t-1)\exp[-(t-1)] & \text{if } t > 1 \\ 0 & \text{otherwise} \end{cases}$$

so that the boundary conditions are satisfied, and that on the other hand $\varphi'(1^-) = 0$ whereas $\varphi'(1^+) = 1$ and thus

$$\frac{d^2\varphi}{dt^2} = \delta_{\{1\}} - \Upsilon_{\{1\}}(t)[4\cos 2(t-1) + 3\sin 2(t-1)]\frac{\exp[-(t-1)]}{2}$$

$$= \delta_{\{1\}} - 2\varphi' + 5\varphi$$

(c)
$$\frac{d^2x}{dt^2} + x = \sin t + \delta_{\{1\}}$$

Then $\varphi(0^+) = \varphi'(0^+) = 0$.

$$\mathscr{L}[\varphi](p) = \frac{1}{(1+p^2)^2} + \frac{\exp(-p)}{1+p^2}$$

$$= \mathscr{L}\left[\int_0^t \sin u \sin(t-u)\,du\right] + \mathscr{L}[\sin(t-1)\Upsilon_1(t)]$$

Hence

$$\varphi(t) = \frac{1}{2}\sin t - \frac{t}{2}\cos t + \Upsilon_{\{1\}}(t)\sin(t-1)$$

We can again verify that φ meets the conditions.

(d) Let T_ε be the distribution

$$\frac{1}{\varepsilon}\delta_{\{t_0+\varepsilon\}} - \frac{1}{\varepsilon}\delta_{\{t_0\}}$$

By definition

$$\forall \varphi \in \mathscr{D} \qquad T_\varepsilon(\varphi) = \frac{\varphi(t_0+\varepsilon) - \varphi(t_0)}{\varepsilon}$$

If $\varepsilon \to 0$, $T_\varepsilon(\varphi)$ has as a limit $\varphi'(t_0) = -\delta'_{\{t_0\}}$; for this reason $\delta'_{\{a\}}$ is called a **doublet** at the point a.

If

$$\frac{d^2x}{dt^2} + x = \delta'_{\{1\}}$$

is to be solved, then $\varphi(0) = \varphi'(0^+) = 0$,

$$\mathscr{L}[\varphi] = \frac{p\exp(-p)}{1+p^2} \quad \text{and} \quad \varphi(t) = \cos(t-1)\Upsilon_{\{1\}}(t)$$

(e)
$$\frac{d^2x}{dt^2} + x = \sum_{0}^{\infty} \delta_{\{2k\pi\}}$$

This is a system with period of free oscillations 2π and subjected to impulses at instants 2π apart. Necessarily $\varphi(0^+) = \varphi'(0^+) = 0$.

$$\mathscr{L}[\varphi](p) = \frac{1}{1+p^2} [1 + \exp(-2\pi p) + \exp(-4\pi p) + \ldots + \exp(-2n\pi p) + \ldots]$$

(convergent series)

$$= \sin t + \Upsilon_{\{2\pi\}}(t)\sin(t-2\pi) + \Upsilon_{\{4\pi\}}(t)\sin(t-4\pi) + \ldots$$

Thus $\varphi(t) = 0$ if $t < 0$, $\varphi(t) = n\sin(t - n\pi)$ if $n\pi \leqslant t < (n+1)\pi$. This is a case of resonance (Fig. 3.6).

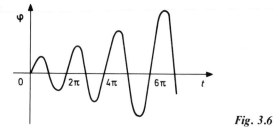

Fig. 3.6

See *Exercise 8* at the end of the chapter.

Symbolic calculus

We note that if we use distributions the formulae are very simple since we multiply the Laplace transform by p each time we differentiate. Equation (6) illustrates this fact by establishing a simple relationship between the Laplace transforms of the output and the input; this relationship is known as soon as we know the particular response to an impulsive input. We shall give an illustration of this.

Thus we note that Theorem 1.1.13 states that integrating f consists in multiplying $\mathscr{L}[f]$ by $1/p$.

Definition 1.4.2 (Transfer function)
Let

$$a_n \frac{\mathrm{d}^n s}{\mathrm{d}t^n} + \ldots + a_0 s = e \tag{7}$$

be a differential equation. We call the expression $G(p) = (a_n p^n + \ldots + a_0)^{-1}$ the transfer function. More generally let

$$a_n \frac{\mathrm{d}^n s}{\mathrm{d}t^n} + \ldots + a_0 s = b_k \frac{\mathrm{d}^k e}{\mathrm{d}t^k} + \ldots + b_0 e \tag{7'}$$

be a relation between an input and an output for a system. We call

$$G(p) = \frac{b_k p^k + \ldots + b_0}{a_n p^n + \ldots + a_0}$$

the transfer function.

Proposition 1.4.3
(a) Let $g(t)$ be the response to the input $e(t) = \delta_{\{0\}} : G(p) = \mathscr{L}[g](p)$.
(b) If e has a Laplace transform it is known when g is known since $\mathscr{L}[s](p) = G(p)\mathscr{L}[e](p)$.
We can also state this as follows: g is the solution of the homogeneous equation

$$a_n \frac{\mathrm{d}^n s}{\mathrm{d}t^n} + \ldots + a_0 s = 0$$

such that $g(0) = g'(0) = \ldots = g^{(n-2)}(0) = 0$ and $g^{(n-1)}(0) = 1$, and thus the continuous response to e, which is zero when $t = 0$, as are its derivatives, is

$$\int_0^t g(u)e(t - u)\,\mathrm{d}u$$

No proof is needed; it is sufficient to note that $\mathscr{L}[\delta](p) = 1$ and thus that $\mathscr{L}[g](p) = G(p)\mathscr{L}[\delta](p) = G(p)$.
Proposition 1.4.3 is a contribution to the solution of the problem of identification of a system and this is where the importance of the transfer function lies.

Part 2: Stability and behaviour of solutions at infinity

2.1. Stability of linear systems

We have already introduced the notion of stability of an equilibrium. We shall now give a general definition of the stability of a solution.

Definition 2.1.1
Let

$$\frac{dx}{dt} = A(t)x + b(t) \tag{8}$$

be a linear differential equation and $\varphi(t, t_0, x_0)$ be the unique solution of (8) such that $x(t_0) = x_0$. We know that this solution is defined $\forall t \in \mathbb{R}$.

(a) We say that $\varphi(t, t_0, x_0)$ is stable if $\forall \varepsilon > 0 \; \exists \eta(\varepsilon, t_0)$ such that

$$\|x_1 - x_0\| \leqslant \eta(\varepsilon, t_0) \Rightarrow \|\varphi(t, t_0, x_0) - \varphi(t, t_0, x_1)\| \leqslant \varepsilon \qquad \forall t \geqslant t_0$$

If there exists T such that, $\forall \varepsilon > 0$, $\exists \eta(\varepsilon)$ independent of t_0 for $t_0 \in [T, \infty[$ and such that

$$\|x_1 - x_0\| \leqslant \eta(\varepsilon) \Rightarrow \|\varphi(t, t_0, x_0) - \varphi(t, t_0, x_1)\| \leqslant \varepsilon \qquad \forall t \geqslant t_0 \geqslant T$$

we say that $\varphi(t, t_0, x_0)$ is **uniformly stable** for $t_0 \geqslant T$.
(b) We say that $\varphi(t, t_0, x_0)$ is **asymptotically stable** if it is stable and if $\exists \eta_1(t_0)$ such that

$$\|x_1 - x_0\| \leqslant \eta_1(t_0) \Rightarrow \lim_{t \to \infty} \|\varphi(t, t_0, x_0) - \varphi(t, t_0, x_1)\| = 0$$

We say that $\varphi(t, t_0, x_0)$ is **uniformly asymptotically stable** for $t_0 \geqslant T$ if it is uniformly stable and if $\exists \eta_1$ independent of t_0 such that for every x_1 satisfying $\|x_0 - x_1\| \leqslant \eta_1$ the following property holds: $\forall \varepsilon$, $\forall t \geqslant T$, $\exists T(\varepsilon)$ such that

$$t \geqslant t_0 + T(\varepsilon) \Rightarrow \|\varphi(t, t_0, x_0) - \varphi(t, t_0, x_1)\| \leqslant \varepsilon$$

i.e. the convergence of $\varphi(t, t_0, x_0) - \varphi(t, t_0, x_1)$ to zero is uniform in t_0 where $\|x_0 - x_1\| \leqslant \eta_1$.
(c) In all other cases φ is called **unstable**.

Scalar equations of order n

We note that in this case the notion of stability means that if initial positions, velocities, accelerations, etc., are close to each other, these positions, velocities, accelerations, etc., remain close to each other for all $t \geqslant t_0$.

We note that in (b) we demand on the one hand stability and on the other hand convergence to zero, the second not necessarily implying the first as can be seen from Fig. 3.7.

Asymptotic stability and systems theory

We shall see that in systems theory the fundamental notion is that of

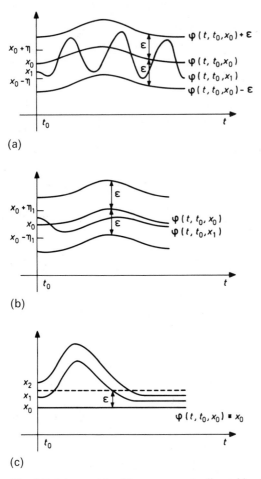

Fig. 3.7. (a) φ stable; (b) φ asymptotically stable; (c) φ unstable although

$$\lim_{t \to \infty} \varphi(t, t_0, x_1) - \varphi(t, t_0, x_0) = 0$$

asymptotic stability: it is this property which ensures that for every bounded input the output is bounded, and that after the disappearance of a temporary input we obtain, at the end of a certain time, a steady state output; in short, it is the situation least sensitive to perturbations.

FUNDAMENTAL THEOREM

The fundamental phenomenon is that all solutions are of the same nature for linear systems. We can speak of stable or unstable systems.

Theorem 2.1.2
Let

$$\frac{dx}{dt} = A(t)x \qquad (9)$$

be a homogeneous system.

(a) From the point of view of stability all solutions of (9) are of the same nature: stable, uniformly stable or asymptotically or uniformly asymptotically stable, or unstable. It is sufficient therefore to know the nature of the solution $x(t) \equiv 0$.

(b) The system is stable for t_0 iff $\exists K(t_0)$ such that $\|\Phi(t, t_0)\| \leqslant K(t_0) \ \forall t \geqslant t_0$; it is uniformly stable for $t_0 \geqslant T$ iff $\exists K(T): \forall T \leqslant s \leqslant t \ \|\Phi(t, s)\| \leqslant K(T)$. The system is stable iff all solutions are bounded.

(c) The system is asymptotically stable for t_0 iff

$$\lim_{t \to \infty} \Phi(t, t_0) = 0$$

It is uniformly asymptotically stable for $t_0 \geqslant T$ iff $\exists K(T)$ and $\alpha(T) > 0$ such that $\forall T \leqslant s \leqslant t < \infty \ \ \|\Phi(t, s)\| \leqslant K(T) \exp[-\alpha(T)(t - s)]$. The system is asymptotically stable iff all solutions have zero as limit when $t \to \infty$.

(d) The system is unstable iff $\Phi(t, t_0)$ is not bounded if $t \to \infty$.

For a system with constant coefficients we know that cases (b), (c) and (d) correspond respectively to eigenvalues λ_i with

$$\text{Re } \lambda_i \leqslant 0 \quad \text{and} \quad \text{Re } \lambda_i < 0 \quad \text{if} \quad v(\lambda_i) > 1 \qquad \forall i$$

$$\text{Re } \lambda_i < 0 \quad \forall i \quad \text{and} \quad \exists i_0 \ \text{Re } \lambda_{i_0} > 0 \quad \text{or} \quad \text{Re } \lambda_{i_0} = 0 \quad \text{and} \quad v(\lambda_{i_0}) > 1$$

In this case the stability is always uniform; in the former case we have already established (c) where

$$-\alpha(T) = \sigma = \inf_i (\text{Re } \lambda_i)$$

For systems with variable coefficients there is no criterion relative to the eigenvalues; even if the latter are fixed and have negative real parts the system can be unstable.

Example

$$\frac{dx}{dt} = \begin{pmatrix} -1 + \frac{3}{2}\cos^2 t & 1 - \frac{3}{2}\sin t \cos t \\ -1 - \frac{3}{2}\sin t \cos t & -1 + \frac{3}{2}\sin^2 t \end{pmatrix} x = A(t)x$$

The eigenvalues of $A(t)$ are $-1/4 \pm i\sqrt{7}/4$; they are fixed, their real part is strictly negative, and yet the system is unstable since a solution is the vector

$$\begin{cases} \exp(-t/2)\cos t \\ \exp(t/2)\sin t \end{cases}$$

as can easily be shown.

Proof of Theorem 2.1.2 (a) Let $\varphi(t, t_0, x_0)$ and $\varphi(t, t_0, y_0)$ be two solutions. $\varphi(t, t_0, x_0) = \Phi(t, t_0)x_0$ so that the latter is stable iff $\exists \eta(t_0)$ such that $\|x_0 - x_1\| \leqslant \eta(t_0) \Rightarrow \|\Phi(t, t_0)(x_0 - x_1)\| \leqslant \varepsilon$. However, if then $\|y_0 - y_1\| \leqslant \eta(t_0)$ we have also $\|\Phi(t, t_0)(y_0 - y_1)\| \leqslant \varepsilon$ and the second solution is stable.

In the same way we can prove the equivalence for the other types of stability.

(b) It is sufficient to study the stability of the solution $x(t) \equiv 0$. The latter is stable iff $\forall \varepsilon > 0 \ \exists \eta(t_0)$ such that $\|x_0\| < \eta(t_0)$ implies that $\|\Phi(t, t_0)x_0\| \leqslant \varepsilon$ $\forall t \geqslant t_0$.

If $\forall t \geqslant t_0 \ \|\Phi(t, t_0)\| \leqslant K(t_0)$, then $\|\Phi(t, t_0)x_0\| \leqslant K(t_0)\|x\|$ which is smaller than ε when

$$\|x_0\| \leqslant \frac{\varepsilon}{K(t_0)} = \eta(t_0)$$

Conversely, if $x(t) \equiv 0$ is stable

$$\|\Phi(t, t_0)\| = \sup_{W:\|W\|=\eta(t_0)} \left\| \Phi(t, t_0)\frac{W}{\eta(t_0)} \right\| \leqslant \frac{\varepsilon}{\eta(t_0)} \qquad \forall t \geqslant t_0$$

which establishes that

$$\|\Phi(t, t_0)\| \leqslant K(t_0) = \frac{\varepsilon}{\eta(t_0)}$$

For uniform stability we write that the preceding relations are true $\forall t_0 \geqslant T$.

(c) Suppose that

$$\lim_{t \to \infty} \Phi(t, t_0) = 0$$

$x(t) \equiv 0$ is then stable. Moreover $\forall x_0 \ \|\varphi(t, t_0, x_0)\|$ has zero for limit and $x(t) \equiv 0$ is asymptotically stable.

If $\|\Phi(t, s)\| \leqslant K(T)\exp[-\alpha(T)(t-s)]$ for $t \geqslant s \geqslant t_0 > T$, for all $\varepsilon > 0$

$$\|\varphi(t, t_0, x_0) - \varphi(t, t_0, x_1)\| \leqslant K(T)\exp[-\alpha(T)(t-t_0)]\|x_0 - x_1\| \leqslant \varepsilon$$

$$\text{for } t \geqslant t_0 + T(\varepsilon)$$

and the system is uniformly asymptotically stable for $t_0 \geq T$.

We can establish conversely that if the system is uniformly asymptotically stable for $t_0 \geq T$ we can find α and K having the stated property.

Since the system is uniformly stable, we know already that $\|\Phi(t, s)\| \leq M(T)$ for all $T \leq s \leq t < \infty$. Moreover

$$\exists \eta_1(T): \|x_0\| \leq \eta_1(T)$$

$$\forall \varepsilon \quad \exists S(\varepsilon): t \geq t_0 + S(\varepsilon) \Rightarrow \|\Phi(t, t_0)x_0\| \leq \varepsilon$$

Then

$$\sup_{W: \|W\| = \eta_1(T)} \|\Phi(t, t_0)W\| = \|\Phi(t, t_0)\|\eta_1(T) \leq \varepsilon$$

Thus

$$\forall \varepsilon \quad \exists S(\varepsilon): \forall t \geq t_0 + S(\varepsilon)$$

$$\|\Phi(t, t_0)\| \leq \frac{\varepsilon}{\eta_1(T)}$$

Let us choose $\varepsilon < \eta_1(T)$ and note that $\Phi[t + 2S(\varepsilon), t]$ is equal to $\Phi[t + 2S(\varepsilon), t + S(\varepsilon)]\Phi[t + S(\varepsilon), t]$ and therefore that

$$\|\Phi[t + 2S(\varepsilon), t]\| \leq \left[\frac{\varepsilon}{\eta_1(T)}\right]^2$$

More generally

$$\|\Phi[t + nS(\varepsilon), t]\| \leq \left[\frac{\varepsilon}{\eta_1(T)}\right]^n M(T)$$

Let

$$\alpha(T) = -\frac{1}{S(\varepsilon)} \log\left[\frac{\varepsilon}{\eta_1(T)}\right] \quad \alpha(T) > 0 \quad \text{and} \quad K(T) = M(T)\exp[\alpha(T)S(\varepsilon)]$$

Also let

$$t_0 + nS(\varepsilon) \leq t < t_0 + (n + 1)S(\varepsilon)$$

Then

$$\|\Phi(t, t_0)\| \leq \Phi[t, t_0 + nS(\varepsilon)]\| \ \|\Phi[t_0 + nS(\varepsilon), t_0]\|$$

$$\leq M(T)\exp[-nS(\varepsilon)\alpha(T)]$$

$$= K(T)\exp[-\alpha(T)S(\varepsilon)]\exp[-\alpha(T)nS(\varepsilon)]$$

$$\leq K(T)\exp[-\alpha(T)(t - t_0)]$$

(d) This point is obvious.

For a system with a right-hand side there is no automatic connection between the facts that the system is stable and that solutions remain bounded; everything depends on the right-hand side.

The essential result is the following.

Theorem 2.1.3

Let

$$\frac{dx}{dt} = A(t)x \tag{10}$$

and

$$\frac{dx}{dt} = A(t)x + b(t) \tag{11}$$

(a) All solutions of (11) are of the same nature from the point of view of stability; we can speak of a system as stable or unstable, etc.

(b) The nature of system (11) is the same as that of system (10).

Proof We shall denote by capital letters the solutions of (11) and by small letters those of (10). Then

$$X(t) = x(t) + \int_{t_0}^{t} \Phi(t, s)b(s) \, ds$$

where X and x have the same initial value.

If X_0 and Y_0 are two initial values and X, x and Y, y the corresponding solutions, then $X(t) - Y(t) = x(t) - y(t)$. The behaviour of $X - Y$ is therefore the same as that of $x - y$ when $t \to \infty$, which establishes at the same time points (a) and (b) of the theorem.

CRITERIA FOR STABILITY FOR EQUATIONS WITH CONSTANT COEFFICIENTS

To study the stability of systems with constant coefficients we must study the real part of the roots of the characteristic polynomial of A.

Definition 2.1.4

If $F(z) = a_0 z^n + a_1 z^{n-1} + \ldots + a_n$ is a polynomial with real coefficients we shall say that it is **uniformly asymptotically stable** (UAS) if, for every root λ, Re $\lambda < 0$.

Proposition 2.1.5

For $F(z) = z_0 z^n + \ldots + a_n$ to be UAS it is necessary that the coefficients be non-zero and all of the same sign.

Proof Let $\lambda_1, \ldots, \lambda_n$ be all the roots of F, distinct or identical, real or not:

$$F(z) = a_0 \prod_1^n (z - \lambda_i)$$

If λ_i is real, $\lambda_i < 0$ and $z - \lambda_i$ is a polynomial with positive coefficients.

If λ_i is complex, $\lambda_i = \alpha + i\beta$ and $\bar{\lambda}_i = \alpha - i\beta$ are roots, so that $[z - (\alpha + i\beta)][z - (\alpha - i\beta)] = z^2 - 2\alpha z + \alpha^2 + \beta^2$ is a factor and its coefficients are strictly positive since $\alpha < 0$; thus all the coefficients have the sign of a_0 when $\text{Re } \lambda < 0 \ \forall \lambda$, root of F.

Proposition 2.1.6

Let $F(z) = z^3 + a_1 z^2 + a_2 z + a_3$; for F to be UAS it is necessary and sufficient that $a_1 > 0$, $a_2 > 0$, $a_3 > 0$ and $a_1 a_2 > a_3$.

Proof We note that $a_3 > 0$ already implies that zero is not a root; $a_1 a_2 \neq a_3$ implies that ib is not a root for we would have $-ib^3 - a_1 b^2 + ia_2 b + a_3 = 0$, $b^2 = a_2$ and $a_1 b^2 = a_3$.

We shall first examine two particular cases.

(a) $z^3 + a_3 = 0$. The roots are as in Fig. 3.8 and F is not UAS; in this case $a_1 a_2 = 0 < a_3$.

(b) $z^3 + a_1 z^2 + a_2 z = 0$. The roots are $z = 0$, $z = \alpha$ and $z = \beta$ where $\alpha + \beta = -a_1 < 0$ whereas $\alpha\beta = a_2 > 0$. Both are negative if they are real and, if not, $\alpha + \bar{\alpha} = 2 \text{ Re } \alpha < 0$; in any case F is UAS, and $a_1 a_2 > 0 = a_3$.

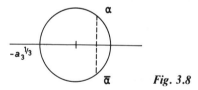

Fig. 3.8

We shall now use the fact that the roots are continuous functions of the coefficients (a_1, a_2, a_3), so that, if a_1, a_2 and a_3 vary without $a_1 a_2 - a_3$ changing sign, the roots remain on the same side of the imaginary axis and therefore their real parts do not change sign.

Thus if $a_1 a_2 < a_3$, we do not change the sign of this inequality by making a_1 and a_2 tend to zero; in the limit we arrive at case (a) and F is not UAS.

However, if $a_1 a_2 > a_3$, by making a_3 tend to zero we arrive at case (b) and F is UAS.

Definition 2.1.7

Let $F(z) = a_0 z^n + a_1 z^{n-1} + \ldots + a_n$. We define the Hurwitz determinant of F as the determinant H obtained in the following way: the diagonal is

$a_1, a_2, \ldots, a_n$ and the columns are formed, starting from the top, by the coefficients a_n in descending order; when we reach a_0 we continue by inserting zeros.

Thus H has the form

$$H = \begin{vmatrix} a_1 & a_3 & a_5 & a_7 & \cdots & & 0 \\ a_0 & a_2 & a_4 & a_6 & & & \vdots \\ 0 & a_1 & a_3 & a_5 & & & \\ 0 & a_0 & a_2 & a_4 & & & \\ 0 & 0 & a_1 & a_3 & \ddots & & \\ 0 & 0 & a_0 & & & & 0 \\ 0 & 0 & 0 & & \cdots & & a_n \end{vmatrix}$$

Theorem 2.1.8 (Routh–Hurwitz criterion)

Let F be a polynomial such that $a_0 > 0$. For F to be UAS it is necessary and sufficient that all the principal minors be strictly positive.

Example   If $n = 4$

$$H = \begin{vmatrix} a_1 & a_3 & 0 & 0 \\ a_0 & a_2 & a_4 & 0 \\ 0 & a_1 & a_3 & 0 \\ 0 & a_0 & a_2 & a_4 \end{vmatrix} \qquad \begin{matrix} D_1 = a_1 > 0 \\ D_2 = a_1 a_2 - a_0 a_3 > 0 \\ D_3 = a_3 D_2 - a_1{}^2 a_4 > 0 \\ D_4 = a_4 D_3 > 0 \end{matrix}$$

If $n = 3$ we get the criterion of Proposition 2.1.6 again.

We shall not prove this criterion.

If

$$a_0 \frac{d^n s}{dt^n} + a_1 \frac{d^{n-1} s}{dt^{n-1}} + \ldots + a_n s = e(s)$$

is a linear system and F the polynomial $a_0 z^n + a_1 z^{n-1} + \ldots + a_n$, $1/F(p)$ is then the transfer function of the system; $1/F(i\omega)$ is called the frequency response and $\omega \rightsquigarrow F(i\omega)$ is called the transfer locus. There are geometrical criteria which enable us to determine whether F is UAS, starting from the transfer locus.

We shall not demonstrate these criteria; they all rest on the same theorem which involves counting the number of zeros of a polynomial which lie in a given domain in the complex plane $\mathbb{C}$.

Theorem 2.1.9 (Leonard's criterion)

Let $F(z) = a_0 z^n + \ldots + a_n$ be a polynomial with positive coefficients and $\omega \rightsquigarrow F(i\omega)$ where $0 < \omega < \infty$ be the transfer locus traversed in the direction of increasing ω; if the locus passes through exactly n quadrants in encircling the origin, F is UAS, and conversely.

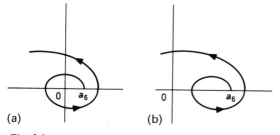

Fig. 3.9. (*a*) *F is UAS;* (*b*) *F is unstable.*

Example $n = 6$ (see Fig. 3.9).

Theorem 2.1.10 (Mihaïlov's criterion)
Let $F(z) = a_0 z^n + \ldots + a_n$ and

$$F_1(x) = a_n - a_{n-2} x^2 + a_{n-4} x^4 + \ldots \quad \text{defined by } F_1(\omega) = \text{Re } F(i\omega)$$

$$F_2(x) = a_{n-1} - a_{n-3} x^2 + \ldots \qquad \text{defined by } \omega F_2(\omega) = \text{Im } F(i\omega)$$

where $x \in \mathbb{R}^+$. Let $u_1, u_2, \ldots$ be the roots of F_1 and $v_1, v_2, \ldots$ be those of F_2. For F to be UAS it is necessary and sufficient that these roots be real and that $0 < u_1 < v_1 < u_2 < v_2 \ldots$.

Example (See Fig. 3.10.)

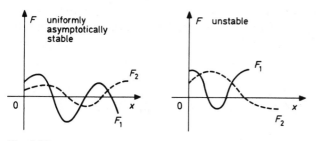

Fig. 3.10

Particular case: tracking (cf. p. 126)
To hold the output equal to e we use as input $\varepsilon = e - s$ (Fig. 3.11). If G is the transfer function of the open loop $\mathcal{L}[s](p) = G(p)\mathcal{L}[\varepsilon](p)$, i.e. $\mathcal{L}[s](p) =$

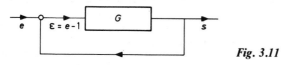

Fig. 3.11

$G(p)\{\mathscr{L}[e](p) - \mathscr{L}[s](p)\}$, the new transfer function is

$$H(p) = \frac{G(p)}{1 + G(p)}$$

and the transfer locus is

$$\omega \rightsquigarrow \frac{G(i\omega)}{1 + G(i\omega)}$$

Leonard's criterion then becomes the following criterion.

Theorem 2.1.11 (Nyquist's criterion)

If the open-loop system is UAS and if the transfer locus $G(i\omega)/[1 + G(i\omega)]$ traversed in the direction of increasing ω keeps the point -1 on the left, the tracking is UAS (Fig. 3.12).

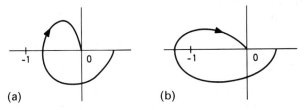

(a)   (b)

Fig. 3.12. (a) *Tracking UAS; (b) tracking unstable.*

Domain of stability

Let $F(z) = a_0 z^n + \ldots + a_n$; the nature of F depends on the coefficients $a_0, \ldots, a_n$. The portion of $\mathbb{R}^{n+1}$ in which the coefficients $a_0, \ldots, a_n$ are such that F is UAS is called the domain of stability; we can in principle find it with the help of Routh's criterion. If it is empty the system is always unstable; we shall say that it is intrinsically unstable.

Example

$$\frac{dx}{dt} = ax + by$$

$$F(z) = z^2 - (a + d)z + ad - bc$$

$$\frac{dy}{dt} = cx + dy$$

The domain of stability is the domain of $\mathbb{R}^4$ defined by $a + d < 0$, $ad - bc < 0$.

Margin of stability

In practice we do not know the parameters of a system exactly and it is not sufficient to suppose that $\mathrm{Re}\,\lambda < 0$ for every root of the characteristic polynomial. In general we give ourselves a margin and say $\mathrm{Re}\,\lambda < \sigma < 0$. By doing this we know that we are obtaining a 'minimum damping velocity' of the transient regime since $\|\Phi(t, s)\| \leqslant a \exp[-\sigma(t - s)]$ and a transient is of the form

$$\int_0^t \Phi(t, s)b(s)\,ds$$

LINEAR PERTURBATIONS

Theorem 2.1.12

Let $dx/dt = A(t)x$ be a system which is UAS for $t_0 \geqslant T$; there exist a and $\sigma > 0$ such that

$$\forall (t, u) \geqslant T \qquad \|\Phi(t, u)\| \leqslant a \exp[-\sigma(t - u)]$$

Let $B \in \mathscr{L}(\mathbb{R}^d, \mathbb{R}^d)$ be such that

$$\exists S: \|B(t)\| \leqslant \eta \leqslant \frac{\sigma}{a} \qquad \forall t \geqslant S$$

Then the system $dx/dt = [A(t) + B(t)]x$ is UAS for all $t_0 \geqslant T$. In particular, if A is UAS, then, if $\forall t \geqslant S$ $\|A(t) - A\| \leqslant \eta < \sigma/a$ (or if simply $\lim_{t \to \infty} A(t) = A$), $dx/dt = A(t)x$ is UAS (for t sufficiently large).

Proof Every solution is of the form

$$x(t) = \Phi(t, t_0)x_0 + \int_{t_0}^t \Phi(t, s)B(s)x(s)\,ds$$

If $t_0 \geqslant S$

$$\|x(t)\| \leqslant a \exp[-\sigma(t - t_0)]\|x_0\| + \int_{t_0}^t a \exp[-\sigma(t - s)]\eta\|x(s)\|\,ds$$

and if Gronwall's lemma is applied to $\exp(\sigma t)\|x(t)\|$

$$\|x(t)\| \leqslant a\|x_0\| \exp[-(\sigma - a\eta)(t - t_0)]$$

whose limit is zero.

If $t_0 < S$ $\varphi(t, t_0, x_0) = \varphi(t, S, x_S)$ and we can apply the preceding result since $S \geqslant S$.

If the system is only uniformly stable, to maintain this stability we require a much stronger condition.

Theorem 2.1.13

Let $dx/dt = A(t)x$ be a system which is uniformly stable for $t_0 \geqslant T$ and let $B(t) \in \mathscr{L}(\mathbb{R}^d, \mathbb{R}^d)$ be such that

$$\int_T^\infty \|B(u)\|\,du < \infty$$

Then $dx/dt = [A(t) + B(t)]x$ is uniformly stable for $t_0 \geqslant T$.

Proof This is again an application of Gronwall's lemma.

$$x(t) = \Phi(t, t_0)x(t_0) + \int_{t_0}^t \Phi(t, s)B(s)x(s)\,ds$$

Here $\|\Phi(t, s)\| \leqslant K$ for $t, s \geqslant t_0 \geqslant T$. Then

$$\|x(t)\| \leqslant K\|x(t_0)\| + \int_{t_0}^{t} K\|B(s)\| \|x(s)\| \, ds$$

and from Gronwall's generalized formula

$$\|x(t)\| \leqslant K\|x(t_0)\| + \int_{t_0}^{t} K\|x(t_0)\|K\|B(s)\| \left[\exp \int_{s}^{t} K\|B(u)\| \, du \right] ds$$

Now

$$\int_{s}^{t} \|B(u)\| \, du \leqslant \int_{T}^{\infty} \|B(u)\| \, du = A < \infty$$

and therefore

$$\|x(t)\| \leqslant K\|x(t_0)\|[1 + KA \exp(KA)]$$

The solutions are then bounded and the system is uniformly stable since $\|x(t)\| \leqslant \varepsilon$ $\forall t \geqslant t_0$ if $\|x(t_0)\| \leqslant \varepsilon K[1 + KA \exp(KA)]^{-1}$.

Important remark   The condition

$$\int_{T}^{\infty} \|B(u)\| \, du < \infty$$

is very strong, and so a system which is only uniformly stable can very easily become unstable.

Example   The system

$$\frac{d^2u}{dt^2} + u = 0$$

is uniformly stable.
  Now the system

$$\frac{d^2u}{dt^2} - \frac{2}{t}\frac{du}{dt} + u = 0$$

is unstable (for $t \cos t - \sin t$ is a solution) whereas

$$\frac{d^2u}{dt^2} + \frac{2}{t}\frac{du}{dt} + u = 0$$

is stable since a fundamental system of independent solutions is formed from

$$\frac{1}{t}\cos t \quad \text{and} \quad \frac{1}{t}\sin t$$

i.e. the system is even asymptotically stable.

For both the perturbed systems

$$B(t) = \begin{pmatrix} 0 & 0 \\ 0 & -\dfrac{2}{t} \end{pmatrix} \quad \text{or} \quad \begin{pmatrix} 0 & 0 \\ 0 & \dfrac{2}{t} \end{pmatrix}$$

$\|B(t)\| = 2/t$ the limit of which is zero but which is not integrable at infinity. See *Exercise 9* at the end of the chapter.

2.2. Response of a uniformly asymptotically stable system

We have just seen that a UAS system has a certain robustness relative to linear perturbations; this is an important property for the automation engineer. We shall establish another, more important, property relating to the responses of a UAS system to bounded stimuli and to transient regimes.

Theorem 2.2.1
Let $dx/dt = A(t)x + b(t)$ be a UAS system.

(a) For every bounded input b, x is bounded.
(b) If $b(t) = 0$ $\forall t \geqslant T$ the response has zero for limit as $t \to \infty$.

Proof

(a)
$$x(t) = \Phi(t, 0)x(0) + \int_0^t \Phi(t, s)b(s)\,ds$$

Let
$$\|b(s)\| \leqslant M \quad \forall s \geqslant 0$$
$$\|\Phi(t, s)\| \leqslant a\exp[-\sigma(t - s)]$$

Then
$$\|x(t)\| \leqslant a\exp(-\sigma t)\,\|x(0)\| + \frac{aM}{\sigma}[1 - \exp(-\sigma t)]$$

is bounded.

(b)
$$\int_0^t \Phi(t, s)b(s)\,ds = \int_0^T \Phi(t, s)b(s)\,ds$$
$$= \Phi(t, T)\int_0^T \Phi(T, s)b(s)\,ds$$

and the norm of this expression is smaller than

$$a\exp[-\sigma(t - T)]\frac{aM}{\sigma}[1 - \exp(-\sigma T)]$$

whose limit is zero.

We shall see that property (a) of the preceding theorem is in practice a property characteristic of UAS systems; this is what constitutes the importance of this notion, which is often denoted simply by the term 'stability' by systems theory specialists.

We recall that a system is UAS iff all the solutions of the homogeneous system have limit zero, with uniform convergence.

Theorem 2.2.2 (Perron)
Let

$$\frac{\mathrm{d}x}{\mathrm{d}t} = A(t)x + b(t) \tag{12}$$

where

$$\sup_{0 \leqslant t \leqslant \infty} \|A(t)\| < \infty$$

and

$$\sup_{0 \leqslant t \leqslant \infty} \|b(t)\| < \infty$$

Every solution is bounded iff A is UAS, which means that the only systems for which all outputs x are bounded for every bounded input b are UAS systems.

Proof   The sufficient condition was established in Theorem 2.2.1. We shall only give indications of the proof of the necessary condition.

If all solutions are bounded

$$\left\| \int_0^t \Phi(t, s)b(s)\,\mathrm{d}s \right\| \leqslant M(b) < \infty$$

A theorem—called the theorem of the uniform bound—states that in these conditions

$$\left\| \int_0^t \Phi(t, s)b(s)\,\mathrm{d}s \right\| \leqslant k \sup_{0 \leqslant t < \infty} \|b(t)\|$$

and that

$$\int_0^t \|\Phi(t, s)\|\,\mathrm{d}s \leqslant A$$

and finally that

$$\|\Phi(t, s)\| \leqslant B \qquad \forall 0 \leqslant s \leqslant t < \infty$$

since

$$\Phi(t, s) = I + \int_s^t A(u)\Phi(u, s)\,\mathrm{d}u$$

The rest of the proof is elementary:

$$(t - u)\Phi(t, u) = \int_u^t \Phi(t, s)\Phi(s, u)\,ds$$

Therefore

$$|t - u|\,\|\Phi(t, u)\| \leqslant \int_u^t \|\Phi(t, s)\|\,\|\Phi(s, u)\|\,ds \leqslant AB$$

Hence

$$\|\Phi(t, u)\| \leqslant \frac{AB}{|t - u|}$$

Consequently $\|\Phi(t, u)x_0\|$ tends uniformly to zero if $t \to \infty$, which ensures that the system is UAS (since $x(t) \equiv 0$ is a UAS solution).

We shall end with a very simple theorem concerning the relation between stability and boundedness of solutions of a non-homogeneous system.

Theorem 2.2.3

Let

$$\frac{dx}{dt} = A(t)x + b(t) \tag{13}$$

be a non-homogeneous system.

(a) If the system is stable and one solution is bounded all solutions are bounded.

(b) If two solutions of (13) are bounded, the system is stable and all solutions are therefore bounded.

Proof (a) Let x be any solution of (13) and let y be a bounded solution. $x(t) - y(t) = \Phi(t, t_0)[x(t_0) - y(t_0)]$ is a solution of the homogeneous system, which is stable; this solution is therefore bounded and therefore so is x.

(b) If x and y are two bounded solutions of (13), $x - y$ is a bounded solution of the homogeneous system which is therefore stable.

2.3. Limits of solutions—comparison of two systems

We know already that if a homogeneous system is UAS all the solutions have limit zero. We shall see that for certain uniformly stable systems the solutions also have limits if t tends to infinity.

Theorem 2.3.1

(a) Let

$$\frac{dx}{dt} = B(t)x \tag{14}$$

be a system in which

$$\int_0^\infty \|B(u)\| \, du = K < \infty$$

Then every solution ψ of (14) has a finite limit if $t \to \infty$ and $\forall V \in \mathbb{R}^d$ there exists a unique solution ψ of (14) such that

$$\lim_{t \to \infty} \psi(t) = V$$

(b) Let

$$\frac{dx}{dt} = B(t)x + b(t) \tag{15}$$

be a system in which

$$\int_0^\infty \|B(u)\| \, du = K < \infty \quad \text{and} \quad \int_0^\infty \|b(u)\| \, du = K_1 < \infty$$

Every solution of (15) then has a finite limit if $t \to \infty$.

Proof (a) We shall show first that every solution of (14) is bounded.

$$\psi(t) = \psi(0) + \int_0^t B(s)\psi(s) \, ds$$

Therefore from Gronwall's generalized formula

$$\|\psi(t)\| \leqslant \|\psi(0)\| + \int_0^t \|\psi(0)\| \, \|B(s)\| \exp\left[\int_s^t \|B(u)\| \, du\right] ds$$

Now

$$\int_s^t \|B(u)\| \, du \leqslant K$$

and therefore

$$\|\psi(t)\| \leqslant \|\psi(0)\|[1 + K \exp(K)]$$

Thus (14) is stable. Then let A be such that $\|\psi(s)\| \leqslant A \ \forall s \in [0, \infty[$:

$$\psi(t_2) - \psi(t_1) = \int_{t_1}^{t_2} B(s)\psi(s) \, ds$$

and therefore

$$\|\psi(t_2) - \psi(t_1)\| \leqslant A \int_{t_1}^{t_2} \|B(s)\| \, ds$$

$\psi(t)$ is a Cauchy sequence like

$$\int_0^t \| B(u) \| \, du$$

Therefore ψ has a finite limit.

We shall show now that there exists only one solution whose limit is zero; this will imply that there is only one solution whose limit is V.

Let

$$\| B(u) \| = \sum_{i,j} |B_{i,j}(u)|$$

and

$$\mathrm{tr}\, B(u) = \sum_i B_{i,i}(u)$$

Then

$$\int_0^t |\mathrm{tr}\, B(u)| \, du \leqslant \int_0^t \| B(u) \| \, du \leqslant K < \infty$$

Therefore

$$\exp \int_0^t \mathrm{tr}\, B(u) \, du \geqslant \sigma > 0$$

and, according to Liouville's theorem, the volume determined by solutions whose initial values form independent vectors cannot have zero as limit (cf. p. (69)), which implies the result stated. If $G(t)$ is a fundamental matrix $\| G(t) \|$ is bounded since all solutions are bounded, but so is $\| G(s)^{-1} \|$ since the determinant of $G(s)$ is bounded below. Now $\Phi(t, s) = G(t)G(s)^{-1}$ is the resolvent of B; the latter is therefore maximized (for $t \in [0, \infty]$ and $s \in [0, \infty]$) by C and the equation is not only stable but uniformly stable.

Then

$$\left\| \int_0^T \Phi(t, s)B(s) \, ds \right\| \leqslant C \int_0^T \| B(s) \| \, ds < CK$$

and

$$\int_0^\infty \Phi(t, s)B(s) \, ds$$

exists so that

$$\lim_{t \to \infty} \int_t^\infty \Phi(t, s)B(s)V \, ds = 0 \quad \forall V$$

If

$$\psi(t) = V - \int_t^\infty \Phi(t, s)B(s)V\,ds$$

$\lim \psi(t) = V$ and ψ is a solution because

$$\frac{d\psi}{dt} = B(t)V - \int_t^\infty B(t)\Phi(t, s)B(s)V\,ds = B(t)\psi$$

(b) For every x_0 $\Phi(t, 0)x_0$ is the solution of (14) with initial value x_0. It therefore has a limit. The solution of (15) with the same initial value is

$$\Phi(t, 0)x_0 + \int_0^t \Phi(t, s)b(s)\,ds$$

It therefore has a limit if

$$\int_0^t \Phi(t, s)b(s)\,ds$$

has a limit. Since the system is uniformly stable, there exists M such that

$$\forall 0 \leqslant s \leqslant t < \infty \qquad \|\Phi(t, s)\| \leqslant M$$

Then

$$\left\| \int_u^t \Phi(t, s)b(s)\,ds \right\| \leqslant M \int_u^t \|b(s)\|\,ds$$

has limit zero if t and u tend to infinity;

$$\int_0^t \Phi(t, s)b(s)\,ds$$

is a Cauchy sequence like

$$\int_0^t b(u)\,du$$

and therefore has a limit.

Remarks   (a) The condition

$$\int_0^\infty \|B(u)\|\,du < \infty$$

implies that the theorem does not apply to differential equations in $\mathbb{R}$ since the first row of $B(u)$ is then $(0\ 0\ 0\ \dots\ 1)$.

(b) Naturally, the preceding system is not asymptotically stable.

(c) The preceding theorem can be obtained by comparison between (14) and $dx/dt = 0$ (cf. Theorem 2.1.13). The following result generalizes this situation.

Theorem 2.3.2

Let

$$A: \exp\left[\int_0^t \text{tr } A(u)\, du\right] > \sigma > 0 \quad \forall t \geqslant 0$$

and

$$\frac{dx}{dt} = A(t)x \tag{16}$$

be a system which is uniformly stable for $t_0 > 0$. Let

$$\frac{dx}{dt} = B(t)x \tag{17}$$

We shall assume that

$$\int_0^\infty \|B(u) - A(u)\|\, du < \infty$$

(a) Every solution ψ of (17) is bounded and if $\varphi(t_0)$ is the initial value of a solution of (16) such that $\varphi(t_0) = \psi(t_0)$, $\varphi(t) - \psi(t)$ has a finite limit.

(b) With every solution ψ of (17) we can associate a solution φ of (16) such that

$$\lim_{t \to \infty} [\varphi(t) - \psi(t)] = 0$$

and conversely.

(This does not mean that φ and ψ have limits.)

We can deduce Theorem 2.3.1 by considering $A(u) \equiv 0$.

Proof Let Φ_A be the resolvent of (16). If φ and ψ have the same initial value

$$\psi(t) - \varphi(t) = \int_0^t \Phi_A(t, s)[B(s) - A(s)]\psi(s)\, ds$$

and the proof is analogous to that of the preceding theorem (it is slightly more intricate).

If $\psi - \varphi$ has a limit

$$\int_0^t \Phi_A(t, s)[B(s) - A(s)]\psi(s)\, ds$$

also has a limit. If ψ is fixed, we can put

$$\varphi(t) = \psi(t) + \int_t^\infty \Phi_A(t, s)[B(s) - A(s)]\psi(s)\, ds$$

$$\frac{d\varphi}{dt} = B(t)\psi(t) + A(t)\int_t^\infty \Phi_A(t, s)[B(s) - A(s)]\psi(s)\, ds - [B(t) - A(t)]\psi(t)$$

$$= A(t)\varphi(t)$$

φ is then the solution of (16) such that $\lim[\varphi(t) - \psi(t)] = 0$. It is unique, for

$$\exp\left[\int_0^t \operatorname{tr} A(u)\, du\right] > a > 0$$

Naturally we could construct ψ, starting from φ, in the same way.

Example Let

$$\frac{d^2 x}{dt^2} + [1 + f(t)]x = 0 \quad \text{where} \quad \int_0^\infty |f(u)|\, du = K < \infty$$

If φ is a solution of (17) with orbit γ and with coordinates $x(t)$, $y(t)$, there exist A and Φ such that

$$\lim_{t \to \infty} \left[x(t) - A\cos(t + \Phi)\right] = 0 = \lim_{t \to \infty} \left[y(t) + A\sin(t + \Phi)\right]$$

and the orbit γ approaches the circle of radius A (Fig. 3.13). (It is sufficient to consider

$$\frac{d^2 x}{dt^2} + x = 0$$

and to apply the preceding theorem.) At the 'limit', $[x(t), y(t)]$ traverses Γ following the law $x(t) = A\cos(t + \Phi)$.

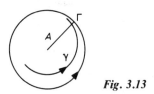

Fig. 3.13

Part 3: Elementary results in control theory

3.1. Optimization with a finite horizon with respect to a quadratic criterion

Let

$$\frac{dX}{dt} = F(t)X(t) + G(t)u(t) + v(t) \tag{18}$$

be a system whose output we wish to affect through the control u.

We shall assume throughout that $X \in \mathbb{R}^n$, $F(t) \in \mathscr{L}(\mathbb{R}^n, \mathbb{R}^n)$ and is a continuous function of t, that $v(t) \in \mathbb{R}^n$ and is also continuous, that $u(t)$ is piecewise continuous and finally that $u(t) \in \mathbb{R}^k$, $G(t) \in \mathcal{L}(\mathbb{R}^k, \mathbb{R}^n)$ and is continuous in t.

For every fixed (t_0, X_0) there exists for every control u a unique solution to the Cauchy problem relative to (t_0, X_0); we shall denote it $\varphi(t, t_0, X_0, u)$. Often when t_0 and X_0 are fixed, we shall abbreviate this notation by writing $\varphi(t, t_0, X_0, u) = X_u(t)$.

The simplest problem we can consider is that of the **linear regulator**; we assume that $v(t) \equiv 0$ and that it is required that, starting from (t_0, X_0), $X(t)$ be as close as possible to zero while the quantity $\{u(s)\}^2$ is kept as small as possible—we can think of this as representing the cost of control. We can formulate the problem in the following way: find a control u which minimizes the quantity

$$\int_{t_0}^t [\langle X_u(s), X_u(s) \rangle + \|u(s)\|^2] \, ds$$

We shall then consider the following slightly more general problem. Let $T < \infty$ be a fixed final instant (**finite horizon**) and let $Q(t)$ and $R(t)$ be two continuous symmetric matrices where $Q(t) \in \mathcal{L}(\mathbb{R}^n, \mathbb{R}^n)$ and $R(t) \in \mathcal{L}(\mathbb{R}^k, \mathbb{R}^k)$; we shall assume moreover that $R(t)$ is invertible $\forall t \in (t_0, T)$. We shall call the following quantity the **quadratic cost** in (t_0, X_0):

$$J_u(t_0, X_0) = \int_{t_0}^T [\langle X_u(t), Q(t)X_u(t) \rangle + \langle u(t), R(t)u(t) \rangle] \, dt$$

Definition 3.1.1

A control u such that, for every control v, $J_u(t_0, X_0) \leqslant J_v(t_0, X_0)$ is called an optimal control for the criterion J and the point (t_0, X_0).

We can choose the control u 'in advance', i.e. we can choose a function u which will serve as a control over (t_0, T), and the system is then called an **open-loop system** (Fig. 3.14(a)). Alternatively, we can choose a fixed function K such that $\forall t \in (t_0, T)$ $K(t) \in \mathcal{L}(\mathbb{R}^n, \mathbb{R}^n)$ and use the output $X(t)$ to construct a control of the form $K(t)X(t)$; the system is then called a **closed-loop system** and the control is called a **feedback** control. In this case we denote the output by X_K and the corresponding cost by $J_K(t_0, X_0)$ (Fig. 3.14(b)).

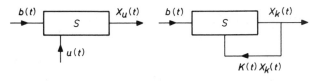

(a)   (b)

Fig. 3.14. (a) Open loop, cost $J_u(t_0, X_0)$; (b) closed loop, cost $J_K(t_0, X_0)$.

Various types of control

If $u(t) \in \mathbb{R}$ then $G(t) \in \mathcal{L}(\mathbb{R}, \mathbb{R}^d)$ is a vector; if the latter is constant, say g, the control $u(t)g$ depends only on a single scalar function u; the system is said to be under simple control.

If, however, $u(t) \in \mathbb{R}^p$ ($p > 1$) and its coordinates are not all identical, and $G(t)$ is such that several scalar functions appear in the control, the system is said to be under multiple control.

Notation   If $M \in \mathcal{L}(\mathbb{R}^n, \mathbb{R}^n)$, $\tilde{M}$ denotes the transpose of M, defined in the following manner: $\forall V, W \in \mathbb{R}^n$, $\langle \tilde{M}V, W \rangle = \langle V, MW \rangle$ (where $\langle U_1, U_2 \rangle$ is the usual scalar product in $\mathbb{R}^n$ of the vectors U_1 and U_2). If V is a vector of $\mathbb{R}^n$, $\tilde{V}$ denotes the transposed (horizontal) vector and $\langle \tilde{M}V, W \rangle = \tilde{V}MW$.

If M is symmetric ($M = \tilde{M}$), it is said to be positive if, $\forall V$, $\tilde{V}MV \geqslant 0$, and strictly positive if, $\forall V \neq 0$, $\tilde{V}MV > 0$.

CALCULATION OF AN OPTIMAL CONTROL

Theorem 3.1.2

Let

$$\frac{dX}{dt} = F(t)X(t) + G(t)u(t) \tag{19}$$

Let Q and R be symmetric (R invertible) and

$$J_u(t_0, X_0) = \int_{t_0}^T \left[\langle X_u(t), Q(t)X_u(t) \rangle + \langle u(t), R(t)u(t) \rangle \right] dt$$

Let

$$-\frac{dP}{dt} = Q(t) + P(t)F(t) + \tilde{F}(t)P(t) - P(t)G(t)R^{-1}(t)\tilde{G}(t)P(t) \tag{20}$$

be the 'Riccati' equation in $\mathcal{L}(\mathbb{R}^n, \mathbb{R}^n)$. Then there exists a symmetric solution of (20) defined over (t_0, T) which is unique and such that $P(T) = 0$, and the feedback control associated with $K(t) = -R^{-1}(t)\tilde{G}(t)P(t)$ is optimal for J and (t_0, X_0), the cost being $J_K(t_0, X_0) = \langle X_0, P(t_0)X_0 \rangle$.

X_K is then the solution of

$$\frac{dX}{dt} = [F(t) + G(t)K(t)]X(t) = [F(t) - G(t)R^{-1}(t)\tilde{G}(t)P(t)]X(t)$$

Proof   We note first that Eqn (20) is equivalent to an equation in $\mathbb{R}^n \times \mathbb{R}^n$ whose right-hand side is locally Lipschitzian; there exists then an interval J containing T over which we can find a unique solution such that $P(T) = 0$. We shall consider for the moment that this solution is defined over $[t_0, T]$.

Let P be a solution of (20). By transposing the equality (20) we find that

$$-\frac{d\tilde{P}}{dt} = Q(t) + \tilde{F}(t)\tilde{P}(t) + \tilde{P}(t)F(t) - \tilde{P}(t)G(t)R^{-1}(t)\tilde{G}(t)\tilde{P}(t)$$

$\tilde{P}(t)$ is thus also a solution of (20) and, since $P(T) = \tilde{P}(T) = 0$, $P(t) = \tilde{P}(t)$. Thus $P(t)$ is symmetric.

Let u be an arbitrary control and $H_u(t) = \langle X_u(t), P(t)X_u(t)\rangle$:

$$\frac{d}{dt} H_u(t) = \langle F(t)X_u(t) + G(t)u(t), P(t)X_u(t)\rangle$$

$$+ \langle X_u(t), P(t)[F(t)X_u(t) + G(t)u(t)]\rangle + \left\langle X_u(t), \frac{dP}{dt} X_u(t)\right\rangle$$

$$= \left\langle X_u(t), \left[\tilde{F}(t)P(t) + P(t)F(t) + \frac{dP}{dt}\right]X_u(t)\right\rangle + \langle u(t), \tilde{G}(t)P(t)X_u(t)\rangle$$

$$+ \langle X_u(t), P(t)G(t)u(t)\rangle$$

Thus $H_u(T) - H_u(t_0) = -\langle X_0, P(t_0)X_0\rangle$, which is equal to (by simplifying the expressions under the integral sign)

$$\int_{t_0}^T \left[\left\langle X_u, \left(\tilde{F}P + PF + \frac{dP}{dt}\right)X_u\right\rangle + \langle u, \tilde{G}PX_u\rangle + \langle X_u, PGu\rangle\right](t)\,dt$$

Then

$$J_u(t_0, X_0) = J_u(t_0, X_0) - \langle X_0, P(t_0)X_0\rangle + \langle X_0, P(t_0)X_0\rangle$$

$$= \langle X_0, P(t_0)X_0\rangle + \int_{t_0}^T \left[\left\langle X_u, \left(Q + \tilde{F}P + PF + \frac{dP}{dt}\right)X_u\right\rangle\right.$$

$$\left. + \langle u, Ru\rangle + \langle u, \tilde{G}PX_u\rangle + \langle X_u, PGu\rangle\right](t)\,dt$$

and by the definition of P

$$J_u(t_0, X_0) = \langle X_0, P(t_0)X_0\rangle + \int_{t_0}^T (\langle u, Ru\rangle + \langle X_u, PGR^{-1}\tilde{G}PX_u\rangle$$

$$+ \langle X_u, PGu\rangle + \langle u, \tilde{G}PX_u\rangle)(t)\,dt$$

Since R is symmetric, we can diagonalize it:

$$R = \tilde{A}\begin{pmatrix} \lambda_1 & & \\ & \ddots & \\ & & \lambda_k \end{pmatrix}A$$

The eigenvalues λ_i are positive since R is positive. Let

$$R^{1/2} = \begin{pmatrix} \lambda_1^{1/2} & & \\ & \ddots & \\ & & \lambda_k^{1/2} \end{pmatrix} A$$

(A is orthogonal). We can verify immediately that $R = \tilde{R}^{1/2}R^{1/2}$ and $R^{-1} = R^{-1/2}\tilde{R}^{-1/2}$. Then

$$0 \leqslant \langle R^{1/2}u + \tilde{R}^{-1/2}\tilde{G}PX_u, R^{1/2}u + \tilde{R}^{-1/2}\tilde{G}PX_u \rangle(t)$$
$$= (\langle u, Ru \rangle + \langle X_u, PGu \rangle + \langle u, \tilde{G}PX_u \rangle + \langle X_u, PGR^{-1}\tilde{G}PX_u \rangle)(t)$$

Thus $J_u(t_0, X_0)$ is a minimum if this quantity is zero $\forall t \in (t_0, T)$, i.e. if $R^{1/2}u + \tilde{R}^{-1/2}\tilde{G}PX_u = 0$ or $u(t) = -R^{-1}(t)\tilde{G}(t)P(t)X_u(t)$, which is then an optimal control, the cost $J_K(t_0, X_0)$ being $\langle X_0, P(t_0)X_0 \rangle$.

Case of non-zero input: tracking

If it is required that $x(t)$ be as close as possible to a given function $f(t)$ which we assume to be differentiable, we put $y(t) = x(t) - f(t)$ so that

$$\frac{dy}{dt} = F(t)y(t) + G(t)u(t) + v(t) \quad \text{where } v(t) = F(t)f(t) - \frac{df}{dt}$$

In addition, it is often necessary to take account of the value $y(T)$.

We can prove the following theorem in the same way as Theorem 3.1.2.

Theorem 3.1.3

Let $dy/dt = F(t)y(t) + G(t)u(t) + v(t)$ and

$$J_u(t_0, X_0) = \int_{t_0}^T [\langle y(t), Q(t)y(t) \rangle + \langle u(t), R(t)u(t) \rangle] \, dt + \langle y(T), My(T) \rangle$$

where M is a symmetric positive matrix. Let $P(t)$ be the solution of the Riccati equation (20) such that $P(T) = M$. Let g be the solution such that $g(T) = 0$ of

$$\frac{dg}{dt} = P(t)v(t) + [P(t)G(t)R^{-1}(t)\tilde{G}(t) - F(t)]g$$

Then the control $u(t) = h(t) + K(t)y(t)$ is optimal, h and K being defined by $h(t) = \tilde{R}^{-1}(t)\tilde{G}(t)g(t)$ and $K(t) = -R^{-1}(t)\tilde{G}(t)P(t)$.

Study of the Riccati equation

We shall give an algorithm which enables us to calculate $P(t)$. At the same time as proving the convergence of this algorithm we shall prove that there exists a solution defined over (t_0, T) such that $P(T) = 0$ (or $P(T) = M$ in the situation of Theorem 3.1.3).

Writing throughout $K(t) = -R^{-1}(t)\tilde{G}(t)P(t)$, we can rewrite Eqn (20):

$$\frac{dP}{dt} + Q(t) + \tilde{K}(t)R(t)K(t) + [\tilde{F}(t) + \tilde{K}(t)\tilde{G}(t)]P(t)$$

$$+ P(t)[F(t) + G(t)K(t)] = 0 \qquad (20')$$

The form $(20')$ suggests the following algorithm.

Algorithm
Let $P_0(t)$ be a positive symmetric matrix such that $P_0(T) = 0$. Construct recursively the sequence P_n from P_{n-1} and $K_{n-1} = -R^{-1}\tilde{G}P_{n-1}$ as a solution of

$$\frac{dP_n}{dt} + Q(t) + \tilde{K}_{n-1}(t)R(t)K_{n-1}(t)$$

$$+ [\tilde{F}(t) + \tilde{K}_{n-1}(t)\tilde{G}(t)]P_n(t) + P_n(t)[F(t) + G(t)K_{n-1}(t)] = 0 \quad (20'')$$

such that $P_n(T) = 0$.
The limit of $P_n(t)$ when $n \to \infty$ is defined over $[t_0, T]$, is a solution of $(20')$ and satisfies $P(T) = 0$; this is the matrix required.

Construction of the algorithm The stages are as follows. Since $P_0(t) \geqslant 0$ we see that $P_1(t) \geqslant 0$ and by recurrence that $P_n(t) \geqslant 0$ $\forall n$. We then establish that the matrix $P_{n+1}(t) - P_n(t)$ is also positive or zero and that there exists a fixed matrix $A \geqslant 0$ such that $A - P_n(t) \geqslant 0$.
A classical theorem, analogous to the theorem of monotone bounded convergence for real numbers, shows that $P_n(t)$ has a limit $P(t)$ which is itself greater than or equal to zero; since $P_n(T) = 0$ $\forall n$, $P(T) = 0$.
Finally we show that the relation $(20'')$ implies the limit $(20')$.

Such an algorithm leads also to numerical calculation of P.

Fundamental remarks: feedback control and open-loop control; noise
(1) We shall first establish that there exists an open-loop control as good as the preceding one; this is due to the very particular forms of the equation to be controlled and of the cost function.

Proposition 3.1.4
Let K be the matrix defined in Theorem 3.1.2 and let $\Phi(t, s)$ be the resolvent of $(F + GK)(t)$; the open-loop control $u_0(t) = K(t)\Phi(t, t_0)X_0$ is also an optimal control for (t_0, X_0).

Proof Let $Y(t) = \varphi(t, t_0, X_0, u_0)$; by definition $Y(t_0) = X_0$ and

$$\frac{dY}{dt} = F(t)Y(t) + G(t)u_0(t) = F(t)Y(t) + G(t)K(t)\Phi(t, t_0)X_0$$

Now $\Phi(t, t_0)X_0$ is also a solution of this last equation so that $Y(t) = \Phi(t, t_0)X_0$ and $dY/dt = [F(t) + G(t)K(t)]Y(t)$, which implies that $Y(t) = X_K(t)$ and consequently that $J_{u_0}(t_0, X_0) = J_K(t_0, X_0)$, which is the minimum cost.

(2) In general an optimal feedback control is strictly better than the best open-loop control, i.e. its cost $J_K(t_0, X_0) \leqslant J_V(t_0, X_0)$ for all open-loop controls V. This is natural since a feedback control uses updated information provided by the output.

(3) Let us consider the preceding linear system with noise added; symbolically we shall represent this as

$$\frac{dX}{dt} = F(t)X + G(t)u(t) + \beta(t)$$

The noise β is a random function of 'zero mean' and of 'non-zero energy'.

This can be defined rigorously, and it can be proved that the optimal feedback control in the absence of noise is still optimal in the presence of noise and that it is therefore strictly better than every open-loop control.

This fact has considerable practical importance. One of its most important consequences is the following separation principle.

Separation principle

Although the following statement is rather vague (because we have not defined the necessary concepts) it seems to us to be quite significant.

Let $dx/dt = Fx + Gu + \beta$ be a deterministic system $Fx + Gu$ excited by 'white noise' β. Suppose that we do not know x but only $y = Hx + \gamma$ where γ is another white noise, independent of β. If we wish to minimize the mean value of

$$\int_{t_0}^{T} (\langle x, Q(x) \rangle + \langle u, Ru \rangle) \, dt$$

we can

(a) estimate x as a function of y (let the result be $\hat{x}$) and
(b) calculate the control $u_0 = -Kx$ which minimizes, for the deterministic system, the quantity

$$\int_{t_0}^{T} (\langle x, Qx \rangle + \langle u, Ru \rangle) \, dt$$

The stochastic control $u = -K\hat{x}$ is then optimal.

DYNAMIC PROGRAMMING

Feedback controls use information provided by the output of the system; this property appears particularly clearly in the following.

Consider the minimum cost from t_0 to T as a function of (t_0, X_0); let V be this function. Let u be any control used between the instants t_0 and $t_0 + \varepsilon$; in this way let the point $(t_0 + \varepsilon, X_u(t_0 + \varepsilon))$ be attained, and let $\delta J_u(\varepsilon)$ be the corresponding cost. Then

$$V(t_0, X_0) \leqslant \delta J_u(\varepsilon) + V[t_0 + \varepsilon, X_u(t_0 + \varepsilon)] \quad \forall u$$

This inequality enables us, under quite general conditions, to obtain an equation which we shall now describe.

First we shall recall results concerning convex sets and functions.

A part Ω of a vector space E is called **convex** if, when X_1 and X_2 are two points of Ω, Ω contains the whole segment (X_1, X_2), in other words $X_1 \in \Omega$ and $X_2 \in \Omega \Rightarrow \forall \lambda \in [0, 1], \lambda X_1 + (1 - \lambda)X_2 \in \Omega$.

A function f defined over Ω with values in $\mathbb{R}$ is called convex if the set U formed from pairs (a, X) where $a \geqslant f(X)$ is a convex set of $\mathbb{R} \times \Omega$ (Ω is assumed to be convex).

Over $\mathbb{R}$ the function f is convex if the set U represented in Fig. 3.15 is convex.

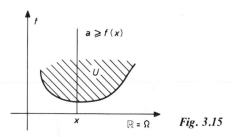

Fig. 3.15

We define the **boundary** $\partial \Omega$ of a set Ω as follows: $X \in \partial \Omega$ if every ball centred at X has a non-empty intersection with $\mathscr{E} - \Omega$ and Ω. For simplicity we shall suppose that $\partial \Omega$ is of class C^1.

We define the **interior** $\mathring{\Omega}$ of a set Ω as follows: $X \in \mathring{\Omega}$ if there exists a ball centred at X all of whose points are in Ω.

$\mathring{\Omega}$ can be empty, e.g. if Ω is a straight line. If $\mathring{\Omega}$ is not empty, we can define at every point of $\partial \Omega$ a tangent plane to Ω and an **outward normal** to Ω, because Ω is entirely on one side of its tangent plane (Fig. 3.16).

If $\forall X \in \Omega$ the tangent plane at X to Ω has only the point X in common with Ω, Ω is called **strictly convex**. A sphere is an example of a convex set the interior of which is non-empty and which is strictly convex.

In many cases it can be shown that, if a control is optimal from 0 to T, it is optimal overy every subinterval of $[0, T]$; under these conditions, for every ε,

$$V(t_0, X_0) = \min_u \{\delta J_u(\varepsilon) + V[t_0 + \varepsilon, X_u(t_0 + \varepsilon)]\}$$

where u runs through the set of all possible controls.

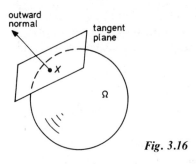

outward
normal

tangent
plane

X

Ω

Fig. 3.16

'Intuitive' proof of Bellman's equation

What follows is not claimed to be rigorous; it is intended to be merely an indication designed to clarify the statement which follows.

Let $dX/dt = F(t, X, u)$ be a differential equation where $X \in \mathbb{R}^n$, $u \in \mathbb{R}^k$; we suppose if necessary that u is restricted for every t in $[0, T]$, the constraint being $u(t) \in K$ where K is closed and bounded in $\mathbb{R}^k$.

For (t_0, X_0) the cost corresponding to the control u is taken to be of the form

$$J_u(t_0, X_0) = \int_{t_0}^{T} L[s, X_u(s), u(s)] \, ds$$

Between t_0 and $t_0 + \varepsilon$ the cost $\delta J_u(\varepsilon)$ is

$$\int_{t_0}^{t_0 + \varepsilon} L[s, X_u(s), u(s)] \, ds$$

which is a quantity equivalent for small ε to $\varepsilon L[t_0, X_0, u(t_0)]$. The cost $V[t_0 + \varepsilon, X_u(t_0 + \varepsilon)]$ is equivalent to

$$V(t_0, X_0) + \varepsilon \frac{\partial}{\partial \varepsilon} \{V[t_0 + \varepsilon, X_u(t_0 + \varepsilon)]\}$$

$$= V(t_0, X_0) + \varepsilon \frac{\partial V}{\partial t}(t_0, X_0) + \varepsilon \left\langle \frac{\partial V}{\partial X}(t_0, X_0), \frac{dX_u}{dt}(t_0, X_0) \right\rangle$$

The relation

$$V(t_0, X_0) = \min_u \{\delta J_u(\varepsilon) + V[t_0 + \varepsilon, X_u(t_0 + \varepsilon)]\}$$

then becomes

$$0 = \min_{u \in K} \left[L(t_0, X_0, u) + \left\langle \frac{\partial V}{\partial X}(t_0, X_0), F(t_0, X_0, u) \right\rangle \right] + \frac{\partial V}{\partial t}(t_0, X_0)$$

If there is no constraint we take the minimum over $u \in \mathbb{R}^k$.

Naturally we are assuming that all the functions which occur are sufficiently regular to permit the operations indicated; this is the case in the situation of Theorem 3.1.5 which follows.

Theorem 3.1.5 (Bellman's equation)

Let $dX/dt = F(t, X, u)$ be a differential equation with $X \in \mathbb{R}^n$, $u \in \mathbb{R}^k$ and possibly restricted by the constraint $u \in K$ (K is either a closed bounded set or $\mathbb{R}^n$) and K is convex.

Let $L(s, X, u)$ be a function such that $L(s, X, u) \geqslant a|u|^{1+\alpha}$ with $\alpha > 0$. To simplify the statement we shall assume that

$$L(s, X, u) = g(t, x) + h(t, u)$$

where $u \rightsquigarrow h(t, u)$ is convex. We assume that F is of class C^1.

Let

$$J_u(t_0, X_0) = \int_{t_0}^T L[s, X_u(s), u(s)]\, ds$$

be the cost associated with u for (t_0, X_0).

Then there exists an optimal control u_0 attaining the minimum cost $V(t_0, X_0)$.

The function V is a solution of the equation—called Bellman's equation—

$$V(T, X) = 0$$

$$\frac{\partial V}{\partial t}(t, X) + \inf_{y \in K}\left[\left\langle \frac{\partial V}{\partial X}(t, X), F(t, X, y)\right\rangle + L(t, X, y)\right] = 0 \qquad (21)$$

Equation (21) is also called the **equation of dynamic programming**. It enables us to calculate $u_0(t)$ by calculating at each instant the value realizing the minimum indicated; this calculation can sometimes be carried out exactly but more often must be done by an approximate method.

We shall prove this theorem in the particular case of Theorem 3.1.2, i.e.

$$F(t, X, u) = F(t)X + G(t)u$$

and

$$L(s, X, u) = \langle X, Q(s), X\rangle + \langle u, R(s)u\rangle$$

Then

$$J_V(t, X) = \int_t^T [\langle \varphi(s, t, X, v), Q(s)\varphi(s, t, X, v)\rangle + \langle v(s), R(s)v(s)\rangle]\, ds$$

Both

$$\frac{\partial J_v}{\partial t}(t, X) \quad \text{and} \quad \frac{\partial J_v}{\partial X}(t, X)$$

exist.

Lemma 3.1.6

For every control v, for 'almost every t' of $[t_0, T]$,

$$0 = \frac{\partial J_v}{\partial t}(t, X) + \left\langle \frac{\partial J_v}{\partial X}(t, X), F(t)X_v(t) + G(t)v(t, X)\right\rangle + L[t, X, v(t, X)]$$

We say that a relation is true for 'almost every t' if the set A in which this relation is not satisfied is such that

$$\int_{t_0}^{T} 1_A(s) \, ds \equiv 0$$

By definition $1_A(s)$ is the indicator function of A given by

$$1_A(s) = \begin{cases} 1 & \text{if } s \in A \\ 0 & \text{otherwise} \end{cases}$$

Proof

$$J_v(t_0, X_0) = -J_v[T, X_v(T)] + J_v[t_0, X_v(t_0)]$$

$$= -\int_{t_0}^{T} \frac{d}{ds} J_v[s, X_v(s)] \, ds$$

$$= -\int_{t_0}^{T} \left\{ \frac{\partial J_v}{\partial s} [s, X_v(s)] \right.$$

$$\left. + \left\langle \frac{\partial J_v}{\partial X}^\dagger [s, X_v(s)], F(s)X_v(s) + G(s)v[s, X_v(s)] \right\rangle \right\} ds$$

Moreover

$$J_v(t_0, X_0) = \int_{t_0}^{T} L\{s, X_v(s), v[s, X_v(s)]\} \, ds$$

Since these two integrals are equal for all t, we obtain the result.

Lemma 3.1.7

A control u which satisfies, for every X and almost every t, the relation

$$\forall y \quad \left\langle \frac{\partial J_u}{\partial X} (t, X), F(t)X + G(t)u(t, X) \right\rangle + L[t, X, u(t, X)]$$

$$\leqslant \left\langle \frac{\partial J_u}{\partial X} (t, X), F(t)X + G(t)y \right\rangle + L(t, X, y) \quad (22)$$

is optimal.

Proof   Let v be any control. Then

$$J_u(t_0, X_0) = -\int_{t_0}^{T} \left(\frac{\partial J_u}{\partial t} [s, X_v(s)] + \left\langle \frac{\partial J_u}{\partial X} [s, X_v(s)], F(s)X_v(s) + G(s)v(s) \right\rangle \right.$$

$$\left. + L\{s, X_v(s), v[s, X_v(s)]\} \right) ds + \int_{t_0}^{T} L\{s, X_v(s), v[s, X_v(s)]\} \, ds$$

$$\leqslant -\int_{t_0}^{T} \left(\frac{\partial J_u}{\partial t}[s, X_v(s)] + \left\langle \frac{\partial J_u}{\partial X}[s, X_v(s)], F(s)X_v(s) + G(s)u[s, X_v(s)] \right\rangle \right.$$

$$\left. + L\{s, X_v(s), u[s, X_v(s)]\} \right) ds$$

$$+ \int_{t_0}^{T} L\{s, X_v(s), v[s, X_v(s)]\} ds$$

Then from Lemma 3.1.6 the last line is equal to

$$\int_{t_0}^{T} L\{s, X_v(s), v[s, X_v(s)]\} ds = J_v(t_0, X_0)$$

Thus

$$J_u(t_0, X_0) \leqslant J_v(t_0, X_0)$$

Suppose then that $J_u(t, X)$ is a solution of Bellman's equation (21):

$$\left\langle \frac{\partial J_u}{\partial X}(t, X), F(t)X + G(t)u(t, X) \right\rangle + L[t, X, u(t, X)]$$

$$= -\frac{\partial J_u}{\partial t}(t, X) \quad \text{(from Lemma 3.1.6)}$$

$$= \inf_{y \in K} \left[\left\langle \frac{\partial J_u}{\partial X}(t, X), F(t)X + G(t)y \right\rangle + L(t, X, y) \right] \quad \text{(from Eqn (21))}$$

$$\leqslant \left\langle \frac{\partial J_u}{\partial X}(t, X), F(t)X + G(t)y \right\rangle + L(t, x, y) \quad \forall y$$

u is therefore an optimal control, from Lemma 3.1.7.

If we can find the solution V of (21) we can find u by going back to Lemma 3.1.7.

See *Exercises 10 to 16* at the end of the chapter.

3.2. Optimization with an infinite horizon: controllability; stability

We now consider a problem analogous to that of Section 3.1 but this time we wish to minimize the cost

$$\int_{t_0}^{\infty} [\langle X_u(t), Q(t)X_u(t) \rangle + \langle u(t), R(t)u(t) \rangle] dt$$

Naturally the first question which arises is whether X_u remains bounded or more precisely whether we can find controls u such that $X_u(t)$ is bounded $\forall t \geqslant t_0$.

Let us examine two examples.

(a) Let

$$\frac{dX}{dt} = \begin{pmatrix} 1 & 0 \\ 1 & -1 \end{pmatrix} X + \begin{pmatrix} 0 & 0 \\ 0 & 1 \end{pmatrix} \begin{pmatrix} u \\ v \end{pmatrix}$$

We shall write

$$X = \begin{pmatrix} x \\ y \end{pmatrix}$$

Whatever the control u, $x(t) = x_0 \exp(t)$ does not remain bounded; *a fortiori*, starting from a point (x_0, y_0), it is impossible to find a control taking the system to the origin. These two facts arise from what cannot be done with x; we say that the system is not completely controllable.

(b) Let

$$\frac{dX}{dt} = \begin{pmatrix} 1 & 0 \\ 1 & -1 \end{pmatrix} X + \begin{pmatrix} 1 & 0 \\ 0 & 0 \end{pmatrix} \begin{pmatrix} u \\ v \end{pmatrix}$$

i.e.

$$\frac{dx}{dt} = x + u \qquad \frac{dy}{dt} = x - y$$

If we choose $u(t) = -2x(t)$ the system becomes

$$\frac{dX}{dt} = \begin{pmatrix} -1 & 0 \\ 1 & -1 \end{pmatrix} X$$

and all solutions are bounded. It can be established also (as an exercise) that for every $\vartheta > 0$ and for every point $(x(0), y(0))$ a control u can be found such that the solution starting from $(x(0), y(0))$ at the instant 0 reaches the origin at the instant ϑ.

The matrix F is the same in the two examples and we see that the possibility of controlling the system is in fact a characteristic of the pair (F, G). We shall make these ideas more precise. First of all we shall introduce the following notation.

Notation 3.2.1

The notation $(t_0, X_0) \rightsquigarrow (t_1, X_1)$ signifies that there exists a control u such that $\varphi(t_1, t_0, X_0, u) = X_1$, i.e. u leads from (t_0, X_0) to (t_1, X_1).

CONTROLLABILITY OF A SYSTEM

Definition 3.2.2

A point X is called controllable at the instant t, starting from the instant τ prior to t $(t > \tau)$, if $(\tau, X) \rightsquigarrow (t, 0)$. It is then controllable at every instant subsequent to t. We say also that the pair (τ, X) is **controllable** at the instant t. We say that (τ, X) is controllable if there exists t such that $(\tau, X) \leadsto (t, 0)$. A system is called **completely controllable** if *all* couples (τ, X) are controllable.

In the particular case of a differential equation of order n in $\mathbb{R}$ a system is completely controllable if, starting at any instant τ whatever from a point x_0 with velocity x_0', acceleration x_0'', etc., one can arrive at the origin at time t with zero velocity, zero acceleration, etc.

Before describing the set of controllable points we shall make two remarks.

Proposition 3.2.3

Let $dX/dt = FX + Gu$ where F and G are constant. Then $(t_0, X_0) \leadsto (t_1, X) \Leftrightarrow (0, X_0) \leadsto (t_1 - t_0, X_1)$. We shall speak then of controllable points X without specifying the instant of departure.

Proof $(t_0, X_0) \leadsto (t_1, X_1)$ iff there exists a control u such that

$$X_1 = \exp[F(t_1 - t_0)]X_0 + \int_{t_0}^{t_1} \exp[F(t_1 - s)]Gu(s)\,ds$$

Then let v be defined over $[0, t_1 - t_0]$ by $v(s) = u(s + t_0)$; the preceding relation becomes

$$X_1 = \exp[F(t_1 - t_0)]X_0 + \int_0^{t_1 - t_0} \exp[F(t_1 - t_0 - s)]Gv(s)\,ds$$

and states that through the use of v $(0, X_0) \leadsto (t_1 - t_0, X_1)$. Conversely if v effects $(0, X_0) \leadsto (t_1 - t_0, X_1)$ we define u over $[t_0, t_1]$ by $u(s) = v(s + t_0)$ and u effects $(t_0, X_0) \leadsto (t_1, X_1)$.

Naturally, in general, $(t_0, X_0) \leadsto (t_1, X_1)$ is not equivalent to $(t_0, X_0 - X_1) \leadsto (t_1, 0)$ even for systems with constant coefficients. For example, let

$$\frac{dX}{dt} = \begin{pmatrix} 1 & 0 \\ 0 & 1 \end{pmatrix} X + \begin{pmatrix} 0 \\ u \end{pmatrix} \qquad X_0 = \begin{pmatrix} 1 \\ 1 \end{pmatrix} \qquad X_1 = \begin{pmatrix} 1 \\ 0 \end{pmatrix}$$

By choosing $u(s) \equiv [\exp(-t) - 1]^{-1}$ we see that $(0, X_0 - X_1) \leadsto (t, 0)$ whereas there exists no control enabling us to join X_0 to X_1.

We shall see that for a completely controllable system with constant coefficients, one can always join two points.

We shall use some elementary results in linear algebra which we shall now recall.

Résumé of linear algebra

If $M \in \mathscr{L}(\mathbb{R}^n, \mathbb{R}^n)$ we shall denote by $\mathscr{K}(M)$ the kernel of M, i.e.

$$\mathscr{K}(M) = \{V : MV = 0\} \qquad \text{(cf. p. 84)}$$

$\mathscr{K}(M) = 0$ iff M^{-1} exists.

We call the image of M the vector space of vectors W of the form MV. We denote this space $R(M)$, so that

$$R(M) = \{W; \exists V: MV = W\}$$

$R(M) = \mathbb{R}^n$ iff M^{-1} exists.

Let U be a vector orthogonal to all the elements of $R(M)$ so that

$$\forall W \in R(M) \qquad \langle U, W \rangle = 0$$

i.e.

$$\forall V \in \mathbb{R}^n \qquad \langle U, MV \rangle = 0$$

Therefore $\langle \tilde{M}U, V \rangle = 0$; consequently $\tilde{M}U = 0$ and $U \in \mathcal{K}(\tilde{M})$. Thus

$$\mathbb{R}^n = R(M) \oplus \mathcal{K}(\tilde{M}) \quad \text{and} \quad R(M) \perp \mathcal{K}(\tilde{M})$$

(cf. p. 85). If M is symmetric

$$\mathbb{R}^n = R(M) \oplus \mathcal{K}(M) \quad \text{and} \quad R(M) \perp \mathcal{K}(M)$$

Theorem 3.2.4

Let $dX/dt = F(t)X + G(t)u$. Let $\Phi(t, s)$ be the resolvent of $F(t)$ and

$$W(\tau, t) = \int_\tau^t \Phi(\tau, s)G(s)\tilde{G}(s)\tilde{\Phi}(\tau, s)\,ds$$

(τ, X) is controllable at the instant t iff $\exists t > \tau$ such that $X \in R[W(\tau, t)]$. The set of controllable points is then a vector space.

Proof Let $X = W(\tau, t)Y$ and $u = -\tilde{G}(t)\tilde{\Phi}(t, \tau)Y$.

$$\Phi(t, \tau)X + \int_\tau^t \Phi(t, s)G(s)u(s)\,ds$$

$$= \Phi(t, \tau)W(\tau, t)Y - \int_\tau^t \Phi(t, s)G(s)\tilde{G}(s)\tilde{\Phi}(\tau, s)Y\,ds$$

$$= 0$$

Every point of $R[W(\tau, t)]$ can thus be controlled at the instant t. To complete the proof, it remains then to verify that every point X of $\mathcal{K}[W(\tau, t)]$ is not controllable at the instant t. If $X \neq 0$ is such a point and if we assume that X is controllable at instant t, then there would exist a control u such that

$$0 = \Phi(t, \tau)X + \int_\tau^t \Phi(t, s)G(s)u(s)\,ds$$

and therefore

$$0 = X + \int_\tau^t \Phi(\tau, s)G(s)u(s)\,ds$$

so that

$$0 < \|X\|^2 = \langle X, X \rangle$$

$$= - \int_\tau^t \langle X, \Phi(\tau, s)G(s)u(s) \rangle \, ds$$

$$= - \int_\tau^t \langle \tilde{G}(s)\tilde{\Phi}(\tau, s)X, u(s) \rangle \, ds$$

Now $W(\tau, t)X = 0$ and therefore

$$0 = \langle X, W(\tau, t)X \rangle$$

$$= \int_\tau^t \langle X, \Phi(\tau, s)G(s)\tilde{G}(s)\tilde{\Phi}(\tau, s)X \rangle \, ds$$

$$= \int_\tau^t \|\tilde{G}(s)\tilde{\Phi}(\tau, s)X\|^2 \, ds$$

Thus $\tilde{G}(s)\tilde{\Phi}(\tau, s)X = 0$ for almost all s, which contradicts the relation

$$0 < - \int_\tau^t \langle \tilde{G}(s)\tilde{\Phi}(\tau, s)X, u(s) \rangle \, ds$$

Corollary 3.2.5

If $W(\tau, t)$ is invertible every pair (τ, X) is controllable at the instant t using the control

$$u(t) = - \tilde{G}(t)\tilde{\Phi}(\tau, t)W(\tau, t)^{-1}X$$

Theorem 3.2.6

Let F and G be constants and consider the system $dX/dt = FX + Gu$. The vector space of controllable points is generated by the columns of the matrix $\mathscr{C}$ defined by

$$\mathscr{C} = [G, FG, \ldots, F^{n-1}G]$$

This is an invariant subspace of the system. Thus the system is completely controllable if the rank of $\mathscr{C}$ is n. X is then controllable for every τ ($X \in \mathbb{R}^n$).

Proof According to the Cayley–Hamilton theorem there exist constants a_i such that $F^n = a_1 F^{n-1} + \ldots + a_{n-1}F + a_n I$. Thus $F^n G = a_1 F^{n-1}G + \ldots + a_n G$ so that, $\forall p \geqslant n$, $F^p G$ can be expressed as a function of $F^{n-1}G, \ldots, FG$ and G.

If X is a point which is not controllable starting from the instant 0 (e.g. because the controllability of X does not depend on the initial instant) then there exists t such that $X \in \mathscr{K}[W(0, t)]$, i.e.

$$0 = \langle X, W(0, t)X \rangle = \int_0^t \|\tilde{G} \exp(-\tilde{F}s)X\| \, ds$$

Therefore $\tilde{G}\exp(-\tilde{F}s)X$, which is a continuous function, is zero over $[0, t]$, and by differentiating it $n-1$ times we establish that

$$0 = \tilde{G}\tilde{F}\exp(-\tilde{F}s)X = \tilde{G}\tilde{F}^2\exp(-\tilde{F}s) = \ldots = \tilde{G}\tilde{F}^{n-1}\exp(-\tilde{F}s)$$

so that, for $s = 0$, $0 = \tilde{G}X = \tilde{G}\tilde{F}X = \ldots = \tilde{G}\tilde{F}^{n-1}X$. Now if M is any matrix, $\tilde{M}X = 0$ is equivalent to stating that X is orthogonal to the rows of $\tilde{M}$, i.e. to the columns of M. Thus X is orthogonal to the columns of $[G, FG, \ldots, F^{n-1}G] = \mathscr{C}$.

We shall show that, conversely, every vector orthogonal to the columns of $\mathscr{C}$ is uncontrollable; in fact such an element is orthogonal to the rows of the matrices $\tilde{G}, \tilde{G}\tilde{F}, \ldots, \tilde{G}\tilde{F}^{n-1}$ and therefore also to $\tilde{G}\tilde{F}^p$ $\forall p$, according to the argument at the beginning of this proof; it is therefore orthogonal to $\tilde{G}\exp(-\tilde{F}s)$ for all s and consequently belongs to $\mathscr{K}[W(0, t)]$ for all t.

We note finally that the vector space generated by $\mathscr{C}$ is an invariant of the system; if X_0 belongs to the space generated by $G, FG, \ldots, F^{n-1}G$, $X(t)$ belongs to this space for every control u and for t since

$$X(t) = \exp(Ft)\,X_0 + \int_0^t \exp[F(t-s)]\,Gu(s)\,\mathrm{d}s$$

and since

$$\exp(Ft)\,X_0 = X_0 + tFX_0 + \frac{t^2}{2!}F^2X_0 + \ldots + \frac{t^p}{p!}F^pX_0 + \ldots$$

can again be expressed as a function of $G, FG, \ldots, F^{n-1}G$.

Example

$$F = \begin{pmatrix} 1 & 1 \\ 0 & 3 \end{pmatrix} \qquad G = \begin{pmatrix} a \\ b \end{pmatrix} \qquad \mathscr{C} = \begin{pmatrix} a & a+b \\ b & 3b \end{pmatrix}$$

and the system is completely controllable iff $b(2a - b) \neq 0$.

Instead of speaking of a completely controllable system, we often speak of a **completely controllable pair** (F, G).

Comments

(i) Naturally, if X can be joined to 0 in an arbitrarily short time, we must expect the control u to be very large. It can be verified as an exercise, and we shall see in the following section that if we impose on u a constraint which limits its modulus (e.g. $\|u(t)\| \leqslant A \; \forall t$) a minimum time will then be needed to control a point.

(ii) We note that if we make a change of coordinates in $\mathbb{R}^n$ with transformation matrix P and in $\mathbb{R}^k$ with transformation matrix Q, then F becomes PFP^{-1}, G becomes PGQ^{-1} and $\mathscr{C}$ becomes $P\mathscr{C}Q^{-1}$. The rank therefore does not change. The notion of controllability is thus independent of the coordinate system (fortunately); it is related to the corresponding linear functions.

(iii) We can easily show that if (F_0, G_0) is completely controllable $\exists \varepsilon : \|F - F_0\| \leqslant \varepsilon$ and $\|G - G_0\| \leqslant \varepsilon$ implies that (F, G) is also completely controllable: every pair sufficiently close to a completely controllable pair is itself also completely controllable. (We are speaking throughout of constant pairs.)

(iv) When G is of rank n the system is completely controllable. This corresponds to the fact that n scalar controls $u_1, \ldots, u_n$ can then operate on the coordinates. It is unnecessary to take $k > n$; we should gain nothing as regards controllability.

The following corollary is fundamental.

Theorem 3.2.7

A scalar equation of order n with constant coefficients is always completely controllable with simple control, i.e. a system

$$\frac{d^n x}{dt^n} = a_0 x + a_1 \frac{dx}{dt} + \ldots + a_{n-1} \frac{d^{n-1}x}{dt^{n-1}} + u$$

is always completely controllable.

Proof The proof consists of verifying that the controllability matrix $\mathscr{C}$ is always of rank n. The equation is equivalent to the system

$$\frac{dX}{dt} = \begin{pmatrix} 0 & 1 & 0 & \cdots & \\ & & 1 & & \\ & & & \ddots & \\ & & & & 1 \\ a_0 & a_1 & & & a_{n-1} \end{pmatrix} X + \begin{pmatrix} 0 \\ \vdots \\ 0 \\ 1 \end{pmatrix} u = FX + Gu$$

It is easy to calculate $\mathscr{C} = (G, FG, \ldots, F^{n-1}G)$ directly since

$$FG = \begin{pmatrix} 0 \\ \vdots \\ 0 \\ 1 \\ a_{n-1} \end{pmatrix} \quad F^2 G = \begin{pmatrix} 0 \\ \vdots \\ 0 \\ 1 \\ a_{n-1} \\ a_{n-2} + a_{n-1}a_{n-2} \end{pmatrix} \quad \cdots \quad F^{n-1}G = \begin{pmatrix} 1 \\ a_{n-1} \\ \vdots \end{pmatrix}$$

Thus

$$\mathscr{C} = \begin{pmatrix} & & & & & 1 \\ 0 & & & & \cdot^{\cdot^{\cdot}} & \\ & & & 1 & & \\ & & 1 & & & \\ & 1 & & \cdot & & X \\ 1 & & & & & \end{pmatrix}$$

which is always of rank n.

We now introduce a notion related to that of controllability as the examples at the beginning of the section predicted.

Definition 3.2.8
A pair (τ, X) is called **reachable** if there exist a control u and an instant $s > \tau$ such that $X = \varphi(\tau, s, 0, u)$. A system is called **completely reachable** if every pair (τ, X) is reachable.

We can prove the following result by using a method analogous to the one we used to prove Theorem 3.2.4.

Theorem 3.2.9
Let

$$\tilde{W}(s, \tau) = \int_{s}^{\tau} \Phi(t, u)G(u)\tilde{G}(u)\tilde{\Phi}(\tau, u)\, du$$

and let $dX/dt = FX + Gu$ be a system with constant coefficients whose resolvent is Φ. A couple (τ, X) is reachable iff

$$\exists s < \tau: X \in R[\hat{W}(s, \tau)]$$

We can then deduce from this the following corollary analogous to Theorem 3.2.6.

Corollary 3.2.10
Let $dX/dt = FX + Gu$ be a system with constant coefficients where $X \in \mathbb{R}^n$. Let $\mathscr{C}$ be the controllability matrix. The vector space of reachable points is generated by the columns of $\mathscr{C}$; this is the same as the vector space of controllable points. A system with constant coefficients is completely controllable iff it is completely reachable.

Proof We note first that the columns of the controllability matrices of the two following systems obviously generate the same space of controllable points:

$$\frac{dX}{dt} = FX + Gu \tag{23}$$

$$\frac{dX}{dt} = -FX - Gu \tag{24}$$

For (23) let u be a control such that $(0, 0) \rightsquigarrow (t, X)$, i.e. such that

$$X = \int_0^t \exp[F(t-s)] \, Gu(s) \, ds$$

which can also be written

$$X = -\int_0^t \exp(Fv) \, Gu(t-v) \, dv$$

or

$$0 = \exp(-Ft) X + \int_0^t \exp[-F(t-v)] \, Gu(t-v) \, dv$$

If $u_1(v) = -u(t-v)$, defined over $[0, t]$, then

$$0 = \exp[(-F)t] \, X + \int_0^t \exp[(-F)(t-v)] \, (-G)u_1(v) \, dv$$

and u_1 is a control for (24) such that $(0, X) \rightsquigarrow (t, 0)$; in other words if X is reachable for (23) it is controllable for (24) and therefore also for (23), which gives the required result.

See *Exercises 17 to 21* at the end of the chapter.

STABILIZATION OF A SYSTEM WITH CONSTANT COEFFICIENTS

To be able to control the solutions of $dX/dt = FX + Gu$ over an infinite time, it is useful to find controls u such that X_u remains bounded; if F is asymptotically stable, we know that X_u is certainly bounded for every function u. We shall show that, if the system (F, G) is completely controllable, we can stabilize the system by means of a feedback control.

Definition 3.2.11

A pair (F, G) is called **stabilizable** if there exists a linear function K such that $F - GK$ is asymptotically stable.

The following theorem states not only that a completely controllable system can be stabilized by means of a feedback control, but that we can choose this control so that the eigenvalues of the matrix $F - GK$ take any values that we wish, a fact which enables us to obtain a convenient 'stability margin' and 'damping velocity'.

Theorem 3.2.12
Let $\Lambda = \bar{\Lambda}$ be a set of n points in the complex plane. For every completely controllable pair (F, G) in $\mathbb{R}^n$ there exists K such that the eigenvalues of $F - GK$ are the points of Λ. In particular we can then choose K so that the eigenvalues λ of $F - GK$ satisfy $\mathrm{Re}\, \lambda < \sigma < 0$; (F, G) is then stabilizable.*

Proof We know that if we make an invertible linear transformation P the eigenvalues of a matrix do not change; they are those of the corresponding linear function. The theorem rests on the fact that we can find P such that the new system has a particularly simple form, starting from which we can easily prove the result.

This theory of the 'canonical form' of controllable systems is quite an intricate subject in linear algebra; we shall use without proof the following fundamental result.

Theorem 3.2.13
Let (F, G) be a completely controllable system in which $G \in \mathbb{R}^n$ is a vector (and therefore u is a scalar); we can find an invertible linear transformation P such that the new system is a scalar differential equation of order n controlled by a simple control and which we know in addition to be completely controllable.
In other words, the transformed forms F_1 and G_1 of F and G are of the form

$$F_1 = PFP^{-1} = \begin{pmatrix} 0 & 1 & \cdots & \\ \vdots & & \ddots & \\ & & & 1 \\ a_0 & a_1 & & a_{n-1} \end{pmatrix}$$

$$G_1 = PG = \begin{pmatrix} 0 \\ \vdots \\ 0 \\ 1 \end{pmatrix}$$

A completely controllable system in which G is a vector is linearly equivalent to a scalar equation of order n with simple control.

We can then verify very easily that we can find $K_1 = (b_0, b_1, \ldots, b_{n-1})$ such that $F_1 - G_1 K_1$ has Λ as its set of eigenvalues.

---

* Naturally it is necessary that the points of Λ be the n roots of an equation of order n with real coefficients.

The characteristic equation is in fact

$$\det \begin{vmatrix} z & -1 & 0 & 0 & & 0 \\ 0 & z & -1 & 0 & & 0 \\ 0 & 0 & z & -1 & & \\ b_0 - a_0 & b_1 - a_1 & \cdots & & & z + b_{n-1} - a_{n-1} \end{vmatrix} = 0$$

We can show by recurrence, working with reference to the first column, that this equation is

$$z^{n+1} - (a_n - b_n)z^n + (b_{n-1} - a_{n-1})z^{n-1} + \ldots + (-1)^{n+1}(b_0 - a_0) = 0$$

(to order n)

The differences $a_n - b_n$ then represent symmetric functions of the roots and enable us to find b_k as a function of a_k for $k = 1, \ldots, n$.

For example in $\mathbb{R}^3$ this equation is

$$z^3 - z^2(a_2 - b_2) + z(b_1 - a_1) - (a_0 - b_0) = 0$$

and if λ_1, λ_2 and λ_3 are the points of Λ

$$a_2 - b_2 = \lambda_1 + \lambda_2 + \lambda_3$$
$$b_1 - a_1 = \lambda_1\lambda_2 + \lambda_2\lambda_3 + \lambda_1\lambda_3$$
$$a_0 - b_0 = \lambda_1\lambda_2\lambda_3$$

which naturally gives the values of b_0, b_1 and b_2.

We note that a stabilizable system is not necessarily controllable. For example if F is asymptotically stable and G is nul, the pair (F, G) is clearly stabilizable but it is not controllable.

We can prove in fact the following result. A system is controllable iff we can find K_+ and K_- such that $F - GK_-$ has eigenvalues with a (strictly) negative real part while those of $F - GK_+$ have a (strictly) positive real part.

We can interpret this result as follows: because of the equivalence of controllability and reachability, a couple is completely controllable iff it is stabilizable for $t \to +\infty$ and $t \to -\infty$.

Remark We can also prove the following less complete result. Let (F, G) be completely controllable and

$$W(0, T) = \int_0^T \exp(-Fs)\, G\tilde{G} \exp(-\tilde{F}s)\, \mathrm{d}s$$

Then $W(0, T)^{-1}$ exists, in view of Corollary 3.2.5. If $K = \tilde{G} \exp(\tilde{F}T)\, W(0, T)^{-1} \exp(FT)$, then the eigenvalues of $F - GK$ have a strictly negative real part.

See *Exercises 22 and 23* at the end of the chapter.

CONTROL, OVER AN INFINITE HORIZON, OF A SYSTEM WITH CONSTANT COEFFICIENTS

If $dX/dt = FX + Gu$ is a system with constant coefficients, we shall be able to solve the problem posed on p. 183, under certain conditions at least. We shall consider the problem of control over $[0, \infty]$ as the limit of control over $[0, T]$ when $T \to \infty$.

We shall suppose the matrices Q and R to be constant and we shall seek to minimize the expression

$$\int_0^\infty \langle X_u(t), Q X_u(t) \rangle + \langle u(t), Ru(t) \rangle \, dt$$

If $P_T(t)$ is the solution of the Riccati equation with respect to T, we shall show that it has a limit independent of t when $T \to \infty$ and that this limit enables us to construct an asymptotically stable system which minimizes the cost

$$\int_0^\infty \langle X_u(t), Q X_u(t) \rangle + \langle u(t), Ru(t) \rangle \, dt$$

Theorem 3.2.14

Consider the equation $dX/dt = FX + Gu$ in $\mathbb{R}^n$ and the cost

$$\int_0^\infty \langle X_u(t), Q X_u(t) \rangle + \langle u(t), Ru(t) \rangle \, dt = J_\infty(u, X_0)$$

We shall assume that F, G, Q and R are constant, Q and R are symmetric and R is strictly positive. Finally we shall assume that (F, G) is stabilizable (it is sufficient that it be controllable).

Consider the equation

$$-\frac{dP}{dt} = Q + PF + \tilde{F}P - PGR^{-1}\tilde{G}P \tag{25}$$

and denote by $P_T(t)$ the solution of (25) such that $P_T(T) = 0$. If $T \to \infty$, $P_T(t)$ has a limit P_∞ independent of t; this limit is a solution of the equation

$$E_\infty : Q + PF + \tilde{F}P - PGR^{-1}\tilde{G}P = 0$$

Then the system $dX/dt = (F - GR^{-1}\tilde{G}P_\infty)X$ is UAS and the control $u_\infty(t) = -R^{-1}\tilde{G}P_\infty X(t)$ minimizes the cost $J_\infty(u, X_0)$.

Finally $J_\infty(u, X_0) = \langle X_0, P_\infty X_0 \rangle$.

Proof Let $\hat{K}$ be such that $F - G\hat{K}$ is asymptotically stable ($\hat{K}$ exists, by hypothesis). Let $dX/dt = (F - G\hat{K})X$; the solution of this equation which equals X_0 at the point t_0 is

$$X_{\hat{K}}(s) = \exp[(F - G\hat{K})(s - t_0)] X_0$$

and the cost over $[0, T]$ corresponding to the control $\hat{K}X$ is thus

$$J_{\hat{K}}(t_0, X_0) = \int_{t_0}^{T} \langle X_0, \exp[(F - G\hat{K})(s - t_0)] (Q + \tilde{\hat{K}}RK)$$

$$\times \exp[(F - G\hat{K})(s - t_0)] X_0 \rangle \, ds$$

We shall denote this expression by $\langle X_0, \hat{P}_T(t_0)X_0 \rangle$.

Let $K(t)$ be the optimal control for (t_0, X_0) over (t_0, T); it entails a cost $\langle X_0, P_T(t_0)X_0 \rangle$ which is smaller than or equal to the preceding one:

$$\langle X_0, P_T(t_0)X_0 \rangle \leqslant \langle X_0, \hat{P}_T(t_0)X_0 \rangle$$

P_T and $\hat{P}_T$ do not depend on X_0, for which $P_T(t_0) \leqslant \hat{P}_T(t_0)$ in the sense that $\hat{P}_T(t_0) - P_T(t_0)$ is a positive symmetric matrix (cf. p. 174).

As $F - G\hat{K}$ is asymptotically stable, $\hat{P}_T(t_0)$ has a finite limit if $T \to \infty$. Thus $P_T(t_0)$ is bounded if $T \to \infty$ (in the sense that there exists $A > 0$ such that $A - P_T(t_0) > 0 \quad \forall T$).

We next verify that $P_T(t_0)$ is an increasing function of T; in fact, from the principle of dynamic programming

$$\langle X_0, P_T(t_0)X_0 \rangle \leqslant V_{T+h}(t_0, X_0) = \langle X_0, P_{T+h}X_0 \rangle \quad \text{for } h > 0$$

so that $P_T(t_0) \leqslant P_{T+h}(t_0)$. We know then (cf. p. 177) that

$$\lim_{T \to \infty} P_T(t) = P_\infty(t) \leqslant A$$

If $s \leadsto P_T(t + s)$, a function of s, this function is also a solution of (25) since Q, F, R and G are constants. Its value for $s = T - t$ is zero. This solution is therefore P_{T-t}. Thus

$$P_{T-t}(s) = P_T(t + s)$$

Therefore $P_{T-t}(0) = P_T(t)$ and

$$\lim_{T \to \infty} P_T(t) = \lim_{T \to \infty} P_{T-t}(0) = P_\infty(0)$$

which proves that $P_\infty(t) \equiv P_\infty(0)$ and therefore that P_∞ is a solution of E_∞.

We must now verify that $F_\infty = F - GR^{-1}\tilde{G}P_\infty$ is asymptotically stable. E_∞ can also be written $\tilde{F}_\infty P_\infty + P_\infty F_\infty + Q + P_\infty GR^{-1}GP_\infty = 0$.

If $S = Q + P_\infty GR^{-1}GP_\infty$, S is symmetric and $S > 0$. Thus the equation $\tilde{F}_\infty Z + ZF_\infty + S = 0$ has a solution $Z = P_\infty > 0$. We shall see that this implies that F_∞ is asymptotically stable. That will be the object of Lemma 3.2.15 which follows this proof.

It remains now to verify that u_∞ is optimal; this results again from the dynamic programming principle. In fact, if there were a better control, this would induce over every interval $(0, T)$ a smaller or equal cost and over at least one of these intervals a strictly smaller cost, which is contrary to the actual principle of construction of P_∞.

It remains to establish the lemma referred to.

Lemma 3.2.15

Let $Q > 0$ and symmetric. The equation $\tilde{F}Z + ZF + Q = 0$ has a symmetric solution $Z > 0$ iff F is asymptotically stable.

Proof

(a) The condition is sufficient. In fact, let

$$Z = \int_0^\infty \exp(\tilde{F}T) Q \exp(FT) \, dt$$

$Z = \tilde{Z}$ and Z is strictly positive because

$$\langle V, ZV \rangle = \int_0^\infty \exp(\tilde{F}T) \langle V, QV \rangle \exp(FT) \, dt > 0$$

Now

$$Z = \int_s^\infty \exp[\tilde{F}(u - s)] Q \exp[F(u - s)] \, ds$$

and

$$\frac{dZ}{ds} = 0 = -Q - \tilde{F}Z - ZF$$

(b) Let $Z = \tilde{Z} > 0$ be a solution of $\tilde{F}Z + ZF + Q = 0$. Let

$$dX/dt = FX \tag{26}$$

and $U(X) = \langle Z, ZX \rangle$, $X \in \mathbb{R}^n$. Since Z is symmetric and strictly positive, the surfaces $\{U(X) = a^2\}$ are closed surfaces containing 0, whose limit if $a \downarrow 0$ is $\{0\}$. (If $Z = I$ these surfaces are spheres of radius a.)

If $X(t)$ is a solution of (26)

$$\frac{dU[X(t)]}{dt} = \left\langle \frac{dX}{dt}, ZX \right\rangle + \left\langle X, Z\frac{dX}{dt} \right\rangle$$

$$= \langle X, (\tilde{F}Z + ZF)X \rangle$$

$$= -\langle X, QX \rangle$$

(which is strictly negative if $X \neq 0$) so that

$$\lim_{t \to \infty} U[X(t)] = 0 \quad \text{and} \quad \lim_{t \to \infty} X(t) = 0$$

Thus, every solution of (26) has 0 as limit, which means that F is asymptotically stable.

Study of the equation E_∞

In general the equation E_∞ has several solutions and it is necessary to know which one to choose. In the scalar case the situation is simple for E_∞ may be written

$$p^2 \frac{g^2}{r} - 2pf - q = 0$$

Since q and r are two positive numbers, there is one positive solution and a negative solution; clearly we choose the former.

In $\mathbb{R}^n$ the situation is in general more complicated. We shall indicate a situation in which we can nevertheless find

$$P_\infty = \lim_{T \to \infty} P_T(t)$$

We recall that, Q being symmetric and diagonalizable and also positive, its eigenvalues are positive, so that

$$Q = \tilde{A} \begin{pmatrix} \lambda_1^{1/2} & & \\ & \ddots & \\ & & \lambda_n^{1/2} \end{pmatrix} \begin{pmatrix} \lambda_1^{1/2} & & \\ & \ddots & \\ & & \lambda_n^{1/2} \end{pmatrix} A$$

Thus we can write $Q = \tilde{H}H$, and we can then prove the following theorem.

Theorem 3.2.16
If $(\tilde{F}, \tilde{H})$ is stabilizable (i.e. if there exists L such that $F + LH$ is asymptotically stable—we say that (H, F) is detectable) P_∞ is the only positive solution of E_∞.

Complements
(a) By definition (H, F) is detectable if $(\tilde{F}, \tilde{H})$ is stabilizable. It is therefore sufficient for this that $(\tilde{F}, \tilde{H})$ be controllable, which is equivalent to the fact that the rank of the following matrix be n:

$$\mathcal{S} = \begin{pmatrix} H \\ HF \\ \vdots \\ HF^{n-1} \end{pmatrix}$$

(b) Let P_0 be symmetric and positive and P_T be the solution of E such that $P(T) = P_0$. We can show that if (F, G) is stabilizable and (H, F) is detectable, the limit of $P_T(t)$ does not depend on P_0 and is always equal to P_∞. We can then construct an algorithm for the solution of E comparable with that on page 177.

If (F, G) is stabilizable and (H, F) is not detectable, we know that

$$\lim_{T \to \infty} P_T(t)$$

always exists, but it depends on P_0. P_∞ is the limit of $P_T(t)$ for $P_0 = 0$. We can then again construct an algorithm for the limit, but we must realize that there exist other positive symmetric solutions of E_∞. If (F, G) is not stabilizable, the situation is still more complicated. If (H, F) is detectable

$$\lim_{T \to \infty} P_T(t)$$

does not exist and E_∞ does not have positive symmetric solutions, whereas if (H, F) is not detectable it can happen that

$$\lim_{T \to \infty} P_T(t)$$

exists or does not exist. It can also happen that positive symmetric solutions of E_∞ exist or do not exist.

See *Exercises 24 to 26* at the end of the chapter.

3.3. Adjoint equation; optimality criterion; bounded controls

In real problems we can clearly use only bounded controls. Under these conditions, starting from a specified point we can reach only a bounded set of points in a prescribed time. In the same way, if two points can be joined by a solution, this can be done only in a time greater than or equal to some non-zero minimum time.

We shall study controls u such that $\forall t\ u(t) \in \Omega$ where Ω is a closed bounded set. We shall denote by $K(t)$ the set of points which can be reached, starting from (t_0, X_0), at the end of time t and we shall study these sets. This study will lead us to the notions of adjoint state and adjoint equation.

This will enable us, as a first application, to study the minimum time necessary to reach a point and to determine the corresponding control.

We shall state first some general theorems which allow the determination of optimal controls in many cases.

SET OF POINTS REACHABLE WITH u BOUNDED; ADJOINT STATE

Theorem 3.3.1

Let $dX/dt = F(t)X + G(t)u + v(t)$ be a system in $\mathbb{R}^n$ where F, G and v are continuous, controlled by functions u such that $u(t) \in \Omega$ where Ω is a convex bounded closed set. Let (t_0, X_0) be fixed and $K(t)$ be the set of points reachable, starting from (t_0, X_0), at the end of time t using a control u. Then $K(t)$ is a convex closed bounded set which varies continuously with t.

Proof Let $\Phi(t, s)$ be the resolvent of F:

$$X(t) = \Phi(t, t_0)X_0 + \int_{t_0}^{t} \Phi(t, s)[G(s)u(s) + v(s)]\, ds$$

With T fixed, we shall consider the equation over the interval $[t_0, T]$. Let

$$M = \sup_{t_0 \leqslant s \leqslant T} \|F(s)\| \qquad N = \sup_{t_0 \leqslant s \leqslant T} \|G(s)\|$$

$$V = \sup_{t_0 \leqslant s \leqslant T} \|v(s)\| \qquad K = \sup_{t_0 \leqslant s \leqslant T} \|u(s)\|$$

Then

$$\|X(t)\| \leqslant \exp[M(T - t_0)]\|X_0\| + \int_{t_0}^{T} \exp[M(t - s)](NK + V)\,\mathrm{d}s = B < \infty$$

$K(t)$ is therefore bounded.

We shall show that $K(t)$ is convex. In fact if X_1 and X_2 are two points of $K(t)$, by definition there exist u_1 and u_2 such that

$$X_1 = \Phi(t, t_0)X_0 + \int_{t_0}^{t} \Phi(t, s)[G(s)u_1(s) + v(s)]\,\mathrm{d}s$$

$$X_2 = \Phi(t, t_0)X_0 + \int_{t_0}^{t} \Phi(t, s)[G(s)u_2(s) + v(s)]\,\mathrm{d}s$$

Therefore

$$\lambda X_1 + (1 - \lambda)X_2$$

$$= \Phi(t, t_0)X_0 + \int_{t_0}^{t} \Phi(t, s)\{G(s)[\lambda u_1(s) + (1 - \lambda)u_2(s)] + v(s)\}\,\mathrm{d}s$$

Since Ω is convex, $\lambda u_1 + (1 - \lambda)u_2$ is a control with values in Ω for $\lambda \in [0, 1]$.

To verify that $K(t)$ is closed, we must verify that the limit of a convergent sequence $\{X_n\}_{n \in \mathbb{N}}$ of points of $K(t)$ belongs to $K(t)$. If $\{X_n\}_{n \in \mathbb{N}}$ is such a sequence, determined by a sequence $\{u_n\}_{n \in \mathbb{N}}$ of controls,

$$X_n = \Phi(t, t_0)X_0 + \int_{t_0}^{t} \Phi(t, s)[G(s)u_n(s) + v(s)]\,\mathrm{d}s$$

We can show that, Ω being bounded and closed and $u_n(s)$ belonging to Ω for $t_0 \leqslant s \leqslant T$, there exists a subsequence $\{u_{n_k}\}_{k \in \mathbb{N}}$ extracted from the sequence u_n and a function u such that $u(s) \in \Omega$ and that

$$\lim_{k \to \infty} \int_{t_0}^{t} h(s)u_{n_k}(s)\,\mathrm{d}s = \int_{t_0}^{t} h(s)u(s)\,\mathrm{d}s$$

for every bounded continuous function h (we say that the sequence u_n is weakly compact).

Then

$$X = \lim_{n \to \infty} X_n = \lim_{k \to \infty} X_{n_k}$$

and this limit is

$$\Phi(t, t_0)X_0 + \int_{t_0}^t \Phi(t, s)[G(s)u(s) + v(s)]\, ds$$

which belongs to $K(t)$.

Let us now define what is meant by the statement that $K(t)$ varies continuously with t. This phrase states that $\forall \varepsilon > 0$ there exists $\eta: |t_1 - t_2| \leqslant \eta$ implies that the distance between $K(t_1)$ and $K(t_2)$ is smaller than or equal to ε, i.e. that

$$\forall X_1 \in K(t_1) \quad \exists X_2 \in K(t_2): \|X_1 - X_2\| \leqslant \varepsilon$$

and conversely. Let

$$X_1 \in K(t_1): X_1 = \Phi(t_1, t_0)X_0 + \int_{t_0}^{t_1} \Phi(t, s)[G(s)u_1(s) + v(s)]\, ds$$

Then let

$$X_2 = \Phi(t_2, t_1)X_1 + \int_{t_1}^{t_2} \Phi(t_2, s)[G(s)u_1(s) + v(s)]\, ds \qquad X_2 \in K(t_2)$$

since X_2 is the point reached at the instant t_2 by continuing to use the control u_1. Then

$$\|X_1 - X_2\| \leqslant \|X_1 - \Phi(t_2, t_1)X_1\| + \|X_2 - \Phi(t_2, t_1)X_1\|$$

$$\leqslant \|I - \Phi(t_2, t_1)\| \, \|X_1\| + \int_{t_1}^{t_2} \|\Phi(t_2, s)\|(NK + V)\, ds$$

Since the resolvent is uniformly continuous over $(t_0, T) \times (t_0, T)$, the first term can be made less than or equal to $\varepsilon/2$ for $|t_2 - t_1| \leqslant \eta_1$, and the second term can be made less than or equal to $\varepsilon/2$ for $|t_2 - t_1| \leqslant \eta_2$, which establishes the result.

We shall now introduce the notion of adjoint state.

Suppose we are seeking a control which will allow us to join (t_0, X_0) to the origin in a minimum time—call this time ϑ. It is clear that $0 \in \partial K(\vartheta)$, the boundary of $K(\vartheta)$. Controls which lead to points on the boundary $\partial K(t)$ of $K(t)$ thus play a particular role; in most cases they are optimal controls. We shall characterize them by introducing the adjoint equation in which the control does not appear, a fact which is fundamental.

Theorem 3.3.2

We consider, throughout, the system

$$\frac{dX}{dt} = F(t)X + G(t)u + v(t)$$

We assume that $u(t) \in \Omega$ which is convex, closed and bounded. Let u_0 be a control such that $K(s)$ is strictly convex (cf. p. 179).

(a) Let $\eta(s)$, $s \in [t_0, T]$, be a vector of $\mathbb{R}^d$ such that

$$\forall s \in [t_0, T] \qquad \frac{d\eta}{ds} = -\tilde{F}(s)\eta$$

and

$$\langle \eta(s), G(s)u_0(s) \rangle = \max_{u \in \Omega} \langle \eta(s), G(s)u \rangle$$

Then $X_{u_0}(s) \in \partial K(s)$ and $\eta(s)$ is an outward normal to $\partial K(s)$ at $X_{u_0}(s)$ $\forall s \in [t_0, T]$.

(b) Conversely, if $X_{u_0}(s) \in \partial K(s)$ $\forall s \in [t_0, T]$ there exists a non-zero solution η of

$$\frac{d\eta}{ds} = -\tilde{F}(s)\eta$$

and

$$\langle \eta(s), G(s)u_0(s) \rangle = \max_{u \in \Omega} \langle \eta(s), G(s)u \rangle$$

The equation $d\eta/ds = -\tilde{F}(s)\eta$ is called the **adjoint equation** and η is the **adjoint state**.

Proof
(a) Let $X_{u_0}(t_0) = X_0$ and $\eta(t_0) = \eta_0$; then

$$X_{u_0}(s) = \Phi(s, t_0)X_0 + \int_{t_0}^s \Phi(s, \vartheta)[G(\vartheta)u_0(\vartheta) + v(\vartheta)] \, d\vartheta$$

and

$$\eta(s) = -\tilde{\Phi}(t_0, s)\eta_0$$

(where Φ always denotes the resolvent of F). Thus

$$\langle \eta(s, X_{u_0}(s) - X_u(s) \rangle = \left\langle \tilde{\Phi}(t_0, s)\eta_0, \int_{t_0}^s \Phi(s, \vartheta)G(\vartheta)[u_0(\vartheta) - u(\vartheta)] \, d\vartheta \right\rangle$$

$$= \int_{t_0}^s \langle \eta(\vartheta), G(\vartheta)[u_0(\vartheta) - u(\vartheta)] \rangle \, d\vartheta$$

If

$$\langle \eta(\vartheta), G(\vartheta)u_0(\vartheta) \rangle = \max_{u \in \Omega} \langle \eta(\vartheta), G(\vartheta)u \rangle$$

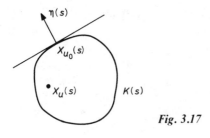

Fig. 3.17

the last integral is positive or zero for every control u, so that

$$\langle \eta(s), X_{u_0}(s) - X_u(s) \rangle \geqslant 0$$

which is possible only if $X_{u_0}(s) \in \partial K(s)$ and $\eta(s)$ is an outward normal to $K(s)$ at this point (Fig. 3.17). In fact

(i)   if $X_{u_0}(s)$ is within $K(s)$, whatever the vector V may be, the scalar product $\langle V, X_{u_0}(s) - X_u(s) \rangle$ does not have fixed sign as X_u goes through $K(s)$;
(ii)   if $M \in \partial K(s)$, an outward normal η is characterized exactly by the fact that the sign of $\langle \eta, M - P \rangle$ is fixed when P goes through $K(s)$.

(b) Let u_0 be a control such that $X_{u_0}(s) \in \partial K(s)$ $\forall s \in [t_0, T]$, let $t_1 \in]t_0, T[$ and η_1 be the normal to $K(t_1)$ at the point $X_{u_0}(t_1)$ of $\partial K(t_1)$ and let $\eta(t)$ be the solution of $d\eta/dt = -\tilde{F}(t)\eta$ such that $\eta(t_1) = \eta_1$. In addition let us define the control $\hat{u}(s)$ by

$$\langle \eta(s), G(s)\hat{u}(s) \rangle = \max_{u \in \Omega} \langle \eta(s), G(s)u \rangle$$

(It would be necessary to verify that this equation correctly defines an admissible control $\hat{u}$—we shall accept this.)
According to the preceding calculation

$$\langle \eta(t_1), X_{u_0}(t_1) - X_{\hat{u}}(t_1) \rangle = \int_{t_0}^{t_1} \langle \eta(\vartheta), G(\vartheta)[u_0(\vartheta) - \hat{u}(\vartheta)] \rangle \, d\vartheta$$

The left-hand side is negative or zero since $\eta(t_1)$ is an outward normal, and the right-hand side is positive or zero by the definition of $\hat{u}$.
Necessarily $X_{u_0}(t_1) = X_{\hat{u}}(t_1)$ for all t_1 of $[t_0, T]$ and $u_0(\vartheta) = \hat{u}(\vartheta)$ over $[t_0, t_1]$, which proves that

$$\langle \eta(s), G(s)u_0(s) \rangle = \max_{u \in \Omega} \langle \eta(s), G(s)u \rangle$$

We have shown that an optimal control leads in general to the boundary $\partial K(t)$ of $K(t)$; it is important to know in what case there exists a single control leading to a given point on the boundary. For this purpose, we introduce the following notion.

Definition 3.3.3

The control problem associated with the equation

$$\frac{dX}{dt} = F(t)X + G(t)u + v(t)$$

with the initial pair (t_0, X_0) and the set of controls $u: u(s) \in \Omega \;\; \forall s \in [t_0, t_1]$ (where Ω is bounded and closed) is called **normal** if two controls u_1 and u_2 which lead to the same point X_1 on the boundary $\partial K(t_1)$ are equal for almost every t in $[t_0, t_1]$.

We shall give, almost without proof, the two following results.

Theorem 3.3.4

The preceding problem is normal iff for every non-zero solution of $d\eta/dt = -\tilde{F}(t)\eta$ and every pair of controls u_1 and u_2 such that

$$\langle \eta(t), G(t)u_1(t) \rangle = \langle \eta(t), G(t)u_2(t) \rangle$$

$$= \max_{u \in \Omega} \langle \eta(t), G(t)u \rangle$$

$u_1(s)$ and $u_2(s)$ are equal for almost every s in $[t_0, t_1]$.

Theorem 3.3.5

If a problem is normal over $[t_0, t_1]$ it is normal over $[t_0, \tau] \;\; \forall \tau \in [t_0, t_1]$; if Ω contains more than one point and if the problem is normal $K(t_1)$ is strictly convex.

We recall that in the statement of Theorem 3.3.2 we assume precisely that $K(s)$ is strictly convex for all s. This condition is then satisfied when the problem is normal. We shall see that for equations with constant coefficients there exist sufficient conditions for a problem to be normal.

We shall now establish that we can remove the hypothesis that Ω is convex.

Definition 3.3.6

Let Ω be a bounded set. We define the **convex closed envelope of Ω** as the smallest convex closed set containing Ω. We shall denote this envelope by $H(\Omega)$.

Examples The convex closed envelope of a set of three points in $\mathbb{R}^2$ is the triangle whose vertices are the points shown in Fig. 3.18.

The following theorem is of great importance. It states for example that controls taking their values at the vertices A, B and C of the triangle $H(\Omega)$ suffice to describe $K_{H(\Omega)}(t)$; in particular, optimal controls which lead to the boundary of $K(t)$ take their values at boundary points of Ω.

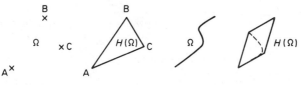

Fig. 3.18

Theorem 3.3.7
Let us consider a control problem which is normal and in which Ω is closed and bounded; if $K(t)$ is the set of points reached at instant t by controls taking their values in Ω, and if $K_{H(\Omega)}(t)$ is the set of points reached at instant t by controls taking their values in $H(\Omega)$, then

$$K(t) = K_{H(\Omega)}(t)$$

This theorem is often called the **bang-bang control principle**, a control taking its values at 'extreme' points being called a bang-bang control.
See *Exercises* **27** *and* **28** at the end of the chapter.

REACHING A POINT IN MINIMUM TIME

We are concerned with the following problem. Given (t_0, X_0), find a control u taking its values in Ω and joining X_0 to the origin in the shortest possible time. We shall establish that this is an immediate application of the preceding results.

We note at once that, if the origin can be reached, there actually exists a minimum time; in fact let $\mathcal{T}$ be the set of instants t such that $K(t)$ contains 0. Since $X_0 \neq 0$ this set is bounded below, and since $K(t)$ varies continuously it is closed. $\mathcal{T}$ therefore has a minimal element, which is the time required.

We shall call the set

$$\partial K(t) \cap_{s<t} K(s)^c$$

the 'new boundary of $K(t)$'. This is then the set of points on the boundary of $K(t)$ which do not belong to any set $K(s)$ for $s < t$. Figure 3.19 shows such a set.

Every point of $K(t)$ which does not belong to the new boundary belongs to $K(s)$ when s is sufficiently near to t ($s < t$). Consequently if T denotes the minimum time required to reach 0, 0 is necessarily a point on the new

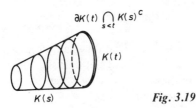

$\partial K(t) \cap_{s<t} K(s)^c$

$K(t)$

$K(s)$

Fig. 3.19

boundary of $K(T)$. The corresponding optimal control then satisfies the maximum principle stated in Theorem 3.3.2.

Without giving a proof we shall state the following theorem.

Theorem 3.3.8

Let $dX/dt = F(t)X + G(t)u + v(t)$ be a differential equation in which u is a control such that $u(t) \in \Omega$ (Ω closed and bounded). Let u_0 be a control allowing us to reach the origin, starting from (t_0, X_0), in a minimum time T. Let $X_{u_0}(t)$ be the corresponding solution. We shall assume that the problem is normal and shall denote by $\eta(t)$ the solution of $d\eta/dt = -\tilde{F}(t)\eta$ such that $\eta(T)$ is the external normal to $K(T)$ at the point 0. Then

$$\forall t_0 \leqslant t \leqslant T \quad 0 \leqslant M(t) = \langle \eta(t), F(t)X_0(t) + G(t)u_0(t) + v(t) \rangle$$
$$= \max_{u \in \Omega} \langle \eta(t), F(t)X_0(t) + G(t)u + v(t) \rangle$$

If Ω has a particular form it is often possible to determine the maximum in question, particularly if $\Omega = \{u = (u_1, \ldots, u_k) \ \forall p \leqslant k, |u_p| \leqslant 1\}$ for which u_0 will necessarily take its values at the vertices.

In this connection we shall state without proof the following theorem.

Proposition 3.3.9

Let $dX/dt = FX + Gu + v$ be a system with constant coefficients. Let Ω be a convex polyhedron with the following property: every vector W parallel to one of its edges is such that the rank of $[GW, FGW, \ldots, F^{n-1}GW]$ is n. The problem is then normal. There exists one and only one optimal control: it is piecewise constant and its values are given by the vertices of Ω.

Hence we deduce the following theorem.

Theorem 3.3.10

Let (F, G) be a constant completely controllable pair. Let $\Omega = \{u: |u_i| \leqslant 1 \ \forall i\}$ be the cube of side 2 centred at the origin of $\mathbb{R}^n$.

Let

$$\frac{dX}{dt} = FX + Gu \tag{27}$$

be an equation in $\mathbb{R}^n$. There exists a control u_0 enabling us to reach the origin, starting from a point X_0, in a minimum time; $u_0(s)$ is piecewise continuous and its coordinates have absolute value 1. This control and the corresponding adjoint state are solutions of the system

$$\frac{d\eta}{ds} = -\tilde{F}\eta \tag{28}$$

$$\langle \eta(s), Gu_0(s) \rangle = \max_{u \in \Omega} \langle \eta(s), Gu \rangle \tag{29}$$

Practical calculation

Equation (28) permits us to calculate η as a function of $\eta(t_0)$ and Eqn (29) gives u and, if $u \in \mathbb{R}$ for example, the changes of sign in u.

We determine in every case the curves corresponding to $u(s) = $ constant, since we necessarily arrive at zero along one of these curves. Thus if $u \in \mathbb{R}$ we have two curves, Γ^+ corresponding to $u \equiv +1$ and Γ^- corresponding to $u \equiv -1$.

In some cases we can construct the control u in the following way. Let T be the time at which the origin is reached starting from X_0 and $v(s) = \eta(T-s)$, $Y(s) = X(T-s)$, $V_0(s) = u_0(T-s)$. Then $dv/ds = \tilde{F}v$, $dY/ds = -FY - GV_0$ and

$$\langle \eta(s), Gu_0(s) \rangle = \langle v(T-s), GV_0(t-s) \rangle$$

$$= \max_{u \in \Omega} \langle v(T-s), Gu \rangle$$

We seek v first, and then u. Then we seek $v(0)$ such that $Y(T) = X_0$.

Example   Let $d^2x/dt^2 = u$ and $\Omega = [-1, +1]$. Let $X_0 = (x_0, y_0)$. The corresponding system is

$$\frac{dX}{dt} = \begin{pmatrix} 0 & 1 \\ 0 & 0 \end{pmatrix} X + \begin{pmatrix} 0 \\ 1 \end{pmatrix} u(t)$$

which is completely controllable.

(a) After we reverse the time the adjoint equation is

$$\frac{dv}{dt} = \begin{pmatrix} 0 & 0 \\ 1 & 0 \end{pmatrix} v$$

whose solutions are

$$v_1(t) = v_1(0)$$

$$v_2(t) = v_2(0) + tv_1(0)$$

$\langle v(t), Gu(t) \rangle = v_2(t)u(t)$, whose maximum over all possible values of u is $|v_2(t)|$ which corresponds to $u(t) = \text{sgn } v_2(t)$.

$u(t)$ has absolute value 1 and changes value at most once.

(b) Consider $dY/dt = -FY - Gu(t)$; let

$$Y = \begin{pmatrix} \tilde{x} \\ \tilde{y} \end{pmatrix}$$

This equation is

$$\frac{d\tilde{x}}{dt} = -\tilde{y} \qquad \frac{d\tilde{y}}{dt} = -u(t) = -\text{sgn } v_2(t)$$

The initial value of Y is 0. First let us examine the set of points reached when v_2 does not change sign.

Let Γ^+ be the set of points with the control $u(t) \equiv +1$. These are the points

$$\tilde{y}(t) = -t \qquad \tilde{x}(t) = \frac{t^2}{2} = \frac{\tilde{y}^2}{2} \; (t)$$

i.e. the points $\tilde{x} = \tilde{y}^2/2$, $\tilde{y} < 0$: Γ^+.

In the same way, let Γ^- be the set of points reached when $u(t) \equiv -1$, i.e. $\tilde{x} = -\tilde{y}^2/2$, $\tilde{y} > 0$: Γ^-.

Let S^+ be the set of points reached with $u(t) \equiv +1$ and then $u(t) \equiv -1$. The corresponding solutions terminate in zero along Γ^-. Before this, there are curves which do not pass through the origin, and consequently

$$\tilde{y} = a - t \qquad \tilde{x} = -at + \frac{t^2}{2}$$

$$\tilde{x} = \frac{\tilde{y}^2}{2} - \frac{a^2}{2} : S^+ \qquad (a \in \mathbb{R} \text{ and variable})$$

Also let S^- be the set of points reached with $u(t) \equiv -1$ and then $u(t) \equiv +1$; then

$$\tilde{x} = \frac{\tilde{y}^2}{2} + \frac{a^2}{2} : S^- \qquad (a \in \mathbb{R} \text{ and variable})$$

(c) The solutions are obtained by reversing the time and are represented in Fig. 3.20; we start with a curve S^+ or S^-; when we meet Γ^- or Γ^+ we change the sign of u and reach the origin by following this last curve.

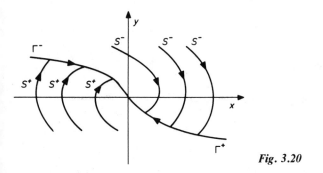

Fig. 3.20

The curve $\Gamma = \Gamma^+ \cup \Gamma^-$ is the curve where u eventually changes sign. It is called the 'switching locus' (Fig. 3.20). Since (x_0, y_0) are fixed, we choose $a^2 = 2x_0 - y_0^2$ or $a^2 = y_0^2 - 2x_0$ according to the sign. Then we calculate T.

See **Exercises 29 to 33** at the end of the chapter.

INTEGRAL COST AND GENERAL STATEMENT

We shall study systems for which the cost to be minimized is of the form

$$g[X(T)] + \int_{t_0}^{T} f(t, x) + h(t, u)\, dt$$

This is an example of an integral cost.

We shall give only one proof, that of the simplest theorem. It will enable us to introduce what is called a '**variational method**'.

It is significant that the general statement which is the subject of Theorem 3.3.11 is of the same type as that of Theorem 3.3.8, i.e. it involves the adjoint equation and a maximum principle. This is a result of the fact that optimal controls, in this situation, and even in general situations, lead to points on the boundary $\partial K(t)$ of $K(t)$.

The only theorem which we shall prove is the following.

Theorem 3.3.11
Let

$$\frac{dX}{dt} = FX + Gu \tag{30}$$

be a differential equation in $\mathbb{R}^n$ with constant coefficients and let

$$J_u(0, X_0) = \int_0^1 h[X_u(s), u(s)]\, ds$$

be an integral cost. Let

$$\frac{d\eta}{dt} = -\tilde{F}\eta + \frac{\partial \tilde{h}}{\partial X}(X_u, u) \tag{31}$$

be the adjoint equation in which $\eta \in \mathbb{R}^n$, $\partial h/\partial X \in \mathcal{L}(\mathbb{R}^n, \mathbb{R})$ and therefore $\partial \tilde{h}/\partial X \in \mathbb{R}^n$. Let u_0 be an optimal control, X_{u_0} be the corresponding solution and η_0 be the solution of (31) such that $\eta_0(1) = 0$. Let $H(\eta, X, u) = \langle \eta, FX + Gu \rangle - h(X, u)$. Then

$$H[\eta_0(t), X_0(t), u_0(t)] = \max_{u \in \mathbb{R}^n} H[\eta_0(t), X_0(t), u]$$

This is called the **maximum principle**.

Comments
(i) If T is the minimum time necessary to reach the origin starting from a point X_0, Theorem 3.3.8 can be stated in the same terms as this, by considering the cost

$$\int_0^T 1(s)\, ds$$

(ii) With the notation indicated, Eqns (30) and (31) may be written

$$\frac{dX}{dt} = \frac{\partial \tilde{H}}{\partial \eta} (X_u, u) \tag{30'}$$

$$\frac{d\eta}{dt} = -\frac{\partial \tilde{H}}{\partial X} (X_u, u) \tag{31'}$$

Proof   The variational method here consists in stating that the cost related to an optimal control $u_0(t)$ is smaller than that related to controls $u_\varepsilon(t) = u_0(t) + \varepsilon f(t)$ where f is a fixed continuous function over $[0, 1]$.

We shall denote the cost related to u_ε by J_ε. We shall calculate $(\partial J_\varepsilon/\partial \varepsilon)_{\varepsilon = 0}$ and equate this derivative to zero; this condition is necessary if J_{u_0} is to be an extremum. We shall accept that it is sufficient for a minimum.

Let $X_\varepsilon(t)$ be the solution corresponding to $u_0 + \varepsilon f$:

$$\frac{dX_\varepsilon}{dt} = FX_\varepsilon + Gu_0 + \varepsilon Gf(t)$$

We know that $(\partial X_\varepsilon/\partial \varepsilon)_{\varepsilon = 0}$ is the solution of the variational equation $dz/dt = Fz + Gf(t)$ such that $z(0) = 0$ (cf. Chapter 1, p. 50). Thus

$$\left. \frac{\partial X_\varepsilon}{\partial \varepsilon} \right|_{\varepsilon = 0} = \int_0^t \exp[F(t - s)] Gf(s) \, ds$$

Then

$$\left. \frac{\partial J_\varepsilon}{\partial \varepsilon} \right|_{\varepsilon = 0} = \int_0^1 \frac{\partial}{\partial \varepsilon} h[X_\varepsilon(s), u_\varepsilon(s)] \bigg|_{\varepsilon = 0} ds$$

$$= \int_0^1 \left\langle \frac{\partial \tilde{h}}{\partial X} [X_0(s), u_0(s)], \frac{\partial X_\varepsilon}{\partial \varepsilon}(s) \right\rangle ds$$

$$+ \int_0^1 \left\langle \frac{\partial \tilde{h}}{\partial u} [X_0(s), u_0(s)], f(s) \right\rangle ds$$

Let I be the first integral:

$$I = \int_0^1 \left\langle \frac{\partial \tilde{h}}{\partial X} [X_0(s), u_0(s)], \int_0^s \exp[F(s - v)] Gf(v) \, dv \right\rangle ds$$

$$= \int_0^1 \left\langle \exp(\tilde{F}s) \frac{\partial \tilde{h}}{\partial X} [X_0(s), u_0(s)], \int_0^s \exp(-Fv) Gf(v) \, dv \right\rangle ds$$

Let

$$\varphi(t) = \int_0^t \exp(\tilde{F}s) \frac{\partial \tilde{h}}{\partial X} [X_0(s), u_0(s)] \, ds$$

and

$$\psi(t) = \int_0^t \exp(-Fv)\, Gf(v)\, dv$$

The preceding quantity is

$$\int_0^1 \langle \varphi'(s), \psi(s) \rangle\, ds$$

$$= [\langle \varphi(t), \psi(t) \rangle]_0^1 - \int_0^1 \langle \varphi(t), \psi'(t) \rangle\, dt$$

$$= \left\langle \int_0^1 \exp(\tilde{F}v)\frac{\partial \tilde{h}}{\partial X}[X_0(v), u_0(v)]\, dv, \int_0^1 \exp(-Fv)\, Gf(v)\, dv \right\rangle$$

$$- \int_0^1 \left\langle \int_0^t \exp(\tilde{F}s)\frac{\partial \tilde{h}}{\partial X}[X_0(s), u_0(s)]\, ds, \exp(-Ft)\, Gf(t) \right\rangle dt$$

Now

$$\eta_0(s) = -\exp(\tilde{F}s)\int_0^1 \exp(\tilde{F}v)\frac{\partial \tilde{h}}{\partial X}[X_0(v), u_0(v)]\, dv$$

$$+ \int_0^s \exp[-\tilde{F}(s-v)]\frac{\partial \tilde{h}}{\partial X}[X_0(s), u_0(s)]\, ds$$

The integral I then equals

$$- \int_0^1 \langle \tilde{G}\eta_0(t), f(t) \rangle\, dt$$

so that

$$\left.\frac{\partial J_\varepsilon}{\partial \varepsilon}\right|_{\varepsilon=0} = - \int_0^1 \left\langle \tilde{G}\eta_0(s) - \frac{\partial \tilde{h}}{\partial u}[X_0(s), u_0(s)], f(s) \right\rangle ds$$

Now

$$\frac{\partial H}{\partial u}(\eta_0, X_0, u_0) = \tilde{G}\eta_0 - \frac{\partial \tilde{h}}{\partial u}(X_0, u_0)$$

The condition

$$\left.\frac{\partial J_\varepsilon}{\partial \varepsilon}\right|_{\varepsilon=0} = 0$$

is then

$$\frac{\partial H}{\partial u}(\eta_0, X_0, u) = 0$$

and we see that this states that $H[\eta_0(t), X_0(t), u_0(t)]$ is the maximum of $H[\eta_0(t), X_0(t), u]$, which is the principle stated.

We can easily see how this theorem can be extended to more general cases (naturally at the price of certain technical difficulties).

We shall state one of these generalizations but shall confine ourselves to linear equations.

Theorem 3.3.12

Let

$$\frac{dX}{dt} = F(t)X + G(t)u \tag{32}$$

be a differential equation in $\mathbb{R}^n$ and let

$$J_u(t_0, X_0) = g[X(T)] + \int_{t_0}^{T} f(t, X) + h(t, u)\, dt$$

where g is convex and of class C^1 and h is strictly convex. We shall consider controls such that $u(t) \in \Omega$ where Ω is a closed bounded set in $\mathbb{R}^k$.

Let

$$\frac{d\eta}{dt} = -\tilde{F}(t)\eta + \frac{\partial \tilde{f}}{\partial X}(t, X) \tag{33}$$

be the adjoint equation and

$$H(\eta, X, u) = \langle \eta, FX + Gu \rangle - h(t, u)$$

Then there exists an optimal control u_0, a solution X_0 and an adjoint state η_0 such that

$$X_{u_0}(t_0, X_0) = X_0 \qquad \eta_0(T) = -\operatorname{grad} g[X_{u_0}(T)]$$

and

$$H[\eta_0(t), X_0(t), u_0(t)] = \max_{u \in \Omega} H[\eta_0(t), X_0(t), u]$$

Particular case   Let $Q(t)$ and $R(t)$ be positive symmetric matrices and $R(t)$ be invertible and let $f(t, X) = \langle X, Q(t)X \rangle$ and $h(t, u) = \langle u, Ru \rangle$. We then obtain the following result which generalizes Theorems 3.1.2 and 3.1.3.

Corollary 3.3.13

Let $dX/dt = F(t)X + G(t)u$ be a differential equation in $\mathbb{R}^n$ and

$$J_u(t_0, X_0) = g[X(T)] + \int_{t_0}^{T} \langle X(t), Q(t)X(t) \rangle + \langle u(t), R(t)u(t) \rangle\, dt$$

where Q and R are symmetric and positive and R is invertible. We suppose g strictly convex and of class C^1.

Then there exists an optimal control $u_0(t) = R(t)^{-1}\tilde{G}(t)\eta_0(t)$ where the optimal solution X_0 and the adjoint state η_0 satisfy

$$\frac{dX}{dt} = F(t)X + G(t)R(t)^{-1}\tilde{G}(t)\eta(t)$$

$$\frac{d\eta}{dt} = Q(t)X - \tilde{F}(t)\eta$$

and

$$X(t_0) = x_0 \qquad \eta(T) = -\tfrac{1}{2}\,\mathrm{grad}\,g[X(T)]$$

Proof   We use Theorem 3.3.12 with $f(t, X) = \langle X, Q(t)X\rangle$ and $h(t, u) = \langle u, R(t)u\rangle$. Thus

$$\frac{\partial f}{\partial X}(t, X) = \left(\frac{\partial f}{\partial x_1}, \ldots, \frac{\partial f}{\partial x_n}\right) = 2\tilde{X}Q(t)$$

and

$$\frac{\partial \tilde{f}}{\partial X} = 2Q(t)X$$

Taking for η half of the vector η_0 occurring in Theorem 3.3.12 we obtain the adjoint equation $d\eta/dt = Q(t)X - \tilde{F}(t)\eta$ and $\eta(T) = -\tfrac{1}{2}\,\mathrm{grad}\,g[X(T)]$.

Still with the notation of Theorem 3.3.12 and $\eta = \tfrac{1}{2}\eta_0$

$$H(\eta_0, X, u) = 2\langle \eta, FX + Gu\rangle - \langle u, Ru\rangle$$

The maximum is attained at the point u at which the tangent function $\partial H/\partial u$ is zero, i.e. $2\tilde{G}\eta - 2Ru = 0$. Therefore $u = R^{-1}(t)\tilde{G}(t)\eta(t)$, which is the stated result.

Connection between Theorems 3.1.2 and 3.3.13

Suppose $g \equiv 0$ and therefore $\eta(T) = 0$ and let us solve the adjoint equation, denoting the resolvent of $-\tilde{F}(t)$ by $\psi(t, s)$:

$$\eta(t) = \psi(t, 0)\eta(0) + \int_0^t \psi(t, s)Q(s)X_0(s)\,ds$$

Since $\eta_0(T) = 0$

$$\eta_0(0) = -\int_0^T \psi(0, s)Q(s)X_0(s)\,ds$$

and

$$\eta_0(t) = -\int_t^T \psi(t, s)Q(s)X_0(s)\,ds$$

then

$$u_0(t) = - \int_t^T R^{-1}(t)\tilde{G}(t)\psi(t, s)Q(s)X_0(s)\,ds$$

If

$$P(t)X_0(t) = \int_t^T \psi(t, s)Q(s)X_0(s)\,ds$$

then

$$u_0(t) = - R^{-1}(t)\tilde{G}(t)P(t)X_0(t)$$

Moreover, by differentiating the expression defining P we obtain

$$\frac{dP}{dt} X_0(t) + P(t)[F(t)X_0(t) - G(t)R^{-1}(t)\tilde{G}(t)P(t)X_0(t)]$$

$$= -\tilde{F}(t)P(t)X_0(t) - Q(t)X_0(t)$$

and

$$X_0(t)\left[\frac{dP}{dt} + Q(t) + \tilde{P}(t)P(t) + P(t)F(t) - G(t)R^{-1}(t)\tilde{G}(t)P(t)\right] = 0$$

so that $P(T)X_0(T) = 0$. If $P(T) = 0$ and if P is a solution of the Riccati equation we find the preceding result and the optimal control is the same.

Supplementary constraints
We shall state two results in current use.

Proposition 3.3.14 (Target problems)
Let

$$\frac{dX}{dt} = F(t)X + G(t)u \tag{34}$$

and

$$J_u(t_0, X_0) = \int_{t_0}^T f(t, X) + h(t, u)\,dt$$

in the same conditions as in Theorem 3.3.13.

We can show that there exists a unique optimal control which is a solution of the two preceding equations but in which $X(T)$ and $\eta(T)$ satisfy the following relations.

(a) If we impose $X(T) = 0$, we calculate u and η and write $X(T) = 0$.
(b) If we impose $X(T) \in \{x: \varphi(x) \leqslant 0\}$ we shall write $\eta(T) = k \operatorname{grad} \varphi[X(T)]$ where $k < 0$, which states the orthogonality of $\eta(T)$ to the target.

Examples
(i) For

$$\frac{d^2x}{dt^2} = u \qquad \begin{cases} x(0) = 0 \\ x'(0) = -3 \end{cases}$$

$$J_u = \int_0^1 [u(s)]^2 \, ds$$

we require that at time 1 the solution starting from $[x(0), x'(0)]$ is $(0, 0)$, with J minimized.

The equivalent system is

$$\frac{dX}{dt} = \begin{pmatrix} 0 & 1 \\ 0 & 0 \end{pmatrix} X + \begin{pmatrix} 0 \\ 1 \end{pmatrix} u$$

and the adjoint equation is

$$\frac{d\eta}{dt} = -\begin{pmatrix} 0 & 0 \\ 1 & 0 \end{pmatrix} \eta$$

whose solutions are $\eta_1(t) = b$, $\eta_2(t) = a - bt$ where a and b are constants.

The optimal control $u_0(t) = R(t)^{-1}\tilde{G}(t)\eta_0(t) = \eta_2(t) = a - bt$ and therefore

$$x(t) = -3t + a\frac{t^2}{2} - b\frac{t^3}{6}$$

and

$$y(t) = -3 + at - b\frac{t^2}{6}$$

The condition $x(1) = y(1) = 0$ imposes values on a and b and we find

$$u(t) = 12 - 18t \qquad x(t) = -3t + 6t^2 - 3t^3$$

$$y(t) = -3 + 12t - 9t^2 \qquad J(u) = 36$$

We note that, since the cost involves only u, in the absence of constraints we would naturally have found $u \equiv 0$, which is the minimum of $J(u)$ corresponding to $a = b = 0$; by solving the Riccati equation we would have found $P \equiv 0$.

(ii) Over $\mathbb{R}$ let

$$\frac{dx}{dt} = x + u \qquad x(0) = x_0 \qquad x(1) = 0 \qquad (35)$$

$$J(u) = \frac{1}{4}\int_0^1 u^4(t) \, dt$$

We seek u which will take the system from x_0 to 0 between times 0 and 1, minimizing J.

The adjoint equation is $d\eta/dt = -\eta$ and $\eta = a \exp(-t)$;

$$H(\eta, x, u) = (x + u)\eta - \frac{u^4}{4}$$

According to the maximum principle, $u_0(t) = [\eta_0(t)]^{1/3}$.
Then (35) becomes $dx/dt = x + a^{1/3} \exp(-t/3)$ with solutions

$$x_0 \exp(t) + \int_0^t \exp(t - s) \exp(-s/3) a^{1/3} ds$$

and $x(1) = 0$ implies that

$$a^{1/3} = -x_0 \left[\int_0^1 \exp(-4s/3) ds \right]^{-1} = \frac{4}{3} x_0 [\exp(-4/3) - 1]^{-1}$$

and then

$$u_0(t) = \frac{4}{3} x_0 [\exp(-4/3) - 1]^{-1} \exp(-t/3)$$

$$x_0(t) = x_0 \exp(t) \left[1 - \frac{1 - \exp(-4t/3)}{1 - \exp(-4/3)} \right]$$

and

$$J_{u_0} = 16 \frac{x_0^4}{27} [1 - \exp(-4/3)]^{-3}$$

Proposition 3.3.15 (Minimum time with integral constraint)
Let

$$\frac{dX}{dt} = F(t)X + G(t)u \tag{36}$$

$$J_u(t_0, X_0) = \int_{t_0}^T f(t, X) + h(t, u) dt$$

f and h satisfy the same conditions as in Corollary 3.3.13 but T is not fixed, and it is required to join the point (t_0, X_0) to the origin in a minimum time T_0, with the imposed constraint

$$\int_{t_0}^{T_0} [f(t, X) + h(t, u)] dt \leqslant A$$

We can show that there exists a unique optimal control obtained in the following manner. We calculate as a function of T the control which minimizes

$$\int_{t_0}^T f(t, X) + h(t, u) dt$$

and we choose for T_0 the smallest number for which the constraint is satisfied.

Example   We consider, as on page 214, the system $d^2x/dt^2 = u$ with initial point $x(0) = 0$, $x'(0) = -3$, and we impose the constraint

$$J(u) = \int_0^T u^2(t)\, dt \leqslant 1$$

It is still required to join the initial point to the origin $(0, 0)$.
We have established that u is of the form $a - bt$, so that

$$J(u) = \int_0^T (a - bt)^2\, dt = a^2 T - abT^2 + b^2 \frac{T^3}{3}$$

Moreover

$$x(t) = -3t + a\frac{t^2}{2} - b\frac{t^3}{6}$$

and

$$y(t) = -3 + at - b\frac{t^2}{6}$$

Therefore $x(T) = y(T) = 0$ imposes $a = 12/T$ and $b = 18/T^2$; then $J(u) = 36/T$. The minimum time T_0 meeting the constraint $J(u) \leqslant 1$ is then $T_0 = 36$.

See **Exercises 34 to 37** at the end of the chapter.

Exercises

1. Prove Theorem 1.1.6 if f has exponential growth.

2. Let $f_1(t) = \cos \omega t$ and $f_2(t) = \sin \omega t$. Calculate $f_1 * f_2$ and

$$\mathcal{L}\left[\frac{x}{2}\cos \omega x\right]$$

3. Find $\mathcal{L}[f]$ for $f = \sinh(kt)$, $\cosh(kt)$, $t\sinh(kt)$, $t\cosh(kt)$, $t^{1/2}$ and $(\sin at)/t$ and for f defined in the following three ways:

$$f(t) = \begin{cases} t & \text{if } 0 < t \leqslant T \\ 2T - t & \text{if } T < 2t < 2T \end{cases} \quad \text{and} \quad f(t) = f(t + 2T)$$

$$f(t) = naT \quad \text{for } nT \leqslant t < (n+1)T$$

$$f(t) = \begin{cases} \sin t & \text{if } 0 \leqslant t \leqslant \pi \\ 0 & \text{if } \pi \leqslant t < 2\pi \end{cases} \quad \text{and} \quad f(t) = f(t + 2\pi)$$

4. Find $\mathcal{L}^{-1}[F]$ for the following functions F:

$$F = \frac{p}{(p^2 + 1)(p^2 + 2)} \qquad F = \frac{1}{p(p + 1)^2} \qquad F = \frac{p - 4}{(p^2 + 4)^2}$$

$$F = \frac{1}{(p - 4)^3} \qquad F = \frac{1}{p^2 + 4p + 9}$$

5. Find directly the result of Example 3 (Section 1.2).

6. Show in Example 4 (Section 1.2) that every solution is twice differentiable; hence deduce an equation satisfied by y and find the result shown.

7. Let

$$A = \begin{pmatrix} -1 & -1 & -2 \\ 1 & 1 & 1 \\ 2 & 1 & 3 \end{pmatrix} \quad \text{and} \quad b = \begin{pmatrix} \exp(t) \\ 0 \\ 0 \end{pmatrix}$$

Find the solution φ of $dX/dt = Ax + b(t)$ such that $\varphi(0) = 0$.

8. Solve

$$\frac{d^2 x}{dt^2} + x = \sum_0^\infty \delta_{\{k\pi\}}$$

and verify that the solutions are bounded.

9. Let

$$A(t) = \begin{pmatrix} -a & 0 \\ 0 & \sin \log t + \cos \log t - 2a \end{pmatrix}$$

where $1 < 2a < 1 + \exp(-\pi)$, and

$$B(t) = \begin{pmatrix} 0 & 0 \\ \exp(-at) & 0 \end{pmatrix}$$

(a) Verify that $dx/dt = A(t)x$ is asymptotically stable.
(b) Verify that $dx/dt = [A(t) + B(t)]x$ is unstable although

$$\int_T^\infty \| B(u) \| \, du < \infty$$

10. Consider the situation of Theorem 3.1.2 with $X \in \mathbb{R}$. Find the optimal feedback control using Eqn (21).

11. If $M \geqslant 0$ is a symmetric matrix, show, in the situation of Theorem 3.1.2, with a cost

$$\int_{t_0}^T \langle X_u(t), Q(t)X_u(t) \rangle + \langle u(t), R(t)u(t) \rangle \, dt + \langle X_u(T), MX_u(T) \rangle$$

that the optimal control is obtained in the same way, modifying only the boundary condition of the Riccati equation: $P(T) = M$ in place of $P(T) = 0$.

12. If $S(t) \geqslant 0$ is a symmetric matrix $\forall t$, replace the cost appearing in the statement of Theorem 3.1.2 by

$$J(t, X, u) = \int_{t_0}^{T} \langle X_u(t), Q(t)X_u(t) \rangle + \langle u(t), R(t)u(t) \rangle + \langle u'(t), S(t)u'(t) \rangle \, dt$$

where u' denotes the derivative of u (which exists except at a finite number of points). Putting

$$Z = \begin{pmatrix} X \\ u \end{pmatrix}$$

show that we are led to Theorem 3.1.2 and find the corresponding Riccati equation.

13. Let F and G be two fixed numbers, u a real function and $dx/dt = Fx + Gu$ a scalar equation; let

$$J_u(t_0, x_0) = \int_{t_0}^{T} \{ax_u^2(s) + \exp[u(s)]\} \, ds \qquad (a > 0)$$

Find the Eqn (21) satisfied by the minimum cost $V(t, x)$.

14. Consider in $\mathbb{R}$ the scalar equation over $[0, \pi/2]$ $dx/dt = x + u$ and the function $f(t) = \cos t$; it is required that x should track $f(t)$, minimizing the expression

$$\int_{0}^{\pi/2} [x_u(s) - \cos s]^2 + u^2(s) \, ds + x\left(\frac{\pi}{2}\right)^2$$

Find the optimal control and the corresponding cost.

15. Consider in $\mathbb{R}$ the equation $dx/dt = u$ which connects the angular momentum about an axis and a control pair. Suppose x_0 is given for $t = 0$ and it is required to reduce $x(1)$, minimizing the expression

$$x(1)^2 + \int_{0}^{1} u^2(s) \, ds$$

Find the optimal control and the corresponding cost.

16. Consider in $\mathbb{R}^2$ the following system controlled by

$$u = \begin{pmatrix} u_1 \\ u_2 \end{pmatrix}$$

$$\left. \begin{array}{l} \dfrac{dx}{dt} = ax + by + u_1(t) \\[2mm] \dfrac{dy}{dt} = cx + dy + u_2(t) \end{array} \right\} \tag{37}$$

Let $\alpha \geqslant 0$, $\beta \geqslant 0$, $\lambda > 0$ and $\mu > 0$; consider the cost

$$\int_0^T \alpha x_u^{\,2}(t) + \beta y_u^{\,2}(t) + \lambda u_1^{\,2}(t) + \mu u_2^{\,2}(t)\, dt$$

Write the equations for finding u_1 and u_2 (cf. Exercise 24). System (37) is a predator–prey system which we are trying to control.

17. Let

$$\frac{dX}{dt} = \begin{pmatrix} 1 & 0 \\ 1 & -1 \end{pmatrix} X + \begin{pmatrix} 1 \\ 0 \end{pmatrix} u$$

Let $A = (1, 1)$ and T be given.

Find a control u (which can be sought in the form $\alpha \exp(-t) + \beta \exp(t)$) leading from A to 0 in time T. What happens if T tends to zero?

18. Let

$$\frac{dX}{dt} = \begin{pmatrix} 1 & 0 \\ 0 & 2 \end{pmatrix} X + \begin{pmatrix} 1 & 0 \\ 0 & 1 \end{pmatrix} \begin{pmatrix} u_1 \\ u_2 \end{pmatrix}$$

Let $A = (1, 1)$.

(a) Verify that the system is completely controllable. Verify that every control u leading from 0 to A in time T becomes infinite if $T \to 0$.

(b) Find (u_1, u_2) leading from 0 to A in time T.

19. Let $dX/dt = FX + gu$, where $F \in \mathcal{L}(\mathbb{R}^n, \mathbb{R}^n)$, $g \in \mathbb{R}^n$ and $u \in \mathbb{R}^1$, be a system with constant coefficients. Suppose that the system is completely controllable and that the eigenvalues of F are purely imaginary and not oscillatory. Show that $\forall \varepsilon > 0$ we can join two points X_0 and X_1 by using a control u such that $\forall s$ $|u(s)| \leqslant \varepsilon$.

20. Let

$$\frac{dX}{dt} = \begin{pmatrix} 2 & \lambda - 3 \\ 0 & 2 \end{pmatrix} X + \begin{pmatrix} 1 & 1 \\ \lambda^2 - \lambda & 0 \end{pmatrix} \begin{pmatrix} u_1 \\ u_2 \end{pmatrix}$$

(a) For what values of λ is the system completely controllable?

(b) For what values of λ can we control it by using a single control (i.e. $u_1 = u_2$)?

21. Let $dX/dt = F(t)X + G(t)u$. Suppose that $(t_0, X_0 - X_1) \rightsquigarrow (t_1, 0)$. Give a sufficient condition for $(t_0, X_0) \rightsquigarrow (t_1, X_1)$. Verify that this condition is met if $G(t) \equiv I$ or $G(t)^{-1}$ exists.

If the system is a scalar equation controlled by simple control show that $\forall (t_0, t_1, x_0 \text{ and } x_1)$

$$(t_0, x_0 - x_1) \to (t_1, 0) \Leftrightarrow (t_0, x_0) \to (t_1, x_1)$$

Consider

$$X_0 = \begin{pmatrix} x_0 \\ x_0' \\ \vdots \\ x_0^{(n-1)} \end{pmatrix} \quad \text{and} \quad X_1 = \begin{pmatrix} x_1 \\ 0 \\ \vdots \\ 0 \end{pmatrix}$$

22. Let

$$F = \begin{pmatrix} -1 & 0 \\ 0 & 1 \end{pmatrix} \qquad G = \begin{pmatrix} 1 \\ 1 \end{pmatrix}$$

Calculate $W(0, T)$ for a given instant T, and then $K = \tilde{G} \exp(\tilde{F}T) W(0, T)^{-1} \exp(FT)$. Verify the remark following Theorem 3.2.13.

23. Let

$$F = \begin{pmatrix} -1 & 2 & 1 \\ -2 & 3 & -1 \\ 1 & -1 & 0 \end{pmatrix} \qquad G = \begin{pmatrix} 0 \\ 1 \\ 0 \end{pmatrix}$$

(a) Verify that (F, G) is completely controllable and F is unstable.

(b) Transform (F, G) into a system (F_1, G_1) corresponding to a scalar equation and find K_1 so that the eigenvalues of $F_1 - G_1 K_1$ are $(-1, -2, -3)$.

(c) Find a feedback control such that the eigenvalues of $F - GK$ are $(-1, -2, -3)$.

24. In $\mathbb{R}^2$ let the system be

$$\frac{dx}{dt} = ax + by + u_1(t) \qquad \frac{dy}{dt} = cx + dy + u_2(t)$$

Let $\alpha \geqslant 0$, $\beta \geqslant 0$, $\lambda > 0$ and $\mu > 0$ and

$$J = \int_0^\infty \alpha x^2(t) + \beta y^2(t) + \lambda u_1^2(t) + \mu u_2^2(t)\, dt$$

(a) Verify that (F, G) is stabilizable.

(b) In what conditions is (H, F) detectable?

(c) Let

$$F = \begin{pmatrix} 1 & 1 \\ 1 & -1 \end{pmatrix}$$

$\alpha = 1$, $\beta = 2$, $\lambda = \mu = 1$. Find the optimal control relative to J (cf. Exercise 16).

25. Let $dx/dt = -x + u$ be a scalar system. Let

$$J = \int_0^\infty x_u^2(s) + u^2(s)\, ds$$

Find the optimal control and solution.

26. Let

$$\frac{dx}{dt} = -x + u \qquad \frac{dy}{dt} = y$$

and

$$J = \int_0^\infty x_u^2(s) + u^2(s)\, ds$$

Does an optimal control exist? What does the problem become if

$$G = \begin{pmatrix} 1 & 0 \\ 0 & 1 \end{pmatrix} \quad \text{and} \quad u = \begin{pmatrix} u_1 \\ u_2 \end{pmatrix}$$

27. Let

$$\frac{dx}{dt} + \lambda x = u$$

be a scalar equation. Let $\Omega = \{u : |u| \leqslant 1\}$. Find the set of points from which the system can be brought to the origin using a control taking its values in Ω and in a time less than or equal to T.

28. For the problem defined in Exercise 27, find the control u_0 leading from a given point to the origin in the minimum time. Calculate this time.

29. Let $d^2x/dt^2 = u$ and $\Omega = [-1, +1]$ as in the example studied on pages 206 and 207. Let $(x_0, y_0) = (-1, \sqrt{7})$; construct explicitly the control u which enables the system to reach the origin in the minimum time. Calculate this minimum time.

30. Let

$$\frac{d^2x}{dt^2} + x = u$$

and $\Omega = [-1, +1]$. Study the problem of the minimum time to reach the origin starting from a point (x_0, y_0).

31. Let

$$\frac{dx}{dt} + \lambda x = u$$

and $\Omega = [-1, +1]$ as in Exercises 27 and 28. Find the control joining a point to the origin in the minimum time.

32. Let $d^3x/dt^3 = u$ and $\Omega = [-1, +1]$. Study the control leading the system in minimum time from the point (x_0, y_0, z_0) to the origin.

33. Study the normality conditions for the system

$$\begin{pmatrix} \dfrac{dx}{dt} \\[2mm] \dfrac{dy}{dt} \end{pmatrix} = \begin{pmatrix} -2 & \lambda \\ 0 & -1 \end{pmatrix}\begin{pmatrix} x \\ y \end{pmatrix} + \begin{pmatrix} 1 & 0 \\ \lambda & \lambda \end{pmatrix}\begin{pmatrix} u_1 \\ u_2 \end{pmatrix}$$

and

$$\Omega: \begin{cases} |u_1(t)| \leqslant 1 \\ |u_2(t)| \leqslant 1 \end{cases}$$

34. Let $d^2x/dt^2 = 0$, $x(0) = 0$, $x'(0) = -3$ and

$$J(u) = \int_0^1 [u(t)]^2 \, dt$$

It is required to control the system minimizing $J(u)$ and taking it into the circle $x^2 + y^2 = 1$.

(a) Show that $\eta(1) = -kX(1)$. (Remember that an optimal control arrives at a target set along the normal.)

(b) Hence deduce the optimal control, the optimal solution and the cost.

35. Let $d^2x/dt^2 = u$, (x_0, y_0) be an initial point and

$$J(u) = \int_0^{2\pi} [u(s)]^2 \, ds$$

It is required to bring the solution to the origin, minimizing J. Find the optimal control, the optimal solution and the cost.

36. Let $dx/dt = u$ and x_0 be an initial point.

(a) Let

$$J(u) = x^2(1) + \int_0^1 [u(s)]^2 \, ds$$

Find the control which minimizes J, the optimal solution and the cost.

(b) It is required to take x_0 to zero, minimizing

$$\int_0^1 [u(s)]^2 \, ds$$

Find the minimizing control and minimum cost.

37. Consider a pendulum with friction controlled by a function u; let

$$\frac{d^2x}{dt^2} + \frac{dx}{dt} + x = u \tag{38}$$

Let $(x_0 = 0, x_0' = 0)$; it is required that for $t = 2$, $x(t) = dx/dt = 1$, while

$$J(u) = \int_0^2 [u(s)]^2 \, ds$$

is minimized. Find the optimal control, the optimal solution and the cost.

Chapter 4

Stability, Lyapunov theory

We have already introduced on two occasions the notion of stability. In the first place in Chapter 2 (p. 104) we have defined the stability of an equilibrium. We have established that the stability of an equilibrium a of $dx/dt = f(x)$ is connected with the stability of this equilibrium relative to the linearized system $dx/dt = Df(a)(x - a)$ (Chapter 2, Theorem 3.1.3, p. 106). We have held back part of the proof, however, until the present chapter.

We next established that, if a dynamic system is structurally stable and a is a stable equilibrium of such a system, every system which is sufficiently close has itself a stable equilibrium b close to a (Chapter 2, Theorem 3.2.12, p. 117).

In Chapter 3 we defined the notion of stability of a solution (Chapter 3, Definition 2.1.1, p. 153). The essence of the results which we established on that occasion can be summarized thus. Firstly, all solutions of a homogeneous linear system are of the same nature, stable or not, uniformly or asymptotically, etc. This stability is a function of the form of the resolvent. Secondly, we showed that the nature of a system with a second term is the same as that of the homogeneous system. Moreover, we studied uniformly asymptotically stable systems from the point of view of the automation engineer: in short, this stability persists for 'moderately perturbed' systems (Chapter 3, Theorems 2.1.2, 2.1.3, 2.1.12 and 2.2.2).

In this chapter we shall first recall the definition of stability and indicate certain phenomena which can occur with non-linear systems. Next we shall show, and this will be the main theme of the chapter, how the use of auxiliary functions enables us to obtain information on the stability of certain solutions.

The reader will recall that we have already used auxiliary functions in a similar context; as early as Chapter 1 (p. 57) the use of a particular first integral enabled us to infer the periodicity of all solutions of a 'predator–prey' equation.

We used auxiliary functions also in the study of Van der Pol's equation (p. 18) and in an exercise describing the orbits of solutions of the pendulum equation; finally, the results in Chapter 3 (p. 196) were themselves also proved by this method.

The first part of this chapter will be devoted to generalities, the second to certain particular systems: systems depending on a gradient and Hamiltonian systems. The third part will be kept for the study, properly so-called, of the Lyapunov method.

Part 1: Generalities

1.1. Stability: definitions

Let

$$\frac{dx}{dt} = f(t, x) \tag{1}$$

be a differential equation in $\mathbb{R}^d$; we assume that for every element (t_0, x_0), $t_0 \in \mathbb{R}$, $x_0 \in \mathbb{R}^d$, of the domain of definition of f there exists a unique solution $\varphi(t, t_0, x_0)$ of (1) defined for all $t \geq t_0$ and such that $\varphi(t_0, t_0, x_0) = x_0$.

Definition 1.1.1 (cf. Chapter 3, Definition 2.1.1, p. 153)

(a) We say that $\varphi(t, t_0, x_0)$ is stable if $\forall \varepsilon > 0 \ \exists \eta(t_0, \varepsilon)$ such that if $\|x_1 - x_0\| \leq \eta(\varepsilon, t_0)$ then $\|\varphi(t, t_0, x_0) - \varphi(t, t_0, x_1)\| \leq \varepsilon \ \forall t \geq t_0$. If there exists T such that $\forall \varepsilon > 0 \ \exists \eta(\varepsilon)$ independent of t_0 so that if $\|x_1 - x_0\| \leq \eta(\varepsilon)$ then

$$\|\varphi(t, t_0, x_0) - \varphi(t, t_0, x_1)\| \leq \varepsilon \qquad \forall t \geq t_0 > T$$

we say that φ is uniformly stable for $t_0 \geq T$.

(b) We say that $\varphi(t, t_0, x_0)$ is asymptotically stable if it is stable and if $\forall t_0$ $\exists \eta_1(t_0)$ such that if $\|x_1 - x_0\| \leq \eta_1(t_0)$ then

$$\lim_{t \to \infty} \|\varphi(t, t_0, x_0) - \varphi(t, t_0, x_1)\| = 0$$

We say that $\varphi(t, t_0, x_0)$ is uniformly asymptotically stable for $t_0 \geq T$ if it is uniformly stable and if $\exists \eta_1$, independent of t_0, such that for every x_1 satisfying $\|x_0 - x_1\| \leq \eta_1$ the following property holds:

$$\forall \varepsilon, \forall t \geq T \quad \exists T(\varepsilon): t \geq t_0 + T(\varepsilon) \Rightarrow \|\varphi(t, t_0, x_0) - \varphi(t, t_0, x_1)\| \leq \varepsilon$$

(c) In all other cases we say that φ is unstable.

We draw the reader's attention to the comments following this definition (p. 153).

In this part we shall give only some indications of the differences which appear when a system is not linear.

STABLE AND UNSTABLE SOLUTIONS OF ONE AND THE SAME EQUATION

Contrary to what happens in the case of linear systems, it is possible that an equation possesses at the same time stable and unstable solutions.

The following equation is an example of this.

Example   Let

$$\frac{dX}{dt} = f(X) \quad \text{where } X = \begin{pmatrix} x \\ y \end{pmatrix} \tag{2}$$

be the system defined by

$$\frac{dx}{dt} = -y(x^2 + y^2)^{1/2} \qquad \frac{dy}{dt} = x(x^2 + y^2)^{1/2}$$

Let $X_0 = (x_0, y_0)$ be a given point and $r_0{}^2 = x_0{}^2 + y_0{}^2$. The orbits of (2) are $x^2 + y^2 = x_0{}^2 + y_0{}^2 = r_0{}^2$ and the solution $\varphi(t, t_0, x_0)$ is defined by its coordinates:

$$x(t) = x_0 \cos r_0(t - t_0) - y_0 \sin r_0(t - t_0)$$

$$y(t) = x_0 \sin r_0(t - t_0) + y_0 \cos r_0(t - t_0)$$

The solution $x(t) \equiv 0 \equiv y(t)$ is uniformly stable since $\forall \varepsilon$ if $r_0{}^2 \leqslant \eta(\varepsilon) = \varepsilon$ then $x^2(t) + y^2(t) = \|\varphi(t, t_0, X_0) - 0\|^2 = r_0 \leqslant \varepsilon$.

However, $\|\varphi(t, t_0, X_0) - \varphi(t, t_0, X_1)\|$ represents the distance between two points which go round two concentric circles at different velocities; this could not then be made smaller than ε $\forall \varepsilon$.

It must be noted, nevertheless, that the distance between the two circles, centred at 0 and with radii r_0 and r_1 respectively, can be made arbitrarily small; we have here a situation in which we must carefully distinguish stability of solutions from proximity of orbits. We shall return in Chapter 8 to this phenomenon. We shall introduce for this purpose the notion of **orbital stability** which characterizes solutions whose orbits have the above property of proximity.

STABILITY AND CHANGE OF FUNCTIONS OR OF VARIABLES

We have seen in Chapter 1 (p. 24) how a differential equation is transformed by application of a differentiable transformation F from $\mathbb{R}^p$ into itself. If $\varphi(t)$ is the solution of $dx/dt = f(x)$, this solution becomes $\psi(t)$, the tangent vector W at the point $F(a)$ being the vector $DF(a)V$ if V is the tangent vector to φ at a (Fig. 4.1). Thus, if DF^{-1} exists in a neighbourhood of the solution ψ we shall be able to deduce from it properties of the solution φ.

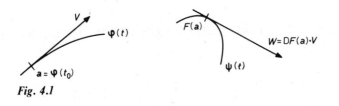

Fig. 4.1

Suppose that F is a change of coordinates in $\mathbb{R}^p$ and that the equation

$$\frac{dX}{dt} = f(X) \tag{3}$$

becomes

$$\frac{dY}{dt} = g(Y) \tag{4}$$

Let $\varphi(t, t_0, X_0)$ be the solution of (3) such that $\varphi(t_0) = X_0$ and let $Y_0 = F(X_0)$ and $F[\varphi(t, t_0, X_0)] = \psi(t, t_0, Y_0)$ (Fig. 4.2). Assume that ψ is

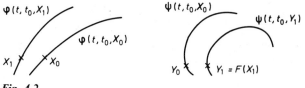

Fig. 4.2

stable; this means that $\forall \alpha \ \exists \beta$ such that if $\| Y_1 - Y_0 \| \leqslant \beta$ then $\| \psi(t, t_0, Y_1) - \psi(t, t_0, Y_0) \| \leqslant \alpha \ \forall t \geqslant t_0$. What can we deduce from this about $\varphi = F^{-1}\psi$? If F^{-1} is continuous

$$\forall \varepsilon \quad \exists \alpha(t) : \| \psi(t, t_0, Y_1) - \psi(t, t_0, Y_0) \| \leqslant \alpha(t) \Rightarrow \| \varphi(t, t_0, X_0) - \varphi(t, t_0, X_1) \| \leqslant \varepsilon$$

We can reach a conclusion only if F^{-1} is uniformly continuous in the neighbourhood of $\psi(t, t_0, Y_0)$ for then $\forall \varepsilon$ we can choose successively α, then β and then η so that

$$\| X_1 - X_0 \| \leqslant \eta \Rightarrow \| Y_1 - Y_0 \| \leqslant \beta$$
$$\Rightarrow \| \psi(t, t_0, Y_1) - \psi(t, t_0, Y_0) \| \leqslant \alpha \quad \forall t \geqslant t_0$$
$$\Rightarrow \| \varphi(t, t_0, X_0) - \varphi(t, t_0, X_1) \| \leqslant \varepsilon \quad \forall t \geqslant t_0$$

Thus the solution φ is of the same nature as the solution $\psi = F(\varphi)$ if, in a neighbourhood of ψ, F^{-1} is uniformly continuous; this will be so if $(DF)^{-1} = D(F^{-1})$ is bounded.

Polar coordinates
Let

$$X = \begin{pmatrix} x \\ y \end{pmatrix} \quad \text{and} \quad \begin{pmatrix} r \\ \vartheta \end{pmatrix} = F(X)$$

where $r \geqslant 0$, $0 \leqslant \vartheta < 2\pi$, $r = (x^2 + y^2)^{1/2}$ and $\vartheta = \arctan(y/x)$. Then

$X = F^{-1}(r, \vartheta)$ is defined by

$$x = r \cos \vartheta \qquad y = r \sin \vartheta$$

$$DF^{-1} = \begin{pmatrix} \cos \vartheta & \sin \vartheta \\ -r \sin \vartheta & r \cos \vartheta \end{pmatrix}$$

Thus we shall be able to study the stability of a solution by converting to polar coordinates if ψ remains within a domain $0 < R_0 \leqslant r \leqslant R_1 < \infty$.

Example   We return to Eqn (2); converting to polar coordinates we get

$$\left. \begin{aligned} \frac{dr}{dt} &= 0 \\[2mm] \frac{d\vartheta}{dt} &= r \end{aligned} \right\} \tag{5}$$

whose solution $\psi[t, t_0, (r_0, \vartheta_0)]$ is

$$\psi(t) = \begin{cases} r_0 \\ \vartheta_0 + r_0 t \quad \text{(modulo } 2\pi) \end{cases}$$

If (r_1, ϑ_1) is a point in a neighbourhood of (r_0, ϑ_0) (Fig. 4.3)

$$\|\psi[t, t_0, (r_0, \vartheta_0)] - \psi[t, t_0, (r_1, \vartheta_1)]\|^2 = |r_1 - r_0|^2 + |\vartheta_1 - \vartheta_0 + (r_1 - r_0)t|^2$$

$$\text{(modulo } 2\pi)$$

a quantity which cannot be made less than or equal to ε unless $r_1 = r_0$.

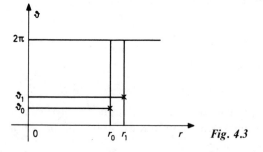

Fig. 4.3

Thus, all solutions of (5)—including the solution $r \equiv 0$—are unstable. Hence we deduce that all solutions of (2)—except possibly the solution $x \equiv 0 \equiv y$—are also unstable. For the solution $r = 0$ we can deduce nothing about its behaviour in polar coordinates relative to the equilibrium $x \equiv 0 \equiv y$ of (2). We have verified, moreover, that the latter is stable.

Other types of functional changes

It is necessary to be extremely careful if we make a change which is not a change of coordinates: it is possible that the behaviour of solutions of the transformed equation bears no relation to that of solutions of the initial equation.

Example Throughout let

$$\left.\begin{array}{l} \dfrac{\mathrm{d}x}{\mathrm{d}t} = -y(x^2 + y^2)^{1/2} \\[3mm] \dfrac{\mathrm{d}y}{\mathrm{d}t} = x(x^2 + y^2)^{1/2} \end{array}\right\} \tag{6}$$

Let $r = (x^2 + y^2)^{1/2}$ and $\alpha = \arctan(y/x) - rt$, i.e. $x = r\cos(rt + \alpha)$ and $y = r\sin(rt + \alpha)$ where r and α are unknown functions of t. Equations (6) become

$$\frac{\mathrm{d}r}{\mathrm{d}t} = 0 \qquad \frac{\mathrm{d}\alpha}{\mathrm{d}t} = 0 \tag{7}$$

The solution with initial value (r_0, α_0) is $r \equiv r_0$, $\alpha \equiv \alpha_0$. In this system all solutions are stable.

We note in fact that we have a family of changes of coordinates depending on t and that

$$DF(t)^{-1} = \begin{pmatrix} \cos(rt + \alpha) - rt\sin(rt + \alpha) & \sin(rt + \alpha) + rt\cos(rt + \alpha) \\ -r\sin(rt + \alpha) & r\cos(rt + \alpha) \end{pmatrix}$$

STABILITY AND STATIONARY POINTS

We can always reduce the study of the stability of a solution of a differential equation to that of the stability, for another equation, of the identically null solution.

Let

$$\frac{\mathrm{d}x}{\mathrm{d}t} = f(t, x) \tag{8}$$

and let (x_0, t_0) be an initial point and $\varphi(t, t_0, x_0)$ the solution such that $x(t_0) = x_0$.

With every solution x we can associate a function y by the relation

$$y(t) = x(t) - \varphi(t, t_0, x_0)$$

If x_1 is the value of $x(t)$ for $t = t_0$, the corresponding value for $t = t_0$ of the function $y(t)$ is $x_1 - x_0$. Moreover y is a solution of the equation

$$\frac{\mathrm{d}y}{\mathrm{d}t} = f[t, y + \varphi(t, t_0, x_0)] - f[t, \varphi(t, t_0, x_0)] \tag{9}$$

Conversely, if y is a solution of (9), $x(t) = y(t) + \varphi(t, t_0, x_0)$ is the solution of (8) such that $x(t_0) = y(t_0) + x_0$.

The solution y which corresponds to $x = \varphi(t, t_0, x_0)$ is $y(t) \equiv 0$. We shall denote by $\psi(t, t_0, y_0)$ the solution of (9) such that $y(t_0) = y_0$.

Proposition 1.1.2

The solution $y \equiv 0$ of (9) is stable, uniformly stable, asymptotically or uniformly asymptotically stable for t_0 iff the solution $\varphi(t, t_0, x_0)$ is itself stable, uniformly stable, asymptotically or uniformly asymptotically stable respectively for t_0.

Proof   The fact that $y \equiv 0$ is stable for t_0 is equivalent to

$$\forall \varepsilon > 0 \quad \exists \eta(t_0): \|y_0\| \leqslant \eta(t_0) \Rightarrow \|\psi(t, t_0, y_0)\| \leqslant \varepsilon \quad \forall t \geqslant t_0$$

or equally

$$\|x_1 - x_0\| \leqslant \eta(t_0) \Rightarrow \|\varphi(t, t_0, x_1) - \varphi(t, t_0, x_0)\| \leqslant \varepsilon \quad \forall t \geqslant t_0$$

This states that $\varphi(t, t_0, x_0)$ is stable for t_0.

Remark   It should not be deduced from the preceding proposition that we can content ourselves with studying the stability of stationary points; in fact the study of the equilibrium $y \equiv 0$ of Eqn (9) necessitates knowledge of Eqn (9), i.e. of $\varphi(t, t_0, x_0)$, which is not always the case.

Partial stability

We can strengthen the condition $\|x_1 - x_0\| \leqslant \eta$ of Definition 1.1.1 and require that $\|\varphi(t, t_0, x_0) - \varphi(t, t_0, x_1)\| \leqslant \varepsilon$ if $\|x_1 - x_0\| \leqslant \eta$ and $x_1 - x_0$ belongs to a particular set of $\mathbb{R}^d$.

Definition 1.1.3

The solution $\varphi(t, t_0, x_0)$ of $dx/dt = f(t, x)$ is partially stable with respect to E if

$$\forall \varepsilon \quad \exists \eta(t_0): x_1 - x_0 \in E \qquad \|x_1 - x_0\| \leqslant \eta(t_0)$$

Then

$$\|\varphi(t, t_0, x_0) - \varphi(t, t_0, x_1)\| \leqslant \varepsilon \qquad \forall t \geqslant t_0$$

Example   The stable subspaces E_λ introduced on page 84, defined as the proper generalized subspaces relative to the eigenvalues λ with strictly negative real part, are in fact composed of initial values x_0 of solutions which are partially stable with respect to E_λ. The invariant stable sets introduced on page 53 are another example.

See **Exercises 1 to 3** at the end of the chapter.

1.2. Linear variational equation and stability

In Chapter 2 we studied the stability of an equilibrium by studying the stability of the linearized equation in the neighbourhood of that equilibrium, and in Chapter 3 we studied the stability of solutions of a linear system. In this section we shall examine up to what point the variational equation in the neighbourhood of a solution (not necessarily constant) of a non-linear equation gives information about the stability of that solution.

The following theorem is the fundamental one on which rests the partial results which we shall show.

Theorem 1.2.1
Let

$$\frac{\mathrm{d}x}{\mathrm{d}t} = A(t)x \tag{10}$$

be a linear system in $\mathbb{R}^n$, uniformly asymptotically stable for $t_0 \geqslant T$. Let

$$f: \begin{array}{c} \mathbb{R} \times \mathbb{R}^d \to \mathbb{R}^d \\ (t, x) \rightsquigarrow f(t, x) \end{array}$$

be a continuous function such that

$$\forall \varepsilon > 0 \quad \exists \eta > 0: \|x\| \leqslant \eta \Rightarrow \|f(t, x)\| \leqslant \varepsilon \|x\| \qquad \forall t \geqslant T$$

Let

$$\frac{\mathrm{d}x}{\mathrm{d}t} = A(t)x + f(t, x) \tag{11}$$

Then the solution $x \equiv 0$ of (11) is uniformly asymptotically stable for $t_0 \geqslant T$.

Proof We can show that the assumptions imply that the Cauchy problem for (11) relative to (t_0, x_0), for $t_0 \geqslant T$ and x_0 sufficiently small, has a unique solution which is defined $\forall t \geqslant t_0 \geqslant T$. We shall accept this.

Let Φ be the resolvent of (10) and x such a solution of (11):

$$x(t) = \Phi(t, t_0)x_0 + \int_{t_0}^t \Phi(t, s)f[s, x(s)] \, \mathrm{d}s$$

We take T as fixed and we know that there exist α and K such that $\|\Phi(t, s)\| \leqslant K \exp[-\alpha(t - s)]$. We can assume $K > 1$. If $\varepsilon \leqslant \alpha/K$ and η is the corresponding value, we shall show that if $\|x_0\| \leqslant \eta/K < \eta$ then $\|x(t)\| \leqslant \eta \ \forall t \geqslant t_0$, which will establish the result.

We know already, through continuity, that, in the neighbourhood of t_0, $\|x(s)\| \leqslant \eta$; then we put $\vartheta = \inf(s, \|x(s)\| > \eta)$. Naturally $t_0 < \vartheta$ and if $\vartheta < \infty$ then $\|x(\vartheta)\| = \eta$. Let $t_0 \leqslant s < t \leqslant \vartheta$: $\|f[s, x(s)]\| \leqslant \varepsilon \|x(s)\|$ and

$$\|x(t)\| \leqslant K \exp[-\alpha(t - t_0)]\|x_0\| + \int_{t_0}^t \varepsilon K \exp[-\alpha(t - s)] \|x(s)\| \, \mathrm{d}s$$

Therefore, from Gronwall's lemma

$$\exp(\alpha t)\|x(t)\| \leqslant K \exp(\alpha t_0) \|x_0\| \exp[\varepsilon K(t - t_0)]$$

or, put another way,

$$\|x(t)\| \leqslant K \exp(\varepsilon K - \alpha)(t - t_0)]\|x_0\| \leqslant K\|x_0\| < \eta$$

This relation shows that if $\|x_0\| \leqslant \eta/K$ then $\forall t < \vartheta$ $\|x(t)\| < \eta$. Then, if $\vartheta < \infty$, by continuity $\|x(\vartheta)\| < \eta$, which is absurd. Thus $\vartheta = \infty$, which implies that if $\|x_0\| < \eta$ then $\|x(t)\| < \eta$ $\forall t \geqslant T$. Now if ε_0 is fixed we can always choose $\eta < \varepsilon_0$ and then $x(t) \equiv 0$ is a uniformly stable solution of (11). Moreover its limit as $t \to \infty$ is zero, which completes the proof.

Now let $dx/dt = F(t, x)$ and let φ be a solution of this equation. To study the stability of φ we write $dz/dt = F(t, z + \varphi) - F(t, \varphi)$ which we linearize in the form

$$\frac{dz}{dt} = \frac{\partial F}{\partial x}(t, \varphi)z + f(t, z)$$

In general $f(t, z)$ does not satisfy the conditions of the preceding theorem (unless for example

$$\left|\frac{\partial^2 F_k}{\partial x_i \partial x_j}(t, x)\right| \leqslant M \qquad \forall t \geqslant T, \forall x$$

and for all the coordinates).

The following result is limited but is nevertheless useful. (It is a corollary of Theorem 1.2.1.)

Theorem 1.2.2
Let

$$\frac{dx}{dt} = Ax + f(t, x) \tag{12}$$

be a differential equation in which $\forall \varepsilon, \exists \eta: \|x\| \leqslant \eta \Rightarrow \|f(t, x)\| \leqslant \varepsilon\|x\|$ $\forall t$ and $A \in \mathscr{L}(\mathbb{R}^d, \mathbb{R}^d)$.

(a) If $\forall i$, Re $\lambda_i < \sigma < 0$ for all the eigenvalues λ_i of A, $x \equiv 0$ is a uniformly asymptotically stable solution of (12).

(b) If $\exists i_0$: Re $\lambda_{i_0} > 0$, $x \equiv 0$ is an unstable solution of (12).

Comment   Through linearization we can study satisfactorily the stability of an equilibrium but, in general, we cannot study the stability of other solutions.

Part 2: Systems deriving from a gradient; Hamiltonian systems

2.1. Systems deriving from a gradient

If f is a differentiable function from $\mathbb{R}^p$ into $\mathbb{R}$, we have defined (Chapter 1, p. 15) the gradient vector of f; this is, at each point a, a vector in $\mathbb{R}^p$ such that

$$\forall V \in \mathbb{R}^p \qquad Df(a) \cdot V = \langle \operatorname{grad} f(a), V \rangle$$

Definition 2.1.1

Let U be a mapping, of class C^2, from $\mathbb{R}^p$ into $\mathbb{R}$. We define a system deriving from the gradient of U, or a system deriving from the potential U, to be a differential equation of the form

$$\frac{dx}{dt} = -\operatorname{grad} U(x) \tag{13}$$

Since U is assumed to be of class C^2, (13) satisfies the existence and uniqueness conditions for the Cauchy problem relative to (t_0, x_0) for every x_0 in the domain of definition of U.

We shall assume that the solution $\varphi(t, t_0, x_0)$ is defined $\forall t \geq t_0$.

We can calculate the variation in $U(x)$ when x describes a solution φ of (13); from Chapter 1, Theorem 2.2.9 (p. 15), we know that if φ is a solution of (13)

$$\frac{d}{dt} U[\varphi(t)] = \langle \operatorname{grad} U[\varphi(t)], -\operatorname{grad} U[\varphi(t)] \rangle$$

$$= -\|\operatorname{grad} U[\varphi(t)]\|^2$$

Thus $U[\varphi(t)] \leq U[\varphi(t_0)]$ $\forall t \geq t_0$, which often enables us to deduce information about the behaviour of φ.

Suppose for example that $U(x, y) > 0$ for all $(x, y) \in \mathbb{R}^2$, that $DU(0, 0) = 0$ and that $D^2 U(0, 0) > 0$, U having the form sketched in Fig. 4.4. We see that

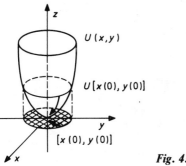

Fig. 4.4

$\lim \varphi[t, t_0, (x_0, y_0)] = (0, 0)$ and that $(0, 0)$ is a uniformly asymptotically stable equilibrium for every initial value (x_0, y_0).

In general, we shall associate with U the surface in $\mathbb{R}^{n+1}$ defined by $z = U(x)$ and we shall denote the **surface in $\mathbb{R}^n$** $U(x) = a$ by the **curve of level** a or the **horizontal section at altitude** a.

The following property is easy to verify.

Proposition 2.1.2

If a is a point of $\mathbb{R}^n$ such that grad $U(a) \neq 0$, this vector is then perpendicular to the level curve $U(x) = U(a)$.

Proof If $a = (a_1, a_2, \ldots, a_n)$, we have verified (Chapter 1, p. 23) that if grad $U(a) \neq 0$ the equation of the tangent plane to the surface $U(x) = U(a)$ at the point a is

$$\frac{\partial U}{\partial x_1}(a)(x_1 - a_1) + \ldots + \frac{\partial U}{\partial x_n}(a)(x_n - a_n) = \langle \text{grad } U(a), x - a \rangle = 0$$

This states that grad $U(a)$ is carried by the normal at a to $U(x) = U(a)$.

The sketch above suggests that knowledge of U enables us not only to verify that the origin is uniformly asymptotically stable but also to find the set of starting points of all solutions which tend to the origin; this is what is called its domain of attraction, of which the definition is the following.

Definition 2.1.3

If $x(t) \equiv a$ is an asymptotically stable equilibrium of an equation $dx/dt = f(x)$, we define the **domain of attraction** of that equilibrium as the set of points x_0 such that

$$\lim_{t \to \infty} \varphi(t, t_0, x_0) = a$$

Theorem 2.1.4

Let $U : \mathbb{R}^n \to \mathbb{R}$ be a function of class C^2 and

$$\frac{dx}{dt} = -\text{grad } U(x) \tag{14}$$

a system deriving from U.

(a) If φ is a solution of (14)

$$\frac{d}{dt} U[\varphi(t)] = -\|\text{grad } U[\varphi(t)]\|^2 \leqslant 0$$

$$\frac{d}{dt}\{U[\varphi(t)]\} = 0$$

iff there exists $a : \varphi(t) \equiv a$.

(b) At every point x at which grad $U(x) \neq 0$ the trajectory of (14) is perpendicular to the level curve passing through that point.

(c) Every isolated minimum of U is a uniformly asymptotically stable equilibrium whose domain of attraction can be found by examining, as h varies, the level curves $U(x) = h$.

Comment We have already proved the points (a) and (b) of the theorem. As far as point (c) is concerned, we shall make it clear by an example. We show that, if a is an isolated minimum, the sets

$$A_n = \left\{ x : U(x) \leqslant U(a) + \frac{1}{n} \right\}$$

by hypothesis, for n sufficiently large, are sets surrounding a, of which an appropriate component has limit a as $n \to \infty$. These sets have the level curves as boundaries.

Theorem 2.1.4 gives information regarding the form of the solutions of systems depending on a gradient, and more particularly on the equilibria; we shall complement it with the following result.

Proposition 2.1.5

Let

$$\frac{dx}{dt} = -\operatorname{grad} U(x) \tag{15}$$

be a system deriving from a gradient. Let a be an equilibrium of (15) and $H[U(a)] = -A$ be the Hessian matrix. The eigenvalues of A are real, a fact which enables us to determine the nature of the equilibrium a if 0 is not one of the eigenvalues.

Proof It suffices to note that every Hessian matrix (cf. p. 15) is symmetric. A is thus diagonalizable; the linearized system of (15) in the neighbourhood of $\varphi(t) \equiv a$ is precisely $dz/dt = A(z - a)$.

Example Let $U(x, y) = x^2 + y^2(y - 1)^2$. The gradient system is $dx/dt = -2x$, $dy/dt = -2y(y - 1)(2y - 1)$. The points O, A and B in Fig. 4.5(a) are equilibria: $O = (0, 0)$, $A = (0, 1)$, $B = (0, \frac{1}{2})$. O and A are asymptotically stable. Their domains of attraction are $y < \frac{1}{2}$ and $y > \frac{1}{2}$. B is a saddle point. The arrowed orbits in Fig. 4.5(b) are orthogonal to the level curves.

See *Exercises 4 to 7* at the end of the chapter.

2.2. Hamiltonian systems

We have already introduced Hamiltonian systems (p. 70); we are often led to transform a system deriving from a gradient in $\mathbb{R}^d$ into a system in $\mathbb{R}^{2d}$ in the

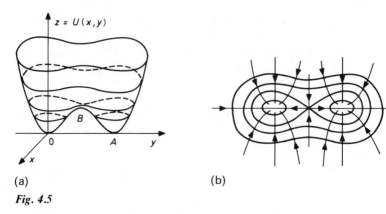

(a)   (b)

Fig. 4.5

following way. Let

$$\frac{d^2 q}{dt^2} = -\text{grad}[U(q)] \quad \text{where } q \in \mathbb{R}^d, \ q = (q_1, \ldots, q_d)$$

Let

$$p_i = \frac{d}{dt} q_i \quad \text{and} \quad p \in \mathbb{R}^d, \ p = (p_1, \ldots, p_d)$$

We put

$$T(p) = \sum_{i=1}^{d} \frac{p_i{}^2}{2}$$

and we denote by $H(p, q)$ a quantity, called the **Hamiltonian** of the system, which is given by

$$H(p, q) = T(p) + U(q)$$

We then consider the system

$$\left.\begin{aligned}
\frac{dq_i}{dt} &= \frac{\partial H}{\partial p_i} & i = 1, \ldots, d \\[2mm]
\frac{dp_i}{dt} &= -\frac{\partial H}{\partial q_i} & i = 1, \ldots, d
\end{aligned}\right\} \tag{16}$$

In Chapter 2 we proved Liouville's theorem which states that the volume determined by the solutions of (16) in $\mathbb{R}^{2d}$ is constant. Another fundamental property of these systems is the following.

Theorem 2.2.1
$H(p, q)$ is a first integral of (16).

Proof It suffices to calculate

$$\frac{d}{dt}H[p(t), q(t)] = \sum_{i=1}^{d}\frac{\partial H}{\partial p_i}\frac{dp_i}{dt} + \sum_{i=1}^{d}\frac{\partial H}{\partial q_i}\frac{dq_i}{dt} = 0$$

Instead of using U we can use H in the same way to demonstrate the following result.

Theorem 2.2.2
If the origin is an isolated minimim of the potential U it is a stable equilibrium which is not asymptotically stable for (16).

Proof The proof consists first in verifying that if 0 is an isolated minimum of U it is in fact an isolated minimum of H. We next consider the solution φ starting from (p_0, q_0) (Fig. 4.6); the orbit of φ is the projection on the subspace $z = 0$ of $\mathbb{R}^{2d+1}$, which establishes the stability of the solution $\varphi(t) \equiv 0$ and the fact that no solution can have the origin as limit.

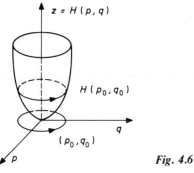

Fig. 4.6

Example: Orbits of a pendulum in the phase space (p, q)
Let $x = q$ and $dx/dt = p$; the equation

$$\frac{d^2x}{dt^2} + \sin x = 0$$

can be written

$$\frac{dp}{dt} = -\frac{\partial H}{\partial q}$$

$$\frac{dq}{dt} = \frac{\partial H}{\partial p}$$

for $H = \frac{p^2}{2} + (1 - \cos q)$

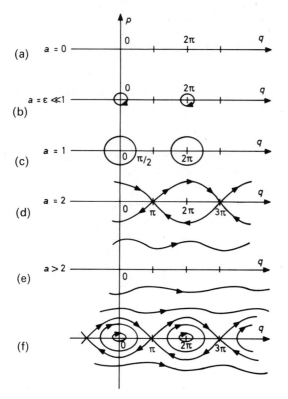

Fig. 4.7

The orbits are the curves

$$a = \frac{p^2}{2} + (1 - \cos q)$$

Figure 4.7 illustrates the following properties.

(a) If $a = 0$, $p = 0$, $q = 2k\pi$ (Fig. 4.7(a)).
(b) If $a = \varepsilon \ll 1$, $p^2 + q^2 \sim 2\varepsilon$: small oscillations (Fig. 4.7(b)).
(c) If $a = 1$, $p = \pm(2\cos q)^{1/2}$ ($\cos q \geqslant 0$): periodic oscillations (Fig. 4.7(c)).
(d) If $a = 2$, $p^2 = 4\cos^2(q/2)$: the system goes from one unstable equilibrium to another in infinite time; aperiodic case (Fig. 4.7(d)).
(e) If $a > 2$, complete indefinite rotations (Fig. 4.7(e)).
(f) Recapitulative figure: the curves $p = \pm 2\cos(q/2)$ are called **separatrices** (Fig. 4.7(f)).

We have indicated that the equilibria $(2k + 1)\pi$ are unstable; the proof is simple since, at such a point,

$$Df = \begin{pmatrix} 0 & 1 \\ 1 & 0 \end{pmatrix}$$

This is a matrix whose eigenvalues are $+1$ and -1, which corresponds to a saddle point.

Definition 2.2.3

We shall define a **conservative system** of the second order as a system whose equation is the following: if $x \in \mathbb{R}$, there exists f of class C^1 and

$$\frac{d^2x}{dt^2} + f(x) = 0 \tag{17}$$

Let

$$F(q) = \int_0^q f(s)\, ds$$

In the phase plane such a system is a Hamiltonian system whose Hamiltonian H is

$$H(p, q) = \frac{1}{2} p^2 + F(q) \qquad \left(q = x, p = \frac{dx}{dt} \right)$$

The orbits of (17) are $p^2 + 2F(q) = \text{constant}$, a point of equilibrium a is such that $f(a) = 0$, and there exist, as in the case of the pendulum, **separatrices** separating the periodic solutions from the others; their equation is $p^2 + 2F(q) = 2F(a)$ (where a is an equilibrium of (17)).

Period of the orbits of a conservative system

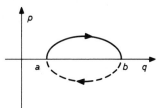

Fig. 4.8. Orbit $H(p, q) = h$; period $T = 2 \int_a^b \dfrac{dq}{\{2[h - F(q)]\}^{1/2}}$

Proposition 2.2.4
Let

$$\frac{d^2x}{dt^2} + f(x) = 0 \tag{18}$$

be a conservative system and

$$H(p, q) = \frac{p^2}{2} + F(q)$$

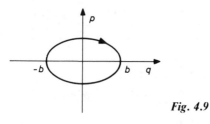

Fig. 4.9

its Hamiltonian. Let $H(p, q) = h$ be a periodic orbit whose equation is $p = \pm\{2[h - F(q)]\}^{1/2}$; let a and b be the extreme points

$$T = 2 \int_a^b \frac{dq}{\{2[h - F(q)]\}^{1/2}}$$

If F is odd and if $b = -a$ (Fig. 4.9)

$$\frac{T}{4} = \int_0^b \frac{dq}{\{2[h - F(q)]\}^{1/2}}$$

Proof The proof is immediate. If q_0 is fixed, the velocity $dq/dt = p$, at the time of the two passages through q_0, has the same absolute value $\{2[h - F(q_0)]\}^{1/2}$. The time required to cover (a, b) is thus the same as that to cover (b, a) and is the half-period. As

$$\frac{dt}{dq} = \frac{1}{\{2[h - F(q)]\}^{1/2}}$$

for $0 < t < T$, then

$$\frac{T}{2} = \int_a^b \frac{dt}{dq} \, dq = \int_a^b \frac{dq}{\{2[h - F(q)]\}^{1/2}}$$

Example: *period of a pendulum*
   Let

$$\frac{d^2x}{dt^2} + k^2 \sin x = 0$$

$$H(p, q) = \frac{p^2}{2} + k^2(1 - \cos q) = \frac{p^2}{2} + F(q)$$

Let $q(0) = b < \pi, p(0) = 0$ be the initial position and velocity of the pendulum in a periodic oscillation $h = H[p(0), q(0)] = k^2(1 - \cos b)$.

$$T = 4 \int_0^b \frac{du}{[2k^2(\cos u - \cos b)]^{1/2}}$$

$$= \frac{2}{k} \int_0^b \frac{du}{\{[\sin(b/2)]^2 - [\sin(u/2)]^2\}^{1/2}}$$

Let ψ be defined by $\sin(u/2) = \sin\psi \sin(b/2)$ where $\psi \in [0, \pi/2]$. Then

$$T = \frac{4}{k} \int_0^{\pi/2} \frac{d\psi}{\{1 - [\sin(b/2)]^2(\sin\psi)^2\}^{1/2}}$$

an integral which cannot be calculated.

We shall give an expression for T by development in series. We know that for $|x| < 1$

$$(1 + x)^{-1/2} = 1 - \frac{1}{2}x + \frac{-\frac{1}{2}(-\frac{1}{2} - 1)}{2!}x^2 + \dots$$

$$+ \frac{-\frac{1}{2}(-\frac{1}{2} - 1)\dots(-\frac{1}{2} - n)}{n!}x^n \dots$$

Thus

$$\left[1 - \left(\sin\frac{b}{2}\right)^2 (\sin\psi)^2\right]^{-1/2} = 1 + \frac{1}{2}\left(\sin\frac{b}{2}\right)^2 \sin^2\psi + \frac{3}{8}\left(\sin\frac{b}{2}\right)^4 \sin^4\psi + \dots$$

We can integrate this series term by term so that

$$T = \frac{4}{k}\left[\frac{\pi}{2} + \frac{1}{2}\left(\sin\frac{b}{2}\right)^2 \int_0^{\pi/2} \sin^2\psi \, d\psi + \frac{3}{8}\left(\sin\frac{b}{2}\right)^4 \int_0^{\pi/4} \sin^4\psi \, d\psi + \dots\right]$$

$$T = \frac{2\pi}{k}\left(1 + \frac{b^2}{16} + 3\frac{b^4}{2^{10}} + \dots\right)$$

(T depends on the initial position b). If $b = \pi$, the series has an infinite sum, but the calculation is not valid because we do not know that there is a periodic solution.

In this case we can calculate the time to go from $q = 0$ to $q = \pi$ directly as

$$\int_0^\pi \frac{dt}{dq} dq = \frac{1}{k}\int_0^\pi \frac{dq}{[2(1 + \cos q)]^{1/2}} = \frac{8}{k}\int_0^{\pi/2} \frac{du}{\cos u} = \infty$$

This case, which is not realistic, corresponds to such a slowing-down when q tends to π (the velocity being $1 + \cos q \sim (\pi - q)^2/2$) that an infinite time is needed to reach it. We shall see later what in fact happens.

We shall show finally how the existence of a potential sometimes enables us to establish the instability of an equilibrium.

Theorem 2.2.5 (Case of instability)
Let

$$\frac{d^2x}{dt^2} + f(x) = 0$$

be a conservative system (cf. Definition 2.2.3).

Assume that $f(0) = 0$ and that there exists $\alpha: 0 \leqslant q \leqslant \alpha \Rightarrow F(q) < 0$. $q(t) \equiv 0$ (or $x \equiv 0$) is then an unstable equilibrium.

Proof Since $f(0) = 0$, $x \equiv 0$ is clearly a solution. Let $p_0 > 0$ and $(0, p_0)$ be an initial point in the neighbourhood of $(0, 0)$; on the trajectory issuing from this point

$$\frac{p^2}{2}(t) + F[q(t)] = \frac{p_0^2}{2}$$

and, as long as $q(t) \leqslant \alpha$, $p^2(t)/2 > p_0^2/2$; therefore $dq/dt > p_0$ and $q(t) > p_0 t$, which implies that

$$q\left(\frac{\alpha}{p_0}\right) > \alpha$$

However small p_0 may be, the distance from it is at least α.

See *Exercise 8* at the end of the chapter.

Part 3: Lyapunov theory

3.1. Autonomous systems

PRINCIPAL THEOREMS

As we indicated in the introduction, the Lyapunov method consists of using auxiliary functions whose form and behaviour along the integral curves enable us to study the stability of certain solutions in a manner similar to that used in the preceding section.

Definition 3.1.1
Let 0 be the origin of $\mathbb{R}^d$ and Ω a neighbourhood of 0. Let U be a real-valued function, defined over Ω, such that $U(0) = 0$ and $U(x) > 0$ if $x \neq 0$: we say that U is **positive definite** in Ω.

We shall denote by $\mathcal{K}$ the family of functions $a: \mathbb{R}^+ \to \mathbb{R}$ which are strictly increasing and such that $a(0) = 0$. We can prove easily that U is positive definite in Ω iff $\exists a \in \mathcal{K}: U(x) \geqslant a(\|x\|)$.

We recall that with every symmetric matrix A of $\mathcal{L}(\mathbb{R}^d, \mathbb{R}^d)$ we can associate a **quadratic form** $U: \mathbb{R}^d \to \mathbb{R}$ defined in the following way:

$$U(x) = x^t A x = \langle x, Ax \rangle = \sum_{i,j=1}^{d} A_{i,j} x_i x_j$$

where we have denoted by x the vector

$$\begin{pmatrix} x_1 \\ \vdots \\ x_d \end{pmatrix}$$

by x^t the transpose of x, and by $A_{i,j}$ the element of A in the ith row and jth column.

We know that U is positive definite iff all the principal minors of A are strictly positive.

Definition 3.1.2

Let

$$\frac{dx}{dt} = f(x) \tag{19}$$

be an autonomous system such that $f(0) = 0$. A function U of class C^1, positive definite in a neighbourhood Ω of 0, is called a **Lyapunov function** for (19) if for every solution φ of (19)

$$\frac{d}{dt} U[\varphi(t)] \leqslant 0$$

We know that

$$\frac{d}{dt} U[\varphi(t)] = \langle \operatorname{grad} U[\varphi(t)], f[\varphi(t)] \rangle$$

$$= \sum_{i=1}^{d} \frac{\partial U}{\partial x_i} [\varphi(t)] f_i[\varphi(t)]$$

Notation We shall write

$$\mathring{U}(x) = \langle \operatorname{grad} U(x), f(x) \rangle$$

$$= \sum_{i=1}^{d} \frac{\partial U}{\partial x_i} (x) f_i(x)$$

A function U, positive definite, such that $\mathring{U}(x) \leqslant 0$ is a Lyapunov function for (19)—we note that, although f does not appear in the notation $\mathring{U}$, this quantity naturally depends on f.

The following elementary theorem is the starting point of the theory. In its exposition we shall denote by B_ρ the ball $\{x : \|x\| \leqslant \rho\}$.

Theorem 3.1.3 (Lyapunov)

Let

$$\frac{dx}{dt} = f(x) \tag{20}$$

be an autonomous system of which 0 is an equilibrium.

(a) If there exists a function U defined over a neighbourhood Ω of 0 in which it is a Lyapunov function of (20), $x \equiv 0$ is a uniformly stable equilibrium.

(b) If, in addition, $-\overset{\circ}{U}(x)$ is positive definite in Ω, $x \equiv 0$ is a uniformly asymptotically stable equilibrium whose **domain of attraction** contains B_ρ when $B_\rho \subset \Omega$.

(c) If there exists in Ω a function V of class C^1 and a number δ such that, on the one hand, $V(0) = 0$ and $V(x) > 0$ for all x such that $0 < \|x\| < \delta$ and, on the other hand, $\overset{\circ}{V}(x)$ is positive definite, then $x \equiv 0$ is an unstable equilibrium.

Proof

(a) Since Ω is a neighbourhood of 0, there exists $\varepsilon > 0$ such that

$$S_\varepsilon = \{x: \|x\| = \varepsilon\} \subset \Omega$$

(see Fig. 4.10). If

$$m = \inf_{x \in S_\varepsilon} U(x)$$

then $\forall \varepsilon > 0$ m is strictly positive ($m > 0$). Since U is continuous, there exists

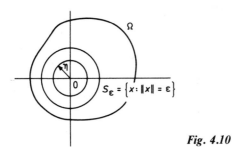

Fig. 4.10

$\eta < \varepsilon$ such that $\|x\| \leqslant \eta \Rightarrow U(x) < m$. If $\|x_0\| \leqslant \eta$ then $U[\varphi(t, t_0, x_0)] \leqslant U(x_0) < m$ $\forall t > t_0$ since $U[\varphi(t)]$ is a decreasing function of t for every trajectory φ.

The trajectory $\varphi(t, t_0, x_0)$ then cannot reach S_ε for $t > t_0$ since at a point x_1, at which it reached it, $U(x_1)$ would have to be at least equal to m; if $\|x_0\| \leqslant \eta \Rightarrow \|\varphi(t, t_0, x_0)\| < \varepsilon$ $\forall t \geqslant t_0$, which establishes the result.

(b) By hypothesis $U(x) \geqslant a(\|x\|)$, and over every bounded set we can also find $b \in \mathscr{K}$ such that $U(x) \leqslant b(\|x\|)$. Thus

$$U(x) \leqslant a(\lambda) \Rightarrow a(\|x\|) \leqslant a(\lambda)$$

Therefore

$$\|x\| \leqslant \lambda$$

and

$$\|x\| \leqslant \mu \Rightarrow b(\|x\|) \leqslant b(\mu)$$

Therefore

$$U(x) \leqslant b(\mu)$$

If λ and μ are positive

$$U(x) \leqslant a(\lambda) \Rightarrow \|x\| \leqslant \lambda$$

$$\|x\| \leqslant \mu \Rightarrow U(x) \leqslant b(\mu)$$

We know also that there exists $c \in \mathcal{K} : \overset{\circ}{U}(x) \leqslant -c(\|x\|) < 0$.

Let ρ be such that $B_\rho = \{\|x\| \leqslant \rho\} \in \Omega$ and M such that $U(x) \leqslant M \; \forall x \in B_\rho$. Finally, if ε is any fixed number, there exists ε' such that $a(\varepsilon) = b(\varepsilon')$. If $x_0 \in B_\rho$

$$U[\varphi(s + t_0, t_0, x_0)] = U(x_0) + \int_{t_0}^{t_0 + s} \overset{\circ}{U}[\varphi(v + t_0, t_0, x_0)] \, dv \qquad (21)$$

If

$$T(\varepsilon) = \frac{|M - a(\varepsilon')|}{c(\varepsilon')}$$

we shall verify that

$$\forall \vartheta > T(\varepsilon) \qquad \|\varphi(\vartheta + t_0, t_0, x_0)\| \leqslant \varepsilon$$

which will establish the result.

Suppose that $\forall v \leqslant s \leqslant T(\varepsilon)$

$$\|\varphi(v + t_0, t_0, x_0)\| > \varepsilon' \qquad (22)$$

then

$$\overset{\circ}{U}[\varphi(v + t_0, t_0, x_0)] \leqslant -c(\varepsilon')$$

and

$$U[\varphi(s + t_0, t_0, x_0)] < M - c(\varepsilon')T(\varepsilon) \leqslant a(\varepsilon')$$

from (21) and the definition of $T(\varepsilon)$. We know that then $\|\varphi(s + t_0, t_0, x_0)\| < \varepsilon'$, which contradicts (22).

Therefore there exists at least s_0 such that (22) is not satisfied and thus such that $U[\varphi(s_0 + t_0, t_0, x_0)] \leqslant b(\varepsilon')$ and since $U[\varphi(t, t_0, x_0)]$ is a decreasing function of t

$$U[\varphi(\vartheta + t_0, t_0, x_0)] \leqslant b(\varepsilon') \qquad \forall \vartheta \geqslant T(\varepsilon) \geqslant s_0$$

Now $b(\varepsilon') = a(\varepsilon)$ and therefore

$$U[\varphi(\varepsilon + t_0, t_0, x_0)] \leqslant a(\varepsilon) \qquad \forall \vartheta \geqslant T(\varepsilon)$$

and

$$\|\varphi(\vartheta + t_0, t_0, x_0)\| \leqslant \varepsilon \qquad \forall \vartheta \geqslant T(\varepsilon)$$

(c) Let $0 < \|x_0\| < \delta$ and $B_\delta = \{x: \|x\| \leqslant \delta\}$ (Fig. 4.11). Let $A = \{x: \|x\| < \delta$

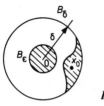

Fig. 4.11

and $V(x) \geqslant V(x_0)\}$. As $V(0) = 0$, $0 \notin A$, there even exists a ball B_ε such that $B_\varepsilon \cap A = \emptyset$ and $B_\varepsilon \subset \{x: V(x) < V(x_0)/2\}$. Thus

$$\alpha = \inf_{x \in A} [\overset{\circ}{V}(x)]$$

is strictly positive $(\alpha > 0)$. As $V[\varphi(t)]$ is an increasing function of t for every solution φ, $\varphi(t, t_0, x_0)$ can only remain in A or leave B_δ.

Let

$$\vartheta = \inf[t: \varphi(t, t_0, x_0] = \delta$$

$$V[\varphi(t, t_0, x_0)] = V(x_0) + \int_{t_0}^{t} \overset{\circ}{V}[\varphi(s, t_0, x_0)] \, ds$$

If $\vartheta > t$ this expression is greater than or equal to $V(x_0) + \alpha t$ and, as V is bounded over B_ρ, ϑ is necessarily finite. For every $\|x_0\|$ (however small), there exists $\vartheta < \infty$ such that $\varphi(t, t_0, x_0)$ reaches the set B_δ at time ϑ; 0 is thus unstable.

This theorem, for which the assumptions are very restrictive, has been generalized in many ways. We shall quote two results which are often applied.

We give two simple examples of the application of Theorem 3.1.3.

Examples
(a)

$$\frac{dx}{dt} = 2y(z - 1)$$

$$\frac{dy}{dt} = -x(z - 1) \tag{23}$$

$$\frac{dz}{dt} = -z^3$$

$0 \equiv x = y = z$ is a solution of (23). The linearized system in the neighbourhood of the origin has matrix

$$A = \begin{pmatrix} 0 & -2 & 0 \\ 1 & 0 & 0 \\ 0 & 0 & 0 \end{pmatrix}$$

This linearized system is stable but, since its eigenvalues 0, $i\sqrt{2}$ and $-i\sqrt{2}$ have zero real part, the stability of the linearized system implies nothing about the initial system.

Now let $U(x, y, z) = x^2 + 2y^2 + z^2$; U is positive definite in $\mathbb{R}^3$. It is a quadratic form of the matrix

$$\begin{pmatrix} 1 & 0 & 0 \\ 0 & 2 & 0 \\ 0 & 0 & 1 \end{pmatrix}$$

$$\overset{\circ}{U}(x, y, z) = 2x2y(z - 1) + 4y(-x)(z - 1) + 2z(-z^3)$$

$$= -2z^4 \leqslant 0$$

The origin is thus a stable equilibrium.

It is easy to see that it is not an asymptotically stable equilibrium since the orbits of solutions whose initial values are in $z \equiv 0$ are $x^2 + 2y^2 = k^2$ where $k^2 = x_0{}^2 + 2y_0{}^2$.

We can solve (23), however:

$$z(t) = \frac{z(0)}{[2z(0)^2 t + 1]^{1/2}}$$

If

$$x^2 + 2y^2 = k^2$$

we put

$$x(t) = k \cos \vartheta(t)$$

and

$$y(t) = \frac{k}{\sqrt{2}} \sin \vartheta(t)$$

Then

$$\frac{d\vartheta}{dt} = \sqrt{2(1 - z)}$$

and

$$\vartheta(t) = \vartheta_0 + \sqrt{2}\, t - \frac{\sqrt{2}}{z_0}[(1 + 2z_0{}^2 t)^{1/2} - 1]$$

$$\lim_{t \to +\infty} z(t) = 0 \qquad \forall z_0$$

(see Fig. 4.12).

We shall be able to try to apply Lasalle's theorem, which is stated below, to this situation.

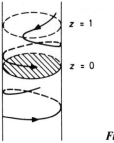

$z = 1$

$z = 0$

Fig. 4.12

(b)

$$\frac{dx}{dt} = x(y^2 - 1)$$

(24)

$$\frac{dy}{dt} = y(x^2 - 1)$$

$x = y = z = 0$ is a solution of (24). The linearized system has matrix

$$A = \begin{pmatrix} -1 & 0 \\ 0 & -1 \end{pmatrix}$$

It is uniformly asymptotically stable; the origin is thus a uniformly asymptotically stable equilibrium of (24), according to Theorem 3.1.3 (Chapter 2, p. 105).

The existence of a Lyapunov function here will give us information about the domain of attraction of the origin.

Let

$$U(x, y) = \tfrac{1}{2}(x^2 + y^2)$$

$$\mathring{U}(x, y) = x^2(y^2 - 1) + y^2(x^2 - 1)$$

Let $\Omega = \{x^2 + y^2 < 1\}$; in Ω, $\mathring{U}(x, y)$ is negative definite; we again find that 0 is uniformly asymptotically stable but additionally that $\forall (x_0, y_0) \in \Omega$ the solution starting from that point has 0 as limit.

We now give—without proof—statements of the two theorems which are frequently applied.

Theorem 3.1.4 (Lasalle: 'Invariant sets')
(a) If

$$\frac{dx}{dt} = f(x)$$

(25)

is an autonomous system in $\mathbb{R}^d$, suppose that there exists a closed bounded set

G, a positive function U of class C^1 over G and $a < \infty$ such that

$$G = \{0 \leqslant U(x) \leqslant a\} \quad \text{and} \quad \mathring{U}(x) \leqslant 0 \quad \text{in } G$$

If $E = \{x \in G, \mathring{U}(x) = 0\}$ and M is the largest subset invariant for (25) contained in E, then

$$\lim_{t \to \infty} \text{distance} \left[\varphi(t_0, x_0), M \right] = 0 \quad \forall x_0 \in G$$

(b) If $f(0) = 0$ and if 0 is the only subset of E invariant for (25), 0 is uniformly asymptotically stable and its domain of attraction contains G.

(c) If U is not bounded while being defined in the entire space $\mathbb{R}^d$ and if $U(x) \geqslant 0$ and $\mathring{U}(x) \leqslant 0 \, \forall x \in \mathbb{R}^n$, all the solutions are bounded and tend to M.

Theorem 3.1.5 (Cetaev: 'Instability')
Let

$$\frac{dx}{dt} = f(x) \tag{26}$$

be an autonomous system for which the origin is an equilibrium $f(0) = 0$.

Let $\rho > 0$ and $B_\rho = \{\|x\| \leqslant \rho\}$, and let Ω be a subset of B_ρ whose boundary F contains 0 (see Fig. 4.13). Suppose that there exists U of class C^1 such that, if $x \in \Omega$ and $x \notin F$, then $U(x) > 0$ and $\mathring{U}(x) > 0$, whereas if $x \in F$ $U(x) \equiv 0$. In these conditions 0 is an unstable equilibrium.

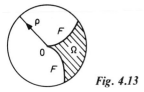

Fig. 4.13

EXAMPLES AND APPLICATIONS

Example 1: Van der Pol's equation (cf. Chapter 1, p. 18)
Let

$$\frac{d^2x}{dt^2} + \lambda(1 - x^2)\frac{dx}{dt} + x = 0 \tag{27}$$

be an equation where $\lambda > 0$. It is equivalent to the following system, obtained by putting $y = dx/dt$:

$$\frac{dx}{dt} = y$$

$$\frac{dy}{dt} = \lambda(x^2 - 1)y - x$$

If $U(x, y) = x^2 + y^2$ then $\overset{\circ}{U}(x, y) = 2\lambda y^2 (x^2 - 1)$. Let $G = \{0 \leqslant U(x, y) \leqslant 1\} = \{0 \leqslant x^2 + y^2 \leqslant 1\}$: $\overset{\circ}{U}(x, y) \leqslant 0$ in G. Let $E = \{x \in G: \overset{\circ}{U}(x) = 0\}$: $E = \{y = 0\}$ and the only invariant set contained in E is $\{x = 0 = y\}$.

According to Lasalle's theorem

$$\lim_{t \to \infty} \varphi[t, (x_0, y_0)] = 0$$

if $x_0^2 + y_0^2 \leqslant 1$. The origin is thus asymptotically stable and its domain of attraction contains G.

The general Van der Pol equation is

$$\frac{d^2 x}{dt^2} + \lambda(x^2 - 1)\frac{dx}{dt} + x = 0 \qquad \text{where } \lambda > 0 \qquad (28)$$

If ψ is a solution of (28) we put $\varphi(t) = \psi(-t)$; then

$$\frac{d\varphi}{dt}(t) = -\frac{d\psi}{dt}(-t) \quad \text{and} \quad \frac{d^2\varphi}{dt^2} = \frac{d^2\psi}{dt^2}(-t)$$

Thus φ is a solution of (27) and the trajectories of (27) and (28) have the same orbits but they are traversed in opposite directions, which is enough to show that 0 is an unstable equilibrium; we could also show, but this is not so simple, that every trajectory of (28) whose initial value is such that $x(0)^2 + y(0)^2 < 1$ leaves the set $x^2 + y^2 < 1$.

We can in fact establish the result directly by using the same function U since, for (F), $\overset{\circ}{U}(x, y) = -2\lambda y^2 (x^2 - 1)$; therefore, if $y_0 \neq 0$, from Cetaev's theorem the trajectory leaves the set $x^2 + y^2 \leqslant 1 - \varepsilon \ \forall \varepsilon > 0$.

If $y_0 = 0$ and $x_0 \neq 0$, the trajectory immediately enters a set in which $\overset{\circ}{U}(x, y) > 0$ and therefore leaves the circle $x^2 + y^2 \leqslant 1 - \varepsilon \ \forall \varepsilon > 0$.

Example 2: Rotation of a solid about its axes of inertia

Consider, in a given system of reference, a solid having one point fixed. Denote by A, B, C the principal moments of inertia and by p, q, r the components of angular velocity. Suppose the system is subjected to no external force. We shall suppose $A \leqslant B \leqslant C$. We shall study the stability of stationary rotations about each of the axes of inertia.

We can show that the equations of motion are

$$A\frac{dp}{dt} = (B - C)qr \qquad B\frac{dq}{dt} = (C - A)rp \qquad C\frac{dr}{dt} = (A - B)pq$$

(1) Let $A < B$ and $p \equiv p_0$ be the stationary rotation about the axis of smallest inertia. Putting $x = p - p_0$, $y = q$ and $z = r$, we obtain the system

$$A\frac{dx}{dt} = (B - C)yz \qquad B\frac{dy}{dt} = (C - A)z(p_0 + x) \qquad C\frac{dz}{dt} = (A - B)(p_0 + x)y$$

Let

$$U(x, y, z) = B(B - A)y^2 + C(C - A)z^2 + (Ax^2 + By^2 + Cz^2 + 2Ap_0x)^2$$

Since $A < B \leqslant C$, U is positive definite (this is not true if $A = B$); since $\mathring{U}(x, y, z) = 0$, the rotation $p \equiv p_0$ is stable (but not asymptotically stable).

(2) Let $B < C$ and $r \equiv r_0$ be the stationary rotation about the axis of greatest inertia. Consider the analogous system

$$A\frac{dx}{dt} = (B - C)y(z + r_0) \qquad B\frac{dy}{dt} = (C - A)(z + r_0)x \qquad C\frac{dz}{dt} = (A - B)xy$$

Consider also

$$U(x, y, z) = B(C - B)y^2 + A(C - A)x^2 + (Ay^2 + Bx^2 + Cz^2 + 2Cr_0z)^2$$

Since $A \leqslant B < C$, U is positive definite; we see that $\mathring{U}(x, y, z) = 0$ and reach the same conclusion: the rotation $r \equiv r_0$ is stable (but not asymptotically).

(3) Let $A < B \leqslant C$ or $A \leqslant B < C$ and $q \equiv q_0$ be the stationary rotation about the intermediate axis. We study the system

$$A\frac{dx}{dt} = (B - C)(y + q_0)z \qquad B\frac{dy}{dt} = (C - A)zx \qquad C\frac{dz}{dt} = (A - B)x(y + q_0)$$

Let $\Omega = \{x^2 + y^2 + z^2 < q_0{}^2\} \cap \{x > 0\} \cap \{z > 0\}$. This is the intersection of a sphere with the first octant (Fig. 4.14).

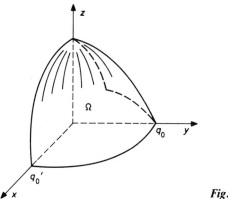

Fig. 4.14

Let $U(x, y, z) = xz$.

$$\mathring{U}(x, y, z) = (y + q_0)\left(\frac{B - C}{A}z^2 + \frac{A - B}{C}x^2\right)$$

We can orient the axis so that $q_0 < 0$; then, in Ω, $U > 0$, $\mathring{U} > 0$ and $0 \in F$: from Cetaev's theorem the rotation is unstable.

Rotations about the extreme axes of inertia are stable, whereas those about intermediate axes are unstable.

Example 3: Pendulum with constant friction

Let

$$\frac{d^2 x}{dt^2} + \frac{dx}{dt} + \sin x = 0$$

The equivalent system is

$$\frac{dx}{dt} = y$$

$$\frac{dy}{dt} = -y - \sin x$$

Let

$$U(x, y) = \frac{y^2}{2} + (1 - \cos x)$$

Then $\overset{\circ}{U}(x, y) = -y^2$. If

$$H = \left\{ x, y : \frac{y^2}{2} + (1 - \cos x) \leqslant 2 \right\}$$

H is a closed set whose boundary is $y^2 = 4\cos^2(x/2)$; we consider the part of this set contained between $-\pi$ and $+\pi$. Let

$$G = \left\{ x, y : \frac{y^2}{2} + (1 - \cos x) \leqslant 2 - \varepsilon \right\}$$

We take the part G_0 of G contained between $-\pi$ and $+\pi$; this is the shaded region in Fig. 4.15. If $E = \{G \cap \overset{\circ}{U}(x, y) = 0\}$, the only invariant subset of G is $x = y = 0$.

From Lasalle's theorem the solutions whose initial value is in G_0 have 0 as limit.

If we take the part of H contained between $(2k - 1)\pi$ and $(2k + 1)\pi$ and the corresponding part G_k of G, we can establish that all the solutions issuing from this part G_k have as limit $x = 2k\pi$, $y = 0$.

The equilibria $y = 0$, $x = 2k\pi$ are asymptotically stable and we know their domain of attraction.

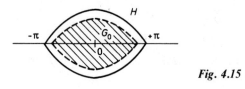

Fig. 4.15

The equilibria $y = 0$, $x = (2k + 1)\pi$ are unstable since the linearized system has matrix

$$Df\{(2k + 1)\pi\} = \begin{pmatrix} 0 & 1 \\ 1 & -1 \end{pmatrix}$$

whose eigenvalues are $(-1 + \sqrt{5})/2$ and $(-1 - \sqrt{5})/2$; these are saddle points.

We can then sketch the orbits of a pendulum with friction and the comparison with those of a pendulum without friction (Fig. 4.16).

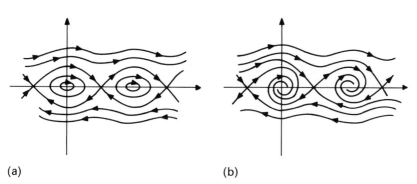

(a)   (b)

Fig. 4.16. (a) *Pendulum without friction; (b) pendulum with friction.*

Example 4: Linear equations with constant coefficients and linearization

If

$$\frac{dx}{dt} = Ax \tag{29}$$

we shall seek U in the form $x^t B x$ where B is a symmetric positive definite matrix. In this case $\overset{\circ}{U}(x) = x^t(A^t B + BA)x$.

If we can find B such that $C = -(A^t B + BA)$ is positive definite we shall have shown that (29) is uniformly asymptotically stable. Now Lemma 3.2.15 (Chapter 3, p. 196) states that if A is an asymptotically stable matrix for every positive symmetric matrix C there exists a unique matrix B such that $C = -(A^t B + BA)$; B is symmetric and positive definite. This gives again the result of Theorem 3.1.2 (Chapter 2, p. 105).

We shall now demonstrate exactly the result relating to the instability of the equilibrium when there exists an eigenvalue of $Df(0)$ with positive real part.

Let

$$\frac{dx}{dt} = Ax + f(x) \tag{30}$$

where $\forall \varepsilon \ \exists \eta \colon \|x\| \leqslant \eta \Rightarrow \|f(x)\| \leqslant \varepsilon \|x\|$. Suppose there exists an eigenvalue λ of A such that $\mathrm{Re}\, \lambda > 0$. We can show that for every symmetric positive definite matrix C there exists B (also symmetric and positive definite) and $a > 0$ such that, if $V(x) = x^t B x$, for Eqn (29) $\mathring{V}(x) = aV(x) + x^t C x$, while for Eqn (30) $\mathring{V}(x) = aV(x) + x^t C x + \langle \text{grad } V, f(x) \rangle$.

In a sufficiently small neighbourhood of the origin $x^t C x + \langle \text{grad } V, f(x) \rangle = W(x)$ is non-negative; therefore $\mathring{V}(x) > 0$. From Cetaev's theorem, $x = 0$ is an unstable equilibrium of (30).

See *Exercises 9 to 12* at the end of the chapter.

3.2. Systems $dx/dt = f(t, x)$ depending on t

Consider a function

$$f : \begin{array}{c} (t, x) \rightsquigarrow f(t, x) \\ \mathbb{R} \times \mathbb{R}^d \to \mathbb{R}^d \end{array}$$

Suppose that $f(t, 0) \equiv 0$ so that $x \equiv 0$ is an equilibrium of the equation

$$\frac{dx}{dt} = f(t, x) \tag{31}$$

In this section we shall only state three theorems which generalize those of the preceding section and indicate examples of applications.

Definition 3.2.1

If Ω is a set containing the origin 0 of $\mathbb{R}^d$, a function $V : \mathbb{R}^+ \times \Omega \to \mathbb{R}$ is called positive definite over $\mathbb{R}^+ \times \Omega$ if

(a) $V(t, 0) = 0 \quad \forall t \in \mathbb{R}^+$,

(b)   there exists U, positive definite over Ω, such that $V(t, x) \geqslant U(x)$; this is equivalent to supposing that there exists $a \in \mathscr{K}$ such that

(b') $V(t, x) \geqslant a(\|x\|)$ over Ω

If φ is a solution of $dx/dt = f(t, x)$ and $V(t, x)$ is a function of class C^1 with values in $\mathbb{R}$, we are interested, as in the preceding section, in the variation of V along the trajectories, i.e. in the function $V[t, \varphi(t)]$; we recall that

$$\frac{d}{dt}\{V[t, \varphi(t)]\} = \frac{\partial V}{\partial t}[t, \varphi(t)] + \left\langle \frac{\partial V}{\partial x}[t, \varphi(t)], f[t, \varphi(t)] \right\rangle$$

Notation   We shall write

$$\mathring{V}(t, x) = \frac{\partial V}{\partial t}(t, x) + \left\langle \frac{\partial V}{\partial x}(t, x), f(t, x) \right\rangle$$

The following theorem is the generalization of Theorem 3.1.3 to the situation in which $f(t, x)$ depends on t.

Theorem 3.2.2 (Lyapunov)
Let

$$\frac{dx}{dt} = f(t, x) \tag{32}$$

be an equation such that $f(t, 0) = 0$.

(a) Suppose there exist a neighbourhood Ω of the origin 0 of $\mathbb{R}^d$ and a function V of class C^1, positive definite over $(\mathbb{R}^+ \times \Omega)$, such that $\dot{V}(t, x) \leqslant 0$ over $\mathbb{R}^+ \times \Omega$. Then $x \equiv 0$ is a stable solution of (32). If $\Omega = \mathbb{R}^n$, all the solutions of (32) are bounded.

(b) Suppose there exist V of class C^1 over $\mathbb{R}^+ \times \Omega$ and a, b, two functions of $\mathscr{K}$ (cf. p. 242), such that over $\mathbb{R}^+ \times \Omega$

$$V(t, 0) \equiv 0 \qquad a(\|x\|) \leqslant V(t, x) \leqslant b(\|x\|) \qquad \dot{V}(t, x) \leqslant 0$$

Then $x \equiv 0$ is a uniformly stable solution of (32).

(c) Suppose there exist V of class C^1 over $\mathbb{R}^+ \times \Omega$ and a, b, c three functions of $\mathscr{K}$, such that over $\mathbb{R}^+ \times \Omega$

$$V(t, 0) \equiv 0 \qquad a(\|x\|) \leqslant V(t, x) \leqslant b(\|x\|) \qquad \dot{V}(t, x) \leqslant -c(\|x\|)$$

Then $x \equiv 0$ is a uniformly asymptotically stable solution of (32). If

$$\rho: \{\|x\| \leqslant \rho\} \subset \Omega \quad \text{and} \quad x_0 \in \{x, V(t_0, x) \leqslant a(\rho)\}$$

then

$$\lim_{t \to \infty} \varphi(t, t_0, x_0) = 0$$

In particular if $\Omega = \mathbb{R}^n$ and if

$$\lim_{\rho \to \infty} a(\rho) = \infty$$

all the solutions of (32) have limit zero.

(d) Suppose there exist V of class C^1 over $\mathbb{R}^+ \times \Omega$ and a and c, two functions of $\mathscr{K}$, such that over $\mathbb{R}^+ \times \Omega$

$$V(t, 0) = 0 \qquad a(\|x\|) \leqslant V(t, x) \qquad \dot{V}(t, x) \leqslant -c(\|x\|)$$

If $f(t, x)$ is bounded for $\|x\| \leqslant A$, $x \equiv 0$ is an asymptotically stable solution of (32). If $\rho: \{\|x\| \leqslant \rho\} \subset \Omega$ and $x_0: V(t_0, x_0) \leqslant a(\rho)$ then

$$\lim_{t \to \infty} \varphi(t, t_0, x_0) = 0$$

Example   Let

$$\frac{d^2x}{dt^2} + a\frac{dx}{dt} + b[1 + \varepsilon f(t)]x = 0 \tag{33}$$

Suppose that a and b are two positive numbers and $f(t)$ is a bounded positive function $(a > 0, b > 0, f(t) \geqslant 0 \ \forall t)$. Let

$$V(x, y) = \frac{1}{2}\left(y + \frac{ax}{2}\right)^2 + \left(\frac{a^2}{4} + b\right)\frac{x^2}{2}$$

$$\mathring{V}(t, x, y) = -\frac{a}{2}y^2 - \frac{ab}{2}[1 + \varepsilon f(t)]x^2 - b\varepsilon f(t)xy$$

If $b\varepsilon^2 f^2(t) - a^2[1 + \varepsilon f(t)] \leqslant c < 0$, $\mathring{V}(t, x, y)$ is negative definite; this happens if $\varepsilon = 0$ and if ε lies between the roots of

$$\varepsilon^2 f^2(t) - \varepsilon\frac{a^2}{b}f(t) - \frac{a^2}{b} = 0$$

In particular if $f(t) \leqslant M \ \forall t$ it suffices that

$$\varepsilon^2 < \frac{a^2}{bM}\left[\left(1 + \frac{4b}{a^2}\right)^{1/2} - 1\right]$$

We can interpret the term $\varepsilon f(t)bx$ as a contribution of energy; if the braking is sufficiently strong the origin is stable, even asymptotically if $\mathring{V}(t, x, y)$ is negative definite.

Like Lyapunov's theorem, Theorem 3.2.2 is often insufficient for applications and it has given rise to numerous generalizations. There follow two very useful ones.

Theorem 3.2.3 (Salvadori)

Let $f(t, x)$ be a function $(\mathbb{R}^+ \times \mathbb{R}^d \leadsto \mathbb{R}^d)$ such that $f(t, 0) \equiv 0$ for every $t \in \mathbb{R}^+$. Let

$$\frac{dx}{dt} = f(t, x) \tag{34}$$

be a differential equation.

Suppose that there exist a neighbourhood Ω of the origin of $\mathbb{R}^d$, two functions V and W defined over $\mathbb{R}^+ \times \Omega$, and three functions a, b, c of $\mathscr{K}$ (cf. p. 242) such that

(a) $V(t, 0) \equiv 0$ $V(t, x) \geqslant a(\|x\|)$ in $\mathbb{R}^+ \times \Omega$
(b) $W(t, 0) \equiv 0$ $W(t, x) \geqslant b(\|x\|)$ in $\mathbb{R}^+ \times \Omega$
(c) $\mathring{V}(t, x) \leqslant -c[W(t, x)]$ and $\exists \alpha, \beta \ \alpha < \mathring{W}(t, x) < \beta$ in $\mathbb{R}^+ \times \Omega$

If $\rho: \{\|x\| \leqslant \rho\} \subset \Omega$ then $\forall x_0: V(t_0, x_0) \leqslant a(\rho)$

$$\lim_{t \to \infty} \varphi(t, t_0, x_0) = 0$$

The origin is therefore asymptotically stable (not necessarily uniformly).

Theorem 3.2.4 (Cetaev)
Let

$$\frac{dx}{dt} = f(t, x) \tag{34}$$

be an equation in which $f: \mathbb{R}^+ \times \mathbb{R}^d \rightsquigarrow \mathbb{R}^d$ and $f(t, 0) = 0$ $\forall t$. If $\rho > 0$ and $\Omega \subset \{\|x\| \leqslant \rho\}$, suppose that the part of the boundary F of Ω such that $\{\|x\| < \rho\}$ contains the point 0 (Ω open) (see Fig. 4.17). Suppose also that there exists V of class C^1 such that

(a) $0 < V(t, x) \leqslant k < \infty$ $\forall t \geqslant t_0$ $x \notin F$ $x \in \Omega$
(b) $\exists a \in \mathscr{K}$ (cf. p. 242): $\mathring{V}(t, x) \geqslant a[V(t, x)]$ $\forall t \geqslant t_0$ $x \in \Omega$
(c) $V(t, x) = 0$ $\forall t \geqslant t_0$ $x \in F$ $\|x\| \leqslant \rho$

Then 0 is an unstable solution of (34).

Fig. 4.17

We can replace (a) and (b) by, $\exists b, c \in \mathscr{K}$,

(a') $0 < V(t, x) \leqslant b(\|x\|)$ $\forall t \geqslant t_0$ $x \in \Omega$ $x \notin F$
(b') $V(t, x) \geqslant c(\|x\|)$ $\forall t \geqslant t_0$ $x \in \Omega$ $x \notin F$

and we can even replace (a') by

(a'') $\exists$ constant k, $k > 0$, $W(t, x)$ continuous over $\mathbb{R}^+ \times \Omega$ with $W(t, x) \geqslant 0$:
$\mathring{V}(t, x) = kV(t, x) + W(t, x)$.

Examples
(1) Let $h(t) > 0$ and

$$\frac{d^2x}{dt^2} + h(t)\frac{dx}{dt} + \sin x = 0 \tag{35}$$

This is the equation of a pendulum subject to friction $h(t)\, dx/dt$ depending on time.
We study (35) by putting $y = dx/dt$. If

$$U(x, y) = \frac{y^2}{2} + (1 - \cos x) \qquad \mathring{U}(t, x, y) = -h(t)y^2$$

$x \equiv 0 \equiv y$ is then a uniformly stable solution, from Theorem 3.2.2. Salvadori's theorem enables us to prove in addition that if

$$V(t, x, y) = \left(\frac{y + \lambda \sin x}{2}\right)^2 + (1 - \cos x)[1 + \lambda h(t) - \lambda^2]$$

$$\overset{\circ}{V}(t, x, y) = -[h(t) - \lambda]y^2 - \lambda\left(2 - \frac{dh}{dt}\right)(1 - \cos x) + g(x, y)$$

where g is independent of t and can be developed in series, the expansion starting with the terms of third order.

If $h(t) \geqslant \alpha > \lambda > 0$, $dh/dt \leqslant \beta < 2$ and

$$W(x, y) = \frac{y^2}{2} + (1 - \cos x)$$

we can verify that $\overset{\circ}{W}(t, x, y) = -h(t)y^2$; $V, -\overset{\circ}{V}$ and W are positive definite and satisfy conditions (a), (b) and (c) of Salvadori's theorem.

Thus 0 is uniformly asymptotically stable and the orbits have behaviour comparable with that sketched on page 253, which corresponds to $h(t) \equiv 1$.

(2) Let $h(t)$ be a real function such that $0 \leqslant h(t) < 2 \quad \forall t \in \mathbb{R}^+$. Let

$$\frac{d^2 x}{dt^2} + h(t)\frac{dx}{dt} - x = 0 \tag{36}$$

To study (36) we put $y = dx/dt$ and $V(x, y) = xy$.

(a) If $h(t) \equiv 0$, $\overset{\circ}{V}(x, y) = x^2 + y^2$ and $\forall \rho$ we leave the circle $\{x^2 + y^2 \leqslant \rho^2\}$, according to Cetaev's Theorem 3.1.5.
(b) If $h(t) \neq 0$, $\overset{\circ}{V}(t, x, y) = x^2 + y^2 - h(t)xy = r^2[1 - h(t)\cos \vartheta \sin \vartheta]$ on passing to polar coordinates. If $x > 0$ and $y > 0$, or if $0 \leqslant \sin \vartheta \cos \vartheta \leqslant \frac{1}{2}$, then $\overset{\circ}{V} \geqslant r^2[1 - \frac{1}{2}h(t)] > 0$ for $r \neq 0$.

Cetaev's Theorem 3.2.4 then enables us to establish that $x \equiv 0 \equiv y$ is an unstable solution of (36).

Comments and complements

The method of Lyapunov which we have presented is extremely powerful provided that we can find Lyapunov functions.

This is a problem to which many mathematicians have addressed themselves and a large number of theorems generalizing the initial ones have been discovered. However, there exists no general method of finding Lyapunov functions.

In cases where first integrals can be found we can often reach a conclusion by combining some of them; the first integral $U(x, y, z)$ used in Example 2(1) (p. 249) is in fact a combination of the following two first integrals:

$$U_1(x, y, z) = A(p_0 + x)^2 + By^2 + Cz^2 \quad \text{(integral of the driving forces)}$$

$$U_2(x, y, z) = A^2(p_0 + x)^2 + B^2 y^2 + C^2 z^2 \quad \text{(modulus of the kinetic energy)}$$

It can be verified that $U(x, y, z) = (U_1 - Ap_0^2)^2 + (U_2 - AU_1)$.

Another method consists of using, for an equation $dx/dt = f(t, x) + g(t, x)$, a Lyapunov function of the equation $dx/dt = f(t, x)$; this is what we did in the two preceding examples. We could say that this method consists in using, for an equation obtained by perturbation from another equation, a Lyapunov function of the initial equation.

In addition, we can obtain results starting from 'reciprocal theorems' which state that, if an equilibrium has certain stability properties, there exist Lyapunov functions, which we can sometimes use without determining them, for equations in the neighbourhood of the initial equation.

Exercises

1. Study the stability of the solutions of the equation $dx/dt = -x(1 - x)$.

2. Repeat Exercise 1 for $dx/dt = 1 - x^2$.

3. Repeat Exercise 1 for

$$\frac{d^2 x}{dt^2} + \frac{x}{2}\left\{x^2 + \left[x^4 + 4\left(\frac{dx}{dt}\right)^2\right]^{1/2}\right\} = 0$$

Put $x = r \cos \vartheta$ and $y = r^2 \sin \vartheta$.

4. Study the following system deriving from the gradient U: $U(x, y) = x^2 + y^2$.

5. Repeat Exercise 4 for $U(x, y) = x^2 - y^2 - 2x - 4y + 5$.

6. Repeat Exercise 4 for $U(x, y) = y \sin x$.

7. Repeat Exercise 4 for $U(x, y, z) = x^2 + y^2 - z$.

8. Study

$$\frac{d^2 x}{dt^2} + f(x) = 0$$

for the following functions f:

(a) $f(x) = x^{2n-1}$ $n \in \mathbb{N}$ $n > 0$
(b) $f(x) = x - x^3$ and $f(x) = x - x^2$
(c) if $\lambda > 0$, $f(x) = x[1 - (1 + \lambda)(1 + x^2)^{-1/2}]$

9. Study the solutions of Eqn (37) below, assuming that the functions h and f are such that $\forall x \neq 0$, $f(x) > 0$, $xh(x) > 0$. We put

$$H(x) = \int_0^x h(u)\, du \quad \text{and} \quad U(x) = \frac{x^2}{2} + H(x)$$

$$\frac{d^2 x}{dt^2} + f(x)\frac{dx}{dt} + h(x) = 0 \tag{37}$$

Verify in particular that all the solutions tend to the origin $(0, 0)$.

10. Consider a series L, R, C circuit and suppose that a current i flowing in the inductance L creates a flux

$$\Phi = Li - \frac{\alpha}{3}i^3$$

α being a positive constant. The equation governing the electric charge q on the capacitor is

$$L\frac{d^2q}{dt^2} - \alpha\left(\frac{dq}{dt}\right)^2\frac{d^2q}{dt^2} + R\frac{dq}{dt} + \frac{1}{C}q = 0$$

Changing the notation, study the following equation in which the constants λ and β are positive:

$$\frac{d^2x}{dt^2} - \beta\left(\frac{dx}{dt}\right)^2\frac{d^2x}{dt^2} + \lambda\frac{dx}{dt} + x = 0 \qquad (38)$$

(a) Trace the orbits corresponding to $\lambda = 0$, using

$$U(x, y) = \frac{1}{2}\left(y^2 - \frac{\beta}{2}y^4 + x^2\right)$$

(b) For $\lambda \neq 0$ prove the asymptotic stability of the origin and find a set contained in the domain of attraction of the origin. Sketch the form of the orbits.

11. Let

$$\frac{dx}{dt} = 2x^3y^2 - x$$

$$\frac{dy}{dt} = -y$$

(a) Prove that the origin is a uniformly asymptotically stable equilibrium.
(b) Find the greatest real number ρ such that

$$\|x_0\| \leqslant \rho \Rightarrow \lim_{t \to \infty} \varphi(t, t_0, x_0) = 0$$

12. This exercise is an illustration of the fact that we can transform a stable equilibrium of an equation $dx/dt = Ax$, where an eigenvalue of A has zero real part, either into an unstable equilibrium or, in contrast, into a uniformly asymptotically stable equilibrium just by a small perturbation. Let

$$\begin{pmatrix} \dfrac{dx}{dt} \\ \dfrac{dy}{dt} \\ \dfrac{dz}{dt} \end{pmatrix} = \begin{pmatrix} 0 & 1 & 0 \\ 0 & 0 & 1 \\ 0 & 0 & 0 \end{pmatrix}\begin{pmatrix} x \\ y \\ z \end{pmatrix} + \begin{pmatrix} g(x, y, z) \\ h(x, y, z) \\ k(x, y, z) \end{pmatrix}$$

where g, h and k have the forms

$$g = -x\varphi(x, y, z)$$
$$h = -y\psi(x, y, z) - 2x^3$$
$$k = -z\eta(x, y, z) - 2y(x^4 + y^2)$$

(If $g \equiv 0 \equiv h \equiv k$, the origin is a stable equilibrium.)

Using the function $U(x, y, z) = z^2 + (y^2 + x^4)^2$, show that we can find functions φ, ψ, η such that the origin is transformed into an unstable equilibrium or, in contrast, into a uniformly asymptotically stable equilibrium.

Chapter 5

First integrals, partial differential equations of the first order

We have already encountered the notion of a first integral: we shall now study this notion more completely and indicate its applications in the study of partial differential equations (PDEs) of the first order.

Part 1: First integrals

1.1. Principal theorem

We know that a first integral is a function which remains constant on the integral curves of a differential equation, so that every function of one or more first integrals is itself a first integral.

We shall establish that for every differential equation there exists locally a certain number of independent first integrals, and all the others can be expressed as functions of these.

The notion of independence was introduced in Chapter 1 (p. 26); we shall recall it and give an equivalent definition.

Definition 1.1.1

Let $f_1, f_2, \ldots, f_n$,

$$f_i: \begin{array}{l} \mathbb{R}^n \longrightarrow \mathbb{R} \\ (x_1, \ldots, x_n) \rightsquigarrow f_i(x_1, \ldots, x_n) \end{array}$$

be n continuously differentiable functions. Let G be an open subset of $\mathbb{R}^n$. We say that the functions $\{f_i\}_{i=1,\ldots,n}$ are independent in G if

$$\det\left[\frac{D(f_1, f_2, \ldots, f_n)}{D(x_1, \ldots, x_n)}\right] \neq 0 \quad \text{in } G$$

Let F be the transformation $x = (x_1, \ldots, x_n) \rightsquigarrow [f_1(x), \ldots, f_n(x)]$ and let $F(G)$ be the image of G; the preceding condition states that F is invertible in G and that the inverse is differentiable at every point of $F(G)$.

If we consider k functions $(k < n)$ $f_1, \ldots, f_k$: $\mathbb{R}^n \to \mathbb{R}$, we say that they are independent in G if the rank of the matrix

$$\frac{D(f_1, \ldots, f_k)}{D(x_1, \ldots, x_n)}$$

is k.

Theorem 1.1.2

n functions of class C^1 $f_1, \ldots, f_n$ from $\mathbb{R}^n$ into $\mathbb{R}$ are independent in G iff there exists no function H from $\mathbb{R}^n$ into $\mathbb{R}$ which is differentiable and not identically zero in $F(G)$ such that $H[f_1(x), \ldots, f_n(x)] \equiv 0$ in G.

Proof

Sufficient condition If there exists no function H which has the properties indicated then the relation

$$\lambda_1 f_1(x) + \lambda_2 f_2(x) + \ldots + \lambda_n f_n(x) = 0 \qquad \forall x \in G$$

implies

$$\lambda_1 = \lambda_2 = \ldots = \lambda_n = 0$$

and in the same way (by *reductio ad absurdum*) the relation

$$\lambda_1 \frac{\partial f_1}{\partial x_1}(x) + \lambda_2 \frac{\partial f_2}{\partial x_1}(x) + \ldots + \lambda_n \frac{\partial f_n}{\partial x_1}(x) = 0 \qquad \forall x \in G$$

Consider then the system

$$\left. \begin{array}{c} 0 = \lambda_1 \dfrac{\partial f_1}{\partial x_1}(x) + \ldots + \lambda_n \dfrac{\partial f_n}{\partial x_1}(x) \\[2ex] 0 = \lambda_1 \dfrac{\partial f_1}{\partial x_2}(x) + \ldots + \lambda_n \dfrac{\partial f_n}{\partial x_2}(x) \\[1ex] \vdots \\[1ex] 0 = \lambda_1 \dfrac{\partial f_1}{\partial x_n}(x) + \ldots + \lambda_n \dfrac{\partial f_n}{\partial x_n}(x) \end{array} \right\} \qquad (1)$$

In G, (1) admits no solution other than $\lambda_1 = \lambda_2 = \ldots = \lambda_n = 0$. The corresponding determinant is therefore null in G and the functions $f_1, \ldots, f_n$ are thus independent in the sense of the preceding definition.

Necessary condition Suppose now that

$$\det\left[\frac{D(f_1, \ldots, f_n)}{D(x_1, \ldots, x_n)}\right] \neq 0 \qquad \forall x \in G$$

and that there exists a differentiable function

$$H: \begin{array}{l} \mathbb{R}^n \to \mathbb{R} \\ (y_1, \ldots, y_n) \rightsquigarrow H(y_1, \ldots, y_n) \end{array}$$

such that $\forall x \in G \;\; H[f_1(x), f_2(x), \ldots, f_n(x)] = 0$. Then

$$\forall x \in G \quad \left\{ \begin{array}{l} \dfrac{\partial H}{\partial y_1} \dfrac{\partial f_1}{\partial x_1}(x) + \ldots + \dfrac{\partial H}{\partial y_n} \dfrac{\partial f_n}{\partial x_1}(x) = 0 \\ \vdots \\ \dfrac{\partial H}{\partial y_1} \dfrac{\partial f_1}{\partial x_n}(x) + \ldots + \dfrac{\partial H}{\partial y_n} \dfrac{\partial f_n}{\partial x_n}(x) = 0 \end{array} \right.$$

Since the preceding system (1) admits no solution other than $\lambda_1 = \ldots = \lambda_n = 0$, then

$$\frac{\partial H}{\partial y_1} = \frac{\partial H}{\partial y_2} = \ldots = \frac{\partial H}{\partial y_n} \equiv 0 \quad \text{in } F(G)$$

Thus H is constant and this constant can only be zero.

We shall now define the notion of first integral.

Definition 1.1.3
(a) If

$$\frac{\mathrm{d}x}{\mathrm{d}t} = f(x) \tag{2}$$

is an autonomous differential equation in $\mathbb{R}^n$, we define a **first integral independent of time** as a differentiable function $U: \mathbb{R}^n \to \mathbb{R}$ such that, for every solution φ of (2),

$$\frac{\mathrm{d}}{\mathrm{d}t} U[\varphi(t)] \equiv 0$$

(b) If

$$\frac{\mathrm{d}x}{\mathrm{d}t} = f(t, x) \tag{3}$$

is a differential equation in which $x \in \mathbb{R}^n$, we define a **first integral dependent on time** as a differentiable function $U: \mathbb{R} \times \mathbb{R}^n \to \mathbb{R}$ such that, for every solution φ of (3),

$$\frac{\mathrm{d}}{\mathrm{d}t} U[t, \varphi(t)] \equiv 0$$

Comments

(1) We can calculate

$$\frac{d}{dt} U[t, \varphi(t)]$$

as

$$\frac{d}{dt} U[t, \varphi(t)] = \frac{\partial U}{\partial t}[t, \varphi(t)] + \langle \text{grad } U[\varphi(t)], f[t, \varphi(t)] \rangle$$

$$= \frac{\partial U}{\partial t}[t, \varphi(t)] + \sum_{i=1}^{n} \frac{\partial U}{\partial x_i}[\varphi(t)] f_i[t, \varphi(t)]$$

denoting by $f_1, \ldots, f_n$ the coordinates of f.

We shall naturally suppose that f is such that in a domain D of $\mathbb{R} \times \mathbb{R}^d$ there is a unique solution passing through every point (t, x) of D, where as usual we denote by $\varphi(t, t_0, x_0)$ the solution such that $\varphi(t_0) = x_0$; $\varphi(t, t, x) \equiv x$ so that, from the preceding calculation, U is a first integral in the domain D iff

$$\frac{\partial U}{\partial t}(t, x) + \sum_{i=1}^{n} \frac{\partial U}{\partial x_i}(t, x) f_i(t, x) \equiv 0$$

It is this fact which establishes the connection between the notion of first integrals and the solution of PDEs of the first order.

(2) In the case of a scalar equation of order n a first integral may be defined by passing to the equivalent first order system. Let

$$\frac{d^n x}{dt^n} = F\left(t, x, \frac{dx}{dt}, \ldots, \frac{d^{n-1}x}{dt^{n-1}}\right) \tag{4}$$

A first integral is a function of $(t, x_1, \ldots, x_n)$ such that for every solution φ of (4)

$$\frac{d}{dt} U\left[t, \varphi(t), \frac{d\varphi}{dt}, \ldots, \frac{d^{n-1}\varphi}{dt^{n-1}}\right] \equiv 0$$

This is a solution of the partial differential equation

$$0 = \frac{\partial}{\partial t} U(t, x_1, \ldots, x_n) + \frac{\partial}{\partial x_1} U(t, x_1, \ldots, x_n) x_2 + \ldots + \frac{\partial}{\partial x_{n-1}} U(t, x_1, \ldots, x_n) x_n$$

$$+ \frac{\partial}{\partial x_n} U(t, x_1, \ldots, x_n) F(t, x_1, \ldots, x_n)$$

obtained by writing the system equivalent to (4).

Let us examine the following very simple particular case:

$$\frac{dy_1}{dt} = 1 \qquad \frac{dy_2}{dt} = 0 \qquad \ldots \qquad \frac{dy_n}{dt} = 0$$

whose solutions are

$$\varphi(t, 0, a) = \begin{pmatrix} a_1 + t \\ a_2 \\ \vdots \\ a_n \end{pmatrix} \quad \text{where } a = (a_1 \ \ldots \ a_n)$$

It is clear that the functions $(y_1, \ldots, y_n) \rightsquigarrow y_i$ for $i \geq 2$ are constant on the integral curves; they form a set of $n - 1$ independent first integrals for

$$\frac{D(y_2, \ldots, y_n)}{D(y_2, \ldots, y_n)} = \text{identity}$$

If U is any first integral of this system, then

$$U[\varphi(t, 0, a)] = U(a_1 + t, a_1, \ldots, a_n)$$

and

$$\frac{d}{dt} U[\varphi(t, 0, a)] = \frac{\partial U}{\partial y_1}(a_1 + t, a_2, \ldots, a_n) \equiv 0$$

which implies that U depends only on $y_2, \ldots, y_n$.

We have thus found $n - 1$ independent first integrals as a function of which all others can be expressed.

Now in the neighbourhood of a point a where $f(a) \neq 0$ an equation $dx/dt = f(x)$ may always be reduced to the preceding particular case according to the straightening theorem (Chapter 1, Theorem 2.2.12, p. 24).

It is not surprising then that what we have established for this particular equation is true in general.

Theorem 1.1.4
(a) Let

$$\frac{dx}{dt} = f(x) \tag{5}$$

be an autonomous system in $\mathbb{R}^n$ where f is differentiable and a such that $f(a) \neq 0$. There then exists a neighbourhood G of a in which we can find $n - 1$ first integrals $U_1, \ldots, U_{n-1}$, independent of time and mutually independent, such that every first integral U independent of time can be expressed in the form

$$U(x) = H[U_1(x), \ldots, U_{n-1}(x)]$$

where H is continuously differentiable in a neighbourhood of $[U_1(a), \ldots, U_{n-1}(a)]$.

(b) Let

$$\frac{dx}{dt} = f(t, x) \tag{6}$$

be a differential equation. For every point (t_0, a) there exists a neighbourhood G in which we can find n first integrals $V_1, \ldots, V_n$, depending on time but mutually independent, such that every first integral V may be expressed in the form $V(t, x) = K[V_1(t, x), \ldots, V_n(t, x)]$ where K is of class C^1 in a neighbourhood of $[V_1(t_0, a), \ldots, V_n(t_0, a)]$.

Proof
(a) Let $(e_1, \ldots, e_n)$ be a basis of $\mathbb{R}^n$, $a \in \mathbb{R}^n$, and f be such that $f(a) \neq 0$ and continuously differentiable. According to the straightening theorem there exist a neighbourhood G of a and a transformation T, differentiable and invertible in G, such that, at every point of G, $[DT] f(x) = e_1$.

If $y = T(x)$, the equation

$$\frac{dx}{dt} = f(x) \tag{7}$$

is transformed into

$$\frac{dy}{dt} = e_1 \tag{8}$$

which can be illustrated as shown in Fig. 5.1.

Let us apply the transformation T to the solutions φ of (7); if $\psi(t) = T[\varphi(t)]$, ψ is a solution of (8). We know that $y_2, \ldots, y_n$ are first integrals of (8); therefore $dy_i[\psi(t)]/dt = 0$. If $U_i(x) = y_i[T(x)]$ for $i = 2, \ldots, n$ so that $U_i[\varphi(t)] = y_i\{T \circ T^{-1}[\psi(t)]\} = y_i[\psi(t)]$, U_i is then a first integral in G. $U_2, \ldots, U_n$ are independent in G, for if there did exist a differentiable H such that

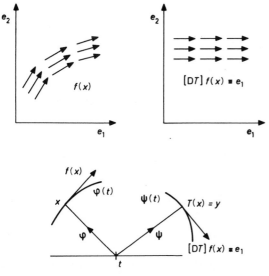

Fig. 5.1

$H(U_2, \ldots, U_n)(x) \equiv 0 \equiv H(y_2, \ldots, y_n)T(x)$, since T is invertible we would have $H(y_2, \ldots, y_n) \equiv 0$.

In the same way if U is a first integral of (7) (not depending on t), if $V(y_1, \ldots, y_n) = U[T^{-1}(y)]$, if ψ is a solution of (8) and $\varphi = T^{-1}(\psi)$ is the corresponding solution of (7), so that

$$\frac{\mathrm{d}}{\mathrm{d}t} V[\psi(t)] = \frac{\mathrm{d}}{\mathrm{d}t} U[\varphi(t)] = 0$$

and if V is a first integral of (8) and can therefore be expressed as a function of $y_2, \ldots, y_n$ only, so that

$$V(y_1, \ldots, y_n) = H(y_2, \ldots, y_n) = H\{U_2[T^{-1}(y)], \ldots, U_n[T^{-1}(y)]\}$$

and if $x = T^{-1}(y)$, then

$$U(x) = H[U_2(x), \ldots, U_n(x)]$$

(b) Let us associate with the equation

$$\frac{\mathrm{d}x}{\mathrm{d}t} = f(t, x) \tag{9}$$

where $x \in \mathbb{R}^n$ an autonomous equation in $\mathbb{R}^{n+1}$

$$\frac{\mathrm{d}\tilde{x}}{\mathrm{d}t} = \tilde{f}(\tilde{x}) \tag{9'}$$

putting $\tilde{x} = (t, x_1, \ldots, x_n)$ and $\tilde{f}(\tilde{x}) = [1, f_1(t, x), \ldots, f_n(t, x)]$. To every solution φ of (9) there corresponds (and conversely) a unique solution $\tilde{\varphi}$ of (9') defined by $\tilde{\varphi} = [t, \varphi_1(t), \ldots, \varphi_n(t)]$. It is clear that, $\forall \tilde{x}$, $\tilde{f}(\tilde{x}) \neq 0$ and that there exist n independent first integrals $U_1(\tilde{x}), \ldots, U_n(\tilde{x})$. Now

$$\frac{\mathrm{d}}{\mathrm{d}t} U_i[\tilde{\varphi}(t)] = \frac{\mathrm{d}}{\mathrm{d}t} U_i[t, \varphi(t)] \equiv 0$$

so that $U_i(t, x)$ is a first integral of (9) dependent on time.

With every first integral $V(t, x)$ of (9) we can associate $U(\tilde{x}) = V(t, x)$, a first integral of (9') which can thus be expressed as a function of the U_i:

$$U(\tilde{x}) = H[U_1(\tilde{x}), \ldots, U_n(\tilde{x})]$$

and

$$V(t, x) = H[U_1(t, x), \ldots, U_n(t, x)]$$

Comments   The essential point of the proof is the following: if $(y_1, \ldots, y_n)$ is given as the value at t of a solution $\varphi(t, 0, a)$ we can recover a as a function of $(t, y_1, \ldots, y_n)$ and each coordinate provides a first integral.

A second method of establishing the theorem consists in verifying that this fact is general when $f(a) \neq 0$. We shall indicate the broad lines of this proof.

Second proof If

$$\frac{dx}{dt} = f(x) \tag{10}$$

and $a = (a_1, \ldots, a_n)$ is a point of $\mathbb{R}^n$ such that $f(a) \neq 0$, we shall assume that the nth coordinate $f_n(a)$ (for example) of $f(a)$ is non-zero ($f_n(a) \neq 0$).

If $\xi = (\xi_1, \ldots, \xi_{n-1})$, $\tilde{\xi} = (\xi_1, \ldots, \xi_{n-1}, a_n)$ and $x = \varphi(t, t_0, \tilde{\xi})$, we shall try to invert this relation, i.e. to determine (t, ξ) in the form $(t, \xi) = H(x)$. We shall denote by $H_1, \ldots, H_n$ the n coordinates of $H(x)$ and by $\varphi_1, \ldots, \varphi_n$ those of φ.

$$\left. \begin{aligned} x_1 &= \varphi_1(t, t_0, \xi_1 \ldots \xi_{n-1}, a_n) \\ &\vdots \\ x_n &= \varphi_n(t, t_0, \xi_1 \ldots \xi_{n-1}, a_n) \end{aligned} \right\} \tag{11}$$

$$\left. \begin{aligned} \xi_1 &= H_1(x_1, \ldots, x_n) \\ &\vdots \\ \xi_{n-1} &= H_{n-1}(x_1, \ldots, x_n) \\ t &= H_n(x_1 \ldots x_n) \end{aligned} \right\} \tag{12}$$

We can write (11) in the equivalent form (12) if

$$\det \frac{D(\varphi_1 \ldots \varphi_n)}{D(\xi_1 \ldots \xi_{n-1}, t)} (t, \xi) \neq 0$$

Now, by hypothesis, $\tilde{\xi} = \varphi(t_0, t_0, \tilde{\xi})$, i.e.

$$\begin{aligned} \xi_1 &= \varphi_1(t_0, t_0, \xi_1 \ldots \xi_{n-1}, a_n) \\ &\vdots \\ \xi_{n-1} &= \varphi_{n-1}(t_0, t_0, \xi_1 \ldots \xi_{n-1}, a_n) \\ a_n &= \varphi_n(t_0, t_0, \xi_1 \ldots \xi_{n-1}, a_n) \end{aligned}$$

and

$$\frac{D(\varphi_1 \ldots \varphi_n)}{D(t_0, \xi_1 \ldots \xi_{n-1})} (t_0, \xi_1 \ldots \xi_{n-1}) = \begin{pmatrix} 1 & & 0 & & f_1(\xi_1) \\ & 1 & & & \\ & & \ddots & & \\ 0 & & & 1 & f_{n-1}(\xi_{n-1}) \\ 0 & & & 0 & f_n(a_n) \end{pmatrix}$$

whose determinant is non-zero; therefore if t is sufficiently close to t_0 this determinant remains non-zero and we can define H so that $H_i[\varphi(t, t_0, \tilde{\xi})] \equiv \xi_i$ is constant. Moreover $H_1, \ldots, H_{n-1}$ are independent, as the following calculation shows:

$$\frac{D(H_1 \ldots H_n)}{D(x_1 \ldots x_n)} (\xi_1 \ldots \xi_{n-1}, a_n) = \left[\frac{D(\varphi_1 \ldots \varphi_n)}{D(\xi_1 \ldots \xi_{n-1}, t)} \right]^{-1} (t_0, \xi_1 \ldots \xi_{n-1})$$

$$= \begin{pmatrix} 1 & & 0 & -\dfrac{f_1(\xi_1)}{f_n(a_n)} \\ & 1 & & \\ 0 & & 1 & -\dfrac{f_{n-1}(\xi_{n-1})}{f_n(a_n)} \\ 0 & 0 & 0 & \dfrac{1}{f_n(a_n)} \end{pmatrix}$$

and therefore

$$\frac{D(H_1, \ldots, H_{n-1})}{D(x_1, \ldots, x_{n-1})} (\xi_1, \ldots, \xi_{n-1}) = \text{identity}$$

We then prove that every first integral U can be expressed as a function of $(H_1, \ldots, H_{n-1})$.

1.2. Geometric interpretation

To limit ourselves to familiar ideas we shall consider the space $\mathbb{R}^3$ but the essence of the results is clearly the same in $\mathbb{R}^n$.

If

$$\frac{dX}{dt} = f(X) \tag{13}$$

is an autonomous equation, we shall write

$$X = \begin{pmatrix} x \\ y \\ z \end{pmatrix} \quad \text{and} \quad F(X) = \begin{pmatrix} P(x, y, z) \\ Q(x, y, z) \\ R(x, y, z) \end{pmatrix}$$

so that Eqn (13) may be written

$$\frac{dx}{dt} = P(x, y, z)$$

$$\frac{dy}{dt} = Q(x, y, z)$$

$$\frac{dz}{dt} = R(x, y, z)$$

A solution will be denoted

$$\Phi(t) = \begin{pmatrix} \varphi(t) \\ \psi(t) \\ \eta(t) \end{pmatrix}$$

Then an orbit is a curve in $\mathbb{R}^3$ tangential to the vector (P, Q, R). Let U be a first integral; the orbit $[\varphi(t), \psi(t), \eta(t)]$ is contained in a surface

$$U(x, y, z) = \text{constant} = U[\varphi(t_0), \psi(t_0), \eta(t_0)]$$

We can determine an orbit as the intersection of two surfaces represented by the equations $U_1(x, y, z) = a_1$ and $U_2(x, y, z) = a_2$, U_1 and U_2 being two independent first integrals.

This result is true when $F(a) \neq 0$ according to the fundamental theorem. If $F(a) = 0$, a is an equilibrium point and we cannot make any statement about the neighbourhood of a.

We shall examine a particular method of expressing the fact that a solution has for its orbit a curve tangential to a given vector.

Study of the equation

$$\frac{dx}{P(x, y, z)} = \frac{dy}{Q(x, y, z)} = \frac{dz}{R(x, y, z)}$$

Definition 1.2.1

Let P, Q, R be three continuous functions from $\mathbb{R}^3$ into $\mathbb{R}$ not simultaneously zero. We define a solution of the equation

$$\frac{dx}{P(x, y, z)} = \frac{dy}{Q(x, y, z)} = \frac{dz}{R(x, y, z)} \tag{14}$$

as a curve whose tangent at each point (x, y, z) is the vector (P, Q, R).

We note that the parametrization of the curve does not appear. If in fact

$$x = \varphi(t)$$
$$y = \psi(t)$$
$$z = \eta(t)$$

is the parametrization, Eqn (14) is equivalent to

$$\frac{d\varphi}{dt} = P[\varphi(t), \psi(t), \eta(t)]$$

$$\frac{d\psi}{dt} = Q[\varphi(t), \psi(t), \eta(t)]$$

$$\frac{d\eta}{dt} = R[\varphi(t), \psi(t), \eta(t)]$$

The curves are the same as the orbits of the solutions of the equations

$$
\left.\begin{aligned}
\frac{dy}{dx} &= \frac{Q}{P}(x, y, z) \\[2mm]
\frac{dz}{dx} &= \frac{R}{P}(x, y, z)
\end{aligned}\right\} \tag{15a}
$$

$$
\left.\begin{aligned}
\frac{dx}{dz} &= \frac{P}{R}(x, y, z) \\[2mm]
\frac{dy}{dz} &= \frac{Q}{R}(x, y, z)
\end{aligned}\right\} \tag{15b}
$$

$$
\left.\begin{aligned}
\frac{dx}{dy} &= \frac{P}{Q}(x, y, z) \\[2mm]
\frac{dz}{dy} &= \frac{R}{Q}(x, y, z)
\end{aligned}\right\} \tag{15c}
$$

at least one of these three equations having a direction at a given point (x, y, z).

Let a, for example, be such that $P(a) \neq 0$ and let G be a neighbourhood of this point where $P(x, y, z) \neq 0$. Let

$$
\Phi = \begin{pmatrix} \varphi \\ \psi \\ \eta \end{pmatrix}
$$

be a solution of (14). Since $P[\varphi(t), \psi(t), \eta(t)] \neq 0$, $(d\varphi/dt)^{-1}$ exists and φ is invertible. Let h be the inverse function $t = h(x) \Leftrightarrow x = \varphi(t)$ and the derivative of h at the point $x_0 = \varphi(t_0)$ is

$$
\frac{dh}{dx_0} = \frac{1}{P[\varphi(t_0), \psi(t_0), \eta(t_0)]} = \frac{1}{P\{x_0, \psi[h(x_0)], \eta[h(x_0)]\}}
$$

If we put $y = \psi[h(x)]$, $z = \eta[h(x)]$, the pair (y, z) is a solution of (15a). Conversely, if $y = f(x)$, $z = g(x)$, is a solution of (15a), the curve described by the point with coordinates

$$
\begin{pmatrix} x \\ f(x) \\ g(x) \end{pmatrix}
$$

has as tangent vector

$$
\left\{\begin{aligned}
& 1 \\[1mm]
& \frac{Q}{P}[x, f(x), g(x)] \\[1mm]
& \frac{R}{P}[x, f(x), g(x)]
\end{aligned}\right.
$$

which, since $P[x, f(x), g(x)]$ is non-zero, is proportional to

$$\begin{pmatrix} P \\ Q \\ R \end{pmatrix} [x, f(x), g(x)]$$

It is then sufficient to change the parametrization by taking for φ a solution of $d\varphi/dt = P\{\varphi(t), f[\varphi(t)], g[\varphi(t)]\}$.

The curve

$$\begin{aligned} x &= \varphi(t) \\ y &= f[\varphi(t)] = \begin{pmatrix} \varphi(t) \\ \psi(t) \\ \eta(t) \end{pmatrix} \\ z &= g[\varphi(t)] \end{aligned}$$

has as its tangent

$$P[\varphi(t), \psi(t), \eta(t)]$$
$$Q[\varphi(t), \psi(t), \eta(t)]$$
$$R[\varphi(t), \psi(t), \eta(t)]$$

and is a solution of (14).

If for example $R(x, y, z) \equiv 0$ the solution of (14) verifies that $z = \text{constant}$; it is convenient to say that, if in (14) a denominator is zero, so is the corresponding numerator.

A first integral of (14) is a function $U(x, y, z)$ constant on the solutions of (14); let (φ, ψ, η) be such a solution.

$$\frac{d}{dt} U[\varphi(t), \psi(t), \eta(t)] = \frac{\partial U}{\partial x} [\varphi(t), \psi(t), \eta(t)] P[\varphi(t), \psi(t), \eta(t)]$$

$$+ \frac{\partial U}{\partial y} [\varphi(t), \psi(t), \eta(t)] Q[\varphi(t), \psi(t), \eta(t)]$$

$$+ \frac{\partial U}{\partial z} [\varphi(t), \psi(t), \eta(t)] R[\varphi(t), \psi(t), \eta(t)]$$

Briefly, U is a first integral of (14) iff

$$0 = P \frac{\partial U}{\partial x} + Q \frac{\partial U}{\partial y} + R \frac{\partial U}{\partial z} \tag{16}$$

and for $P \neq 0$

$$0 = \frac{\partial U}{\partial x} + \frac{Q}{P} \frac{\partial U}{\partial y} + \frac{R}{P} \frac{\partial U}{\partial z}$$

$U(x, y, z)$ is then a first integral of (15a) depending on x.

The relation (16) states that, if U is a first integral, the surface $U(x, y, z) = \text{constant}$ which contains the solution passing through a given point has a tangent plane which contains the vector (P, Q, R) at every point of the solution curve.

Practical method

To find a first integral of (14) we can manipulate the fractions which occur in (14) as in ordinary fractions, assuming that a zero numerator corresponds to a zero denominator.

Suppose for example that

$$\begin{pmatrix} \varphi \\ \psi \end{pmatrix}$$

is a solution of the equation in $\mathbb{R}^2$

$$\frac{dx}{P} = \frac{dy}{Q}$$

This means that

$$\frac{d\varphi/dt}{P[\varphi(t), \psi(t)]} = \frac{d\psi/dt}{Q[\varphi(t), \psi(t)]}$$

which for all functions R and S is again equal to

$$\frac{R[\varphi, \psi](t) \, d\varphi/dt + S[\varphi, \psi](t) \, d\psi/dt}{R(\varphi, \psi)P(\varphi, \psi)(t) + S(\varphi, \psi)Q(\varphi, \psi)(t)}$$

If U exists such that

$$R(x, y) = \frac{\partial U}{\partial x}(x, y) \quad \text{and} \quad S(x, y) = \frac{\partial U}{\partial y}(x, y)$$

U will be a first integral if the denominator $RP + SQ$ is zero.

Instead of writing $d\varphi/dt$ and $d\psi/dt$ throughout, we shall keep to dx and dy. Thus, if

$$\frac{dx}{y} = \frac{dy}{x} \tag{17}$$

we shall write

$$\frac{dx}{y} = \frac{dy}{x} = \frac{x \, dx - y \, dy}{0} = \frac{d(x^2 - y^2)}{0}$$

and $x^2 - y^2$ is a first integral of (17).

Example   Solve the following system in $\mathbb{R}^3$:

$$\frac{dx}{y(x + y) + az} = \frac{dy}{x(x + y) - az} = \frac{dz}{z(x + y)} \tag{18}$$

$$\frac{dz}{z(x + y)} = \frac{dx + dy}{(x + y)^2}$$

Then

$$\frac{dz}{z} = \frac{d(x+y)}{x+y}$$

and $(x+y)/z$ is a first integral.

$$\frac{x\,dx - y\,dy}{az(x+y)} = \frac{dz}{z(x+y)}$$

Therefore $x^2 - y^2 - az$ is another first integral. These two first integrals are clearly independent in every region in which $z \neq 0$. Moreover

$$\frac{D(U_1, U_2)}{D(x, y, z)} = \begin{pmatrix} \dfrac{1}{z} & \dfrac{1}{z} & -\dfrac{x+y}{z^2} \\ 2x & -2y & -a \end{pmatrix}$$

where we denote by $U_1(x, y, z)$ the quantity $(x+y)/z$ and by $U_2(x, y, z)$ the quantity $x^2 - y^2 - az$.

The minors are

$$-\frac{2}{z}(x+y) \qquad -\frac{1}{z^2}[az + 2y(x+y)] \qquad \frac{1}{z^2}[2x(x+y) - az]$$

and cannot all be zero together for $z \neq 0$.

The solutions of (18) are the intersections of the surfaces S_{k_1,k_2}:

$$x + y = k_1 z \qquad x^2 - y^2 - az = k_2$$

We can separately seek the solutions of (18) in the plane $z \equiv 0$. They are the solutions of

$$\frac{dx}{y(x+y)} = \frac{dy}{x(x+y)}$$

i.e. $x^2 - y^2 = \text{constant}$.

See *Exercises 1 and 2* at the end of the chapter.

Part 2: Linear partial differential equations of the first order

We have seen (p. 265) that a function U is a first integral in a domain iff it is a solution of a certain linear PDE of the first order; it is this fact which enables us to solve these equations.

2.1. Solution of equations

Definition 2.1.1

We define a linear PDE of the first order in $\mathbb{R}^n$ by an equation of the form

$$\sum_{i=1}^{n} f_i[x_1, \ldots, x_n, u(x_1 \ldots x_n)] \frac{\partial u}{\partial x_i} (x_1 \ldots x_n) = h[x_1, \ldots, x_n, u(x_1 \ldots x_n)]$$

$$(19)$$

$u: \mathbb{R}^n \to \mathbb{R}, x = (x_1, \ldots, x_n) \rightsquigarrow u(x)$ is a solution of (19) in a domain if it satisfies this relation in this domain.

More generally, we seek a solution in an implicit form: $H(x_1, \ldots, x_n, u) = 0$ where H is a differentiable function (to be determined) such that in the neighbourhood of a point (a, u_0)

$$\frac{\partial H}{\partial u} (x; u) \neq 0$$

Then we know that

$$\frac{\partial u}{\partial x_i} = - \frac{\partial H/\partial x_i}{\partial H/\partial u}$$

so that u is a solution of (19) iff H is a solution of

$$\sum_{i=1}^{n} f_i(x; u) \frac{\partial H}{\partial x_i} (x; u) + h(x; u) \frac{\partial H}{\partial u} (x; u) = 0 \qquad (19')$$

The differences between (19) and (19′) are as follows.

(a) On the one hand, the unknown function u appears in the coefficients f_i of Eqn (19) whereas the unknown function H does not appear in the coefficients of Eqn (19′).

(b) On the other hand, if we separate the terms containing a derivative and those which do not, Eqn (19′) has a zero second term whereas this is not in general the case with Eqn (19).

We note that Eqn (19′) has the same form as Eqn (16) (p. 273)

$$0 = P \frac{\partial U}{\partial x} + Q \frac{\partial U}{\partial y} + R \frac{\partial U}{\partial z}$$

provided that $h(x; u)$ is non-zero. This case is examined separately and we exploit the fact that (16) is the equation of first integrals of the system

$$\frac{dx_1}{f_1} = \frac{dx_2}{f_2} = \ldots = \frac{dx_n}{f_n} = \frac{du}{h} \qquad (20)$$

Definition 2.1.2

The system (20) is called the characteristic system of (19′) (or of (19)).

We thus obtain the following result.

Theorem 2.1.3
Let

$$\sum_{i=1}^{n} f_i[x; u(x)] \frac{\partial u}{\partial x_i} = h[x; u(x)] \quad \text{where } x \in \mathbb{R}^n \text{ and } u: \begin{array}{l} \mathbb{R}^n \to \mathbb{R} \\ x \rightsquigarrow u(x) \end{array}$$

$$(21)$$

Let

$$\sum_{i=1}^{n} f_i(x; u) \frac{\partial H}{\partial x_i} (x; u) + h(x, u) \frac{\partial H}{\partial u} (x; u) = 0 \qquad (21')$$

where

$$H: \begin{array}{l} \mathbb{R}^{n+1} \to \mathbb{R} \\ (x; u) \rightsquigarrow H(x; u) \end{array}$$

A solution of (21) is defined implicitly by a solution of (21'); the solutions of (21') are the first integrals of the characteristic system

$$\frac{\mathrm{d}x_1}{f_1} = \frac{\mathrm{d}x_2}{f_2} = \ldots = \frac{\mathrm{d}x_n}{f_n} = \frac{\mathrm{d}u}{h} \qquad (20)$$

There exist n mutually independent solutions $(H_1, \ldots, H_n)$ of (21'), as a function of which the others may be expressed in the neighbourhood of every point.

Remarks
(1) It often happens that we can find, in the entire space $\mathbb{R}^n$, n independent first integrals; the solutions are then defined as functions of them over the whole of $\mathbb{R}^n$.
(2) If (21) is already in the simplified form

$$\sum_{i=1}^{n} f_i(x) \frac{\partial u}{\partial x_i} = 0 \qquad (21'')$$

we naturally look directly for the solutions as first integrals of the simplified characteristic system

$$\frac{\mathrm{d}x_1}{f_1} = \ldots = \frac{\mathrm{d}x_n}{f_n}$$

There are $n - 1$ independent solutions as a function of which the others may be expressed; in addition there is the obvious solution $u = \text{constant}$.

Examples

(1)

$$y\frac{\partial z}{\partial x} - x\frac{\partial z}{\partial y} = 0 \tag{22}$$

where the unknown is the function $z(x, y)$. Constants are solutions and the system is in simplified form; its characteristic system is $dx/y = -dy/x$; $x^2 + y^2$ is a first integral.

All solutions are then of the form $z = \Phi(x^2 + y^2)$; these are surfaces of revolution about the axis $0z$, horizontal planes being actually particular cases.

This result may also be found directly. In fact if $z = \varphi(x, y)$ is a solution and $\varphi(x, y) = a$ is a horizontal section of that surface, and if (x_0, y_0) is a point of the section, the tangent to the section at the point (x_0, y_0) is

$$(x - x_0)\frac{\partial \varphi}{\partial x}(x_0, y_0) + (y - y_0)\frac{\partial \varphi}{\partial y}(x_0, y_0) = 0$$

Equation (22) states that the vector (x_0, y_0) is orthogonal to this tangent; this is a characteristic property of circles centred at the origin.

(2)

$$z\frac{\partial z}{\partial x} = (1 - z^2)^{1/2} \tag{23}$$

We examine separately the case $0 = (1 - z^2)^{1/2}$, i.e. $z = \pm 1$. These planes are solutions of (23). Let us seek the other solutions in the domain $|z| < 1$ in the form $H(x, y, z) = 0$.

Equation (21′) is

$$z\frac{\partial H}{\partial x} + (1 - z^2)^{1/2}\frac{\partial H}{\partial z} = 0$$

and the characteristic system is

$$\frac{dx}{z} = \frac{dy}{0} = \frac{dz}{(1 - z^2)^{1/2}}$$

y is a first integral and the relation

$$dx = \frac{z\,dz}{(1 - z^2)^{1/2}}$$

provides a second, namely $x + (1 - z^2)^{1/2}$.

The solutions of (21′) are then $H(x, y, z) = \Phi[y, x + (1 - z^2)^{1/2}]$ and those of (21) are $z^2 = 1 - [x - F(y)]^2$ where F is a continuously differentiable function such that $[x - F(y)]^2 < 1$. This is the equation of the surface generated by a circle of radius 1 traced in a plane $y = $ constant, whose centre describes the curve $z = 0$, $x = F(y)$ (Fig. 5.2).

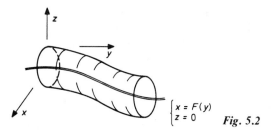

$$\begin{cases} x = F(y) \\ z = 0 \end{cases}$$ *Fig. 5.2*

The planes $z = 1$ and $z = -1$ are the envelope of all these surfaces; we shall meet this situation again later.

See **Exercises 3 and 4** at the end of the chapter.

2.2. Geometric interpretation—the Cauchy problem

We can give, in a natural way, a geometric interpretation in $\mathbb{R}^3$ which can be carried over to $\mathbb{R}^n$ $(n > 3)$ if required. Let

$$f(x, y, z)\frac{\partial z}{\partial x} + g(x, y, z)\frac{\partial z}{\partial y} = h(x, y, z) \tag{24}$$

and

$$f(x, y, z)\frac{\partial H}{\partial x} + g(x, y, z)\frac{\partial H}{\partial y} + h(x, y, z)\frac{\partial H}{\partial z} = 0 \tag{24'}$$

Equation (24') states that a solution of (24) defined in implicit form by $H(x, y, z) = 0$ is a surface whose tangent plane contains at each point the vector $(f, g, h)(x, y, z)$.

The solutions of

$$\frac{\mathrm{d}x}{f} = \frac{\mathrm{d}y}{g} = \frac{\mathrm{d}z}{h}$$

are the characteristic curves which are therefore tangent curves at each point to the vector (f, g, h).

We can construct the characteristic curves as the intersection of two surfaces $F(x, y, z) = k_1$, $G(x, y, z) = k_2$ where F and G are two independent first integrals of the characteristic system (cf. p. 271). Now the solution surfaces of (24) are of the form $H[F(x, y, z), G(x, y, z)] = 0$. Thus we can construct them by eliminating k_1 and k_2 in the relations

$$H(k_1, k_2) = 0 \qquad F(x, y, z) = k_1 \qquad G(x, y, z) = k_2$$

This means that the solution surfaces are loci of characteristic curves.

This is what are established in the earlier examples (p. 278). In Example (1) the horizontal circles, intersections of $z = k_1$ and $x^2 + y^2 = k_2$, are the characteristic curves and the surfaces of revolution about the vertical axis $(x = 0, y = 0)$ are loci of such circles. In Example (2) the circles, intersections of $y = k_1$ and $x + (1 - z^2)^{1/2} = k_2$ (i.e. $z^2 + (x - k_2) = 1$), are the characteristic curves, and the solutions are clearly loci of such circles.

It follows from these properties that an infinity of solution surfaces passes through a given characteristic curve. In fact, if Γ is such a characteristic defined by $F(x, y, z) = \alpha$, $G(x, y, z) = \beta$, every surface $F(x, y, z) = \varphi[G(x, y, z)]$ is a solution surface which contains Γ where $\alpha = \varphi(\beta)$.

The Cauchy problem

For an ordinary differential equation $dy/dx = F(y, x)$ the Cauchy problem is to find solutions passing through a point (x_0, y_0). For a PDE it is the search for a solution passing through a given curve.

Proposition 2.2.1

If (24) is a linear PDE of the first order, then a unique solution of (24) to the Cauchy problem passes through every curve which is not a characteristic.

Proof Let Γ be a curve which we can assume to be defined by the equations $x = h(z)$, $y = k(z)$, and let U_1 and U_2 be two first integrals of the characteristic system of (24); the equations $U_1(x, y, z) = k_1$ and $U_2(x, y, z) = k_2$ define a characteristic curve and every solution of (24) may be obtained by establishing a relation between k_1 and k_2. Let us try to arrange that such a surface contains Γ:

$$U_1[h(z), k(z), z] = k_1$$
$$U_2[h(z), k(z), z] = k_2$$

These two relations must be compatible; this then establishes a relation between k_2 and k_1 unless there exist exactly two constants a_1 and a_2 such that

$$U_1[h(z), k(z), z] \equiv a_1$$
$$U_2[h(z), k(z), z] \equiv a_2$$

which means that Γ is a characteristic curve.

We find this result again in Examples (1) and (2) (p. 278); in the first example there exists only one surface of revolution about the axis $0z$ passing through a given curve Γ, unless Γ is a horizontal curve. The same phenomenon appears again with the vertical circles of Example (2).

Proposition 2.2.2 (Practical method)

The solution passing through a curve Γ is obtained by considering two first integrals of the characteristic system U_1 and U_2 and by eliminating k_1 and k_2 between the equations $U_1 = k_1$, $U_2 = k_2$ and the equations of Γ.

Example Find the solution of

$$y \frac{\partial z}{\partial x} - x \frac{\partial z}{\partial y} = 0$$

which passes through the ellipse Γ: $x^2 + (y - a)^2 = R^2$; $z = my$ (a and $m \neq 0$).

The characteristic curves, as we have shown, are the circles $x^2 + y^2 = k_1$, $z = k_2$; by eliminating k_1 and k_2 we obtain successively $k_2 = my$, $k_1 = R^2 + 2ay - a^2$, and the solution has equation

$$x^2 + y^2 = \frac{2az}{m} + R^2 - a^2$$

(which is a paraboloid of revolution about the axis $0z$).

See **Exercises 5 and 6** at the end of the chapter.

Part 3: Exact differentials and integrating factors

To simplify the statements we shall refer to $\mathbb{R}^3$; the reader will be able to make the changes to $\mathbb{R}^n$ ($n > 3$) in the definitions that follow.

Let $H(x, y, z)$: $\mathbb{R}^3 \to \mathbb{R}$ be a function of class C^1 and DH its differential, i.e. the linear mapping which, corresponding to every vector $V = (V_1, V_2, V_3)$ of $\mathbb{R}^3$, produces

$$DH(x, y, z)V = \frac{\partial H}{\partial x}(x, y, z)V_1 + \frac{\partial H}{\partial y}(x, y, z)V_2 + \frac{\partial H}{\partial z}(x, y, z)V_3$$

We know that if x, y and z are functions of t

$$\frac{dH}{dt}[x(t), y(t), y(t)] = \frac{\partial H}{\partial x}[x(t), y(t), z(t)] \frac{dx}{dt}$$

$$+ \frac{\partial H}{\partial y}[x(t), y(t), z(t)] \frac{dy}{dt} + \frac{\partial H}{\partial z}[x(t), y(t), z(t)] \frac{dz}{dt}$$

We write this symbolically as

$$dH = \frac{\partial H}{\partial x} dx + \frac{\partial H}{\partial y} dy + \frac{\partial H}{\partial z} dz \tag{25}$$

We know that to the second order a function and the tangent linear mapping vary in the same way; we interpret the preceding formula by noting that, for variations δx, δy, δz of the variables, H experiences (to the second order) a variation

$$\delta H \approx \frac{\partial H}{\partial x} \delta x + \frac{\partial H}{\partial y} \delta y + \frac{\partial H}{\partial z} \delta z$$

It is natural to wonder whether we can give meaning to a symbolic expression of the type on the right-hand side of (25); we are led to the following statement.

Definition 3.1.1

We define a **differential form** as an expression of the form $P(x, y, z)\,dx + Q(x, y, z)\,dy + R(x, y, z)\,dz$.

In general there is no function H such that $dH = P\,dx + Q\,dy + R\,dz$. If there is a function H with this property we say that the differential form is an **exact total differential**. We say that a differential form is **integrable** if there exists a function $\mu(x, y, z)$, called an **integrating factor**, such that the form $(\mu P)\,dx + (\mu Q)\,dy + (\mu R)\,dz$ is an **exact total differential**.

In the first case the function H must satisfy $\partial H/\partial x = P$, $\partial H/\partial y = Q$, $\partial H/\partial z = R$. In the second, there must exist K such that $\partial K/\partial x = \mu P$, $\partial K/\partial y = \mu Q$, $\partial K/\partial z = \mu R$.

Definition 3.1.2

We define a **solution** of the equation

$$P\,dx + Q\,dy + R\,dz = 0 \tag{26}$$

as a surface $H(x, y, z) = \text{constant}$ such that the tangent plane at each point is orthogonal to the vector P, Q, R.

Geometrical significance

Since the tangent plane is perpendicular to the vector

$$\left(\frac{\partial H}{\partial x}, \frac{\partial H}{\partial y}, \frac{\partial H}{\partial z}\right)$$

the preceding condition states that the vectors

$$(P, Q, R) \quad \text{and} \quad \left(\frac{\partial H}{\partial x}, \frac{\partial H}{\partial y}, \frac{\partial H}{\partial z}\right)$$

are collinear (i.e. have the same direction) or that (P, Q, R) is a vector perpendicular to the surfaces $H = \text{constant}$.

If $(\delta x, \delta y, \delta z)$ is a displacement in the tangent plane to H at a point, for every function μ the vector $(\mu P, \mu Q, \mu R)$ is orthogonal to this displacement; if μ is an integrating factor $\delta H = (\mu P)\,\delta x + (\mu Q)\,\delta y + (\mu R)\,\delta z$ is then zero.

Conversely if $H = \text{constant}$ is a solution there exists an integrating factor since

$$(P, Q, R) \quad \text{and} \quad \left(\frac{\partial H}{\partial x}, \frac{\partial H}{\partial y}, \frac{\partial H}{\partial z}\right)$$

are proportional.

Proposition 3.1.3

Equation (26) has a solution iff the form $P\,dx + Q\,dy + R\,dz$ is integrable.

We shall examine in $\mathbb{R}^2$ and $\mathbb{R}^3$ under what conditions a differential form is an exact total differential or an integrable form.

3.1. Differential forms in $\mathbb{R}^2$

Theorem 3.1.4

(a) If P and Q are two functions of class C^1 from $\mathbb{R}^2$ into $\mathbb{R}$, the form $P(x, y)\,dx + Q(x, y)\,dy$ is exact iff

$$\frac{\partial P}{\partial y}(x, y) = \frac{\partial Q}{\partial x}(x, y)$$

(b) Every differential form in $\mathbb{R}^2$ is integrable (P and Q being of class C^1).

Proof

(a) The condition is obviously necessary because, if there exists H such that $\partial H/\partial x = P$ and $\partial H/\partial y = Q$, H is of class C^2 and

$$\frac{\partial P}{\partial y} = \frac{\partial^2 H}{\partial x\,\partial y} = \frac{\partial Q}{\partial x}$$

We shall show that it is sufficient. Let

$$F(x, y) = \int_{x_0}^{x} P(u, y)\,du$$

(We suppose naturally that $P(u, y)$ is defined over the segment (x_0, x); for simplification we shall suppose that P and Q are defined at least in a disc.)

P, being of class C^1, is bounded above over the segment (x_0, x) so that we can differentiate the relation defining F:

$$\frac{\partial F}{\partial y}(x, y) = \int_{x_0}^{x} \frac{\partial P}{\partial y}(u, y)\,du$$

$$= \int_{x_0}^{x} \frac{\partial Q}{\partial x}(u, y)\,du$$

$$= Q(x, y) - Q(x_0, y)$$

Thus

$$\frac{\partial F}{\partial x}(x, y) = P(x, y)$$

and

$$\frac{\partial F}{\partial y}(x, y) = Q(x, y) - Q(x_0, y)$$

We can then choose

$$H(x, y) = F(x, y) + \int_{y_0}^{y} Q(x_0, v) \, dv$$

(b) The form $P(x, y) \, dx + Q(x, y) \, dy$ is integrable if there exists a function μ such that

$$\frac{\partial(\mu P)}{\partial y} = \frac{\partial(\mu Q)}{\partial x}$$

from (a). Now this equation is

$$\frac{\partial \mu}{\partial x} Q - \frac{\partial \mu}{\partial y} P = \mu \left(\frac{\partial P}{\partial y} - \frac{\partial Q}{\partial x} \right)$$

The equation has as solution the first integrals of the characteristic system:

$$\frac{dx}{Q} = -\frac{dy}{P} = \frac{d\mu}{\mu(\partial P/\partial y - \partial Q/\partial x)}$$

Solution of the equation $P \, dx + Q \, dy = 0$

The method consists in applying the above: seek an integrating factor and then solve $(\mu P) \, dx + (\mu Q) \, dy = 0$ as in (a).

Remark   Let μ and v be two integrating factors and $H(x, y) = a$ and $G(x, y) = b$ be the solutions obtained starting from μ and v respectively. Since

$$\left| \frac{D(G, H)}{D(x, y)} \right| = \left| \begin{matrix} vP & vQ \\ \mu P & \mu Q \end{matrix} \right| = 0$$

G and H are dependent and there exists a function F of class C^1 such that $G(x, y) = F[H(x, y)]$, which is natural since if $b = F(a)$, $H(x, y) = a \Leftrightarrow G(x, y) = b$ and these two equations represent the same surface. Thus

$$v(x, y)P(x, y) = \frac{\partial G}{\partial x}(x, y)$$

$$= F'[H(x, y)] \frac{\partial H}{\partial y}$$

$$= F'[H(x, y)]\mu(x, y)P(x, y)$$

so that

$$v(x, y) = \mu(x, y)F'[H(x, y)] \quad \text{(if } P = 0, \, Q \neq 0)$$

$F'[H(x, y)] = $ constant is certainly a solution and we have shown that every integrating factor is the product of a fixed integrating factor and an arbitrary solution, or equally, every solution is the quotient of two independent integrating factors.

Example   Solve

$$2y^3\,\mathrm{d}x + (3xy^2 - 1)\,\mathrm{d}y = 0 \qquad (27)$$

An integrating factor μ is a solution of

$$\frac{\partial\mu}{\partial x}(3xy^2 - 1) - 2y^3\frac{\partial\mu}{\partial y} = 3\mu y^2 \qquad (27')$$

This is an equation whose characteristic system is

$$\frac{\mathrm{d}x}{3xy^2 - 1} = \frac{\mathrm{d}y}{-2y^3} = \frac{\mathrm{d}\mu}{3\mu y^2}$$

Thus

$$0 = 3\mu\,\mathrm{d}y + 2y\,\mathrm{d}\mu$$

$$= \frac{3}{2}\mu y^{1/2}\,\mathrm{d}y + y^{3/2}\,\mathrm{d}\mu$$

$$= \mathrm{d}(\mu y^{3/2})$$

If $y \geqslant 0$, $\mu y^{3/2}$ is a first integral; $\mu y^{3/2} = $ constant is a solution of (27′) and $\mu = y^{-3/2}$ is an integrating factor of (27) which becomes

$$2y^{3/2}\,\mathrm{d}x + (3xy^{1/2} - y^{-3/2})\,\mathrm{d}y = 0$$

$$= \mathrm{d}(2xy^{3/2} + 2y^{-1/2})$$

Thus $xy^2 + 1 = ay^{1/2}$ is a solution of (27) for $y > 0$ and in the same way $xy^2 + 1 = b(-y)^{1/2}$ is a solution of (27) for $y < 0$; these may be put together in the form $y = k(xy^2 + 1)^2$ where k is a constant.

Naturally every equation $\mathrm{d}y/\mathrm{d}x = F(x, y)$ can equally well be written

$$\frac{\mathrm{d}y}{\mathrm{d}x} = F(y, x)\frac{\mu(y, x)}{\mu(y, x)}$$

where μ is an integrating factor of $\mathrm{d}y - F(y, x)\,\mathrm{d}x$ or $0 = \mu(y, x)\,\mathrm{d}y - F(y, x)\mu(y, x)\,\mathrm{d}x$. The preceding solution is therefore also a solution of the differential equation

$$\frac{\mathrm{d}y}{\mathrm{d}x} = \frac{2y^3}{1 - 3y^2 x} \qquad (28)$$

Sometimes we solve differential equations by the same method: we replace (28) by (27) and look for an integrating factor; in general, we seek an integrating factor of $\mathrm{d}y - F(y, x)\,\mathrm{d}x$ in order to solve $\mathrm{d}y/\mathrm{d}x = F(y, x)$.

3.2. Differential forms in $\mathbb{R}^3$

The situation is more complicated in $\mathbb{R}^3$ than in $\mathbb{R}^2$ and a differential form is not necessarily integrable.

To express the conditions for integrability we shall use the notion of curl of a vector, whose definition we recall.

If V is of class C^1 and $V = (V_1, V_2, V_3)$

$$\operatorname{curl} V = \left(\frac{\partial V_3}{\partial y} - \frac{\partial V_2}{\partial z}, \frac{\partial V_1}{\partial z} - \frac{\partial V_3}{\partial x}, \frac{\partial V_2}{\partial x} - \frac{\partial V_1}{\partial y} \right)$$

Denoting by i, j, k the unit vectors on the coordinate axes we can write symbolically

$$\operatorname{curl} V = \left(i \frac{\partial}{\partial x} + j \frac{\partial}{\partial y} + k \frac{\partial}{\partial z} \right) \wedge V$$

or again

$$\operatorname{curl} V = \begin{vmatrix} i & j & k \\ \dfrac{\partial}{\partial x} & \dfrac{\partial}{\partial y} & \dfrac{\partial}{\partial z} \\ V_1 & V_2 & V_3 \end{vmatrix}$$

Expansion of the determinant provides the coordinates of curl V. We can easily prove that if μ is of class C^1 $\operatorname{curl}(\mu V) = \mu \operatorname{curl} V + \operatorname{grad} \mu \wedge V$. We can prove also that for every $U : \mathbb{R}^3 \to \mathbb{R}$ of class C^2, $\operatorname{curl}(\operatorname{grad} U) = 0$.

Theorem 3.2.1

Let P, Q, R be three functions, $\mathbb{R}^3 \to \mathbb{R}$, of class C^1, and let V be the vector with components P, Q, R.

(a) The differential form $P(x, y, z) \, dx + Q(x, y, z) \, dy + R(x, y, z) \, dz$ is an exact total differential iff curl $V = 0$.

(b) This form is integrable iff $V \cdot \operatorname{curl} V = 0$.

Proof

(a) The condition is obviously necessary; in fact if there exists U such that $P = \partial U / \partial x$, $Q = \partial U / \partial y$, $R = \partial U / \partial z$, U is then of class C^2 and curl V has components

$$\frac{\partial^2 U}{\partial z \, \partial y} - \frac{\partial^2 U}{\partial y \, \partial z} \quad \frac{\partial^2 U}{\partial x \, \partial z} - \frac{\partial^2 U}{\partial z \, \partial x} \quad \text{and} \quad \frac{\partial^2 U}{\partial y \, \partial x} - \frac{\partial^2 U}{\partial x \, \partial y}$$

which vanish.

To prove that the condition is sufficient we shall proceed as in the case of $\mathbb{R}^2$ and construct the function U; we shall suppose for simplicity that P, Q and R are defined throughout the entire space $\mathbb{R}^3$. Let (x_0, y_0, z_0) be a point and let

$$U(x, y, z) = \int_{x_0}^{x} P(u, y, z) \, du + F(y, z)$$

As before, we can differentiate the integral. We shall assume F (which is to be determined) is of class C^1 and then

$$\frac{\partial U}{\partial x} = P(x, y, z)$$

and

$$\frac{\partial U}{\partial y} = \int_{x_0}^{x} \frac{\partial}{\partial y} P(u, y, z)\, du + \frac{\partial F}{\partial y}(y, z)$$

$$= \int_{x_0}^{x} \frac{\partial}{\partial x} Q(u, y, z)\, du + \frac{\partial F}{\partial y}(y, z)$$

$$= Q(x, y, z) - Q(x_0, y, z) + \frac{\partial F}{\partial y}(y, z)$$

To obtain $\partial U/\partial y = Q(x, y, z)$ we shall put

$$F(y, z) = \int_{y_0}^{y} Q(x_0, v, z)\, dv + G(z)$$

so that

$$\frac{\partial U}{\partial z}(x, y, z) = \int_{y_0}^{y} \frac{\partial R}{\partial y}(x_0, v, z)\, dv + \int_{x_0}^{x} \frac{\partial R}{\partial x}(u, y, z)\, du + \frac{dG}{dz}$$

$$= R(x_0, y, z) - R(x_0, y_0, z) + R(x, y, z) - R(x_0, y, z) + \frac{dG}{dz}$$

$$= R(x, y, z) - R(x_0, y_0, z) + \frac{dG}{dz}$$

We shall then choose

$$U(x, y, z) = \int_{x_0}^{x} P(u, y, z)\, du + \int_{y_0}^{y} Q(x_0, v, z)\, dv + \int_{z_0}^{z} R(x_0, y_0, w)\, dw$$

(to which we can naturally add a constant).

(b) It is easy to establish the necessary condition for integrability; in fact to say that there exists an integrating factor μ is to state, from the results above, that $\operatorname{curl}(\mu V) = 0$.   Now $\operatorname{curl}(\mu V) = \mu(\operatorname{curl} V) + \operatorname{grad} \mu \wedge V$;   therefore $V \cdot \operatorname{curl} V = 0$.

To prove that the condition is sufficient we shall construct a solution of the equation $P\, dx + Q\, dy + R\, dz = 0$.

We note immediately that if $V \cdot \operatorname{curl} V = 0$ then $(fV) \cdot \operatorname{curl}(fV) = 0$ for every differentiable function f, since

$$(fV) \cdot \operatorname{curl}(fV) = f^2 V \cdot \operatorname{curl} V + f(V \cdot \operatorname{grad} f \wedge V)$$

First we shall solve, for z fixed, the equation $P\, dx + Q\, dy = 0$ which is an integrable form; there exist an integrating factor μ (which depends on z) and a solution $U(x, y)$ (which depends on z) such that $U(x, y, z) = C(z)$ is a solution of $P\,dx + Q\,dy = 0$:

$$\frac{\partial U}{\partial x}(x, y, z) = \mu(x, y, z)P(x, y, z)$$

and

$$\frac{\partial U}{\partial y}(x, y, z) = \mu(x, y, z)Q(x, y, z)$$

Now consider the equation $(\mu P)\,dx + (\mu Q)\,dy + (\mu R)\,dz = 0$ which can be written

$$\frac{\partial U}{\partial x}\,dx + \frac{\partial U}{\partial y}\,dy + \frac{\partial U}{\partial z}\,dz + \left(\mu R - \frac{\partial U}{\partial z}\right)dz = 0 \qquad (29)$$

The vector W whose coordinates are the coefficients of this differential form is

$$W = \mu V = \operatorname{grad} U + \begin{pmatrix} 0 \\ 0 \\ \mu R - \dfrac{\partial U}{\partial z} \end{pmatrix}$$

If $S = \mu R - \partial U/\partial z$, the relation $\mu V \cdot \operatorname{curl} \mu V = 0$ becomes, taking account of the fact that $\operatorname{curl}(\operatorname{grad} U) = 0$,

$$\mu V \cdot \left(\frac{\partial S}{\partial y}\,i - \frac{\partial S}{\partial x}\,j\right) = 0$$

i.e.

$$0 = \frac{\partial U}{\partial x}\frac{\partial S}{\partial y} - \frac{\partial S}{\partial x}\frac{\partial U}{\partial y} = \frac{D(U, S)}{D(x, y)}$$

Thus, z being fixed, S is a function of U; let $S = \Phi(U, z)$. Equation (29) then becomes $dU + \Phi(U, z)\,dz = 0$, a differential form in $\mathbb{R}^2$ which is therefore integrable; the solutions are determined implicitly by $H[U(x, y, z), z] = $ constant.

In practice, if we have proved that the form is integrable, we can abbreviate the last stage in the construction of U; having solved $P\,dx + Q\,dy = 0$ for z fixed, we obtain $U(x, y, z) = \varphi(z)$ and we must determine φ to obtain the solution in the form $H(U, z) = $ constant.

We can then fix x (or y) and consider that $U(x_0, y, z) = \varphi(z)$ implicitly defines the same function as $H[U(x_0, y, z), z] = $ constant, i.e. a solution of

$$Q(x_0, y, z)\,dy + R(x_0, y, z)\,dz = 0 \qquad (30)$$

We determine φ by writing that $U(x_0, y, z) = \varphi(z)$ defines the solution which we have found for Eqn (30).

It is necessary to prove beforehand that the form is integrable.

Practical method
(a) Prove that the form is integrable.
(b) Solve $P \, dx + Q \, dy = 0$ for z fixed. We find a solution $U(x, y, z) = \varphi(z)$.
(c) Solve $Q(x_0, y, z) \, dy + R(x_0, y, z) \, dz = 0$ and determine φ by using the fact that the solution is also defined by $U(x_0, y, z) = \varphi(z)$.

Naturally we can make each of the variables play the roles of z and x respectively.

Example Solve

$$z(z + y^2) \, dx + z(z + x^2) \, dy - xy(x + y) \, dz = 0 \qquad (31)$$

(a) We easily verify that (31) is integrable.
(b) We fix y and solve $z(z + y^2) \, dx - xy(x + y) \, dz = 0$. Let

$$\frac{dx}{xy(x + y)} = \frac{dz}{(z + y^2)z} \quad \text{or} \quad \frac{y \, dx}{x(x + y)} = \frac{y^2 \, dz}{z(z + y^2)}$$

or again

$$0 = \left(\frac{1}{x} - \frac{1}{x + y} \right) dx + \left(\frac{1}{z + y^2} - \frac{1}{z} \right) dz = d\left[\frac{x(z + y^2)}{(x + y)z} \right]$$

For y fixed, the solutions of $P \, dx + R \, dz$ are then

$$\frac{x(z + y^2)}{(x + y)z} = \text{constant}$$

(c) We then seek a solution of (31) defined by

$$\varphi(y) = \frac{x(z + y^2)}{z(x + y)}$$

If $z = 1$

$$\varphi(y) = \frac{x(1 + y^2)}{x + y}$$

In addition, for $z = 1$ this solution must satisfy

$$(1 + y^2) \, dx + (1 + x^2) \, dy = 0$$

or

$$\arctan x + \arctan y = \text{constant}$$

or

$$\arctan \frac{x + y}{1 - xy} = \arctan k^{-1}$$

Then $(1 - xy)/(x + y) = k$, which must be equivalent to

$$\varphi(y) = \frac{x(1 + y^2)}{x + y}$$

Eliminating x we find that $\varphi(y) = 1 - ky$. The solutions of (31) are the surfaces $x(z + y^2) = (1 - ky)z(x + y)$, where $k = $ constant.

Remark   A change of variables often simplifies the calculations; if for example P, Q, R are homogeneous, we could put $y = ux$, $z = vx$.

Example $yz(y + z)\,dx + xz(x + z)\,dy + xy(x + y)\,dz = 0$. We verify that $V \cdot \operatorname{curl} V = 0$. We find that

$$uv(u + v)\,dx + v(v + 1)(u\,dx + x\,du) + u(u + 1)(v\,dx + x\,dv) = 0$$

or

$$\frac{dx}{x} + \frac{v(v + 1)\,du + u(u + 1)\,dv}{2uv(1 + u + v)} = 0$$

$$= 2\frac{dx}{x} + \frac{du}{u} + \frac{dv}{v} - \frac{d(1 + u + v)}{1 + u + v}$$

$$= 0$$

$$= d\left(\frac{x^2uv}{1 + u + v}\right)$$

The solutions are then of the form $xyz = k(x + y + z)$ ($k = $ constant).

See *Exercises 7 to 9* at the end of the chapter.

Part 4: Non-linear partial differential equations of the first order

4.1. Envelopes of surfaces

FAMILIES OF SURFACES DEPENDING ON ONE PARAMETER

For simplicity, we shall consider $\mathbb{R}^3$ throughout.

We consider a family of surfaces $\{S_\lambda\}$ depending on a real parameter λ and with equation S_λ: $F(x, y, z, \lambda) = 0$. We assume that F is of class C^2 and that $\partial^2 F/\partial\lambda^2 \neq 0$.

Definition 4.1.1

We define the **characteristic curve of the surface** S_λ as the curve Γ_λ which is the intersection of the surfaces S_λ with equation $F(x, y, z, \lambda) = 0$ and Σ_λ with equation

$$\frac{\partial F}{\partial \lambda}(x, y, z, \lambda) = 0$$

$$\Gamma_\lambda = S_\lambda \cap \Sigma_\lambda$$

Definition 4.1.2

The surface Σ generated by the characteristic curves Γ_λ when λ varies is called the **envelope of the family**.

Along the curve Γ_λ the surfaces Σ and S_λ have the same tangent plane.

This definition contains a statement which we shall prove. If $G(x, y, z) = 0$ is the equation of Σ, we obtain it by eliminating λ between the equations of S_λ and Σ_λ:

$$F(x, y, z, \lambda) = 0$$

$$\frac{\partial F}{\partial \lambda}(x, y, z, \lambda) = 0$$

Since $\partial^2 F / \partial \lambda^2 \neq 0$ the second equation defines $\lambda(x, y, z)$ implicitly and the equation of Σ is $F[x, y, z, \lambda(x, y, z)] = 0$; the tangent plane to this surface at a point (x_0, y_0, z_0) then has equation

$$0 = (x - x_0)\left(\frac{\partial F}{\partial x} + \frac{\partial F}{\partial \lambda}\frac{\partial \lambda}{\partial x}\right)(x_0, y_0, z_0) + (y - y_0)\left(\frac{\partial F}{\partial y} + \frac{\partial F}{\partial \lambda}\frac{\partial \lambda}{\partial y}\right)(x_0, y_0, z_0)$$

$$+ (z - z_0)\left(\frac{\partial F}{\partial z} + \frac{\partial F}{\partial \lambda}\frac{\partial \lambda}{\partial z}\right)(x_0, y_0, z_0)$$

Now, along Γ_λ, $\partial F / \partial \lambda$ is null and if $(x_0, y_0, z_0) \in \Gamma_\lambda$ the equation obtained is that of the tangent plane to S_λ.

Geometrical significance

Let $\Gamma_{\lambda,\mu}$ be the intersection of S_λ and S_μ whose equation is obtained by writing

$$F(x, y, z, \lambda) = 0$$

$$F(x, y, z, \mu) = 0$$

These are the same points as those obtained by writing

$$F(x, y, z, \lambda) = 0$$

$$\frac{F(x, y, z, \lambda) - F(x, y, z, \mu)}{\lambda - \mu} = 0$$

If $\mu \to \lambda$, $\Gamma_{\lambda,\mu}$ then has as limit the characteristic curve Γ_λ. The characteristic curve Γ_λ is the limit of $S_\lambda \cap S_\mu$ if $\mu \to \lambda$; the two tangent planes to S_λ and S_μ along $S_\lambda \cap S_\mu$ have as limit the tangent plane to S_λ which is also the tangent plane to the envelope.

Example   The family of spheres with radius 1 centred on the $0z$ axis with equation $x^2 + y^2 + (z - \lambda)^2 = 1$ has for envelope the cylinder $x^2 + y^2 = 1$ which is tangential to the sphere S_λ along the characteristic Γ_λ with equations $z = \lambda$, $x^2 + y^2 = 1$ obtained as the intersection of $x^2 + y^2 + (z - \lambda)^2 = 1$ and $z - \lambda = 0$ (see Fig. 5.3).

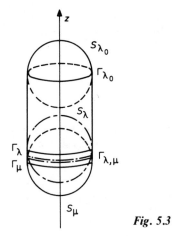

Fig. 5.3

FAMILIES OF SURFACES DEPENDING ON TWO PARAMETERS

We consider a family of surfaces $\{S_{\lambda,\mu}\}$ depending on two real parameters λ and μ and with equation $S_{\lambda,\mu}: F(x, y, z, \lambda, \mu) = 0$. We assume that F is of class C^2 and has a non-zero derivative of order 2 in λ and μ.

If μ is given as a function of class C^2, $\mu = \varphi(\lambda)$, we obtain a one-parameter family whose envelope is obtained by eliminating λ between

$$F[x, y, z, \lambda, \varphi(\lambda)] = 0$$

$$\frac{\partial F}{\partial \lambda} + \frac{\partial F}{\partial \mu} \frac{d\varphi}{d\lambda} = 0$$

We assume that the second equation determines λ as a function of (x, y, z).

If a point simultaneously satisfies

$$F(x, y, z, \lambda, \mu) = 0$$

$$\frac{\partial F}{\partial \lambda}(x, y, z, \lambda, \mu) = 0$$

$$\frac{\partial F}{\partial \mu}(x, y, z, \lambda, \mu) = 0$$

it belongs to the envelope of the one-parameter family whatever the choice of φ; this point generates a surface tangential to all the surfaces of $S_{\lambda,\mu}$.

Definition 4.1.3
We define a **characteristic point** of the surface $S_{\lambda,\mu}$ as a point such that

$$F(x, y, z, \lambda, \mu) = 0$$

$$\frac{\partial F}{\partial \lambda}(x, y, z, \lambda, \mu) = 0$$

$$\frac{\partial F}{\partial \mu}(x, y, z, \lambda, \mu) = 0$$

This point generates a surface Σ whose equation is obtained by eliminating λ and μ between the three preceding equations. This surface is called the **envelope of the two-parameter family** $\{S_{\lambda,\mu}\}$.

All the surfaces $S_{\lambda,\mu}$ are tangential to Σ at their characteristic point. It is clear that an envelope of a two-parameter family does not always exist.

Examples
(1) The family of spheres of radius 1 with centre $z = 0$, $x = \lambda$, $y = \mu$, and with equation $(x - \lambda)^2 + (y - \mu)^2 + z^2 = 1$ has as its envelope the planes $z = \pm 1$; the equation is obtained by eliminating λ and μ between the equations

$$(x - \lambda)^2 + (y - \mu)^2 + z^2 = 1 \qquad x - \lambda = 0 \qquad x - \mu = 0$$

Every one-parameter family $(x - \lambda)^2 + [y - \varphi(\lambda)]^2 + z^2 = 1$ is tangential to these two planes, as is its envelope.
(2) The spheres $(x - 2)^2 + y^2 + z^2 = \mu$ have no two-parameter envelope.

PARTIAL DIFFERENTIAL EQUATIONS ASSOCIATED WITH TWO-PARAMETER FAMILIES

If $F(x, y, z, \lambda, \mu) = 0$ is the equation of a surface $S_{\lambda,\mu}$, we shall assume that $\partial F/\partial z \neq 0$ so that $F(x, y, z, \lambda, \mu) = 0$ defines $z(x, y, \lambda, \mu)$ implicitly. We shall use the standard notation $\partial z/\partial x = p$, $\partial z/\partial y = q$. We know that the function

$z(x, y, \lambda, \mu)$ is of class C^1 and that

$$\frac{\partial F}{\partial x}[x, y, z(x, y, \lambda, \mu), \lambda, \mu] + \frac{\partial F}{\partial z}[x, y, z(x, y, \lambda, \mu), \lambda, \mu]\frac{\partial z}{\partial x}(x, y, \lambda, \mu) = 0$$

and

$$\frac{\partial F}{\partial y}[x, y, z(x, y, \lambda, \mu), \lambda, \mu] + \frac{\partial F}{\partial z}[x, y, z(x, y, \lambda, \mu), \lambda, \mu]\frac{\partial z}{\partial y}(x, y, \lambda, \mu) = 0$$

Definition 4.1.4

The PDE associated with the family $S_{\lambda,\mu}$ is the relation obtained by eliminating λ and μ between the equations

$$F = 0$$

$$\frac{\partial F}{\partial x} + \frac{\partial F}{\partial z}p = 0$$

$$\frac{\partial F}{\partial y} + \frac{\partial F}{\partial z}q = 0$$

The surfaces $S_{\lambda,\mu}$ are solutions of this PDE.

The fundamental result for PDEs of the first order is the following: the family $S_{\lambda,\mu}$ and its (one- or two-parameter) envelopes make up the set of solutions of the equation associated with the family.

Example   The spheres $(x - \lambda)^2 + (y - \mu)^2 + z^2 = 1$ are solutions of the PDE obtained by eliminating λ and μ between

$$(x - \lambda)^2 + (y - \mu)^2 + z^2 = 1$$

$$2(x - \lambda) + 2zp = 0$$

$$2(y - \mu) + 2zq = 0$$

This is $z^2(1 + p^2 + q^2) = 1$. (The planes $z = \pm 1$ are also solutions.)

4.2. The complete integral and solution of the partial differential equation

Definition 4.2.1

Let

$$G(x, y, z, p, q) = 0 \tag{32}$$

be a PDE of the first order and let $F(x, y, z, \lambda, \mu) = 0$ be a two-parameter family of solutions of (32). This family is called a **complete integral** of (32). Every one-parameter envelope is called a **general integral** of (32); if there exists a two-parameter envelope, we call it a **singular integral**.

We recall how Eqn (32) is obtained from F; we define $\lambda(x, y, z, p, q)$ and $\mu(x, y, z, p, q)$ by

$$\frac{\partial F}{\partial x} + p \frac{\partial F}{\partial z} = 0$$

$$\frac{\partial F}{\partial y} + q \frac{\partial F}{\partial z} = 0$$

Then

$$G(x, y, z, p, q) = F[x, y, z, \lambda(x, y, z, p, q), \mu(x, y, z, p, q)] = 0$$

Differentiating this equation with respect to x and y we obtain the relations

$$\frac{\partial F}{\partial x} + \frac{\partial F}{\partial z} p + \frac{\partial F}{\partial \lambda} \frac{\partial \lambda}{\partial x} + \frac{\partial F}{\partial \mu} \frac{\partial \mu}{\partial x} = 0$$

$$= \frac{\partial F}{\partial y} + \frac{\partial F}{\partial y} q + \frac{\partial F}{\partial \lambda} \frac{\partial \lambda}{\partial y} + \frac{\partial F}{\partial \mu} \frac{\partial \mu}{\partial y}$$

so that the equation $G(x, y, z, p, q) = 0$ is equivalent to

$$F(x, y, z, \lambda, \mu) = 0$$

$$\frac{\partial F}{\partial \lambda} \frac{\partial \lambda}{\partial x} + \frac{\partial F}{\partial \mu} \frac{\partial \mu}{\partial x} = 0$$

$$\frac{\partial F}{\partial \lambda} \frac{\partial \lambda}{\partial y} + \frac{\partial F}{\partial \mu} \frac{\partial \mu}{\partial y} = 0$$

The solutions of this system are of three types.

(a) $\lambda = $ constant, $\mu = $ constant, $F(x, y, z, \lambda, \mu) = 0$. This is a surface of the complete integral.

(b) $\dfrac{\partial F}{\partial \lambda} (x, y, z, \lambda, \mu) = 0 = \dfrac{\partial F}{\partial \mu} (x, y, z, \lambda, \mu)$ and $F = 0$

This is the singular integral if it exists.

(c) At least one of the quantities $\partial F/\partial \lambda$ and $\partial F/\partial \mu$ is non-zero; then $D(\lambda, \mu)/D(x, y) = 0$ and $\lambda = \varphi(\mu)$ (if μ is not constant), and the system becomes

$$F[x, y, z, \varphi(\mu), \mu] = 0$$

$$\frac{\partial F}{\partial \lambda} \varphi'(\mu) + \frac{\partial F}{\partial \mu} = 0$$

and this is the equation of a one-parameter envelope.

We have thus demonstrated the following result.

Theorem 4.2.2
Let (32) be a PDE of which we know a complete integral. There is no solution of (32) other than the elements of the complete integral, the general integrals, and possibly the singular integral.

Example (cf. p. 294)   All the solutions of $z^2(1 + p^2 + q^2) = 1$ are the spheres $(x - \lambda)^2 + (y - \mu)^2 + z^2 = 1$ (the complete integral), the envelopes of the one-parameter family (the general integrals), and the planes $z = \pm 1$ (the singular integral).

Naturally there are in general several complete integrals, the elements of one being general integrals of the others.

For example the family $(y - \alpha x - \beta)^2 = (1 + \alpha^2)(1 - z^2)$ is a complete integral of $z^2(1 + p^2 + q^2) = 1$; the complete integral is composed of the cylinders with axis $y = \alpha x + \beta$, $z = 0$ and radius 1 (Fig. 5.4). These cylinders are the envelopes of the following one-parameter family of spheres extracted from the complete integral $(x - \lambda)^2 + (y - \mu)^2 + z^2 = 1$, namely the family $(x - \lambda)^2 + [y - (\alpha\lambda + \beta)]^2 + z^2 = 1$.

To integrate a PDE of the first order is thus a matter of finding a complete integral.

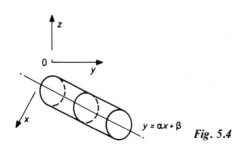

Fig. 5.4

4.3. Charpit's method of obtaining a complete integral

Charpit's method consists in adding to the equation $G(x, y, z, p, q) = 0$ another equation $H(x, y, z, p, q) = 0$ chosen so that the solutions are the same and that the expressions for p and q are such that $-dz + p(x, y, z) dx + q(x, y, z) dy$ is an exact total differential from which the solution can then be extracted.

Definition 4.3.1
We say that the equations $G(x, y, z, p, q) = 0$ and $H(x, y, z, p, q) = 0$ are **compatible** if every solution of one is also a solution of the other.

Suppose that $D(G, H)/D(p, q) \neq 0$; we can then express p and q as functions of x, y, z, say $p = \varphi(x, y, z)$ and $q = \psi(x, y, z)$. If $z(x, y)$ is a common solution of the two equations, then $p = \partial z/\partial x$, $q = \partial z/\partial y$ and $-dz + \varphi \, dx + \psi \, dy$ is an exact total differential. Then, denoting by V the vector $(\varphi, \psi, -1)$, $V \cdot \text{curl } V = 0$, i.e.

$$
\begin{vmatrix}
\varphi & \psi & -1 \\
\dfrac{\partial}{\partial x} & \dfrac{\partial}{\partial y} & \dfrac{\partial}{\partial z} \\
\varphi & \psi & -1
\end{vmatrix}
= -\varphi \frac{\partial \psi}{\partial z} - \frac{\partial \psi}{\partial x} + \psi \frac{\partial \varphi}{\partial z} + \frac{\partial \varphi}{\partial y} = 0
\tag{33}
$$

Conversely, if (33) is satisfied, we can find z such that $\varphi = \partial z/\partial x$ and $\psi = \partial z/\partial y$, and z is a solution of the two equations. Thus the system is compatible iff

$$
\varphi \frac{\partial \psi}{\partial z} + \frac{\partial \psi}{\partial x} = \psi \frac{\partial \varphi}{\partial z} + \frac{\partial \varphi}{\partial y}
$$

Proposition 4.3.2

In order for the system

$$
G(x, y, z, p, q) = 0
$$

$$
H(x, y, z, p, q) = 0
$$

where $D(G, H)/D(p, q) \neq 0$ to be compatible, it is necessary and sufficient that the expression written $[G, H]$ and called the Jacobi bracket should be zero:

$$
0 = [G, H] = \frac{D(G, H)}{D(x, p)} + p \frac{D(G, H)}{D(z, p)} + \frac{D(G, H)}{D(y, q)} + q \frac{D(G, H)}{D(z, q)}
$$

Proof The proof consists in calculating φ and ψ and applying the preceding result. Differentiating G and H with respect to x and z we get

$$
\frac{\partial G}{\partial x} + \frac{\partial G}{\partial p} \frac{\partial \varphi}{\partial x} + \frac{\partial G}{\partial q} \frac{\partial \psi}{\partial x} = 0
$$

and

$$
\frac{\partial G}{\partial z} + \frac{\partial G}{\partial p} \frac{\partial \varphi}{\partial z} + \frac{\partial G}{\partial q} \frac{\partial \psi}{\partial z} = 0
$$

or

$$
\frac{\partial G}{\partial x} + \varphi \frac{\partial G}{\partial z} + \frac{\partial G}{\partial p} \left(\frac{\partial \varphi}{\partial x} + \varphi \frac{\partial \varphi}{\partial z} \right) + \frac{\partial G}{\partial q} \left(\frac{\partial \psi}{\partial x} + \varphi \frac{\partial \psi}{\partial z} \right) = 0
$$

and likewise

$$
\frac{\partial H}{\partial x} + \varphi \frac{\partial H}{\partial z} + \frac{\partial H}{\partial p} \left(\frac{\partial \varphi}{\partial x} + \varphi \frac{\partial \varphi}{\partial z} \right) + \frac{\partial H}{\partial q} \left(\frac{\partial \psi}{\partial x} + \varphi \frac{\partial \psi}{\partial z} \right) = 0
$$

or again

$$\left(\frac{\partial \psi}{\partial x} + \varphi \frac{\partial \psi}{\partial z}\right) \frac{D(G, H)}{D(p, q)} = \frac{D(G, H)}{D(x, p)} + \varphi \frac{D(G, H)}{D(z, p)}$$

and likewise

$$\left(\frac{\partial \varphi}{\partial y} + \psi \frac{\partial \varphi}{\partial z}\right) \frac{D(G, H)}{D(p, q)} = -\left[\frac{D(G, H)}{D(y, q)} + \psi \frac{D(G, H)}{D(z, q)}\right]$$

which establishes the result. The result holds also for the system

$$\left.\begin{array}{l} G = 0 \\ H = a \end{array}\right\} \quad a \text{ is constant}$$

To find a compatible system we then look for H such that $G = 0 \Rightarrow [G, H] = 0$.

Notation $P = \partial G/\partial p, Q = \partial G/\partial q, X = \partial G/\partial x, Y = \partial G/\partial y, Z = \partial G/\partial z$. With this notation

$$0 = P \frac{\partial H}{\partial x} + Q \frac{\partial H}{\partial y} + (Pp + Qq) \frac{\partial H}{\partial z} - (X + pZ) \frac{\partial H}{\partial p} - (Y + qZ) \frac{\partial H}{\partial q}$$

is the required condition; if H is a solution of this PDE, i.e. a first integral of the corresponding characteristic system, the system is compatible and we have proved the following result.

Theorem 4.3.3
Let

$$G(x, y, z, p, q) = 0 \tag{32}$$

and let P, Q, X, Y, Z be the quantities defined above. To find a complete integral of (32) we choose a first integral $H(p, q, x, y, z)$ of the characteristic system

$$\frac{dx}{P} = \frac{dy}{Q} = \frac{dz}{Pp + Qq} = -\frac{dp}{X + pZ} = -\frac{dq}{Y + qZ}$$

The system

$$G = 0$$
$$H = a$$

is compatible for every constant a. The solutions of this system depend on a parameter b and form a complete integral $F(x, y, z, a, b) = 0$.

Remark If G does not depend on z, we look for H, also independent of z, in seeking a first integral of

$$\frac{dx}{P} = \frac{dy}{Q} = -\frac{dp}{X} = -\frac{dq}{Y}$$

(In this case the Jacobi bracket $[G, H]$ may be simplified and becomes

$$\frac{D(G, H)}{D(x, p)} + \frac{D(G, H)}{D(y, q)}$$

which we often call the Poisson bracket.)

Examples
(1) If x and y do not appear in G, p/q is a first integral. Let

$$z^2(1 + p^2 + q^2) = 1 \tag{34}$$

The characteristic system is

$$\frac{dx}{2z^2 p} = \frac{dy}{2z^2 q} = \frac{dz}{2z^2(p^2 + q^2)} = -\frac{dp}{2pz(1 + p^2 + q^2)} = -\frac{dq}{2qz(1 + p^2 + q^2)}$$

The system

$$z^2(1 + p^2 + q^2) = 1$$
$$p = aq$$

is compatible:

$$p^2 = \frac{a^2}{1 + a^2} \frac{1 - z^2}{z^2} \quad \text{and} \quad q^2 = \frac{1}{1 + a^2} \frac{1 - z^2}{z^2}$$

Then

$$dz = \left[\frac{1 - z^2}{z^2(1 + a^2)}\right]^{1/2} (a\,dx + dy)$$

or

$$(1 + a^2)(1 - z^2) = (y - ax - b)^2$$

which is a complete integral of (34) (cf. p. 298).
(2) If x, y and z do not appear in G, then p and q are first integrals.
Let

$$pq = 1 \tag{35}$$

The system is

$$\frac{dx}{q} = \frac{dy}{p} = \frac{dz}{2pq} = \frac{dp}{0} = \frac{dq}{0}$$

The system

$$p = a$$

$$pq = 1$$

is compatible. Thus

$$dz = a\,dx + \frac{1}{a}\,dy$$

so that

$$z - ax - \frac{1}{a}\,y = b$$

is a complete integral of (35).

(3) *Separable equations* $f(x, p) = g(y, q)$: the characteristic system is

$$\frac{dx}{\partial f/\partial p} = \frac{dy}{-\partial g/\partial q} = \frac{dz}{p\,\partial f/\partial p - q\,\partial g/\partial q} = \frac{-dp}{\partial f/\partial x} = \frac{dq}{\partial g/\partial y}$$

$$\frac{\partial f}{\partial p}\,dp + \frac{\partial f}{\partial x}\,dx = 0$$

$f(x, p)$ is a first integral as is also $g(y, q)$. The system $f(x, p) = a = g(y, q)$ then gives p and q. Let

$$p^2 y(1 + x^2) = qx^2 \tag{36}$$

which can also be written

$$\frac{p^2(1 + x^2)}{x^2} = \frac{q}{y}$$

$$p = \frac{ax}{(1 + x^2)^{1/2}} \qquad q = a^2 y$$

Then

$$z = a(1 + x^2)^{1/2} + a^2\,\frac{y^2}{2} + b$$

is a complete integral of (36).

(4) *Clairaut's equation* $z = px + qy + f(p, q)$: the characteristic system is

$$\frac{dx}{x + \partial f/\partial p} = \frac{dy}{y + \partial f/\partial q} = \frac{dz}{px + qy + p\,\partial f/\partial p + q\,\partial f/\partial q} = -\frac{dp}{p - p} = -\frac{dq}{q - q}$$

p and q are first integrals; $z = ax + by + f(a, b)$ is a complete integral of (36), which generalizes the classical result on differential equations.

See **Exercises 10 and 11** at the end of the chapter.

4.4. The Cauchy problem

This is concerned with finding a solution of $G(x, y, z, p, q) = 0$ passing through a given curve Γ.

We shall assume that a complete integral is known and that we have already found out whether Γ belongs to one of the surfaces $S_{\lambda,\mu}$. In the case where the two-parameter family possesses an envelope we shall assume that we have already found out whether Γ is situated on this surface. Finally, we shall find out whether Γ belongs to one of the general integrals of (36).

Definition 4.4.1

If $F(x, y, z, \lambda, \mu) = 0$ is a complete integral of (36), we define the **characteristic curves** of (36) to be the following family of curves depending on three parameters:

$$F(x, y, z, \lambda, \mu) = 0$$

$$\frac{\partial F}{\partial \lambda}(x, y, z, \lambda, \mu) + v\frac{\partial F}{\partial \mu}(x, y, z, \lambda, \mu) = 0$$

Let φ be a differentiable function of λ and let $F[x, y, z, \lambda, \varphi(\lambda)] = 0$ be the corresponding family of solutions of (36); this family has as its envelope a general integral whose equation can be obtained by eliminating λ between the two relations

$$F[x, y, z, \lambda, \varphi(\lambda)] = 0$$

$$\frac{\partial F}{\partial \lambda} + \varphi'(\lambda)\frac{\partial F}{\partial \mu} = 0$$

Thus a general integral is the locus of a one-parameter family of characteristic curves with $\mu = \varphi(\lambda)$ and $v = \varphi'(\lambda)$. Therefore through each of the characteristic curves passes an infinity of general integrals all having the same tangent plane along this curve; the Cauchy problem is indeterminate.

Moreover this fact is sufficient to characterize the characteristic curves, as the following theorem shows.

Theorem 4.4.2

If Γ is a non-characteristic curve of (36), there exists a finite number of solutions to the Cauchy problem relative to Γ which can be obtained in the following way. Let Γ be

$$x = f(t) \qquad y = g(t) \qquad z = h(t)$$

We can determine $\lambda(t)$ and $\mu(t)$ (and therefore $\lambda(\mu)$ or $\mu(\lambda)$) by writing that at the point of intersection of Γ with $S_{\lambda,\mu}$ the tangent to Γ is in the tangent plane to $S_{\lambda,\mu}$. The surface required is then the envelope of the corresponding family depending on a single parameter.

In other words we can derive $\lambda(t)$ and $\mu(t)$ from the system

$$F[f(t), g(t), h(t), \lambda, \mu] = 0$$

$$\frac{\partial F}{\partial x}[(f, g, h, \lambda, \mu)(t)]f'(t) + \frac{\partial F}{\partial y}[(f, g, h, \lambda, \mu)(t)]g'(t)$$

$$+ \frac{\partial F}{\partial z}[(f, g, h, \lambda, \mu)(t)]h'(t) = 0 \qquad (37)$$

Comments

(1) The second equation can be replaced by

$$\frac{\partial F}{\partial t}[f(t), g(t), h(t), \lambda, \mu] = 0$$

The system obtained expresses the fact that the parameter t of the point of intersection of Γ with $S_{\lambda,\mu}$ has a double root, which corresponds to the fact that the tangent to Γ is in the tangent plane to $S_{\lambda,\mu}$ (Fig. 5.5).

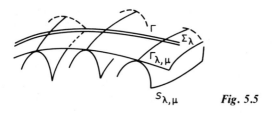

Fig. 5.5

(2) We can find the characteristics by writing that the Cauchy problem is indeterminate, which establishes that they do not depend on the complete integral F.

Proof of Theorem 4.4.2   It is convenient to verify first that the system (37) allows us to calculate $\lambda(t)$ and $\mu(t)$.

We shall write this system more briefly in the form

$$F(t, \lambda, \mu) = 0$$

$$\Phi(t, \lambda, \mu) = 0$$

It determines $\lambda(t)$ and $\mu(t)$ iff

$$\frac{D(F, \Phi)}{D(\lambda, \mu)} \neq 0$$

Let us examine the different possible cases.

(a) If $\partial F/\partial \lambda$ and $\partial F/\partial \mu$ are identically zero, this Jacobian is null but we know that then Γ is the singular integral (which therefore exists).

(b) Suppose, for example, that $\partial F/\partial \mu \neq 0$. In this case the Jacobian is null iff there exists $v(t, \lambda, \mu)$ of class C^1 such that

$$\frac{\partial F/\partial \lambda}{\partial F/\partial \mu} = \frac{\partial \Phi/\partial \lambda}{\partial \Phi/\partial \mu} = -v(t, \lambda, \mu)$$

i.e.

$$\frac{\partial F}{\partial \lambda} + v(t, \lambda, \mu)\,\frac{\partial F}{\partial \lambda} = 0 = \frac{\partial \Phi}{\partial \lambda} + v(t, \lambda, \mu)\,\frac{\partial \Phi}{\partial \mu}$$

Now by differentiating the first relation with respect to t we obtain the equality

$$0 = \frac{\partial^2 F}{\partial \lambda\, \partial x}\, f'(t) + \frac{\partial^2 F}{\partial \lambda\, \partial y}\, g'(t) + \frac{\partial^2 F}{\partial \lambda\, \partial z}\, h'(t) + \frac{\partial v}{\partial t}\frac{\partial F}{\partial \mu}$$

$$+ v\left[\frac{\partial^2 F}{\partial \mu\, \partial x}\, f'(t) + \frac{\partial^2 F}{\partial \mu\, \partial y}\, g'(t) + \frac{\partial^2 F}{\partial \mu\, \partial z}\, h'(t)\right]$$

Then

$$0 = \left(\frac{\partial \Phi}{\partial \lambda} + v\,\frac{\partial \Phi}{\partial \mu}\right)[f(t), g(t), h(t)] + \frac{\partial v}{\partial t}\frac{\partial F}{\partial \mu}$$

and consequently $\partial v/\partial t = 0$. Therefore we can find λ, μ, v such that

$$F[f(t), g(t), h(t), \lambda, \mu] = 0$$

$$\frac{\partial F}{\partial \lambda}\,[f(t), g(t), h(t), \lambda, \mu] + v\,\frac{\partial F}{\partial \mu}\,[f(t), g(t), h(t), \lambda, \mu] = 0$$

Γ is then a characteristic curve.

Thus, if Γ is not a characteristic and if it is not situated on the singular integral, we can find $\lambda(t)$ and $\mu(t)$ or μ as a function of λ: $\mu(\lambda)$ (or $\lambda(\mu)$).

The relation $F[f(t), g(t), h(t), \lambda, \mu(\lambda)] = 0$ then enables us to write, by differentiation, that

$$\frac{\partial F}{\partial \lambda}\,[f(t), g(t), h(t), \lambda, \mu(\lambda)] + \frac{\partial F}{\partial \mu}\,[f(t), g(t), h(t), \lambda, \mu(\lambda)]\,\frac{\partial \mu}{\partial \lambda} = 0$$

The point of intersection of Γ with $S_{\lambda, \mu(\lambda)}$ is on the envelope of the family and therefore on the general integral, which completes the proof.

Example Find a solution of

$$(p^2 + q^2)x = pz \tag{38}$$

passing through the parabola $x = 0$, $z^2 = 4y$.

First we seek a complete integral $F(x, y, z, a, b)$ of (38) whose characteristic system is

$$\frac{dx}{2px - z} = \frac{dy}{2qx} = \frac{dz}{2p^2x - pz + 2q^2x} = \frac{-dp}{(p^2 + q^2) - p^2} = \frac{-dq}{-pq}$$

$p^2 + q^2$ is a first integral. The system

$$p^2 + q^2 = a^2$$
$$(p^2 + q^2)x = pz$$

allows us to verify that $z^2 = a^2x^2 + \varphi(y)$. Therefore $zp = a^2x$ and $zq = \frac{1}{2}\varphi'(y)$. Then

$$z^2(p^2 + q^2) = a^2z^2 = a^4x^2 + \frac{1}{4}[\varphi'(y)]^2 = a^4x^2 + a^2\varphi(y)$$

Therefore $\varphi(y) = (ay + b)^2$ and $F(x, y, z, a, b) = z^2 - a^2x^2 - (ay + b)^2$.
  Let Γ be

$$x = 0, \qquad y = t^2, \qquad z = 2t \tag{39}$$

We form the following system (whose second equation is modified in agreement with the first comment):

$$F[x(t), y(t), z(t), a, b] = 0$$

$$\frac{\partial F}{\partial t}[x(t), y(t), z(t), a, b] = 0$$

or

$$4t^2 = (at^2 + b)^2$$

$$8t = 4at(at^2 + b)$$

Elimination of t gives us $ab = 1$ and hence the one-parameter family

$$z^2 = a^2x^2 + \left(ay + \frac{1}{a}\right)^2$$

which can also be written

$$a^4(x^2 + y^2) + a^2(2y - z^2) + 1 = 0$$

whose envelope $4(x^2 + y^2) = (2y - z^2)^2$ is the required surface.

See *Exercises 12 to 14* at the end of the chapter.

Exercises

1. Solve the following equations:

(a)
$$\frac{dx}{x(y-z)} = \frac{dy}{y(z-x)} = \frac{dz}{z(x-y)}$$

(b)
$$\frac{dx}{x^2(y^3-z^3)} = \frac{dy}{y^2(z^3-x^3)} = \frac{dz}{z^2(x^3-y^3)}$$

2. *Orthogonal trajectories.* Let $F(x, y, z) = 0$ be a surface S in $\mathbb{R}^3$; we are given a family of curves Γ_λ traced on it by the intersection of S with $G(x, y, z) = \lambda$ (λ varies in $\mathbb{R}$). It is required to find curves traced on S orthogonal to the original ones. Let P, Q, R be the direction of the tangent to Γ_λ. Find the direction of the tangents to the orthogonal trajectories

$$P' = R\frac{\partial F}{\partial y} - Q\frac{\partial F}{\partial z} \qquad Q' = P\frac{\partial F}{\partial z} - R\frac{\partial F}{\partial x} \qquad R' = Q\frac{\partial F}{\partial x} - P\frac{\partial F}{\partial y}$$

(Example: $F \equiv x^2 + y^2 + z^2 = 1$; $G(x, y, z) \equiv xy - z$.)

3. Solve the equation

$$y + z\frac{\partial z}{\partial y} = 0$$

4. Solve the equation

$$x^2\frac{\partial z}{\partial x} + y^2\frac{\partial z}{\partial y} = (x + y)z$$

5. Find the solution of the equation

$$(2xy - 1)\frac{\partial z}{\partial x} + (z - 2x^2)\frac{\partial z}{\partial y} = 2(x - yz)$$

passing through the curver Γ with equations

$$x = 1 \qquad y = 0$$

6. Find the solution of the equation

$$(x - y)\frac{\partial z}{\partial x} + (y - x - z)\frac{\partial z}{\partial y} = z$$

which passes through the curve Γ with equations

$$x^2 + y^2 = 1 \qquad z = 0$$

7. Solve the equation

$$2y(\lambda - x)\,dx + [z - y^2 + (\lambda - x)^2]\,dy - y\,dz = 0$$

8. Solve the equation

$$yz\,dx + xz\,dy + xy\,dz = 0$$

9. Solve the equation

$$(1 + yz)\,dx + x(z - x)\,dy - (1 + xy)\,dz = 0$$

10. Find complete integrals for the following equations (the answer is given in parentheses):

(a) $p + q = pq$ $\qquad\qquad \left(z = ax + \dfrac{ay}{a-1} + b\right)$

(b) $zpq = p + q$ $\qquad\qquad \left(z^2 = 2(a + 1)\left(x + \dfrac{y}{a}\right) + b\right)$

(c) $zpq = p^2(p^2 + xq) + q^2(q^2 + yp)$ $\quad \left(z = \dfrac{a^4 + b^4}{ab} + ax + by\right)$

11. Repeat Exercise 10 for the following equations:

(a) $p^2z^2 + q^2 = 1$

$$(az(1 + a^2z^2)^{1/2} + \log[az + (1 + a^2z^2)^{1/2}] = 2a(ax + y + b))$$

(b) $qz = (p^2 + q^2)y$ $\qquad\qquad\qquad ((x + b)^2 + y^2 = az^2)$

(c) $px^5 - 4q^3x^2 + 6x^2z = 2$

$$\left(z = \dfrac{2}{3}(y + a)^{3/2} + \dfrac{1}{9} + \dfrac{1}{3x^2} + b\exp\left(\dfrac{3}{x^2}\right)\right)$$

(d) $2(z + xp + pq) = yp^2$ $\qquad\qquad \left(z = \dfrac{ax}{y^2} + \dfrac{b}{y} - \dfrac{a^2}{4y^3}\right)$

12. Find the solution of $p^2x + qy = z$ which passes through the curve Γ with equations $y = 1$, $x + z = 0$.

13. Find the solution of $pq = z$ which passes through the curve Γ with equations $x = 0$, $y^2 = z$.

14. Find the solution of $z = p^2 - q^2$ which passes through the curve Γ with equations $4z + x^2 = 0$, $y = 0$.
Find directly the characteristics of the equation.

Chapter 6

Linear operators; Green's function; Integral equations; Sturm–Liouville's problem

Introduction

Consider a metal bar of constant section and length L which conducts heat uniformly with a constant coefficient of diffusion k (Fig. 6.1).

At time zero the temperature of the bar at point x is equal to $f(x)$; the end-points are at temperature zero and are maintained at this temperature. The temperature $u(x, t)$ at each point x and at each instant t is then observed.

We shall prove that $u(x, t)$ is a solution of the equation

$$\frac{\partial u}{\partial t}(x, t) = k \frac{\partial^2 u}{\partial x^2}(x, t) \quad \text{for } 0 < t \text{ and } 0 \leqslant x \leqslant L \tag{1}$$

and satisfies the supplementary conditions, called boundary conditions,

$$\left. \begin{array}{ll} u(x, 0) = f(x) & \text{for } 0 \leqslant x \leqslant L \\ u(0, t) = u(L, t) = 0 & \text{for } 0 \leqslant t \end{array} \right\} \tag{2}$$

To solve this problem we can seek solutions of the form $u(x, t) = \varphi(x)T(t)$ where φ and T must then satisfy

$$\varphi(x) \frac{\mathrm{d}T}{\mathrm{d}t}(t) = kT(t) \frac{\mathrm{d}^2 \varphi}{\mathrm{d}x^2}(x)$$

or

$$\frac{1}{kT(t)} \frac{\mathrm{d}T}{\mathrm{d}t}(t) = \frac{1}{\varphi(x)} \frac{\mathrm{d}^2 \varphi}{\mathrm{d}x^2}(x)$$

The term on the left is a function of t, that on the right is a function of x; they can be equal for all x and all t only if they are constant.

Thus we ask if there exist constants λ and functions T and φ, not identically zero, such that

$$\frac{\mathrm{d}T}{\mathrm{d}t} = \lambda kT(t) \qquad \frac{\mathrm{d}^2 \varphi}{\mathrm{d}x^2}(x) = \lambda \varphi(x)$$

For $u(x, t)$ to satisfy the conditions (2) it is then necessary that $T(t)\varphi(0) = 0 =$

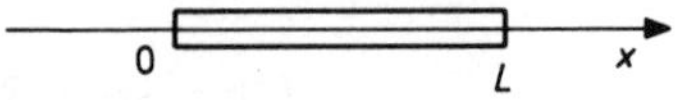

Fig. 6.1

$T(t)\varphi(L)$ and therefore, since T is not identically zero, we have $\varphi(0) = 0 = \varphi(L)$. We are thus led to study the following problem.

(a) Find, on the one hand, numbers λ and functions φ such that

$$\frac{\mathrm{d}^2\varphi}{\mathrm{d}x^2} = \lambda\varphi(x) \quad \text{and} \quad \varphi(0) = 0 = \varphi(L) \tag{3}$$

and, on the other hand, functions T such that

$$\frac{\mathrm{d}T}{\mathrm{d}t} = \lambda T \tag{4}$$

(b) Hence deduce a solution of (1) satisfying (2) and therefore $u(x, 0) = f(x)$. For this purpose we note that if $(\varphi_n, T_n)_{n\in\mathbb{N}}$ is a sequence of solutions of (3) and (4) every sequence

$$u_p(x, t) = \sum_{n=1}^{p} \varphi_n(x)T_n(t)$$

is a solution of (1). We are thus led to ask whether there exists a sequence $\{a_n\}_{n\in\mathbb{N}}$ such that $\sum a_n\varphi_n(x) = f(x)$. We shall then put $T_n(0) = a_n$ which will determine T_n.

Important remark   This method—called the method of separation of variables—cannot always be applied (cf. Exercise 16).

This is the set of problems which we shall examine. First we shall study this type of boundary conditions; it is not the same as occurs in a Cauchy problem—it has particular properties (e.g. 'the Fredholm alternative').

Next we shall see that the search for solutions φ_n and numbers λ_n which satisfy (1) is analogous to the search for eigenfunctions and eigenvalues of linear mappings in $\mathbb{R}^d$, so that the problem of representation of f in the form $f = \sum a_n\varphi_n$ is that of the search for a basis of a certain vector space constructed from eigenvectors of a linear mapping.

The first part of the chapter is devoted to the study of **Green's functions** which characterize these particular boundary conditions.

Part 2 introduces some elements of the study of Hilbert spaces and of linear operators over them, more particularly their eigenvalues and eigenfunctions. We shall illustrate this study by examples, notably spaces of **square integrable functions** and **Fourier series**.

In Part 3 we shall apply the results presented in Part 2 to **integral equations** and to the **Sturm–Liouville problem**. A theorem of Hilbert–Schmidt often allows us to pass from L^2 convergence to absolute and uniform convergence. We shall conclude by indicating how to use certain 'special' functions for the solution of physical problems similar to the one we introduced at the beginning of this chapter.

In the applications to partial differential equations we shall not concern ourselves with problems of uniqueness which could be dealt with by using a theorem on contractions (cf. Proposition 2.2.12).

Part 1: Green's function

1.1. The Fredholm alternative

The Fredholm alternative is a phenomenon related to the particular nature of the boundary conditions which we shall impose on the solutions of a differential equation.

We shall content ourselves with describing the situation in the case of differential equations of the second order. It would be easy to give statements for equations of order n; this is left to the reader.

In this chapter we shall seek solutions of differential equations which are real functions defined over an interval $[a, b]$ of $\mathbb{R}$, in general bounded, and satisfying conditions of the type

$$\left. \begin{array}{l} \alpha_0 f(a) + \alpha_1 \dfrac{df}{dx}(a) = A \\[3mm] \beta_0 f(b) + \beta_1 \dfrac{df}{dx}(b) = B \end{array} \right\} \quad \text{where } \alpha_0, \alpha_1, \beta_0, \beta_1, A \text{ and } B \text{ are constants}$$

$$(5)$$

If $A = 0 = B$ the boundary condition is called **homogeneous**; the principal characteristic of this type of conditions is that the set of functions which satisfies homogeneous conditions is a vector space of functions. This fact plays a central role for it is only over such spaces that we can define linear mappings.

For the moment we shall establish that this type of boundary condition causes a particular phenomenon concerned with the existence and uniqueness of solutions.

Let us examine the following very simple example. Let

$$\frac{d^2f}{dx^2} = 0 \tag{6}$$

$$f(0) = 0 = f(\pi) \tag{7}$$

The only solution of (6) satisfying (7) is $f(x) \equiv 0$. Let

$$\frac{d^2f}{dx^2} + f = 0 \tag{6'}$$

$$f(0) = 0 = f(\pi) \tag{7'}$$

There exists an infinity of solutions of (6') satisfying (7'): $f(x) = A \sin x$.

Now let us consider these two equations with a non-zero right-hand side. Let

$$\frac{d^2 f}{dx^2} = g(x) \tag{8}$$

$$f(0) = \alpha \qquad f(\pi) = \beta \tag{9}$$

There exists only one solution satisfying (9) which is the following, as can be immediately verified:

$$f(x) = \alpha + \frac{x}{\pi}\left[\beta - \alpha - \int_0^\pi (\pi - y)g(y)\,dy\right] + \int_0^x (x - y)g(y)\,dy$$

Such a unique solution exists for every function g continuous over $[0, \pi]$. Let

$$\frac{d^2 f}{dx^2} + f = g(x) \tag{8'}$$

$$f(0) = \alpha \qquad f(\pi) = \beta \tag{9'}$$

The solutions of (8') are

$$f(x) = f(0) \cos x + f'(0) \sin x + \int_0^x \sin(x - y)g(y)\,dy$$

These functions satisfy (9') if

$$f(0) = \alpha \quad \text{and} \quad \beta = f(\pi) = \alpha + \int_0^\pi \sin(\pi - y)g(y)\,dy$$

If g is not such that

$$\beta - \alpha = \int_0^\pi \sin y\, g(y)\,dy$$

there is no solution of (8') which satisfies (9'). If, however,

$$\beta - \alpha = \int_0^\pi g(y) \sin y\,dy$$

there exists an infinity of solutions of (8') which satisfy (9'); these are the functions

$$f(x) = \alpha \cos x + A \sin x + \int_0^x g(y) \sin(x - y)\,dy$$

If we denote by $f_0(x)$ a solution of (6') satisfying (7') (to which all other solutions satisfying (7') are proportional), we note that Eqn (8'), with the function g on the right-hand side, has either no solution or an infinity of solutions, according to whether or not

$$\int_0^\pi g(y)f_0(y)\,\mathrm{d}y = \beta - \alpha$$

This is a general phenomenon which we shall now describe.

Definition 1.1.1

Let p, q and g be three functions continuous over a bounded interval $[a, b]$ of $\mathbb{R}$; consider the equation

$$\frac{\mathrm{d}^2 f}{\mathrm{d}x^2}(x) + p(x)\frac{\mathrm{d}f}{\mathrm{d}x}(x) + q(x)f(x) = g(x) \qquad x \in [a, b] \tag{10}$$

for which we seek solutions satisfying the boundary condition

$$\left.\begin{aligned} a_0 f(a) + a_1 \frac{\mathrm{d}f}{\mathrm{d}x}(a) &= \alpha \\[2mm] b_0 f(b) + b_1 \frac{\mathrm{d}f}{\mathrm{d}x}(b) &= \beta \end{aligned}\right\} \tag{11}$$

where a_0, a_1, b_0, b_1, α and β are real numbers such that $|a_0| + |a_1| > 0$ and $|b_0| + |b_1| > 0$ (so that F does not reduce to the condition $0 \equiv 0$).

We define a **homogeneous problem** associated with (10) and (11) as the search for solutions of the following Eqn (12) satisfying the conditions (13):

$$\frac{\mathrm{d}^2 f}{\mathrm{d}x^2}(x) + p(x)\frac{\mathrm{d}f}{\mathrm{d}x}(x) + q(x)f(x) = 0 \tag{12}$$

$$\left.\begin{aligned} a_0 f(a) + a_1 \frac{\mathrm{d}f}{\mathrm{d}x}(a) &= 0 \\[2mm] b_0 f(b) + b_1 \frac{\mathrm{d}f}{\mathrm{d}x}(b) &= 0 \end{aligned}\right\} \tag{13}$$

Theorem 1.1.2 (The Fredholm alternative)

Equation (10) has for every α and every β a unique solution satisfying (11), iff the homogeneous problem has $f(x) \equiv 0$ as its only solution.

Proof Every solution of (10) is the sum of a particular solution of (10) and the general solution of (12).

Let f_1 be the solution of (12) such that

$$f_1(a) = a_1 \quad \text{and} \quad \frac{\mathrm{d}f_1}{\mathrm{d}x}(a) = -a_0$$

Let f_2 be the solution of (12) such that

$$f_2(b) = -b_1 \quad \text{and} \quad \frac{\mathrm{d}f_2}{\mathrm{d}x}(b) = b_0$$

f_1 and f_2 are non-zero and their Wronskian $W(f_1, f_2)$ is such that

$$W(f_1, f_2)(a) = a_0 f_2(a) + a_1 \frac{df_2}{dx}(a)$$

$$W(f_1, f_2)(b) = b_0 f_1(b) + b_1 \frac{df_1}{dx}(b)$$

We know that the Wronskian of two solutions is either always or never null.

Suppose first that $W(f_1, f_2) \neq 0$. Every solution of (12) is then of the form $f(x) = A_1 f_1(x) + A_2 f_2(x)$; it satisfies (13) iff

$$a_0[A_1 f_1(a) + A_2 f_2(a)] + a_1\left[A_1 \frac{df_1}{dx}(a) + A_2 \frac{df_2}{dx}(a)\right] = 0$$

$$b_0[A_1 f_1(b) + A_2 f_2(b)] + b_1\left[A_1 \frac{df_1}{dx}(b) + A_2 \frac{df_2}{dx}(b)\right] = 0$$

i.e. if $A_2 W(f_1, f_2)(a) = 0 = A_1 W(f_1, f_2)(b)$, i.e. if $A_1 = A_2 = 0$. Thus $W(f_1, f_2) \neq 0$ implies that the only solution of the homogeneous problem is $f \equiv 0$.

Let ψ be the solution of (10) such that

$$\psi(a) = \frac{d\psi}{dx}(a) = 0$$

The general solution of (10) is $B_1 f_1(x) + B_2 f_2(x) + \psi(x)$. Such a solution satisfies (11) if

$$B_2 W(f_1, f_2)(a) = \alpha$$

and

$$B_1 W(f_1, f_2)(b) + b_0 \psi(b) + b_1 \frac{d\psi}{dx}(b) = \beta$$

B_1 and B_2 are unique and completely determined by this system since $W(f_1, f_2) \neq 0$.

Suppose now that $W(f_1, f_2) \equiv 0$. f_1 and f_2 are non-zero and proportional. They are thus both solutions of the homogeneous problem which therefore has many solutions, say $f_2 = k f_1$. Let f_3 be another solution of (12) independent of f_1 (and therefore of f_2). $W(f_2, f_3) = k W(f_1, f_3)$ and every solution of (10) is of the form $B_1 f_1 + B_3 f_3 + \psi$; such a solution satisfies (11) if

$$B_3 W(f_1, f_3)(a) = \alpha$$

$$B_3 W(f_2, f_3)(b) + b_0 \psi(b) + b_1 \frac{d\psi}{dx}(b) = \beta$$

i.e.

$$B_3 W(f_1, f_3)(a) = \alpha$$

$$B_3 k W(f_1, f_3)(a) + b_0 \psi(b) + b_1 \frac{d\psi}{dx}(b) = \beta$$

If

$$b_0\psi(b) + b_1\frac{d\psi}{dx}(b) = \beta - k\alpha$$

B_3 is determinate and B_1 is arbitrary. There then exists an infinity of solutions of (10) $B_1f_1 + B_3f_3 + \psi$ which satisfy (11). If, however,

$$b_0\psi(b) + b_1\frac{d\psi}{dx}(b) \neq \beta - k\alpha$$

there is no solution. This completes the proof.

We shall return later to this situation which we shall interpret in terms of linear operators; this will enable us to state the result a little more precisely.

Geometrical analogy

The functions satisfying (13) form a vector space E. The mapping which, for every element f of E, gives $Lf = f'' + pf' + qf$ is clearly linear so that the solution of $Lf = g$ is analogous to the following problem. Let $A \in \mathscr{L}(\mathbb{R}^n, \mathbb{R}^n)$ and $V \in \mathbb{R}^n$; solve $AU = V$. The corresponding homogeneous equation is $AU = 0$, which only has the solution $U = 0$ iff A^{-1} exists; in this case $A^{-1}V$ is the only solution of $AU = V$.

If, however, 0 is an eigenvalue of A and thus of its transpose A^*, there exist V_0 and W_0 (non-zero) such that $A^*V_0 = 0 = AW_0$. Then $\forall U \in \mathbb{R}^d$ $\langle A^*V_0, U \rangle = \langle V_0, AU \rangle = 0$ so that there exists a solution of $AU = V$ only if $\langle V_0, V \rangle = 0$. If U is such a solution, then there exists an infinity of solutions, of the form $U + kW_0$.

We shall meet this analogy again in Section 1.3.

1.2. Green's function: generalities

We shall first consider an example of Green's function, whose role and existence we shall explain in an intuitive manner.

Consider a beam fixed at its ends, for which we shall suppose the abscissae to be 0 and 1 respectively. Suppose this beam is subjected to a force distributed over each element $[y, y + dy]$ with intensity $g(y)\,dy$ where g is a continuous function. We denote by $f(x)$ the displacement of the beam at the point x and it is shown that f is a solution of the equation

$$\frac{d^2f}{dx^2} = kg(x) \tag{14}$$

where k is a constant characteristic of the beam.

Naturally f satisfies the boundary conditions

$$f(0) = 0 = f(1) \tag{15}$$

Let $G(x, y)$ be the solution of (14) corresponding to $g = \delta_{\{y\}}$, the Dirac delta at the point y (cf. Chapter 3, p. 142). For every $y \in [0, 1]$ $G(x, y)$ has the following properties.

(a) $x \rightsquigarrow G(x, y)$ is a continuous function over $[0, 1]$.

(b) $$\frac{\partial}{\partial x} G(x, y) = a \text{ if } x < y$$

while

$$\frac{\partial}{\partial x} G(x, y) = a + 1 \text{ if } x > y$$

(c) If $x \neq y$

$$\frac{\partial^2 G}{\partial x^2} (x, y) = 0$$

G, a function of x, is a solution of the homogeneous equation $d^2 f/dx^2 = 0$ over the intervals $[0, y[$ and $]y, 1]$.

(d) G satisfies the boundary conditions $G(0, y) = 0 = G(1, y)$.

(e) If

$$f(x) = \int_0^1 G(x, y) kg(y) \, dy$$

then $d^2 f/dx^2 = kg(x)$ and f is the solution of (14) such that $f(0) = 0 = f(1)$.

To prove this we note that

$$f(x) = \int_0^x G(x, y) kg(y) \, dy + \int_x^1 G(x, y) kg(y) \, dy$$

Since $\partial G(x, y)/\partial x$ is bounded above in the interval $[0, 1]$ we can differentiate under the integral sign so that

$$\frac{df}{dx} (x) = G(x, x) kg(x) + \int_0^x (1 + a) kg(y) \, dy - G(x, x) kg(x) + \int_x^1 akg(y) \, dy$$

$$= a \int_0^1 kg(y) \, dy + \int_0^x kg(y) \, dy$$

Thus

$$\frac{d^2 f}{dx^2} = kg(x)$$

We see, therefore, that G enables us to find for every function g the solution of (14) satisfying (15); this is the reason for the importance of Green's function.

In the subject of the strength of materials such a function is also called 'the influence function'.

It is of course absurd to suppose that a force is entirely concentrated at one point and we can consider $G(x, y)$ is the response to a force of intensity

$g_\varepsilon(z) \geqslant 0$, zero for $|z - y| > \varepsilon$ and such that

$$\int_{y-(\varepsilon/2)}^{y+(\varepsilon/2)} g_\varepsilon(z)\, dz = 1$$

Then

$$\frac{\partial^2 G}{\partial x^2}(x, y) = g_\varepsilon(x)$$

is zero for $|x - y| > \varepsilon$ and

$$\int_{y-(\varepsilon/2)}^{y+(\varepsilon/2)} g_\varepsilon(z)\, dz = \int_{y-(\varepsilon/2)}^{y+(\varepsilon/2)} \frac{\partial^2 G}{\partial x^2}(z, y)\, dz$$

$$= \frac{\partial G}{\partial x}\left(y + \frac{\varepsilon}{2}, y\right) - \frac{\partial G}{\partial x}\left(y - \frac{\varepsilon}{2}, y\right)$$

which is equal to 1.

If we consider $\varepsilon \ll 1$, G satisfies for practical purposes the conditions (a)–(d); the use of the Dirac delta is only the mathematical means of expressing this result rigorously.

The fifth relation states that, by superposing at points $\{z_i\}$, a distance $1/\varepsilon$ apart, forces of intensity $a_i g_\varepsilon(z_i)$, the effects are superposed and the resultant is that corresponding to the effect of a function equal to a_i at the different points z_i.

Calculation of $G(x, y)$

It is important to note that the conditions (a), (b), (c) and (d) suffice to define the function G completely.

In fact, since its second derivative is zero there exist two constants a and b such that

$$G(x, y) = ax \qquad \text{if } x < y$$

$$G(x, y) = b(x - 1) \qquad \text{if } x > y$$

since $G(0, y) = 0 = G(1, y)$.

As G is continuous, $ay = b(y - 1)$; since finally $\partial G(x, y)/\partial x = a + 1$, for $x > y$ $b = a + 1$, so $a = y - 1$, $b = y$. Thus $G(x, y) = x(y - 1)$ if $x \leqslant y$, $G(x, y) = y(x - 1)$ if $x \geqslant y$. We note finally that if we let $f_1(x)$ and $f_2(x)$ be the independent solutions of the homogeneous equation $d^2 f/dx^2 = 0$ which are respectively $f_1(x) = (y - 1)x$ and $f_2(x) = y(x - 1)$, we can write G in the form

$$G(x, y) = \begin{cases} \dfrac{f_1(x) f_2(y)}{W(f_1, f_2)} & \text{if } x \leqslant y \\[3mm] \dfrac{f_1(y) f_2(x)}{W(f_1, f_2)} & \text{if } x \geqslant y \end{cases}$$

We shall meet this property again.

We shall now give a general definition of Green's function.

Definition 1.2.1

Consider the homogeneous equation

$$a_0(x)\frac{d^n f}{dx^n}(x) + a_1(x)\frac{d^{n-1}f}{dx^{n-1}}(x) + \ldots + a_n(x)f(x) = 0 \tag{16}$$

with the boundary conditions

$$\alpha_1 f(a) + \alpha_1^{(1)}\frac{df}{dx}(a) + \ldots + \alpha_1^{(n-1)}\frac{d^{n-1}f}{dx^{n-1}}(a) + \beta_1 f(b) + \ldots + \beta_1^{(n-1)}\frac{d^{n-1}f}{dx^{n-1}}(b) = 0$$

$$\alpha_2 f(a) + \alpha_2^{(1)}\frac{df}{dx}(a) + \ldots \qquad\qquad + \beta_2 f(b) + \ldots + \beta_2^{(n-1)}\frac{d^{n-1}f}{dx^{n-1}}(b) = 0$$

$$\vdots$$

$$\alpha_n f(a) + \alpha_n^{(1)}\frac{df}{dx}(a) + \ldots \qquad\qquad + \beta_n f(b) + \ldots + \beta_n^{(n-1)}\frac{d^{n-1}f}{dx^{n-1}}(b) = 0$$

$$\tag{17}$$

We suppose that the rank of the following determinant Δ is n:

$$\Delta = \begin{vmatrix} \alpha_1 & \alpha_1^{(1)}\ldots\alpha_1^{(n-1)} & \beta_1\ldots\beta_1^{(n-1)} \\ \alpha_2 & \alpha_2^{(2)}\ldots & \vdots \\ \vdots & & \\ \alpha_n & \alpha_2^{(n)}\ldots & \ldots\beta_n^{(n-1)} \end{vmatrix}$$

(i.e. the n linear mappings V_p from $\mathbb{R}^{2n}$ into $\mathbb{R}$ defined by

$$V_p(u_1, \ldots, u_{2n}) = \alpha_p u_1 + \alpha_p^{(1)}u_2 + \ldots + \alpha_p^{(n-1)}u_n + \beta_p u_{n+1} + \ldots + \beta_p^{(n-1)}u_{2n}$$

are independent).

Consider the differential equation (16) over the closed interval $[a, b]$. We define a Green function of Eqn (16) with boundary conditions (17) as a function $G(x, y)$, defined for every y of $]a, b[$ and every x of $[a, b]$, satisfying the following conditions.

(a) $G(x, y)$ is continuous and possesses partial derivatives with respect to x up to order $n - 2$ which are continuous in x and y over the square $a \leqslant x \leqslant b$, $a < y < b$.

(b) $\dfrac{\partial^{n-1}G}{\partial x^{n-1}}(y^+, y) - \dfrac{\partial^{n-1}}{\partial x^{n-1}}G(y^-, y) = \dfrac{1}{a_0(y)}$

NB   If h is a real function

$$f(u^+) \text{ denotes } \lim_{\substack{x \to u \\ x > u}} f(x) \quad \text{and} \quad f(u^-) \text{ denotes } \lim_{\substack{x \to u \\ x < u}} f(x)$$

(c) $\forall y \in]a, b[, x \leadsto G(x, y)$ is a solution of (16) in each of the intervals $[a, y[$ and $]y, b]$.

(d) $\forall y \in]a, b[, x \leadsto G(x, y)$ satisfies the boundary conditions (17).

The following theorem is essential.

Theorem 1.2.2

(a) If the homogeneous equation (16) with boundary conditions (17) admits only the solution $f \equiv 0$, there exists for this problem a unique Green function $G(x, y)$. We suppose $a_0(x) \neq 0 \; \forall x \in [a, b]$.

(b) Suppose that the above conditions are satisfied and consider the equation

$$a_0(x) \frac{d^n f}{dx^n}(x) + \ldots + a_n(x)f(x) = g(x) \tag{16'}$$

where the term on the right-hand side is a continuous function g. Then there exists only one solution of (16') satisfying the boundary conditions (17) and this solution is

$$f(x) = \int_a^b G(x, y)g(y) \, dy$$

We shall prove this theorem only in the case of an equation of the second order; it would be easy—but tedious—to extend the proof to the case of an equation of order n.

Proof Consider then a second-order equation

$$p_0(x) \frac{d^2 f}{dx^2}(x) + p_1(x) \frac{df}{dx}(x) + p_2(x)f(x) = 0 \tag{18}$$

where $p_0(x) \neq 0 \; \forall x \in [a, b]$ with the boundary conditions

$$\left. \begin{array}{l} a_1 f(a) + a_2 \dfrac{d^2 f}{dx}(a) + b_1 f(b) + b_2 \dfrac{df}{dx}(b) = 0 \\[2mm] \alpha_1 f(a) + \alpha_2 \dfrac{df}{dx}(a) + \beta_1 f(b) + \beta_2 \dfrac{df}{dx}(b) = 0 \end{array} \right\} \tag{19}$$

We suppose that the rank of

$$\begin{vmatrix} a_1 & a_2 & b_1 & b_2 \\ \alpha_1 & \alpha_2 & \beta_1 & \beta_2 \end{vmatrix}$$

is two, i.e. that the rows are not proportional and that we have certainly two conditions which are not identical, and the functions which satisfy them form a vector space.

The proof consists in fact in constructing the Green function starting from the properties (a), (b), (c) and (d).

(1) Let f_1 and f_2 be two independent solutions of (18). From the third property of the statement, $x \rightsquigarrow G(x, y)$ is a solution in each of the intervals $[a, y[$ and $]y, b[$. Thus there exist four numbers depending on y such that

$$
\left.
\begin{array}{ll}
\text{if } a \leqslant x < y & G(x, y) = \lambda_1(y) f_1(x) + \lambda_2(y) f_2(x) \\
\text{if } y < x \leqslant b & G(x, y) = \mu_1(y) f_1(x) + \mu_2(y) f_2(x)
\end{array}
\right\}
\tag{20}
$$

Since G must be continuous (property (a)) and $\partial G(x, y)/\partial x$ has as a discontinuity $1/p_0(y)$ we must also have the equalities

$$
\left.
\begin{array}{l}
\lambda_1(y) f_1(y) + \lambda_2(y) f_2(y) = \mu_1(y) f_1(y) + \mu_2(y) f_2(y) \\[4pt]
\lambda_1(y) \dfrac{df_1}{dy}(y) + \lambda_2(y) \dfrac{df_2}{dy}(y) = \mu_1(y) \dfrac{df_1}{dy}(y) + \mu_2(y) \dfrac{df_2}{dy}(y) - \dfrac{1}{p_0(y)}
\end{array}
\right\}
\tag{21}
$$

Putting $v_1(y) = \mu_1(y) - \lambda_1(y)$ and $v_2(y) = \mu_2(y) - \lambda_2(y)$, (21) becomes

$$
\left.
\begin{array}{l}
v_1(y) f_1(y) + v_2(y) f_2(y) = 0 \\[4pt]
v_1(y) \dfrac{df_1}{dy}(y) + v_2(y) \dfrac{df_2}{dy}(y) = \dfrac{1}{p_0(y)}
\end{array}
\right\}
\tag{22}
$$

Since $W(f_1, f_2) \neq 0$ for all y the relations (22) define $v_1(y)$ and $v_2(y)$ uniquely.

These being known, it remains to use the fact that $G(x, y)$ satisfies the boundary conditions (19) (property (d)). We can express these conditions as functions of $\lambda_1(y)$ and $\lambda_2(y)$ since $\mu_1(y) = \lambda_1(y) + v_1(y)$ and $\mu_2(y) = \lambda_2(y) + v_2(y)$. Putting the known terms involving $v_1(y)$ and $v_2(y)$ into the right-hand side we obtain the relations

$$
\left.
\begin{array}{l}
\lambda_1(y) \left[a_1 f_1(a) + a_2 \dfrac{df_1}{dx}(a) + b_1 f_1(b) + b_2 \dfrac{df_1}{dx}(b) \right] \\[8pt]
\quad + \lambda_2(y) \left[a_1 f_2(a) + a_2 \dfrac{df_2}{dx}(a) + b_1 f_2(b) + b_2 \dfrac{df_2}{dx}(b) \right] = K_1(y) \\[12pt]
\lambda_1(y) \left[\alpha_1 f_1(a) + \alpha_2 \dfrac{df_1}{dx}(a) + \beta_1 f_1(b) + \beta_2 \dfrac{df_1}{dx}(b) \right] \\[8pt]
\quad + \lambda_2(y) \left[\alpha_1 f_2(a) + \alpha_2 \dfrac{df_2}{dx}(a) + \beta_1 f_2(b) + \beta_2 \dfrac{df_2}{dx}(b) \right] = K_2(y)
\end{array}
\right\}
\tag{23}
$$

We shall prove that the determinant Δ of this system in $\lambda_1(y)$, $\lambda_2(y)$ is not zero, which ensures that $\lambda_1(y)$ and $\lambda_2(y)$ are well defined and unique and that this is also true of $\mu_1(y)$ and $\mu_2(y)$.

We shall do this in the simple case of the boundary conditions of Definition 1.1.1:

$$
a_1 f(a) + a_2 f'(a) = 0
$$
$$
\beta_1 f(b) + \beta_2 f'(b) = 0
$$

Under the same conditions f_1 denotes the solution of the homogeneous equation such that $f_1(a) = a_2$ and $f_1'(a) = -a_1$, while f_2 is that satisfying $f_2(b) = -\beta_2$ and $f_2'(b) = \beta_1$; we know that $W(f_1, f_2) \neq 0$ is a necessary and sufficient condition for the homogeneous problem to have only one solution. Now

$$\Delta = \begin{pmatrix} 0 & W(f_1, f_2)(a) \\ W(f_1, f_2)(b) & 0 \end{pmatrix}$$

Since the homogeneous problem has only one solution, Δ is not zero. We could prove this in a similar (but more complicated) manner for more general conditions in the form of those of the statement.

It is clear that no other function exists which has the properties (a), (b), (c) and (d).

(2) Let

$$f(x) = \int_a^b G(x, y)g(y)\,dy = \int_a^x G(x, y)g(y)\,dy + \int_x^b G(x, y)g(y)\,dy$$

G is differentiable with respect to x in each of the intervals. Therefore

$$\frac{df}{dx} = \int_a^x \frac{\partial G}{\partial x}(x, y)g(y)\,dy + \int_x^b \frac{\partial G}{\partial x}(x, y)g(y)\,dy + G(x, x)g(x) - G(x, x)g(x)$$

Let (z, z) be a point on the diagonal of the square $[a, b] \times [a, b]$.

By hypothesis $\partial G(x, y)/\partial x$ is a continuous function of (x, y) in the set $a \leqslant y \leqslant x \leqslant b$ which is shown hatched in Fig. 6.2 and which contains the

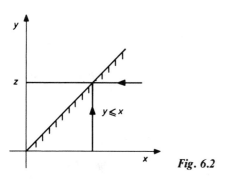

Fig. 6.2

diagonal. Consequently the following two limits are equal:

$$\lim_{\substack{x \to z \\ x > z}} \frac{\partial G}{\partial x}(x, z) = \frac{\partial G}{\partial x}(z^+, z)$$

and

$$\lim_{\substack{y \to z \\ y < z}} \frac{\partial G}{\partial x}(z, y) = \frac{\partial G}{\partial x}(z, z^-)$$

Let us calculate

$$\frac{d^2 f}{dx^2} = \int_a^x \frac{\partial^2 G}{\partial x^2}(x, y)g(y)\,dy + g(x)\frac{\partial G}{\partial x}(x, x^-)$$

$$+ \int_x^b \frac{\partial^2 G}{\partial x^2}(x, y)g(y)\,dy - g(x)\frac{\partial G}{\partial x}(x, x^+)$$

Now

$$\frac{\partial G}{\partial x}(x^+, x) = \frac{\partial G}{\partial x}(x, x^-)$$

and similarly

$$\frac{\partial G}{\partial x}(x^-, x) = \frac{\partial G}{\partial x}(x, x^+)$$

We can establish then, from property (b) of the definition of G, that

$$\frac{d^2 f}{dx^2} = \int_a^b \frac{\partial^2 G}{\partial x^2}(x, y)g(y)\,dy + \frac{g(x)}{p_0(x)}$$

Thus

$$p_0(x)\frac{d^2 f}{dx^2} + p_1(x)\frac{df}{dx} + p_2(x)f$$

$$= g(x) + \int_a^b g(y)\left[p_0(x)\frac{\partial^2 G}{\partial x^2}(x, y) + p_1(x)\frac{\partial G}{\partial x}(x, y) + p_2 G(x, y) \right]dy$$

$$= g(x)$$

Finally since

$$f(a) = \int_a^b G(a, y)g(y)\,dy \qquad \frac{df}{dx}(a) = \int_a^b \frac{\partial G}{\partial x}(a, y)g(y)\,dy$$

and

$$f(b) = \int_a^b G(b, y)g(y)\,dy \qquad \frac{df}{dx}(b) = \int_a^b \frac{\partial G}{\partial x}(b, y)g(y)\,dy$$

f satisfies the conditions (19), which completes the proof.

Example Let

$$\frac{d^4f}{dx^4} = 0 \tag{24}$$

$$f(0) = 0 \qquad f'(0) = 0 \qquad f(1) = 0 \qquad f'(1) = 0 \tag{25}$$

The determinant of the forms V_1, V_2, V_3, V_4 is

$$\begin{vmatrix} 1 & 0 & 0 & 0 & 0 & 0 & 0 & 0 \\ 0 & 1 & 0 & 0 & 0 & 0 & 0 & 0 \\ 0 & 0 & 0 & 0 & 1 & 0 & 0 & 0 \\ 0 & 0 & 0 & 0 & 0 & 1 & 0 & 0 \end{vmatrix}$$

whose rank is 4.

Every solution of (24) is of the form $A + Bx + Cx^2 + Dx^3$; it satisfies (25) if $A = B = C = D = 0$; we are clearly within the conditions of application of Theorem 1.2.2. We shall construct G as in the proof. Let

$$G(x, y) = a_0(y) + a_1(y)x + a_2(y)x^2 + a_3(y)x^3 \quad \text{for } 0 \leqslant x \leqslant y$$
$$G(x, y) = b_0(y) + b_1(y)x + b_2(y)x^2 + b_3(y)x^3 \quad \text{for } y \leqslant x \leqslant 1$$

The conditions (25) imply that

$$\left. \begin{aligned} a_0(y) &= 0 \\ a_1(y) &= 0 \\ b_0(y) + b_1(y) + b_2(y) + b_3(y) &= 0 \\ b_1(y) + 2b_2(y) + 3b_3(y) &= 0 \end{aligned} \right\} \tag{25'}$$

Now let

$$c_0(y) = b_0(y) - a_0(y)$$
$$c_1(y) = b_1(y) - a_1(y)$$
$$c_2(y) = b_2(y) - a_2(y)$$
$$c_3(y) = b_3(y) - a_3(y)$$

From the second property in the definition of G, $G(x, y)$, $\partial G(x, y)/\partial x$ and $\partial^2 G(x, y)/\partial x^2$ are continuous at the point (y, y) and

$$\frac{\partial^3 G}{\partial x^3}(y^+, y) - \frac{\partial^3 G}{\partial x^3}(y^-, y) = 1$$

which gives us the relations

$$c_0(y) + c_1(y)y + c_2(y)y^2 + c_3(y)y^3 = 0$$
$$c_1(y) + 2c_2(y)y + 3c_3(y)y^2 = 0$$
$$2c_2(y) + 6c_3(y)y = 0$$
$$6c_3(y) = 1$$

Thus

$$c_3(y) = \frac{1}{6} \qquad c_2(y) = -\frac{1}{2}y \qquad c_1(y) = \frac{y^2}{2} \qquad c_0(y) = -\frac{y^3}{6}$$

We deduce from this, by using relations (25') that

$$b_1(y) = \frac{y^2}{2} \qquad b_0(y) = -\frac{y^3}{6}$$

$$b_2(y) + b_3(y) = -\frac{y^2}{2} + \frac{y^3}{6}$$

$$2b_2(y) + 3b_3(y) = -\frac{y^2}{2}$$

so that

$$b_2(y) = \tfrac{1}{2}y^3 - y^2 \quad \text{and} \quad b_3(y) = \tfrac{1}{2}y^2 - \tfrac{1}{3}y^3$$

and therefore that

$$a_2(y) = b_2(y) - c_2(y) = \tfrac{1}{2}y^3 - y^2 + \tfrac{1}{2}y$$

$$a_3(y) = b_3(y) - c_3(y) = -\tfrac{1}{3}y^3 + \tfrac{1}{2}y^2 - \tfrac{1}{6}$$

Then

$$G(x, y) = (\tfrac{1}{2}y - y^2 + \tfrac{1}{2}y^3)x^2 - (\tfrac{1}{6} - \tfrac{1}{2}y^3 + \tfrac{1}{3}y^3)x^3 \quad \text{for } 0 \leqslant x \leqslant y$$

$$G(x, y) = (\tfrac{1}{2}x - x^2 + \tfrac{1}{2}x^3)y^2 - (\tfrac{1}{6} - \tfrac{1}{2}x^2 + \tfrac{1}{3}x^3)y^3 \quad \text{for } y \leqslant x \leqslant 1$$

We note that in this case $G(x, y) = G(y, x)$; we shall see the reason for this in the following section.

1.3. Green's function for a self-adjoint equation

We shall concern ourselves with the solution of the following equation over an interval $[a, b]$:

$$\frac{\mathrm{d}}{\mathrm{d}x}\left[p(x) \frac{\mathrm{d}f}{\mathrm{d}x}(x) \right] + q(x)f(x) = g(x) \tag{26}$$

under the boundary conditions—called **separate**—

$$\left.\begin{array}{c} a_0 f(a) + a_1 \dfrac{\mathrm{d}f}{\mathrm{d}x}(a) = \alpha \\[2mm] b_0 f(b) + b_1 \dfrac{\mathrm{d}f}{\mathrm{d}x}(b) = \beta \end{array}\right\} \tag{27}$$

Definition 1.3.1

The preceding problem is called **regular** if $[a, b]$ is a **bounded interval** over which p and q are founded and $p(x) \neq 0$ $\forall x \in [a, b]$. We must note carefully that conditions (27) involve a and b *separately*.

HOMOGENEOUS REGULAR PROBLEMS WITH UNIQUE SOLUTION

Theorem 1.3.2

Consider a regular problem (26), (27) of the type given above. Suppose the associated homogeneous problem (12), (13) (see Definition 1.1.1) has no solution other than $f \equiv 0$. We know that there then exists a unique Green function $G(x, y)$. Let f_1 be the solution of (12) such that

$$f_1(a) = a_1 \qquad \frac{df_1}{dx}(a) = -a_0$$

Let f_2 be the solution of (12) such that

$$f_2(b) = b_1 \qquad \frac{df_2}{dx}(b) = -b_0$$

Then $p(x)W(f_1, f_2)(x)$ is a non-zero constant and the Green function $G(x, y)$ is the symmetric function

$$G(x, y) = \begin{cases} \dfrac{1}{p(x)W(f_1, f_2)(x)} f_1(x)f_2(y) & \text{if } x \leqslant y \\[4mm] \dfrac{1}{p(x)W(f_1, f_2)(x)} f_1(y)f_2(x) & \text{if } x \geqslant y \end{cases}$$

Proof f_1 and f_2 are two non-zero solutions by hypothesis. We know that they are independent since the homogeneous problem has only one solution. Thus $W(f_1, f_2)(x) \neq 0$ $\forall x \in [a, b]$. Moreover

$$\frac{d}{dx}\left\{ p(x)\left[f_1(x)\frac{df_2}{dx}(x) - f_2(x)\frac{df_1}{dx}(x) \right] \right\} = -f_1(x)f_2(x)q(x) + f_1(x)f_2(x)q(x)$$

Consequently $p(x)W(f_1, f_2)(x)$ is a non-zero constant C.

Now let

$$H(x, y) = \begin{cases} \dfrac{1}{C} f_1(x)f_2(y) & \text{if } x \leqslant y \\[4mm] \dfrac{1}{C} f_1(y)f_2(x) & \text{otherwise} \end{cases}$$

We can very easily establish that $H(x, y)$ has all the properties of the Green function and is therefore equal to it.

(a) H is continuous over $[a, b] \times\,]a, b[$.

(b)
$$\frac{\partial H}{\partial x}(x, y) = \frac{1}{C}\frac{df_1(x)}{dx} f_2(y) \quad \text{if } x \leqslant y$$

is continuous over $a \leqslant x \leqslant y \leqslant b$ and

$$\frac{\partial H}{\partial x}(x, y) = \frac{1}{C}\frac{df_2(x)}{dx} f_1(y) \quad \text{if } y \leqslant x$$

is continuous over $a \leqslant y \leqslant x \leqslant b$. In addition

$$\frac{\partial H}{\partial x}(y^+, y) - \frac{\partial H}{\partial x}(y^-, y) = \frac{W(f_1, f_2)(y)}{p(y)W(f_1, f_2)(y)} = \frac{1}{p(y)}$$

(c) Over the intervals $[a, y[$ and $]y, b]$ $x \rightsquigarrow G(x, y)$ is a solution of the homogeneous equation H, as are f_1 and f_2.

(d) $x \rightsquigarrow G(x, y)$ satisfies the boundary conditions (13) because of the choice of f_1 and f_2.

Example   Consider the homogeneous problem

$$\frac{d^2 f}{dx^2}(x) + f(x) = 0 \tag{28}$$

$$f(0) = 0 = f\left(\frac{\pi}{2}\right) \tag{29}$$

This is a regular homogeneous problem whose only solution is $f \equiv 0$. Let f_1 be the solution of (28) such that

$$f(0) = 0 \qquad \frac{df_1}{dx}(0) = -1: f_1(x) = -\sin x$$

Let f_2 be the solution of (28) such that

$$f_2\left(\frac{\pi}{2}\right) = 0 \qquad \frac{df_2}{dx}\left(\frac{\pi}{2}\right) = -1: f_2(x) = \cos x$$

$$p(x)W(f_1, f_2)(x) = \begin{vmatrix} -\sin x & \cos x \\ -\cos x & -\sin x \end{vmatrix} = 1$$

Thus

$$G(x, y) = \begin{cases} -\sin x \cos y & \text{if } 0 \leqslant x \leqslant y \\ -\sin y \cos x & \text{if } y \leqslant x \leqslant \dfrac{\pi}{2} \end{cases}$$

The solution of

$$\frac{d^2 f}{dx^2}(x) + f(x) = -1$$

satisfying (29) is

$$\int_0^{\pi/2} -G(x, y)\,dy = \sin x + \cos x - 1$$

(which is obvious).

In general, it is easy to solve a regular system whose associated homogeneous system has no solution other than $f \equiv 0$.

Theorem 1.3.3

Let

$$\frac{d}{dx}\left[p(x)\frac{df}{dx}(x)\right] + q(x)f(x) = g(x) \tag{30}$$

$$\left.\begin{aligned} a_0 f(a) + a_1 \frac{df}{dx}(a) &= \alpha \\[2mm] b_0 f(b) + b_1 \frac{df}{dx}(b) &= \beta \end{aligned}\right\} \tag{31}$$

be a regular system whose associated homogeneous system possesses only the solution $f \equiv 0$; let $G(x, y)$ be the corresponding Green function, $G_1(x)$ the solution of (12) such that

$$a_0 f(a) + a_1 \frac{df}{dx}(a) = 1$$

$$b_0 f(b) + b_1 \frac{df}{dx}(b) = 0$$

and $G_2(x)$ the solution of (12) such that

$$a_0 f(a) + a_1 \frac{df}{dx}(a) = 0$$

$$b_0 f(b) + b_1 \frac{df}{dx}(b) = 1$$

Then G_1 and G_2 exist and are unique, and the solution of (30) satisfying (31) is unique and equal to

$$\int_a^b G(x, y)g(y)\,dy + \alpha G_1(x) + \beta G_2(x)$$

Proof We know from Theorem 1.2.1 (the Fredholm alternative) that there exists only one solution of (30) satisfying (31); it is sufficient to verify that the proposed solution is correct, which amounts to proving the existence of G_1 and G_2 since the proposed function is a solution of (30) which certainly satisfies (31).

Let φ_2 be the solution of (12) such that $\varphi_2(b) = b_1$ and $d\varphi_2(b)/dx = -b_0$.

$$b_0 \varphi_2(b) + b_1 \frac{d\varphi_2}{dx}(b) = 0$$

and every solution φ such that the relation

$$b_0 \varphi(b) + b_1 \frac{d\varphi}{dx}(b) = 0$$

is true satisfies $W(\varphi, \varphi_2)(x) = 0 \ \forall x \in [a, b]$ and is thus proportional to φ_2. Now

$$a_0 \varphi_2(a) + a_1 \frac{d\varphi_2}{dx}(a)$$

is non-zero since φ_2 is not a solution of the homogeneous problem; thus there exists a constant k such that $G_1(x) = k\varphi_2(x)$.

In the same way we can prove the existence of $G_2(x)$.

GENERALIZED GREEN'S FUNCTION: REGULAR HOMOGENEOUS PROBLEMS WITH NON-ZERO SOLUTION

We shall now suppose that the homogeneous problem has a non-zero solution f_0; naturally all the functions kf_0, where k is any constant, are also solutions of the homogeneous problem. If there existed a solution of the homogeneous problem independent of f_0, all solutions of the homogeneous equation would then satisfy the boundary conditions, which would then no longer have any significance.

To fix f_0 once and for all we shall suppose, for example, that

$$\int_a^b f_0^2(x) \, dx = 1$$

We can then prove the following theorem.

Theorem 1.3.4

Let

$$\frac{d}{dx} \left[p(x) \frac{df}{dx}(x) \right] + q(x)f(x) = g(x) \tag{32}$$

and

$$\left. \begin{array}{l} a_0 f(a) + a_1 \dfrac{df}{dx}(a) = 0 \\[2mm] b_0 f(b) + b_1 \dfrac{df}{dx}(b) = 0 \end{array} \right\} \tag{33}$$

Let $f_0(x)$ be the only solution of the homogeneous problem such that

$$\int_a^b f_0^2(u)\,du = 1$$

(1) Equation (32) has a solution satisfying (33) iff

$$\int_a^b f_0(x)g(x)\,dx = 0$$

(2) There exists a function $G(x, y)$ called the **generalized Green function** characterized by the following properties.

(a) $G(x, y)$ is a continuous function of (x, y) in the square $[a, b] \times]a, b[$.
(b) $\partial G(x, y)/\partial x$ is continuous in (x, y) in each of the triangles $a \leqslant x \leqslant y \leqslant b$ and $a \leqslant y \leqslant x \leqslant b$; moreover

$$\frac{\partial G}{\partial x}(y^+, y) - \frac{\partial G}{\partial x}(y^-, y) = \frac{1}{p(y)}$$

(c) $\forall y \in]a, b[$

$$\frac{d}{dx}\left[p(x)\frac{\partial G}{\partial x}(x, y) \right] + q(x)G(x, y) = -f_0(x)f_0(y)$$

over each of the intervals $[a, y[$ and $]y, b]$.
(d) $\forall y \in]a, b[$ $x \rightsquigarrow G(x, y)$ satisfies conditions (33).
(e) $\forall y \in]a, b[$

$$\int_a^b G(x, y)f_0(x)\,dx = 0$$

(3) G is symmetric and if

$$\int_a^b f_0(x)g(x)\,dx = 0$$

every solution of (32) satisfying (33) is of the form

$$f(x) = \int_a^b G(x, y)g(y)\,dy + kf_0$$

(k is an arbitrary constant), and the only one which satisfies the relation

$$\int_a^b f(x)f_0(x)\,dx = 0$$

is

$$\int_a^b G(x, y)g(y)\,dy$$

Proof

(1) First let us prove that if two functions φ_1 and φ_2 are of class C^2 and satisfy the boundary conditions (33) then

$$\int_a^b \varphi_1(x)\left\{\frac{d}{dx}\left[p(x)\frac{d\varphi_2}{dx}(x)\right] + q(x)\varphi_2(x)\right\}dx$$

$$= \int_a^b \varphi_2(x)\left\{\frac{d}{dx}\left[p(x)\frac{d\varphi_1}{dx}(x)\right] + q(x)\varphi_1(x)\right\}dx \quad (34)$$

Let us calculate in fact

$$\int_a^b \varphi_1(x)\frac{d}{dx}\left[p(x)\frac{d\varphi_2}{dx}(x)\right]dx = \left[\varphi_1(x)p(x)\frac{d\varphi_2}{dx}(x)\right]_a^b$$

$$- \int_a^b \frac{d\varphi_1}{dx}(x)p(x)\frac{d\varphi_2}{dx}(x)\,dx$$

$$= \left\{p(x)\left[\varphi_1(x)\frac{d\varphi_2}{dx}(x) - \varphi_2(x)\frac{d\varphi_1}{dx}(x)\right]\right\}_a^b$$

$$+ \int_a^b \varphi_2(x)\frac{d}{dx}\left[p(x)\frac{d\varphi_1}{dx}(x)\right]dx$$

The first term of the last expression is zero if φ_1 and φ_2 satisfy (33).

Now let f be a solution of (32) satisfying (33), f_0 being a solution of (28):

$$0 = \int_a^b f(x)\left\{\frac{d}{dx}\left[p(x)\frac{df_0}{dx}\right] + q(x)f_0(x)\right\}dx$$

$$= \int_a^b f_0(x)\left\{\frac{d}{dx}\left[p(x)f'(x)\right] + q(x)f(x)\right\}dx$$

$$= \int_a^b f_0(x)g(x)\,dx$$

which is then zero if there exists a solution f of (32) and (33).

We shall construct $G(x, y)$ and prove (3), which will complete the proof.

(2) We construct $G(x, y)$, starting from the properties indicated, in the same way as in the previous case. Let f_1 and f_2 be two independent solutions of the homogeneous equation and let f_3 be a solution of the equation

$$\frac{d}{dx}\left[p(x)\frac{df}{dx}(x)\right] + q(x)f(x) = -f_0(x)f_0(y)$$

Let

$$G(x, y) = \begin{cases} a_1(y)f_1(x) + a_2(y)f_2(x) + f_3(x) & \text{if } a \leqslant x \leqslant y \leqslant b \\ b_1(y)f_1(x) + b_2(y)f_2(x) + f_3(x) & \text{if } a \leqslant y \leqslant x \leqslant b \end{cases}$$

G then satisfies property (c), and we determine a_1, a_2, b_1 and b_2 in such a way that properties (a)–(e) are satisfied. We can in fact show that these properties are compatible and define a_1, a_2, b_1 and b_2 uniquely.

(3) Now let

$$f(x) = \int_a^b G(x, y)g(y)\,dy \qquad \frac{df}{dx} = \int_a^b \frac{\partial G}{\partial x}(x, y)g(y)\,dy$$

and

$$\frac{d^2 f}{dx^2}(x) = \int_a^b \frac{\partial^2 G}{\partial x^2}(x, y)g(y)\,dy + [G(x^+, x) - G(x^-, x)]g(x)$$

as before, so that

$$\frac{d}{dx}\left[p(x)\frac{df}{dx}(x)\right] + q(x)f(x)$$

is equal to

$$\int_a^b \left\{ \frac{d}{dx}\left[p(x)\frac{\partial G}{\partial x}(x, y)\right] + q(x)G(x, y) \right\} g(y)\,dy + g(x)$$

$$= -\int_a^b f_0(x)f_0(y)g(y)\,dy + g(x)$$

$$= g(x)$$

which completes the proof since f satisfies (33).

Example   Let

$$\frac{d^2 f}{dx^2} = g(x) \tag{35}$$

$$\left.\begin{array}{c} f(0) = 0 \\[2mm] f(1) - \dfrac{df}{dx}(1) = 0 \end{array}\right\} \tag{36}$$

The solutions of $d^2f/dx^2 = 0$ are of the form $f(x) = ax + b$ and if $b = 0$ they satisfy (36). Let

$$f_0(x) = x\sqrt{3}: \quad \int_0^1 f_0^2(x)\,dx = 1$$

$x \rightsquigarrow G(x, y)$ is then a solution of

$$\frac{d^2 f}{dx^2}(x) = -3xy$$

All the solutions of this equation are of the form

$$a(y)x + b(y) - \frac{x^3 y}{2}$$

Then let

$$G(x, y) = \begin{cases} -\dfrac{x^3 y}{2} + a_1(y)x + b_1(y) & \text{if } 0 \leqslant x < y \\[3mm] -\dfrac{x^3 y}{2} + a_2(y)x + b_2(y) & \text{if } y < x \leqslant 1 \end{cases}$$

G satisfies the boundary conditions (36) if $b_1(y) = G(0, y) = 0$ and if

$$G(1, y) - \frac{\partial G}{\partial x}(1, y) = -\frac{y}{2} + a_2(y) + b_2(y) + \frac{3}{2}\, y - a_2(y) = 0$$

Thus $b_1(y) = 0$, $b_2(y) = -y$.

Since $G(x, y)$ is continuous

$$G(y, y) = -\frac{y^4}{2} + a_1(y)y = -\frac{y^4}{2} + a_2(y)y - y$$

and $a_1(y) = a_2(y) - 1$.

To complete the determination of G it remains only to satisfy the property (e):

$$\int_0^1 G(x, y)\sqrt{3x}\, dx = 0$$

Let

$$0 = \int_0^y \left\{ -\frac{x^4 y}{2} + [a_2(y) - 1]x^2 \right\} dx + \int_y^1 \left[-\frac{x^4 y}{2} + a_2(y)x^2 - xy \right] dx$$

$$= -\frac{y}{2}\frac{1}{5} + a_2(y)\frac{1}{3} - \frac{y^3}{3} - \frac{y}{2}(1 - y^2)$$

$$= \frac{a_2(y)}{3} + \frac{y^3}{6} - \frac{3y}{5}$$

Thus

$$a_2(y) = -\frac{y^3}{2} + \frac{9y}{5}$$

and

$$a_1(y) = -\frac{y^3}{2} + \frac{9y}{5} - 1$$

Therefore

$$G(x, y) = \begin{cases} -\dfrac{x^3 y}{2} - \dfrac{y^3 x}{2} + \dfrac{9xy}{5} - x & \text{for } 0 \leqslant x \leqslant y \\[3mm] -\dfrac{y^3 x}{2} - \dfrac{x^3 y}{2} + \dfrac{9xy}{5} - y & \text{for } y \leqslant x \leqslant 1 \end{cases}$$

Thus, if

$$\int_0^1 xg(x)\,\mathrm{d}x = 0$$

(35) has a solution which satisfies (36) and it is

$$f(x) = \int_0^1 \left(\frac{9xy}{5} - \frac{x^3y + y^3x}{2} \right) g(y)\,\mathrm{d}y - \int_0^x yg(y)\,\mathrm{d}y - x \int_x^1 g(y)\,\mathrm{d}y$$

If for example

$$g(x) = 2 - 3x \qquad \int_0^1 g(x)f_0(x)\,\mathrm{d}x = 0$$

and

$$f(x) = \int_0^1 -\frac{x}{2}(y^3)(2 - 3y)\,\mathrm{d}y - x \int_x^1 (2 - 3y)\,\mathrm{d}y - \int_0^x y(2 - 3y)\,\mathrm{d}y$$

$$= -\frac{x^3}{2} + x^2 - \frac{9x}{20} + kx$$

Geometrical analogy

We shall pursue the analogy which we have already met on page 313. Let φ_1 and φ_2 be two elements of the vector space of functions satisfying the boundary conditions; the quantity

$$\int_a^b \varphi_1(u)\varphi_2(u)\,\mathrm{d}u$$

denoted by $\langle \varphi_1, \varphi_2 \rangle$ is a scalar product over this space.

Let A be the linear operator defined by

$$A\varphi(x) = \frac{\mathrm{d}}{\mathrm{d}x} \left[p(x)\frac{\mathrm{d}\varphi}{\mathrm{d}x} \right] + q(x)\varphi(x)$$

The relation (34) established at the beginning of Theorem 1.3.4 states that $\langle A\varphi_1, \varphi_2 \rangle = \langle \varphi_1, A\varphi_2 \rangle$, which justifies the adjective 'self-adjoint' describing the form of the equation.

The condition

$$\int_a^b f_0(u)g(u)\,\mathrm{d}u = \langle f_0, g \rangle = 0$$

is the analogue of the relation $\langle V_0, V \rangle = 0$.

Since $A = A^*$ we can identify V_0 and W_0 with f_0 in the analogy, and the form of the solutions, when there are any, is analogous to $V + kV_0$.

In the second part of this chapter, we shall give an account of the theoretical ideas which enable us to explore this analogy.

Complement: an example of a singular problem

A problem which does not satisfy the conditions of Definition 1.3.1 is called singular. There are many problems of this type, e.g. those for which the coefficient of d^2f/dx^2 can vanish, those for which the interval $[a, b]$ is infinite and those for which p or q are not bounded.

We shall not present the theory of these different problems, which is very intricate, but it is advisable to be aware that in many cases we can still define a Green function in general by adjusting the boundary conditions.

We shall still assume that if $p(x) = x$ and if $a = 0$ we are looking for Green's function for a problem in which it is required that $f(x)$ remains bounded if $x \to 0$ or even that $f(x)$ has a limit if $x \to 0$. Similarly for $b = \infty$ it could be required that $f(x)$ remain bounded if $x \to \infty$ or even that

$$\lim_{x \to \infty} f(x)$$

exist.

When Green's function exists its determination can be made directly from its defining properties as in Theorem 1.2.2 or 1.3.4 for a generalized Green function.

Example Find Green's function for the problem

$$x\frac{d^2f}{dx^2}(x) + \frac{df}{dx}(x) = 0 \quad \text{for } x \in \left]0, 1\right] \tag{37}$$

$$\left.\begin{array}{l} f(x) \quad \text{is bounded when } x \to 0 \\ f(1) - kf'(1) = 0 \end{array}\right\} \tag{38}$$

The solutions of (37) are $f(x) = a + b \log x$; they satisfy (38) iff $a = b = 0$. We can then look for a true Green function. Thus let

$$G(x, y) = \begin{cases} a_1(y) + b_1(y) \log x & \text{for } 0 < x \leqslant y \\ a_2(y) + b_2(y) \log x & \text{for } y \leqslant x \leqslant 1 \end{cases}$$

Let us write first that G is continuous:

$$[a_1(y) - a_2(y)] + [b_1(y) - b_2(y)] \log y = 0$$

Putting

$$c(y) = a_1(y) - a_2(y) \quad \text{and} \quad d(y) = b_1(y) - b_2(y)$$

then

$$c(y) + d(y) \log y = 0$$

As

$$\frac{\partial G}{\partial y}(y^+, y) = \frac{b_2(y)}{y} \quad \text{and} \quad \frac{\partial G}{\partial y}(y^-, y) = \frac{b_1(y)}{y}$$

we must have

$$\frac{1}{y} = -\frac{d(y)}{y}$$

Thus $d(y) = -1$ and $c(y) = \log y$.

We must now write that $x \rightsquigarrow G(x, y)$ satisfies (38), which requires us to put $b_1(y) = 0$ on the one hand and $a_2(y) - kb_2(y) = 0$ on the other. Thus $b_1(y) = 0$, $b_2(y) = 1$, $a_2(y) = k$ and $a_1(y) = k + \log y$. Therefore

$$G(x, y) = \begin{cases} k + \log y & \text{for } 0 < x \leqslant y \\ k + \log x & \text{for } y \leqslant x \leqslant 1 \end{cases}$$

The solution of

$$x\frac{d^2 f}{dx^2} + \frac{df}{dx} = 1$$

satisfying (38) is thus

$$f(x) = k + x \log x + \int_x^1 \log y \, dy = x + k - 1$$

which is perfectly clear. However, we must *note carefully* that, even if $G(x, y)$ exists, i.e. if there exists a function satisfying the properties (a), (b), (c) and (d) of Theorem 1.2.2, we cannot without great care construct the expression

$$\int_0^1 G(x, y)g(y) \, dy$$

It is first necessary to ensure that

$$\int_0^1 g(y) \log y \, dy$$

has a meaning. If g satisfies

$$\int_0^1 |g(y) \log y| \, dy < \infty$$

the solution of

$$x\frac{d^2 f}{dx^2} + \frac{df}{dx} = g(x)$$

is then certainly equal to

$$(k + \log x) \int_0^x g(y) \, dy + \int_x^1 (k + \log y)g(y) \, dy$$

See *Exercises 1 to 15* at the end of the chapter.

Part 2: Hilbert spaces; linear operators; eigenfunctions

In the introduction to this chapter we encountered the following problem. Find numbers λ and functions φ (not identically zero) such that

$$\frac{d^2\varphi}{dx^2} = \lambda\varphi(x) \tag{39}$$

and

$$\varphi(0) = \varphi(L) = 0 \tag{40}$$

Then, f being a given function, determine whether it is possible to find a sequence of numbers a_n and of functions φ_n, solutions of (39) satisfying (40), which allow us to write—in a sense to be made precise—

$$f(x) = \sum_{n \in \mathbb{N}} a_n \varphi_n(x)$$

It is easy to see that if $\lambda = \mu^2 \geqslant 0$ the only solution of (39) satisfying the conditions (40) is $\varphi(x) \equiv 0$.

If on the contrary $\lambda = -\mu^2$ ($\mu > 0$), the solutions of (39) such that $\varphi(0) = 0$ are $\varphi(x) = A \sin x$, and if $\mu = n\pi/L$ the function $A \sin(n\pi x/L)$ is a solution of (39) and satisfies (40); if $\mu \neq n\pi/L$ only the function $\varphi \equiv 0$ is suitable.

Thus there exists a sequence $\lambda_n = n\pi/L$ ($n > 0$) and a sequence $\varphi_n = A \sin(n\pi x/L)$ satisfying (39) and (40); every function which is a multiple of φ_n naturally is equally appropriate. To fix φ_n we can impose the condition

$$\int_0^L \varphi_n^{\,2}(x)\,dx = 1$$

Then

$$\varphi_n(x) = \left(\frac{2\pi}{L}\right)^{1/2} \sin\frac{n\pi}{L}x$$

and we can confirm that

$$\int_0^L \varphi_n(x)\varphi_m(x)\,dx = 0$$

for all $n \neq m$.

We are now in a situation quite similar to that which we encounter in the search for an orthonormal basis of eigenvectors of a matrix A; conversely the situation of the Fredholm alternative which is met in certain differential equations can very easily be represented as applying to equations involving matrices and vectors in $\mathbb{R}^n$, as we have seen.

This part is devoted to the exact study of this analogy. We shall introduce linear operators over vector spaces of functions, and we shall establish that when a vector space is not of finite dimension the problems that we wish to solve are much more complex, but that there exists a situation comparable at all points with that of symmetric matrices in $\mathbb{R}^n$—that of compact self-adjoint operators—which is exactly what we meet in the problems of integral equations and differential equations. This will be the subject of the third part of this chapter.

See *Exercise 16* at the end of the chapter.

2.1. Hilbert spaces

It is clear that we can define a linear mapping only on a vector space and that we can speak of an orthonormal basis only on a vector space with a scalar product.

We shall study function spaces on which such notions have been defined. In Chapter 1 (p. 7) we have already introduced the notion of norm and that of a Cauchy sequence. We recall that by a complete space we mean a space in which every Cauchy sequence is convergent.

We have established that in $\mathbb{R}^n$ all norms are equivalent but we know that this is not in general the case for spaces of infinite dimension (cf. Chapter 1, p. 8).

Definition 2.1.1

Let B be a vector space with a norm; if for the distance associated with this norm B is complete, we say that B is a **Banach space**.

We have proved that the space of continuous functions over $[0, 1]$ with the norm

$$\sup_{x \in [0, 1]} |f(x)|$$

is a Banach space.

We shall now introduce the notion of scalar product; this scalar product will allow us to define a norm and thus the notion of limit.

HILBERT SPACES

Definition 2.1.2

Let H be a vector space constructed over $\mathbb{C}$. We define a scalar product over H as a mapping

$$H \times H \to \mathbb{C}$$

$$(u, v) \rightsquigarrow \langle u, v \rangle$$

such that

(a) $\langle u, u \rangle \geqslant 0$ $\langle u, u \rangle = 0 \Leftrightarrow u = 0$ $\forall u \in H$
(b) $\langle u_1 + u_2, v \rangle = \langle u_1, v \rangle + \langle u_2, v \rangle$ $\forall u_1, u_2, v \in H$
     $\langle \lambda u, v \rangle = \lambda \langle u, v \rangle$ $\forall u, v \in H$ and $\forall \lambda \in \mathbb{C}$
(c) $\langle u, v \rangle = \langle \overline{v, u} \rangle$ $\forall u, v \in H$ so that $\langle u, \lambda v \rangle = \bar{\lambda} \langle u, v \rangle$

(2) Let H be a vector space constructed over $\mathbb{R}$. We define a scalar product over H as a mapping

$$H \times H \to \mathbb{R}$$

$$(u, v) \rightsquigarrow \langle u, v \rangle$$

having the properties (a) and (b) for $\lambda \in \mathbb{R}$ as well as

(c') $\langle u, v \rangle = \langle v, u \rangle$

(3) We say that u and v are orthogonal if $\langle u, v \rangle = 0$.

Theorem 2.1.3

(a) The quantity $\langle u, u \rangle^{1/2}$ is a norm over H.

We shall write $\|u\| = \langle u, u \rangle^{1/2}$ for **the norm** of an element u. Moreover, for all u and v of H

(b) $\|u + v\| \leqslant \|u\| + \|v\|$ (triangular inequality).

If $u \perp v$ then $\|u + v\|^2 = \|u\|^2 + \|v\|^2$ (Pythagoras' theorem) and

(c) $|\langle u, v \rangle|^2 \leqslant \langle u, u \rangle \langle v, v \rangle$ (Schwarz's inequality).

Remark   If H is a Hilbert space constructed over $\mathbb{R}$, then

(b') $\|u + v\|^2 = \|u\|^2 + \|v\|^2 \Leftrightarrow u \perp v$.

Definition 2.1.4
If H is complete for this norm, H is called a **Hilbert space**.

Proof of Theorem 2.1.3   Let us first establish the Schwarz inequality:

$$\forall \lambda \in \mathbb{C} \quad \forall u, v \in H$$

$$0 \leqslant \langle u + \lambda v, u + \lambda v \rangle = \langle u, u \rangle + \lambda \langle v, u \rangle + \bar{\lambda} \langle u, v \rangle + |\lambda|^2 \langle v, v \rangle$$

Let $\langle u, v \rangle = R \exp(i\vartheta)$ and $\lambda = x \exp(i\vartheta)$. The preceding equation can be written for one λ in the form $0 \leqslant \langle u, u \rangle + 2xR + x^2 \langle v, v \rangle$. Therefore $|\langle u, v \rangle|^2 = R^2 \leqslant \langle u, u \rangle \langle v, v \rangle$ and there is equality only if $u + \lambda v = 0$.

Now we prove that $\langle u, u \rangle^{1/2}$ is actually a norm. It is clear that $\|u\| \geqslant 0$ and that $\|u\| = 0 \Leftrightarrow u = 0$. It is also clear that $\|\lambda u\| = |\lambda| \|u\|$ $\forall \lambda \in \mathbb{C}$.

It remains to prove that $\|u + v\| \leqslant \|u\| + \|v\|$, i.e. that

$$\langle u + v, u + v \rangle^{1/2} \leqslant \langle u, u \rangle^{1/2} + \langle v, v \rangle^{1/2}$$

or, on squaring,

$$\langle u+v, u+v\rangle \leqslant \langle u, u\rangle + \langle v, v\rangle + 2\langle u, u\rangle^{1/2}\langle v, v\rangle^{1/2}$$

This is true since the first member equals $\langle u, u\rangle + \langle v, v\rangle + 2\operatorname{Re}\langle u, v\rangle$ and since $\operatorname{Re}\langle u, v\rangle \leqslant |\langle u, v\rangle| \leqslant \langle u, u\rangle^{1/2}\langle v, v\rangle^{1/2}$.

The triangular inequality states simply that the length of one side of a triangle is less than or equal to the sum of the lengths of the two other sides since $|\,\|u\| - \|v\|\,| \leqslant \|u - v\| \leqslant \|u\| + \|v\|$.

Basis of a Hilbert space

Definition 2.1.5

We define an orthogonal basis of a Hilbert space as a family $\{e_i\}_{i\in I}$ of elements of this space that have the following two properties:

(a) $\langle e_i, e_j\rangle = 0 \quad \forall i \neq j$
(b) if $\langle u, e_i\rangle = 0 \quad \forall i \quad$ then $u = 0$

If in addition $\|e_i\| = 1$, the basis is called orthonormal (we sometimes use the expression Hilbertian basis). We can prove that every Hilbert space has an **orthonormal basis**.

A Hilbert space is called **separable** if there exists a sequence $\{e_n\}_{n\in\mathbb{N}}$ of elements of H which forms an **orthonormal basis** of H. (All the Hilbert spaces which we shall use will be separable.)

Projection on a subspace of finite dimension

Theorem 2.1.6

Let $e_1, e_2, \ldots, e_p$ be p unit and mutually orthogonal vectors of H, and let K be the vector subspace which they generate:

$$K = \{u \in H : u = a_1 e_1 + a_2 e_2 + \ldots + a_p e_p\}$$

(a) $\forall u \in K$, $a_i = \langle u, e_i\rangle$ and the representation of u in terms of the basis $(e_1, \ldots, e_p)$ is unique and

$$\|u\|^2 = \sum_{i=1}^{p} a_i^2$$

(b) Let $\{u_n\}_{n\in\mathbb{N}}$ be a sequence of elements of K converging in H to u; then $u \in K$ (K is closed).

(c) $\forall v \in H$ there exists a unique element of K, denoted $P^K v$ and called the orthogonal projection of v on K, such that

$$\forall u \in K \quad \langle v - P^K v, u\rangle = 0$$

This element also attains the minimum of $\|v - u\|$ as u runs through K. The minimum is called the distance from v to K.

(d) $H = K \oplus K^{\perp}$, i.e. $\forall v \in H$, $\exists v_1 \in K$, $v_2 \in K^{\perp}$: $v = v_1 + v_2$.

Proof

(a) It is sufficient to calculate $\langle u, e_i \rangle$. We note then that according to Pythagoras' theorem

$$\|u\|^2 = \sum_{i=1}^{p} a_i^2$$

(b) By hypothesis $\{u_n\}_{n \in \mathbb{N}}$ is a Cauchy sequence in H. If $u_n = b_1^{(n)} e_1 + \ldots + b_p^{(n)} e_p$ then $\|u_n - u_m\|^2 = |b_1^{(n)} - b_1^{(m)}|^2 + \ldots + |b_p^{(n)} - b_p^{(m)}|^2$ so that the sequences $\{b_i^{(n)}\}_{n \in \mathbb{N}}$ are Cauchy sequences for $i = 1, \ldots, p$. Thus they converge to b_i and $u = b_1 e_1 + \ldots + b_p e_p \in K$.

(c) Let $P^K v = \langle v, e_1 \rangle e_1 + \ldots + \langle v, e_p \rangle e_p$. It is clear that $P^K v \in K$ and that $\langle v - P^K v, e_i \rangle = \langle v, e_i \rangle - \langle P^K v, e_i \rangle = 0 \ \forall i$; therefore $\langle v - P^K v, u \rangle = 0 \ \forall u \in K$. If w is another element of K having the same property, then

$$\langle v, e_i \rangle = \langle P^K v, e_i \rangle = \langle w, e_i \rangle$$

and

$$w = \sum_{i=1}^{p} \langle w, e_i \rangle e_i = P^K v$$

Finally we construct $\|v - u\|^2 = \|v - P^K v + P^K v - u\|^2 = \|v - P^K v\|^2 + \|P^K v - u\|^2$, since $P^K v - u \in K$; this quantity is a minimum for $u = P^K v$.

In $\mathbb{R}^d$ every element u may be written uniquely as

$$u = \sum_{i=1}^{d} \langle u, e_i \rangle e_i$$

in terms of an orthonormal basis $e_1, \ldots, e_d$ and

$$\|u\|^2 = \sum_{i=1}^{d} |\langle u, e_i \rangle|^2$$

according to Pythagoras' theorem.

In infinite dimensions every element may be expressed as a limit of linear combinations; more precisely, the difference between u and its projection on the space generated by $e_1, \ldots, e_p$ tends to zero as p tends to infinity.

Theorem 2.1.7

Let H be a separable Hilbert space and $\{e_n\}_{n \in \mathbb{N}}$ an orthogonal basis. Then

$$\lim_{n \to \infty} \left\| u - \sum_{1}^{n} \langle u, e_p \rangle e_p \right\| = 0$$

and

$$\|u\|^2 = \sum_{p=1}^{\infty} \langle u, e_p \rangle^2$$

Conversely

$$u_n = \sum_{1}^{n} a_p e_p$$

has a limit u when $n \to \infty$ iff

$$\sum_{p=1}^{\infty} a_p^2 < \infty$$

and in this case

$$\|u\|^2 = \sum_1^\infty a_p{}^2$$

We shall write for every element u of H

$$u = \sum_{p=1}^\infty \langle u, e_p \rangle e_p$$

It must be clearly understood that this involves a limit for the norm v of H.

Proof For every integer n

$$u = \sum_{p=1}^n \langle u, e_p \rangle e_p + \left(u - \sum_{p=1}^n \langle u, e_p \rangle e_p \right)$$

The two terms on the right-hand side of this equality are orthogonal; we can see this by constructing the scalar product

$$\left\langle \sum_{p=1}^n \langle u, e_p \rangle e_p, u - \sum_{q=1}^n \langle u, e_q \rangle e_q \right\rangle$$

$$= \sum_{p=1}^n \langle u, e_p \rangle \langle e_p, u \rangle - \sum_{p=1}^n \sum_{q=1}^n \langle u, e_p \rangle \langle \overline{u, e_q} \rangle \langle e_p, e_q \rangle$$

$$= \sum_{p=1}^n \langle u, e_p \rangle \langle e_p, u \rangle - \sum_{p=1}^n \langle u, e_p \rangle \langle \overline{u, e_p} \rangle = 0$$

Thus

$$\|u\|^2 = \left\| \sum_{p=1}^n \langle u, e_p \rangle e_p \right\|^2 + \left\| u - \sum_1^n \langle u, e_p \rangle e_p \right\|^2$$

$$= \sum_{p=1}^n |\langle u, e_p \rangle|^2 + \left\| u - \sum_1^n \langle u, e_p \rangle e_p \right\|^2$$

Thus

$$\forall n: \sum_{p=1}^n |\langle u, e_p \rangle|^2 \leqslant \|u\|^2$$

then $\sum_{p=1}^\infty |\langle u, e_p \rangle|^2$ exists. Therefore

$$\left\| \sum_1^m \langle u, e_p \rangle e_p - \sum_1^n \langle u, e_q \rangle e_q \right\|^2 = \sum_{n+1}^m |\langle u, e_p \rangle|^2$$

so that

$$\sum_{p=1}^n \langle u, e_p \rangle e_p$$

is a Cauchy sequence in H which is complete. Thus

$$\sum_{p=1}^n \langle u, e_p \rangle e_p$$

has a limit v and

$$\langle v, e_r \rangle = \lim_{n \to \infty} \sum_{p=1}^n \langle u, e_p \rangle \langle e_p, e_r \rangle = \langle u, e_r \rangle$$

Thus $\forall r \in \mathbb{N}$, $\langle u - v, e_r \rangle = 0$. Therefore

$$u = v = \lim_{n \to \infty} \sum_{p=1}^{n} \langle u, e_p \rangle e_p$$

EXAMPLES OF HILBERT SPACES

Euclidean space $\mathbb{R}^n$

The vector space $\mathbb{R}^n$ with the scalar product

$$\langle u, v \rangle = \sum_{1}^{n} u_i v_i$$

for $u = (u_1, \ldots, u_n)$, $v = (v_1, \ldots, v_n)$ is a Hilbert space; the corresponding distance is the ordinary distance, also called the Euclidean distance.

The space l^2

The space l^2 is the vector space formed from sequences $\{a_n\}_{n \in \mathbb{N}}$ of real numbers such that

$$\sum_{n=1}^{\infty} |a_n|^2 < \infty$$

If $A = \{a_n\}_{n \in \mathbb{N}}$ and $B = \{b_n\}_{n \in \mathbb{N}}$ are two such sequences, we must satisfy ourselves that $A + B = \{a_n + b_n\}_{n \in \mathbb{N}} \in l^2$. Now

$$(a_n + b_n)^2 \geqslant 0 \Rightarrow a_n^2 + b_n^2 \geqslant -2a_n b_n$$
$$(a_n - b_n)^2 \geqslant 0 \Rightarrow a_n^2 + b_n^2 \geqslant 2a_n b_n$$

Thus

$$a_n^2 + b_n^2 \geqslant 2|a_n b_n|$$

Then

$$(a_n + b_n)^2 \leqslant 2(a_n^2 + b_n^2) \quad \text{and} \quad \sum (a_n + b_n)^2 \leqslant 2 \sum a_n^2 + 2 \sum b_n^2 \leqslant \infty$$

We can easily verify that l^2 is a Hilbert space for the scalar product

$$\langle A, B \rangle = \sum_{n=1}^{\infty} a_n b_n$$

The space $L_\sigma^2[(a, b)]$

First let $[a, b]$ be a bounded segment of $\mathbb{R}$ and σ a function $\sigma(x) > 0$ $\forall x \in [a, b]$. If f and g are two continuous functions on $[a, b]$ we denote by $\langle f, g \rangle$ the quantity

$$\int_a^b f(x)g(x)\sigma(x)\,\mathrm{d}x$$

It is clear that $\langle f, f \rangle \geqslant 0$ and that $\langle f, f \rangle = 0 \Leftrightarrow f \equiv 0$. It is clear also that $\langle f_1 + f_2, g \rangle = \langle f_1, g \rangle + \langle f_2, g \rangle$ and that $\forall \lambda \in \mathbb{R}\ \lambda \langle f, g \rangle = \langle \lambda f, g \rangle$ and finally $\langle f, g \rangle = \langle g, f \rangle$.

Thus $\langle f, g \rangle$ is a scalar product for the vector space of continuous functions on $[a, b]$. However, it is not complete for the associated norm and we must consider a larger set of functions; these are the integral functions, in the Lebesgue sense, with respect to $\sigma(x)$.

We shall not give the precise definition of this space, but we shall bear in mind that for every continuous function σ, for every segment $[a, b]$ bounded or not, there exists a space $L_\sigma^2[a, b]$, the space of square integrable functions with respect to σ, which is a Hilbert space for the scalar product

$$\langle f, g \rangle = \int_a^b f(x)g(x)\sigma(x)\,\mathrm{d}x$$

and which contains all functions such that

$$\int_a^b f^2(x)\sigma(x)\,\mathrm{d}x < \infty$$

for the usual Riemann integral.

In this space two functions f and g such that

$$\int_a^b [f(x) - g(x)]^2\sigma(x)\,\mathrm{d}x = 0$$

are considered identical. If they are continuous they are actually identical. If $\sigma \equiv 1$ we shall write $L^2[a, b]$ for $L_1^2[a, b]$ and shall note that if $\sigma \geqslant 0$

$$f \in L_\sigma^2[a, b] \Leftrightarrow f\sigma^{1/2} \in L^2[a, b]$$

$$\langle f, g \rangle_{L_\sigma^2[a,b]} = \langle f\sigma^{1/2}, g\sigma^{1/2} \rangle_{L^2[a,b]}$$

so that we are quite often brought back to $L^2[a, b]$.

If we consider functions which can take complex values, we obtain a Hilbert space constructed on $\mathbb{C}$ by choosing as scalar product

$$\langle f, g \rangle = \int_a^b f(x)\bar{g}(x)\sigma(x)\,\mathrm{d}x$$

Example of projection: least squares approximation in $L^2[0, 1]$

Let $H = L^2[0, 1]$; the function $f_1 \equiv 1$ has norm 1. The first-degree polynomial with norm 1 orthogonal to f_1 is $f_2 = \sqrt{3}(2x - 1)$, which we obtain by writing

$$\int_0^1 (ax + b)^2\,\mathrm{d}x = 1 \quad \text{and} \quad \int_0^1 ax + b = 0$$

The second-degree polynomial with norm 1, orthogonal to f_1 and f_2, is

$f_3 = \sqrt{5}(6x^2 - 6x + 1)$ which we obtain in the same way by writing

$$\int_0^1 [f_3(x)]^2 \, dx = 0$$

while

$$\int_0^1 f_1(x)f_3(x) \, dx = 0 = \int_0^1 f_2(x)f_3(x) \, dx$$

Let K be the vector space generated by f_1, f_2, f_3. If $g \in L^2[0, 1]$, $P^K g$, which minimizes the distance from g to K, is called the least squares approximation to g by an element of K. Its calculation is very simple: $P^K g = \langle g, f_1 \rangle f_1 + \langle g, f_2 \rangle f_2 + \langle g, f_3 \rangle f_3$, i.e.

$$P^K g(x) = \int_0^1 g(u) \, du + 3(2x - 1) \int_0^1 g(u)(2u - 1) \, du + 5(6x^2 - 6x + 1)$$

$$\times \int_0^1 g(u)(6u^2 - 6u + 1) \, du$$

If $g(x) \equiv x^3$, the second-degree polynomial which is the closest approximation to $g(x)$ in $L^2[0, 1]$ is thus

$$P^K g(x) = \tfrac{1}{4} + \tfrac{9}{20}(2x - 1) + \tfrac{1}{4}(6x^2 - 6x + 1) = \tfrac{1}{20}(30x^2 - 12x + 1)$$

It is important to define precisely the notion of convergence associated with the norm in a space L^2.

Notation   We shall write $\| f \|_2$ for the norm of a function f belonging to a space $L^2[a, b]$.

Definition 2.1.8
We say that a sequence $\{f_n\}_{n \in \mathbb{N}}$ of functions in $L_\sigma^2[a, b]$ **converges in quadratic mean** with respect to σ, or more simply in $L_\sigma^2[a, b]$, to f if

$$\lim_{n \to \infty} \int_a^b |f_n(x) - f(x)|^2 \sigma(x) \, dx = 0$$

A Cauchy sequence in $L_\sigma^2[a, b]$ is then a sequence such that

$$\forall \varepsilon > 0 \quad \exists N: n, m \geq N \quad \int_a^b |f_n(x) - f_m(x)|^2 \sigma(x) \, dx \leq \varepsilon$$

Since $L_\sigma^2[a, b]$ is complete, such a sequence converges in the sense defined above. This notion of convergence is different from the notion of pointwise convergence; it is possible that

$$f_n \xrightarrow[L^2[a,b]]{} f$$

but that $f_n(x) \not\to f(x)$ for every x in $[a, b]$, and conversely.

However, if a sequence $\{f_n\}_{n\in\mathbb{N}}$ converges at the same time in L^2 and pointwise, the limit is the same function f.

Examples
(1) Let

$$f_n(x) = \begin{cases} 0 & \text{if } x = 0 \\ n^{1/2} & \text{if } 0 < x < 1/n \\ 0 & \text{if } 1/n \leqslant x \leqslant 1 \end{cases}$$

as shown in Fig. 6.3. It is clear that $\forall x$

$$\lim_{n \to \infty} f_n(x) = 0$$

and $f_n \to 0$ pointwise. However, $f_n \nrightarrow 0$ in $L^2[0, 1]$ since

$$\int_0^1 f_n^2(x)\,dx = 1$$

does not tend to zero.

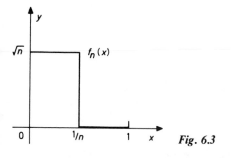

Fig. 6.3

(2) As before, consider $L^2[0, 1]$. We construct a sequence $\{f_n\}_{n\in\mathbb{N}}$ in the following way. We consider the subdivisions of $[0, 1]$ obtained by dividing this segment successively into 2, 4, 8, ..., 2^n, ... equal parts. We enumerate the segments obtained in the following order: $[0\,\frac{1}{2}][\frac{1}{2}\,1]$, then $[0\,\frac{1}{4}][\frac{1}{4}\,\frac{1}{2}][\frac{1}{2}\,\frac{3}{4}][\frac{3}{4}\,1]$, then $[0\,\frac{1}{8}][\frac{1}{8}\,\frac{2}{8}]$ etc.
Consider the sequence defined thus

$$f_0(x) \equiv 1$$

$f_1(x) = 1$ over the first segment, 0 elsewhere

$f_2(x) = 1$ over the second segment, 0 elsewhere

$\vdots$

$f_n(x) = 1$ over the nth segment, 0 elsewhere

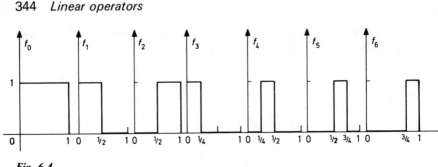

Fig. 6.4

Figure 6.4 represents the graphs of the functions f_0 to f_6. We observe that $\forall x \in [0, 1]$ there exists an infinity of indices n such that $f_n(x) = 0$ and an infinity of indices m such that $f_m(x) = 1$; therefore

$$\lim_{n \to \infty} f_n(x)$$

does not exist.

However,

$$\left[\int_0^1 f_n^2(x)\,dx \right]^{1/2} = \| f_n \|_2 = \frac{1}{2^p} \quad \text{if } 2^p - 1 \leqslant n < 2^{p+1} - 1$$

and therefore

$$\lim_{n \to \infty} (\text{in } L^2[0, 1]) f_n = 0$$

(3) Spaces $H^1[a, b]$ and $H^2[a, b]$.

Definition 2.1.9
We say that a real function is **absolutely continuous** over $[a, b]$ if there exists a function g such that

$$f(x) = f(a) + \int_a^x g(y)\,dy \qquad \forall x \in [a, b]$$

g is not necessarily continuous over $[a, b]$ and therefore f is not necessarily differentiable, but if φ is a function C^∞ such that $\varphi(a) = \varphi(b) = 0$

$$\int_a^b \varphi(s)g(s)\,ds = [\varphi(s)f(s)]_a^b - \int_a^b \frac{d\varphi}{ds} f(s)\,ds = - \int_a^b \frac{d\varphi}{ds} f(s)\,ds$$

which means that g is the derivative of f in the sense of distributions (cf. p. 143).

Definition 2.1.10
The space $H^1[a, b]$ is the vector space of functions absolutely continuous over $[a, b]$ such that $f \in L^2[a, b]$ and $g \in L^2[a, b]$, with the following scalar

product. If

$$f_1(x) = f_1(a) + \int_a^x g_1(y)\,dy$$

$$f_2(x) = f_2(a) + \int_a^x g_2(y)\,dy$$

$$\langle f_1, f_2 \rangle_{H^1[a,b]} = \int_a^b f_1(s)f_2(s)\,ds + \int_a^b g_1(s)g_2(s)\,ds$$

Since $L^2[a, b]$ is a Hilbert space, it is clear that $H^1[a, b]$ is one also.

We define likewise the space $H^2[a, b]$.

Definition 2.1.11
The space $H^2[a, b]$ is the vector space of differentiable functions f whose derivative df/dx is absolutely continuous over $[a, b]$, i.e.

$$\frac{df}{dx}(x) = \frac{df}{dx}(a) + \int_a^x g(y)\,dy$$

such that $f \in L^2[a, b]$, $df/dx \in L^2[a, b]$ and $g \in L^2[a, b]$, with the scalar product

$$\langle f_1, f_2 \rangle_{H^2[a,b]} = \int_a^b f_1(s)f_2(s)\,ds + \int_a^b \frac{df_1}{dx}(u)\frac{df_2}{dx}(u)\,du + \int_a^b g_1(u)g_2(u)\,du$$

$H^2[a, b]$ is a Hilbert space.

EXAMPLES OF BASES OF SEPARABLE HILBERT SPACES

Legendre polynomials
We define over $[-1, +1]$ the Legendre polynomials $\{P_n\}_{n \in \mathbb{N}}$ by the relation

$$P_n = \left(n + \frac{1}{2}\right)^{1/2} \frac{1}{2^n n!} \frac{d^n}{dx^n}[(x^2 - 1)^n]$$

Thus

$$P_0(x) = \frac{1}{\sqrt{2}} \qquad P_1(x) = \left(\frac{3}{2}\right)^{1/2} x \qquad P_2(x) = \left(\frac{5}{2}\right)^{1/2} \frac{3x^2 - 1}{2}$$

We shall study these polynomials in Exercise 17.
We can verify that $\{P_n\}_{n \in \mathbb{N}}$ is an orthonormal basis of $L^2[-1, +1]$. Every function f of $L^2[-1, +1]$ is thus the limit in $L^2[-1, +1]$ of

$$\sum_{n=1}^{p} a_n P_n(x)$$

when $p \to \infty$ with

$$a_n = \int_{-1}^{+1} f(x) P_n(x) \, dx$$

and

$$\sum a_n^{\,2} = \int_{-1}^{+1} |f(x)|^2 \, dx$$

We can verify also that P_n is a solution of the differential equation

$$(1 - x^2) \frac{d^2 f}{dx^2} - 2x \frac{df}{dx} + n(n+1)f = 0$$

with the condition

$$\lim_{|x| \to 1} f(x) < \infty$$

It should be noted that these polynomials are no longer orthogonal over $[0, 1]$ since, for example,

$$\int_0^1 P_0(u) P_1(u) \, du = \frac{\sqrt{3}}{4}$$

We can verify immediately that if f is even its expansion involves only the polynomials P_{2n} ($n \in \mathbb{N}$), while if f is odd it involves only the polynomials P_{2n+1} ($n \in \mathbb{N}$). Hence we can deduce that the polynomials ($n \in \mathbb{N}$)

$$Q_n = (4n+1)^{1/2} \frac{1}{2^{2n}(2n)!} \frac{d^{2n}}{dx^{2n}} [(x^2 - 1)^{2n}] = \sqrt{2} \, P_{2n}$$

form an orthonormal basis of $L^2[0, 1]$ as do the polynomials ($n \in \mathbb{N}$)

$$R_n = (4n+3)^{1/2} \frac{1}{2^{2n+1}(2n+1)!} \frac{d^{2n+1}}{dx^{2n+1}} [(x^2 - 1)^{2n+1}] = \sqrt{2} \, P_{2n+1}$$

We shall meet the Legendre polynomials again in Part 3.

Fourier series

Consider the space $L_\mathbb{C}^2[(0, 2\pi), dx/2\pi]$ of functions with complex values defined over $[0, 2\pi]$ and such that

$$\frac{1}{2\pi} \int_0^{2\pi} |f(x)|^2 \, dx < \infty$$

This is a Hilbert space for the scalar product

$$\langle f, g \rangle = \int_0^{2\pi} f(x) \bar{g}(x) \frac{dx}{2\pi}$$

We can show that the set of functions $\{x \rightsquigarrow \exp(inx)\}_{n \in \mathbb{Z}}$ is an orthonormal basis of $L_\mathbb{C}^2[(0, 2\pi), dx/2\pi]$.

Let $f \in L_{\mathbb{C}}^2([a, b], dx/2\pi)$ and

$$\alpha_p = \frac{1}{2\pi} \int_0^{2\pi} f(y) \exp(-ipy) \, dy \quad \text{for all } p \in \mathbb{Z}$$

Then

$$f(x) = \lim_{L^2} \sum_{-N}^{N} \alpha_p \exp(ipx)$$

If f is real-valued we can regroup the terms $\alpha_p \exp(ipx)$ and $\alpha_{-p} \exp(-ipx)$. Now

$$\alpha_p \exp(ipx) = \frac{1}{2\pi} \left[\int_0^{2\pi} f(y)(\cos py - i \sin py) \, dy \right] [\cos px + i \sin px]$$

$$\alpha_{-p} \exp(-ipx) = \frac{1}{2\pi} \left[\int_0^{2\pi} f(y)(\cos py + i \sin py) \, dy \right] [\cos px - i \sin px]$$

Therefore

$$f(x) = \lim_{N \to \infty} (L^2) \left[a_0 + \sum_1^N a_p \cos px + \sum_1^N b_p \sin px \right]$$

where

$$a_0 = \frac{1}{2\pi} \int_0^{2\pi} f(y) \, dy$$

$$a_p = \frac{1}{\pi} \int_0^{2\pi} f(y) \cos py \, dy$$

$$b_p = \frac{1}{\pi} \int_0^{2\pi} f(y) \sin py \, dy$$

In fact we can prove the following result.

Theorem 2.1.12

If f is a function continuous over $(0, 2\pi)$ whose derivative is piecewise continuous, then f is the limit, uniform over $[0, 2\pi]$, of

$$a_0 + \sum_1^N a_p \cos px + \sum_1^N b_p \sin px$$

which is called a **Fourier series** for f.

This result applies naturally to a periodic function of period 2π, or, by change of variable, to every periodic function.

If f and f' are piecewise continuous we can show that the Fourier series for f converges uniformly and absolutely to $\frac{1}{2}[f(x^+) + f(x^-)]$.

With reference to this result, when f is expanded in terms of a basis $\{e_n\}_{n \in \mathbb{N}}$ of a Hilbert space in the form

$$f = \sum_{n=0}^{\infty} \langle f, e_n \rangle e_n$$

we call this expansion the **Fourier expansion in terms of the basis** $\{e_n\}_{n \in \mathbb{N}}$ and we call $\langle f, e_n \rangle$ the Fourier coefficient corresponding to e_n.

See *Exercises 17 to 21* at the end of the chapter.

2.2. Linear operators

LINEAR OPERATORS IN BANACH SPACES

Definition 2.2.1
Let E and F be two Banach spaces. We shall denote by $\|x\|_E$ the norm of an element x of E and by $\|y\|_F$ that of an element y of F.

We define a linear operator from E into F as a mapping from E into F such that

$$A(\lambda u_1 + \mu u_2) = \lambda A(u_1) + \mu A(u_2)$$

$\forall u_1, u_2 \in E$ and $\lambda, \mu \in \mathbb{C}$ or $\mathbb{R}$ according as E and F are constructed on $\mathbb{C}$ or on $\mathbb{R}$.

A is called bounded if $\exists M$ such that $\|A(u)\|_F \leqslant M \|u\|_E \ \forall u \in E$. We denote by $\mathscr{L}(E, F)$ the vector space of bounded linear operators. If $A \in \mathscr{L}(E, F)$ we define its norm by

$$\|A\|_{\mathscr{L}(E,F)} = \sup_{\|u\|_E = 1} \|A(u)\|_F$$

This number is the smallest of the numbers M such that

$$\|A(u)\|_F \leqslant M \|u\|_E$$

In fact, if M satisfies this relation, $M \geqslant \|A\|_{\mathscr{L}(E,F)}$ on the one hand and on the other hand

$$\|A(u)\|_F = \|u\|_E \left\| \frac{Au}{\|u\|_E} \right\|_F \leqslant \|A\|_{\mathscr{L}(E,F)} \|u\|_E$$

Usually we shall suppress the suffix indicating the space over which we are taking a norm, since there cannot normally be any confusion between an operator from E into F and an element of E or of F; thus we shall write

$\|A(u)\|_F \leqslant \|A\| \|u\|_E$. We can show that with this norm $\mathcal{L}(E, F)$ is a Banach space.

If E and F are of finite dimension every linear operator is bounded; moreover it is represented in a coordinate system in E and F by a matrix and

$$\sup_{\|u\|_E = 1} \|A(u)\|_F$$

is always finite.

In the case of infinite dimensions there exist linear operators which are not bounded, a fact which is the source of many difficulties and which gives rise to phenomena different from those of finite dimensions.

Theorem 2.2.2

A linear operator A is bounded iff

$$\forall \varepsilon > 0 \quad \exists \eta > 0: \forall u_1, u_2 \quad \|u_1 - u_2\|_E \leqslant \eta \Rightarrow \|A(u_1) - A(u_2)\|_F \leqslant \varepsilon$$

We say also that A is continuous; this is the reason for the importance of these operators, especially since

$$\lim_{n \to \infty} u_n = u \Rightarrow \lim_{n \to \infty} A u_n = Au$$

Proof Let A be bounded and ε given:

$$\|A(u_1) - A(u_2)\|_F = \|A(u_1 - u_2)\|_F \leqslant \|A\| \|u_1 - u_2\|_E \leqslant \varepsilon$$

$$\text{if } \|u_1 - u_2\|_E \leqslant \frac{\varepsilon}{\|A\|} = \eta$$

Conversely, if A is continuous, there exists R such that $\|x - 0\|_E \leqslant R$ implies $\|A(x) - A(0)\|_F \leqslant 1$, i.e. $\|x\|_E \leqslant R \Rightarrow \|A(x)\|_F \leqslant 1$.

If $x \in E$ and $y = xR/\|x\|_E$, then $x = \|x\|_E y/R$ and $\|y\|_E = R$, so that

$$A(x) = \frac{\|x\|_E}{R} A(y)$$

and

$$\|A(x)\|_F \leqslant \|x\|_E \frac{1}{R}$$

We can easily prove the following result.

Theorem 2.2.3

$$\|A\| = \sup_{\|x\|_E \leqslant 1} \|Ax\|_F = \sup_{\|x\|_E = 1} \|Ax\|_F = \sup_{x \neq 0} \frac{\|Ax\|_F}{\|x\|_E}$$

$$\|Ax\|_F \leqslant \|A\| \|x\|_E$$

$$\|A \cdot B\| \leqslant \|A\| \|B\| \quad \text{if } A \in \mathcal{L}(E, F), B \in \mathcal{L}(F, G)$$

We write without distinction Ax or $A(x)$ for the image of x through A.
We define the inverse of an operator of $\mathscr{L}(E, F)$ as follows.

Definition 2.2.4

We say that $A \in \mathscr{L}(E, F)$ is invertible if there exists an operator of $\mathscr{L}(F, E)$ (hence linear and bounded) denoted A^{-1} such that

$$\forall u \in F \quad \exists v \text{ unique} \in E: Av = u \quad \text{and} \quad v = A^{-1}u$$

The fact that A^{-1} exists can be split into two results: first, there exists a unique v such that $Av = u$ and we can write $v = Bu$, where B is necessarily linear; next, the fact that A^{-1} exists involves the fact that B is a bounded operator from F into E, which is not implied by the fact that there is only one v such that $Av = u$ (except, of course, with finite dimensions, in which every linear operator is bounded).

Example   If $E = \mathscr{C}(0, 1)$, the set of functions continuous over $[0, 1]$, with the norm

$$\|f\| = \sup_{x \in [0,1]} |f(x)|$$

Let $F = \mathscr{C}_0^1$ be the subspace of E formed of functions null at zero and possessing a continuous derivative, always with the norm

$$\|f\| = \sup_{x \in [0,1]} |f(x)|$$

(F is not complete for this norm.)
We define a bounded linear operator from E into F by putting

$$Af(x) = \int_0^x f(u)\, du = g(x)$$

It is clear that A is a linear operator, that $g \in \mathscr{C}_0^1$ and that

$$|g(x)| \leqslant \int_0^x |f(u)|\, du \leqslant \int_0^1 |f(u)|\, du \leqslant \|f\|_E$$

A has norm 1.

For every function g of $\mathscr{C}_0^1$ there exists a unique function f of $\mathscr{C}(0, 1)$ such that $Af = g$; in fact this equality is equivalent to $g = df/dx$. If $g = Bf$ it is easy to see that B is not bounded, for $\forall M$ there exists a function whose absolute value is bounded by 1 but whose derivative is, at one point at least, greater than M, so that

$$\sup_{\|f\|=1} \frac{df}{dx} \geqslant M \quad \forall M$$

For example,

$$f_\varepsilon(x) = \exp\left[\frac{(x-\varepsilon)^2}{(x-\varepsilon)^2 - \varepsilon^2}\right] 1_{[0 \leqslant x \leqslant 2\varepsilon]}$$

The identity I is certainly invertible. If A is close to I, it is invertible also.

Theorem 2.2.5

Let I be the identity mapping from E, defined by $I \cdot u = u \;\; \forall u \in E$. I is an invertible operator with norm 1. If $B \in \mathscr{L}(E, E)$ is of norm 1, then $I + B$ is invertible, the series $I - B + B^2 + \ldots + (-1)^n B^n + \ldots$ is convergent and

$$(I + B)^{-1} = I - B + B^2 + \ldots + (-1)^n B^n + \ldots$$

More generally, if $A \in \mathscr{L}(E, E)$ is such that A^{-1} exists, then

$$\forall B : \|B\| < \frac{1}{\|A^{-1}\|}$$

the operator $(A + B)^{-1}$ exists.

Proof Let $C_n = I - B + B^2 + \ldots + (-1)^n B^n$.

$$\|C_n - C_m\| \leqslant \|B^{m+1}\| + \|B^{m+2}\| + \ldots + \|B^n\| \leqslant \sum_{m+1}^{\infty} \|B\|^p$$

which is the remainder series of a convergent series; C_n is therefore a Cauchy sequence in $\mathscr{L}(H, H)$.

Let C be its limit:

$$\|(I + B)C - I\| = \lim_{n \to \infty} \|(I + B)C_n - I\| = \lim_{n \to \infty} \|B\|^{n+1} = 0$$

Then $(I + B)C = I = C(I + B)$ and $C = (I + B)^{-1}$.

If now A is invertible, $A + B = A(I + A^{-1}B)$ is invertible when $I + A^{-1}B$ is, which is the case if $\|A^{-1}B\| < 1$; for this it is sufficient that $\|A^{-1}\| \|B\| < 1$. This completes the proof.

We note that if A is invertible $I = AA^{-1}$ and therefore $1 \leqslant \|A\| \|A^{-1}\|$, so that

$$\frac{1}{\|A^{-1}\|} \leqslant \|A\|$$

It suffices then, for $A + B$ to be invertible, that $\|B\| < \|A\|$.

We can easily write $(A + B)^{-1}$ in the form of a series, since

$$A + B = A(I + A^{-1}B)$$

and

$$(A + B)^{-1} = (I + A^{-1}B)^{-1}A^{-1}$$
$$= [I - A^{-1}B + (A^{-1}B)^2 - \ldots]A^{-1}$$
$$= A^{-1} - A^{-1}BA^{-1} + A^{-1}BA^{-1}BA^{-1} - \ldots$$

To prove the existence and uniqueness of solutions of certain equations the following notion is fundamental.

Contracting operator

Definition 2.2.6
Let A be an operator, not necessarily linear, from E into E. We say that it is a contracting operator if there exists $\alpha < 1$ such that, $\forall u, v \in E$, $\|Au - Av\| \leqslant \alpha\|u - v\|$.

Theorem 2.2.7
If A is a contracting operator from E into E, the equation $Au = u$ has a unique solution. This result is still true if there exists an integer n_0 such that A^{n_0} is contracting.

Proof If $u_0 \in E$, we shall show that the sequence defined by $u_1 = Au_0$, $u_2 = Au_1, \ldots, u_{n+1} = Au_n$ converges to the required solution. In fact

$$\|u_{n+1} - u_n\| \leqslant \alpha\|u_n - u_{n-1}\| \leqslant \alpha^2\|u_{n-1} - u_{n-2}\| \ldots$$

so that

$$\|u_{n+1} - u_n\| \leqslant \alpha^n\|u_1 - u_0\|$$

Then

$$\|u_m - u_n\| \leqslant (\alpha^m + \alpha^{m-1} + \ldots + \alpha^n)\|u_1 - u_0\|$$

and $\{u_n\}$ is a Cauchy sequence converging to a limit u. Since A is continuous, $\{Au_n\}$ is a sequence converging to Au; as $u_{n+1} = Au_n$ the sequence $\{u_n\}$ converges at the same time to u and to Au which are therefore equal.

Now let v be another solution of $Av = v$; by hypothesis $\|Au - Av\| \leqslant \alpha\|u - v\|$, but $Au - Av = u - v$. Since $\alpha < 1$, this is absurd unless $u = v$.

The completion of the proof forms part of Exercise 23.

CASE OF HILBERT SPACES; SELF-ADJOINT OPERATORS

We now suppose that E and F each have a scalar product, denoted $\langle u, v \rangle_E$ for E and $\langle u, v \rangle_F$ for F. We suppose that for the corresponding norms E and F are complete and are thus Hilbert spaces.

Definition 2.2.8

If $A \in \mathscr{L}(E, F)$, there exists a unique $A^* \in \mathscr{L}(F, E)$ such that

$$\forall u \in E, \forall v \in F \qquad \langle Au, v \rangle_F = \langle u, A^*v \rangle_E$$

A^* is called the **adjoint operator**— or transposed operator—of A. This notion generalizes that of a transposed matrix.

If v is fixed, the mapping $u \rightsquigarrow \langle Au, v \rangle_F$ is a linear mapping from E into $\mathbb{C}$. According to Schwarz's inequality

$$|\langle Au, v \rangle| \leqslant \|Au\|_F \|v\|_F \leqslant \|A\| \|v\|_F \|u\|_E$$

so that this is a bounded linear mapping. A theorem put forward by Riesz then states that there exists a unique element w of F such that this linear form is equal to $u \rightsquigarrow \langle u, w \rangle_F$. Since w is determined by v, we put $w = A^*v$ and it can easily be verified that $A^* \in \mathscr{L}(F, E)$.

Proposition 2.2.9

If A^{-1} exists $(A^*)^{-1}$ exists and $(A^*)^{-1} = (A^{-1})^*$.

Proof If $u = A^*v$ it is sufficient to write the following sequence of equalities valid for all w in E:

$$\langle u, w \rangle_E = \langle u, A^{-1} \cdot Aw \rangle_E = \langle (A^{-1})^*u, Aw \rangle_E = \langle A^*(A^*)^{-1}u, w \rangle_E$$

In Chapter 3 (p. 186) we have used the very important result that, $\forall M \in \mathscr{L}(\mathbb{R}^n, \mathbb{R}^n)$, $\mathbb{R}^n = R(M) \oplus \mathscr{K}(\tilde{M})$ and $R(M) \perp \mathscr{K}(\tilde{M})$.
We shall generalize this result.

Theorem 2.2.10

If $A \in \mathscr{L}(H, H)$, we denote by $\mathscr{K}(A)$ the kernel of A, i.e. $\mathscr{K}(A) = \{u \in H: Au = 0\}$, and by $\mathscr{R}(A)$ the image of A, i.e. $\mathscr{R}(A) = \{u \in H: \exists v \ Av = u\}$.
Then $H = \mathscr{K}(A) \oplus \mathscr{R}(A^*)$, which means that every element of $\mathscr{K}(A)$ is orthogonal to every element of $\mathscr{R}(A^*)$ and that every element u of H may be written uniquely in the form $u = u_1 + u_2$, $\langle u_1, u_2 \rangle = 0$, $u_1 \in \mathscr{K}(A)$, $u_2 \in \mathscr{R}(A^*)$.

Proof The proof is identical with that on page 186.

$$u \perp \mathscr{R}(A^*) \Leftrightarrow \langle u, A^*v \rangle = 0 \quad \forall v \in H$$

$$\Leftrightarrow \langle Au, v \rangle = 0 \quad \forall v \in H$$

$$\Leftrightarrow Au = 0$$

$$\Leftrightarrow u \in \mathscr{K}(A)$$

Definition 2.2.11

If $A \in \mathscr{L}(H, H)$, A is called self-adjoint if $A = A^*$. In this case $H = \mathscr{K}(A) \oplus \mathscr{R}(A)$.

Proposition 2.2.12

If A is self-adjoint

$$\|A\| = \sup_{\|f\|=1} |\langle Af, f \rangle|$$

Proof   If

$$a = \sup_{\|f\|=1} |\langle Af, f \rangle|$$

we note at once that $|\langle Af, f \rangle| \leqslant a \|f\|^2 \ \forall f \in H$. Moreover

$$a \leqslant \sup_{\|f\|=1} \|Af\| \cdot \|f\|$$

from Schwarz's inequality. Thus $a \leqslant \|A\|$. We shall prove that $\|A\| \leqslant a$ by proving that

$$\forall f \in H \qquad \|Af\| \leqslant a \|f\|$$

$$a\|f+g\|^2 \geqslant \langle A(f+g), f+g \rangle = \langle Af, f \rangle + 2 \operatorname{Re}\langle Af, g \rangle + \langle Ag, g \rangle$$

$$-a\|f-g\|^2 \leqslant \langle A(f-g), f-g \rangle = \langle Af, f \rangle - 2 \operatorname{Re}\langle Af, g \rangle + \langle Ag, g \rangle$$

because

$$\langle Af, g \rangle + \langle Ag, f \rangle = \langle Af, g \rangle + \langle g, Af \rangle = \langle Af, g \rangle + \overline{\langle Af, g \rangle}$$

since A is self-adjoint. Thus

$$4 \operatorname{Re}\langle Af, g \rangle \leqslant 2a(\|f+g\|^2 + \|f-g\|^2) = 2a(\|f\|^2 + \|g\|^2)$$

Let $\|f\| = 1$ and $Af \not\equiv 0$ and let $g = Af/\|Af\|$; for these functions

$$4 \operatorname{Re}\langle Af, g \rangle = \frac{4\langle Af, Af \rangle}{\|Af\|} = 4\|Af\|$$

which is smaller than or equal to $2a(1 + 1)$.

If $\|f\| = 1$ and $Af \neq 1$, $\|Af\| \leqslant 1$

If $\|f\| \neq 1$ and $Af \not\equiv 0$, $\|Af\| = \|f\| \left\| A \frac{f}{\|f\|} \right\| \leqslant a\|f\| \ \forall f$

If $Af = 0$, naturally $\|Af\| \leqslant a\|f\|$

Remark $(A^*)^* = A$ for

$$\langle Au, v \rangle = \langle u, A^*v \rangle$$

$$= \overline{\langle A^*v, u \rangle}$$

$$= \overline{\langle v, (A^*)^*u \rangle}$$

$$= \langle (A^*)^*u, v \rangle$$

THE FREDHOLM ALTERNATIVE

If $A \in \mathscr{L}(H, H)$ and A^* is the adjoint operator, we recall that $H = \mathscr{R}(A) \oplus \mathscr{K}(A^*)$.

Theorem 2.2.13
For the equation $Af = g$ to have a solution it is necessary and sufficient that g be orthogonal to every element of $\mathscr{K}(A^*)$.

Proof This results from the relation $H = \mathscr{R}(A) \oplus \mathscr{K}(A^*)$ since g is the image of an element f of H through A iff $g \in \mathscr{R}(A)$ and thus is orthogonal to $\mathscr{K}(A^*)$.

If A^{-1} exists, $(A^*)^{-1}$ exists also, $\mathscr{K}(A^*) = \{0\}$ and every element g of H is orthogonal to $\mathscr{K}(A^*)$; we find again that for every g of H there exists a solution $A^{-1}g$ which, moreover, is unique.

If A is self-adjoint, $H = \mathscr{K}(A) \oplus \mathscr{R}(A)$; if d is the dimension of $\mathscr{K}(A)$, there exist d independent vectors $f_1, f_2, \ldots, f_d$ such that $Af_i = 0$. If $g \in \mathscr{R}(A)$, there exists f_0 such that $Af_0 = g$; if $\tilde{f}$ satisfies $A\tilde{f} = g$ also, then $A(f - \tilde{f}) = 0$. Therefore $f_0 - \tilde{f} \in \mathscr{K}(A)$ and $f_0 - \tilde{f} = a_1 f_1 + a_2 f_2 + \ldots + a_d f_d$. Thus the set of solutions is of the form $f_0 + a_1 f_1 + \ldots + a_d f_d$; we say that these solutions depend on d parameters. If $g \notin \mathscr{R}(A)$ there are no solutions.

See *Exercises 22 to 24* at the end of the chapter.

2.3. Eigenvalues ; eigenvectors ; compact self-adjoint operators

We shall consider in this section only operators in $\mathscr{L}(H, H)$ where H is a Hilbert space.

Definition 2.3.1
Let $A \in \mathscr{L}(H, H)$ and $\lambda \in \mathbb{C}$ be a complex number. Let

$$E_\lambda(A) = \mathscr{K}(A - \lambda I) = \{u \in H : Au = \lambda u\}$$

$E_\lambda(A)$ is a vector space possibly reducing to $\{0\}$ (a space containing only 0).

If $E_\lambda(A) \neq \{0\}$, we say that λ is an **eigenvalue** of A, and every non-zero element of $E_\lambda(A)$ is called an **eigenvector** of A: $Af = \lambda f, f \neq 0$.

The **dimension** of $E_\lambda(A)$ is called the **multiplicity** of λ.

We note that if λ is an eigenvalue $|\lambda| \leqslant \|A\|$. (In fact $|\lambda| \|u\| = \|Au\| \leqslant \|A\| \|u\|$.)

Note In contrast, in the case of finite dimensions, it can happen that λ is not an eigenvalue, i.e. that $\mathscr{K}(A - \lambda I) = \{0\}$ and that $A - \lambda I$ is not invertible. In other words, if λ is an eigenvalue, $A - \lambda I$ is not invertible, but the converse is not true.

Example Let $H = L^2(0, 1)$ and A be defined by $(Af)(x) = xf(x)$; $A \in \mathcal{L}(H, H)$ obviously.

A number λ is an eigenvalue of A if there exists f, not identically zero and square integrable, such that $\lambda f(x) = Af(x) = xf(x) \, \forall x$, which is impossible; A has no eigenvalue.

Let us examine, then, whether $A - \lambda I$ is invertible, i.e. whether for every f in H there exists g in H such that $(A - \lambda I)g = f$ and whether the operator thus defined is bounded. It would be suitable, in particular, if there existed g such that $(A - \lambda I)g = 1$, i.e. $(x - \lambda)g(x) = 1$; the only function satisfying this would be

$$g(x) = \frac{1}{x - \lambda}$$

which does not belong to $L^2(0, 1)$ if $\lambda \in [0, 1]$.

There exist, however, operators which behave, from the point of view of eigenvalues and eigenvectors, in a manner quite comparable with that of operators in $\mathcal{L}(\mathbb{R}^n, \mathbb{R}^n)$; these are the compact self-adjoint operators.

Definition 2.3.2

Let $B_1 = \{u \in H : \|u\| \leqslant 1\}$ be the unit ball in H. Let $A \in \mathcal{L}(H, H)$ and $U_1 = \{v \in H; \exists u \in B_1 : v = Au\}$ be the image of B_1 through A.

A is called **compact** if from every sequence in U_1 we can extract a subsequence having a limit (which, however, does not necessarily belong to U_1).

The simplest example of a compact operator is the following. Let $A \in \mathcal{L}(H, H)$ and be such that $\mathcal{R}(A)$ is of finite dimension; then A is compact. In fact U_1 is then a bounded closed set in a space of finite dimension, and it is a classical result that in this case we can extract from every sequence in U_1 a convergent subsequence (the Borel–Lebesgue theorem). Such an operator is called an **operator of finite rank**.

We note that, if A is compact, $\forall \alpha \in \mathbb{C} \; \alpha A$ is compact; that, if A and B are compact, $A + B$ is compact; and that, if $\{A_n\}_{n \in \mathbb{N}}$ is a convergent sequence of compact operators, its limit is compact. If A is compact and $B \in \mathcal{L}(H, H)$, then AB and BA are compact, as can easily be verified.

This implies that, if H is of infinite dimension, a compact operator is not invertible. In fact if A^{-1} existed $I = AA^{-1}$ would be compact and the image B of the unit ball B_1 through the identify I would have to have the following property: from every sequence of B_1 a convergent subsequence can be extracted. It can be shown that this is not the case.

If $\{A_n\}_{n \in \mathbb{N}}$ is a sequence of operators of finite rank having a limit, this limit is compact. The converse is exact; every compact operator is the limit of a sequence of finite operators.

This property is widely used in numerical analysis.

We note that if A is compact, A^* is compact. In this case, if λ is an eigenvalue of A, $\bar\lambda$ is also an eigenvalue of A^*.

The fundamental theorem is as follows.

Theorem 2.3.3 (Spectrum of compact self-adjoint operators)

Let $A \in \mathscr{L}(H, H)$ be a compact self-adjoint operator. There exists a real eigenvalue λ_0 such that $|\lambda_0| = \|A\|$. There exists at most a denumerable infinity of eigenvalues. They are **real**, and if there is an infinity of them they can be arranged in order of decreasing absolute value:

$$\|A\| = |\lambda_0| > |\lambda_1| > |\lambda_2| > \ldots > |\lambda_n| > \ldots >$$

and $\lim \lambda_n = 0$. 0 can be an eigenvalue.

Each space E_{λ_i} for $\lambda_i \neq 0$ is of finite dimension; if $\lambda_i \neq \lambda_j$, $E_{\lambda_i} \perp E_{\lambda_j}$ and there exists an orthonormal basis of H formed from eigenvectors of A. (If $\dim H = \infty$, and if there exists only a finite number of non-zero eigenvalues, 0 is an eigenvalue of infinite multiplicity.)

Let $\{\varphi_p\}_{p \in \mathbb{N}}$ be such a basis of eigenvectors: $A\varphi_p = \lambda_p \varphi_p$. If λ is not an eigenvalue $(A - \lambda I)^{-1}$ exists.

$$\forall f \in H \qquad f = \lim_{n \to \infty} \sum_{p=0}^{n} \langle f, \varphi_p \rangle \varphi_p$$

and

$$Af = \lim_{n \to \infty} \sum_{p=0}^{\infty} \lambda_p \langle f, \varphi_p \rangle \varphi_p$$

The convergence of these series naturally takes place in H.

We shall accept this theorem, noting only that the expansion

$$f = \sum_{0}^{\infty} \langle f, \varphi_p \rangle \varphi_p$$

implies that if $\lambda \neq \lambda_p$ $\forall p$, $A - \lambda I$ is invertible. In fact

$$(A - \lambda I)f = g \Leftrightarrow \sum_{0}^{\infty} \lambda_p \langle f, \varphi_p \rangle \varphi_p - \lambda \sum_{0}^{\infty} \langle f, \varphi_p \rangle \varphi_p = \sum_{0}^{\infty} \langle g, \varphi_p \rangle \varphi_p$$

Since an expansion is unique, this implies that

$$\langle f, \varphi_p \rangle = \frac{1}{\lambda - \lambda_p} \langle g, \varphi_p \rangle$$

which defines f by means of

$$f = \sum_{0}^{\infty} \frac{1}{\lambda - \lambda_p} \langle g, \varphi_p \rangle \varphi_p = (A - \lambda I)^{-1} g$$

We shall use the following theorem, which is comparable with the preceding result.

Theorem 2.3.4 (Application)

Let A be a compact self-adjoint operator in $\mathcal{L}(H, H)$. Let $\{\lambda_n\}_{n \in \mathbb{N}}$ be its eigenvalues and let $\{\varphi_n\}_{n \in \mathbb{N}}$ be an orthonormal basis of H formed from the eigenvectors with $A\varphi_n = \lambda_n \varphi_n$ $\forall n \in \mathbb{N}$.

Consider the equation

$$f - \lambda A f = g \tag{41}$$

(a) If $1 - \lambda\lambda_n \neq 0$ $\forall n$, (41) has as a unique solution:

$$f = g + \lambda \sum_{n=0}^{\infty} \frac{\lambda_n}{1 - \lambda\lambda_n} \langle g, \varphi_n \rangle \varphi_n$$

(b) If $1 - \lambda\lambda_{n_0} = 0$, there exists a solution of (41) iff g is orthogonal to every eigenfunction φ_{n_0} corresponding to λ_{n_0}. If this is the case

$$f = g + \lambda \sum_{\substack{n=0 \\ \lambda\lambda_{n_0} \neq 1}}^{\infty} \frac{\lambda_n}{1 - \lambda\lambda_n} \langle g, \varphi_n \rangle \varphi_n + \sum_{\lambda\lambda_{n_0} = 1}^{\infty} c_{n_0} \varphi_{n_0}$$

(The first sum involves eigenvectors other than those corresponding to λ_{n_0}, whereas the second is a sum of arbitrary linear combinations of eigenvectors corresponding to λ_{n_0}.)

This statement is yet another form of the Fredholm alternative in which we can see exactly the parameters of Theorem 2.2.13.

Proof

(a) Let

$$1 - \lambda\lambda_n \neq 0$$

and

$$g = \sum_{n=0}^{\infty} \langle g, \varphi_n \rangle \varphi_n \qquad f = \sum_{n=0}^{\infty} \langle f, \varphi_n \rangle \varphi_n$$

$$\lambda A f = \sum_{n=0}^{\infty} \lambda\lambda_n \langle f, \varphi_n \rangle \varphi_n$$

and

$$f - \lambda A f = \sum_{n=0}^{\infty} (1 - \lambda\lambda_n) \langle f, \varphi_n \rangle \varphi_n$$

which establishes the result.

(b) Let $1 - \lambda\lambda_{n_0} = 0$

$$f - \lambda A f = \sum_{n=0}^{\infty} (1 - \lambda\lambda_n) \langle f, \varphi_n \rangle \varphi_n$$

is an expansion in which the coefficients of eigenvectors corresponding to λ_{n_0}

are zero. It is necessary that they be zero also in the expansion of g, i.e. that $\langle g, \varphi_{n_0} \rangle = 0$ for every eigenvector φ_{n_0} corresponding to λ_{n_0}.

If this is the case the coefficients in the expansion of f corresponding to these vectors φ_{n_0} are arbitrary and the others are given by

$$(1 - \lambda\lambda_n)\langle f, \varphi_n \rangle = \langle g, \varphi_n \rangle$$

Interpretation of the eigenvalues

We know already that

$$|\lambda_0| = \|A\| = \sup_{\|u\|=1} |\langle Au, u \rangle|$$

We can prove the following result.

Theorem 2.3.5

$$|\lambda_0| = \sup_{\|u\|=1} |\langle Au, u \rangle|$$

$$|\lambda_1| = \sup_{\|u\|=1, u \perp E_\lambda{}^0} |\langle Au, u \rangle|$$

$$\vdots$$

$$|\lambda_p| = \sup_{\|u\|=1, u \perp E_{\lambda_0}, u \perp E_{\lambda_2}, \dots, u \perp E_{\lambda_{p-1}}} |\langle Au, u \rangle|$$

Without giving any proof, we shall verify that the result is correct for a symmetric matrix whose properties are analogous to those of a compact self-adjoint operator in infinite dimensions.

Let us choose a basis $e_1, \dots, e_n$ of eigenvectors associated with the eigenvalues $\lambda_1, \dots, \lambda_n$. Let

$$v = \sum_{i=1}^n a_i e_i: Av = \sum_{i=1}^n a_i \lambda_i e_i \quad \text{and} \quad \langle Av, v \rangle = \sum_{i=1}^n a_i^2 \lambda_i$$

Thus

$$\sup_{\|v\|=1} |\langle Av, v \rangle| = \sup_{\{v: \sum_{i=1}^n a_i^2 = 1\}} \sum_{i=1}^n a_i^2 \lambda_i = |\lambda_0| \quad \text{if } |\lambda_0| > |\lambda_1| > \dots > |\lambda_n|$$

If

$$v \perp e_1 \qquad v = \sum_{i=2}^n a_i e_i$$

and

$$\sup_{\|v\|=1, v \perp e_1} |\langle Av, v \rangle| = \sup_{\{v: \sum_{i=2}^n a_i^2 = 1\}} \sum_{i=2}^n a_i^2 \lambda_i = |\lambda_1|$$

These relations are fundamental; they have applications in the calculus of variations but also for the calculation of approximations to eigenvalues and to eigenvectors.

Theorem 2.3.6

(a) *Rayleigh's principle*: let A be a compact self-adjoint operator all of whose eigenvalues are positive:

$$\lambda_0 > \lambda_1 > \lambda_2 > \ldots > \lambda_n > \ldots > 0$$

Let $u \in H$ and $u_1 = Au$, $u_2 = Au_1, \ldots, u_{n+1} = Au_n$. Let

$$\alpha_n = \frac{\|u_{n+1}\|^2}{\langle u_{n+1}, u_n \rangle} \quad \text{and} \quad \beta_n = \frac{\|u_{n+1}\|}{\|u_n\|}$$

Then $0 < \beta_n \leqslant \alpha_n$ and, if u is not orthogonal to an eigenvector associated with λ_0, $\lim \alpha_n = \lim \beta_n = \lambda_0$. If u is orthogonal to $E_{\lambda_0}, \ldots, E_{\lambda_{n_0-1}}$, $\lim \alpha_n = \lim \beta_n = \lambda_{n_0}$.

(b) *The Kryloff–Weinstein upper bound*: if

$$\lim \alpha_n = \lambda = \lim \beta_n$$

then

$$\left(\frac{\alpha_n^{\,2}}{\beta_n^{\,2}} - 1 \right)^{1/2} \geqslant \frac{\lambda - \alpha_n}{\lambda}$$

We shall not give any proof here either but shall verify Rayleigh's principle for symmetric matrices.

Throughout let $e_1, \ldots, e_p$ be a basis of eigenvectors associated with the positive eigenvalues $\lambda_1 > \lambda_2 > \ldots > \lambda_p > 0$. Let

$$v = \sum_{i=1}^{p} a_i e_i : u_n = A^n v = \sum_{i=1}^{p} a_i \lambda_i^{\,n} e_i$$

Let

$$a_1 \neq 0 : \|u_{n+1}\|^2 = a_1^{\,2} \lambda_1^{\,2n+2} \left[1 + \left(\frac{a_2}{a_1} \right)^2 \left(\frac{\lambda_2}{\lambda_1} \right)^{2n+2} + \ldots + \left(\frac{a_p}{a_1} \right)^2 \left(\frac{\lambda_p}{\lambda_1} \right)^{2n+2} \right]$$

$$\langle u_{n+1}, u_n \rangle = a_1^{\,2} \lambda_1^{\,2n+1} \left[1 + \left(\frac{a_2}{a_1} \right)^2 \left(\frac{\lambda_2}{\lambda_1} \right)^{2n+1} + \ldots + \left(\frac{a_p}{a_1} \right)^2 \left(\frac{\lambda_p}{\lambda_1} \right)^{2n+1} \right]$$

Thus

$$\lim_{n \to \infty} \alpha_n = \lambda_1 = \lim_{n \to \infty} \beta_n$$

If

$$v \perp e_1, \ldots, e_{i_0-1} : a_1 = a_2 = \ldots = a_{i_0-1} = 0$$

and if $v \not\perp e_{i_0}$ then $a_{i_0} \neq 0$. Thus

$$\|u_{n+1}\|^2 = a_{i_0}^{\ 2}\lambda_{i_0}^{\ 2n+2}\left[1 + \ldots + \left(\frac{a_p}{a_{i_0}}\right)^2\left(\frac{\lambda_p}{\lambda_{i_0}}\right)^{2n+2}\right]$$

$$\langle u_{n+1}, u_n \rangle = a_{i_0}^{\ 2}\lambda_{i_0}^{\ 2n+1}\left[1 + \ldots + \left(\frac{a_p}{a_{i_0}}\right)^2\left(\frac{\lambda_p}{\lambda_{i_0}}\right)^{2n+2}\right]$$

and

$$\lim_{n\to\infty} \alpha_n = \lambda_{i_0} = \lim_{n\to\infty} \beta_n$$

We note in the same way that if u is not orthogonal to e_1

$$\lim_{n\to\infty} \frac{A^n u}{\|A^n u\|} = e_1$$

which is used to estimate the eigenvectors.

Example Let

$$G(x, y) = \begin{cases} x(y-1) & 0 \leqslant x \leqslant y \\ y(x-1) & y \leqslant x \leqslant 1 \end{cases}$$

and

$$Af(x) = -\int_0^1 G(x, y)f(y)\,dy$$

where $f \in L^2(0, 1)$ and $f(0) = f(1) = 0$.

Let H be the subspace of $L^2(0, 1)$ formed from functions such that $f(0) = f(1) = 0$. We shall verify that A is compact self-adjoint, which can also be verified directly.

$$u_1 \equiv 1 \qquad u_2 = \frac{x}{2}(1 - x) \qquad u_3 = \frac{1}{24}(x^4 - 2x^3 + x)$$

Then

$$\beta_1 = \frac{1}{10.95} \qquad \beta_2 = \frac{1}{9.88} \qquad \alpha_2 = \frac{1}{9.87} \qquad \alpha_1 = \frac{1}{10}$$

If f is an eigenvector

$$\lambda f(x) = (1 - x)\int_0^x yf(y)\,dy + x\int_1^x (y - 1)f(y)\,dy$$

From Schwarz's inequality f is bounded, so that

$$\int_0^x yf(y)\,dy$$

is continuous; therefore so also is f, which is then twice differentiable and $\lambda f''(x) = -f(x)$. Let

$$\frac{d^2 f}{dx^2} = -\frac{1}{\lambda} f \qquad f(0) = f(1) = 0$$

Then

$$\frac{1}{\lambda_n} = (n+1)^2 \pi^2 \qquad \frac{1}{\lambda_0} = \pi^2$$

Thus

$$\frac{1}{\beta_0} > \frac{1}{\alpha_0} > \frac{1}{\beta_1} > \frac{1}{\alpha_1} > \frac{1}{\lambda_0} = \pi^2 = 9.8696$$

2.4. Differential operators

It must not be forgotten that to define an operator we must state precisely the vector space over which this operator is defined; the operator can have different properties according to the space over which it is defined.

Consider for example $H = C^\infty$, the vector space of infinitely differentiable functions over $\mathbb{R}$. If $Af = d^2 f/dx^2$, the eigenvalues λ are the numbers such that $d^2 f/dx^2 = \lambda f$.

If $\lambda > 0$, $\exp(x\lambda^{1/2})$ and $\exp(-x\lambda^{1/2})$ are eigenvectors and λ is an eigenvalue
If $\lambda < 0$, $\cos[x(-\lambda)^{1/2}]$ and $\sin[x(-\lambda)^{1/2}]$ are eigenvectors and λ is an eigenvalue
If $\lambda = 0$, $ax + b$ is an eigenfunction

Thus the operator $d^2 f/dx^2$ defined over C^∞ admits all real numbers as eigenvalues.

If now $H = P$, the vector space of real polynomials, we can still define $Af = d^2 f/dx^2$.

If $\lambda \neq 0$, a polynomial is an eigenfunction if $d^2 f/dx^2 = \lambda f$, which is absurd. If $\lambda = 0$, $ax + b$ is an eigenfunction. Thus the operator $d^2 f/dx^2$ defined over the space of polynomials has only zero as an eigenvalue.

We recall that the study of solutions of a differential equation necessitates the introduction of boundary conditions; according to the nature of the latter, the solutions do or do not form a vector space. The conditions which we studied in Part 1 allow precise definition of linear operators.

To study the problems of expansion in terms of an orthogonal basis of eigenfunctions it is convenient to introduce a scalar product; we then study an operator in a Hilbert space. If we wish to apply Theorem 2.3.3 it is necessary also that $Af \in H$ if $f \in H$; this does not allow the use of the spaces L^2 or H_2 in a simple way.

We then use the *equivalence* of the following two facts.

(a) f satisfies the boundary conditions

$$Af = g \qquad (42)$$

and the homogeneous problem admits only the solution $f \equiv 0$.

(b) G is the Green function of the preceding problem and

$$f = \int G(x, y)g(y)\,dy = Gg$$

G is an operator inverse to A; if $\lambda \neq 0$ and $Af = \lambda f$ then $f = Gf$ and $Gf = f/\lambda$; i.e. the eigenvalues of G and of A are reciprocals. (We note that $\lambda = 0$ is not an eigenvalue of A by hypothesis since the homogeneous problem has only one solution.)

We shall study operators of the type of G and verify that they are exactly compact and self-adjoint.

See **Exercises 25 and 26** at the end of the chapter.

Part 3: Integral equations; the Sturm–Liouville problem and applications

3.1. Integral operators, kernels

We have just seen that it is convenient to study certain differential equations, from the point of view of eigenvalues, by using as an intermediary the Green function $G(x, y)$ and the operator associated with it through

$$f \rightsquigarrow \int G(x, y)f(y)\,dy$$

Such an operator is an example of an integral operator; the function $G(x, y)$ is called the **kernel** of this operator.

We shall study such operators over spaces $L_\sigma^2[a, b]$ where $\sigma(x) > 0$ and $[a, b]$ is an interval of $\mathbb{R}$, bounded or not. We recall that $f \in L^2[a, b]$ if

$$\int_a^b f^2(x)\,dx < \infty$$

and that $f \in L_\sigma^2[a, b]$ if $f\sigma^{1/2} \in L^2[a, b]$; likewise f and g are orthogonal in $L_\sigma^2[a, b]$ iff $f\sigma^{1/2}$ and $g\sigma^{1/2}$ are orthogonal in $L^2[a, b]$.

We say that $K(x, y)$ is a square integrable function over $[a, b] \times [a, b]$ if

$$\int_a^b \int_a^b K^2(x, y)\,dx\,dy = M < \infty$$

We say that $K(x, y)$ is a square integrable function over $[a, b] \times [a, b]$ with respect to σ if

$$\int_a^b \int_a^b K^2(x, y)\sigma(x)\sigma(y)\,dx\,dy = N < \infty$$

i.e. if $K(x, y)[\sigma(x)]^{1/2}[\sigma(y)]^{1/2}$ is a square integrable function over $[a, b] \times [a, b]$.

Finally we say that K is symmetric if $K(x, y) = K(y, x)$.

We shall now define an integral operator with kernel K.

Definition 3.1.1

Let $H = L_\sigma^2[a, b]$ and let K be a square integrable function over $[a, b] \times [a, b]$ with respect to σ.

We define an integral operator associated with the function K as the bounded linear operator from $L_\sigma^2[a, b]$ into itself defined by

$$Kf(x) = \int_a^b K(x, y)f(y)\,dy$$

The norm of this operator, also denoted K, is

$$\|K\| = \sup_{\|f\| = 1} \|Kf\|$$

We say that $K(x, y)$ is the kernel of the operator K. If the kernel K is symmetric, the operator is self-adjoint and consequently

$$\|K\| = \sup_{\|f\| = 1} |\langle Kf, f \rangle|$$

This definition requires a proof.

Proof It is necessary to prove that K is indeed a bounded operator from H into H and that K is self-adjoint if its kernel is symmetric.

For simplicity we shall suppose that there exist two constants such that

$$0 < c_1 \leqslant \sigma(x) \leqslant c_2 < \infty \qquad \forall x \in [a, b]$$

It is sufficient to give the proof for $\sigma \equiv 1$ in view of the remarks we made before the definition.

We prove first that K is self-adjoint if its kernel is symmetric:

$$\langle Kf, g \rangle = \int_a^b \left[\int_a^b K(x, y)f(y)\,dy \right] g(x)\,dx$$

$$= \int_a^b f(y) \left[\int_a^b K(y, x)g(x)\,dx \right] dy$$

(according to a classical theorem on integration (Fubini)

$$= \langle f, Kg \rangle$$

To show that $Kf \in L^2[a, b]$ we apply Schwarz's inequality

$$\left[\int_a^b K(x, y) f(y) \, dy \right]^2 \leqslant \int_a^b |K(x, y)|^2 \, dy \int_a^b |f(y)|^2 \, dy \qquad \forall x \in [a, b]$$

Therefore

$$\int_a^b \left[\int_a^b K(x, y) f(y) \, dy \right]^2 dx \leqslant \int_a^b |f(y)|^2 \, dy \int_a^b \int_a^b |K(x, y)|^2 \, dy \, dx$$

$$\leqslant \| f \|^2_{L^2[a, b]} \cdot M$$

which proves that K is indeed a bounded operator.

Theorem 3.1.2

The operator K defined above is a compact operator (at least if $0 < c_1 \leqslant \sigma(x) \leqslant c_2 < \infty \; \forall x$). It is self-adjoint if $K(x, y) = K(y, x)$.

Proof We shall suppose that $\sigma(x) \equiv 1$.

(a) Let $[a, b] = [0, 1]$ and let $K(x, y)$ be continuous over $[0, 1] \times [0, 1]$ so that $K(x, y) \leqslant A \; \forall (x, y) \in [0, 1] \times [0, 1]$.

Let f_n be a sequence of elements of $L^2(0, 1)$ such that $\| f_n \| = 1$ and let $g_n = Kf_n$:

$$|g_n(x)|^2 \leqslant \left[\int_0^1 |K(x, y)| \, |f_n(y)| \, dy \right]^2 \leqslant A^2$$

always, from Schwarz's inequality.

For y fixed, $x \rightsquigarrow K(x, y)$ is uniformly continuous over $[0, 1]$ and

$$\forall \varepsilon \quad \exists \eta(y) : |x_1 - x_2| \leqslant \eta(y) \Rightarrow |K(x_1, y) - K(x_2, y)| \leqslant \varepsilon$$

If $y \in [0, 1]$, which is a closed bounded set, we can show that there exists an η, not depending on y, which has the same property as $\eta(y)$ so that

$$|g_n(x_1) - g_n(x_2)|^2 \leqslant \left[\int_0^1 |K(x_1, y) - K(x_2, y)| \, |f_n(y)| \, dy \right]^2$$

$$\leqslant \varepsilon \quad \text{if } |x_1 - x_2| \leqslant \eta$$

We have thus proved the following two properties.

(i) $|g_n(x)|^2 \leqslant A^2 \; \forall x \in [0, 1]$: we say that $\{g_n\}_{n \in \mathbb{N}}$ is **uniformly bounded**

(ii) $\forall \varepsilon, \; \exists \eta : |x_1 - x_2| \leqslant \eta \Rightarrow |g_n(x_1) - g_n(x_2)| \leqslant \varepsilon$: we say that $\{g_n\}_{n \in \mathbb{N}}$ is **equicontinuous**

We can show (Ascoli's theorem) that if a sequence has these two properties we can extract from it a sequence $\{g_{n_k}\}_{k \in \mathbb{N}}$ which converges uniformly to a function $g(x)$ (necessarily continuous).

Thus, if $K(x, y)$ is continuous over $[0, 1] \times [0, 1]$, K is compact.

(b) Throughout let $[a, b] = [0, 1]$ and $K \in L^2(0, 1)$. If K is not a continuous function we can show that there exist continuous functions $\{K_n(x, y)\}_{n \in \mathbb{N}}$ such that

$$\lim_{n \to \infty} \int_0^1 \int_0^1 |K_n(x, y) - K(x, y)|^2 \, dx \, dy = 0$$

With each function $K_n(x, y)$ is associated a compact operator K_n. Now

$$\|(K - K_n)f\|^2 = \left\| \int_0^1 [K(x, y) - K_n(x, y)] f(y) \, dy \right\|^2$$

$$\leqslant \left[\int_0^1 \int_0^1 |K(x, y) - K_n(x, y)|^2 \, dy \, dx \right] \cdot \|f\|^2$$

(from Schwarz) so that

$$\lim_{n \to \infty} \|K - K_n\| = 0$$

The operator K, the limit of compact operators, is compact.

(c) Let $[a, b] = [0, \infty]$ or $[-\infty, 0]$ or $[-\infty, \infty]$. Let us take, for example, the first case and let $K_n(x, y)$ be the sequence defined by

$$K_n(x, y) = \begin{cases} K(x, y) & \text{over } [0, n] \times [0, n] \\ 0 & \text{elsewhere} \end{cases}$$

With $K_n(x, y)$ we associate a compact operator K_n and we prove again that

$$\lim_{n \to \infty} \|K - K_n\| = 0$$

so that K is compact.

We proceed in a similar way for the other cases. Hence we deduce the following theorem.

Theorem 3.1.3

Let $K(x, y)$ be a symmetric square integrable kernel over $[a, b] \times [a, b]$ and let K be the associated operator in $L^2[a, b]$.

(a) There exists a real eigenvalue λ_0 of K and

$$|\lambda_0| = \|K\| = \sup_{\|f\| = 1} \langle Kf, f \rangle$$

The other eigenvalues are real and can be arranged in a sequence tending to zero: $|\lambda_0| > |\lambda_1| > |\lambda_2| > \ldots > |\lambda_n| > \ldots \to 0$. In addition

$$\int_a^b \int_a^b K^2(x, y) \, dx \, dy = \sum \lambda_i^2$$

(b) The spaces $E_n = \mathcal{K}(K - \lambda_n I)$ are orthogonal and of finite dimension for $\lambda_n \neq 0$. If $\lambda_n = 0$ is an eigenvalue the space $\mathcal{K}(K)$ can be of infinite dimension.

(c) There exists an orthonormal basis of $L^2[a, b]$ formed from the eigenvectors of K.

(d) If K is square integrable with respect to σ and if

$$0 < c_1 \leqslant \sigma(x) \leqslant c_2 < \infty$$

the conclusions hold in $L_\sigma^2[a, b]$.

These conclusions can hold also in more general cases as we shall see at the end of the chapter.

Example   Let $K(x, y) = \exp(x + y)$ be a kernel over $[0, 1] \times [0, 1]$. λ is an eigenvalue if there exists f, non-zero, such that

$$\int_0^1 \exp(x + y) f(y) \, dy = \lambda f(x)$$

If

$$c(f) = \int_0^1 \exp(u) f(u) \, du$$

then necessarily $\lambda f(x) = c(f) \exp(x)$.

First if $\lambda \neq 0$

$$f(x) = \frac{c(f)}{\lambda} \exp(x)$$

so that by substitution in the expression for $c(f)$

$$c(f) = \frac{c(f)}{\lambda} \int_0^1 \exp(u) \exp(u) \, du = \frac{c(f)}{\lambda} \frac{\exp(2) - 1}{2}$$

Since an eigenfunction is not identically zero, $c(f) \neq 0$ and $\lambda = [\exp(2) - 1]/2$; there exists only one eigenvalue different from zero and the corresponding eigenvectors are $f_\lambda(x) = c \exp(x)$.

We can verify that zero is an eigenvalue of multiplicity infinity since there exists an infinity of non-zero, mutually independent, functions such that

$$\int_0^1 \exp(y) f(y) \, dy = 0$$

We can, for example, construct a sequence of polynomials orthogonal to $\exp(y)$ and of arbitrary degree.

The above theorem states in particular that every function $f \in L^2[a, b]$ can be expressed as a limit in $L^2[a, b]$ of linear combinations of eigenfunctions. We would often wish to have a stronger result and if possible a pointwise limit. The following theorem gives a partial answer.

Theorem 3.1.4 (Hilbert–Schmidt: pointwise limit)

Let $K(x, y)$ be a symmetric kernel over $[a, b] \times [a, b]$, $\{\lambda_n\}_{n \in \mathbb{N}}$ the sequence of its eigenvalues and $\{\varphi_n\}_{n \in \mathbb{N}}$ an orthonormal sequence of eigenfunctions forming a basis of $L^2[a, b]$: $K\varphi_n = \lambda_n \varphi_n \ \forall n \in \mathbb{N}$.

If $g = Kf$ then

$$\sum_{p=1}^{n} \langle g, \varphi_p \rangle \varphi_p$$

converges towards g in $L^2[a, b]$ and also uniformly and absolutely over $[a, b]$.

Proof By hypothesis

$$\langle g, \varphi_n \rangle = \langle Kf, \varphi_n \rangle = \langle f, K\varphi_n \rangle = \lambda_n \langle f, \varphi_n \rangle$$

Then

$$\sum_{n=p}^{q} |\langle Kf, \varphi_n \rangle| |\varphi_n(x)| = \sum_{n=p}^{q} |\langle f, \varphi_n \rangle| |\lambda_n \varphi_n(x)| \qquad (43)$$

The following is a form of the Schwarz inequality for all sequences $\{a_n\}_{n \in \mathbb{N}}$ and $\{b_m\}_{m \in \mathbb{N}}$:

$$\left(\sum_{n=1}^{p} |a_n b_n| \right)^2 \leqslant \left(\sum_{n=1}^{p} a_n^2 \right) \left(\sum_{m=1}^{p} b_m^2 \right)$$

In fact the first term is equal to

$$\sum_{n=1}^{p} a_n^2 b_n^2 + 2 \sum_{m \neq n = 1}^{p} |a_n a_m b_n b_m|$$

The second term is equal to

$$\sum_{n=1}^{p} a_n^2 b_n^2 + \sum_{n \neq m} (a_n^2 b_m^2 + a_m^2 b_n^2)$$

Since $a_n^2 b_m^2 + a_m^2 b_n^2 \geqslant 2|a_n a_m b_n b_m|$ (cf. p. 340) the property is established.
By applying this relationship to Eqn (43) we obtain the following inequality:

$$\sum_{n=p}^{q} |\langle Kf, \varphi_n \rangle| |\varphi_n(x)| \leqslant \left[\sum_{n=p}^{q} \langle f, \varphi_n \rangle^2 \right]^{1/2} \left[\sum_{n=p}^{q} \lambda_n^2 |\varphi_n(x)|^2 \right]^{1/2}$$

Now

$$K(x, z) = \lim_{L^2, p \to \infty} \sum_{n=1}^{p} \langle K(x, z), \varphi_n(z) \rangle \varphi_n(z)$$

$$= \lim_{L^2} \sum_{n=1}^{p} \lambda_n \bar{\varphi}_n(x) \varphi_n(z)$$

Denoting by $K(x, \cdot)$ the function $z \rightsquigarrow K(x, z)$ we see that

$$\|K(x, \cdot)\|_2^2 = \int_a^b |K(x, z)|^2 \, dz \geqslant \sum_{n=1}^{p} \lambda_n^2 |\varphi_n(x)|^2$$

and therefore

$$\sum_{n=p}^{q} |\langle Kf, \varphi_n\rangle| |\varphi_n(x)| \leqslant \|K(x, \cdot)\|_2 \left[\sum_{p}^{q} \langle f, \varphi_n\rangle^2 \right]^{1/2}$$

Since the series on the right is the remainder of a convergent series, the sequence

$$\sum_{1}^{n} |\langle Kf, \varphi_m\rangle| |\varphi_m(x)|$$

is a Cauchy sequence and

$$\sum_{n=1}^{\infty} \langle Kf, \varphi_n\rangle \varphi_n(x)$$

is a uniformly and absolutely convergent series. Its limit is the same as in L^2: it is therefore g.

We shall also use the following very simple proposition.

Proposition 3.1.5

Let $K(x, y)$ be a square integrable function over $[a, b] \times [a, b]$ and let K be the associated compact operator.

The iterative operators $K_2, K_3, \ldots, K_n$ defined by $K_2 f = K(Kf)$, $K_3 f = K(K_2 f), \ldots, K_n f = K(K_{n-1} f)$ are also operators with kernels. The kernel of K_n is

$$K_n(x, y) = \int_a^b K(x, u) K_{n-1}(u, y) \, du$$

The operators K_n are compact and self-adjoint if K is.

Proof By elementary calculation based on Fubini's theorem of inversion of the order of integration

$$K(Kf)(x) = \int_a^b K(x, u) Kf(u) \, du$$

$$= \int_a^b K(x, u) \left[\int_a^b K(u, y) f(y) \, dy \right] du$$

$$= \int_a^b \left[\int_a^b K(x, u) K(u, y) \, du \right] f(y) \, dy$$

$$= \int_a^b K_2(x, y) f(y) \, dy$$

putting

$$K_2(x, y) = \int_a^b K(x, u) K(u, y) \, du$$

If $K(x, y) = K(y, x)$ it is clear that $K_2(x, y) = K_2(y, x)$ and that K_2 is therefore a self-adjoint operator. It is compact, for $K_2(x, y)$ is square integrable over $[a, b] \times [a, b]$.

We can establish in the same way that

$$K_n(x, y) = \int_a^b K(x, u)K_{n-1}(u, y)\, dy$$

We can verify by recurrence that K_n is compact since K_{n-1} is, and that if $K(x, y)$ is symmetric K_n is self-adjoint.

3.2. Volterra integral equations

Consider the linear differential equation of the second order

$$\frac{d^2 f}{dx^2} + a(x)\frac{df}{dx} + b(x)f(x) = g(x) \tag{44}$$

Let F be the solution such that $F(0) = \alpha$ and $dF(0)/dx = \beta$ and let $\varphi(x) = d^2 F/dx^2$. Then

$$\frac{dF}{dx}(y) = \beta + \int_0^y \varphi(u)\, du$$

where β is an arbitrary constant and

$$F(x) = \alpha + \beta x + \int_0^x \left[\int_0^y \varphi(u)\, du \right] dy \quad \text{(where } \alpha \text{ is another constant)}$$

$$= \alpha + \beta x + \int_0^x (x - u)\varphi(u)\, du$$

Substituting in (44) we see that φ is a solution of

$$\varphi(x) + \beta a(x) + \int_0^x a(x)\varphi(y)\, dy + \alpha b(x) + \beta x b(x)$$

$$+ \int_0^x (x - y)\varphi(y)b(x)\, dy = g(x)$$

Putting $-K(x, y) = a(x) + b(x)(x - y)$ and $G(x) = g(x) - \beta a(x) - \alpha b(x) - \beta x b(x)$ the preceding relation becomes

$$\varphi(x) = \int_0^x K(x, y)\varphi(y)\, dy + G(x) \tag{45}$$

The solution of (44) thus reduces to that of (45). We shall naturally obtain the same result starting from an equation of order n by putting $\varphi(x) = d^n F/dx^n$. We shall study equations of the same type as (44).

Definition 3.2.1
We define a **Volterra kernel** over a bounded interval $[a, b]$ as a function $K(x, y)$ such that $K(x, y) = 0$ if $y > x$.

Let K be a square integrable Volterra kernel over $[a, b] \times [a, b]$. Let $\lambda \in \mathbb{R}$ and $f \in L^2[a, b]$. We define a **Volterra equation of the second kind** as an equation of the form

$$\varphi(x) = \lambda \int_a^x K(x, y)\varphi(y)\,\mathrm{d}y + f(x) \tag{46}$$

where the unknown $\varphi \in L^2[a, b]$ and λ is real.

Theorem 3.2.2
Equation (46) has, for every λ, a unique solution

$$\varphi(x) = \lim_{n \to \infty} f(x) + \lambda K(fx) + \ldots + \lambda^n K_n f(x) \tag{47}$$

or

$$\varphi(x) = f(x) + \lambda \int_0^x H(x, y, \lambda)f(y)\,\mathrm{d}y$$

where

$$H(x, y, \lambda) = \sum_{n=0}^{\infty} \lambda^n K_{n+1}(x, y)$$

($K_n(x, y)$ is the kernel of K_n, the nth iterate of K.)

In the equality (47) the convergence is absolute and uniform over $[a, b]$; also the series converges in quadratic mean.

Proof   For simplicity, we shall suppose $[a, b] = [0, 1]$. With the kernel $K(x, y)$ we associate an operator K in $L^2[0, 1]$ and Eqn (46) can be written $(I - \lambda K)\varphi = f$ (we suppose $\lambda \neq 0$).

We know that if $|\lambda| < 1/\|K\|$, $(I - \lambda K)^{-1}$ exists and is precisely $I + \lambda K + \lambda^2 K^2 + \ldots + \lambda^n K^n + \ldots$, so that the equality (47) is true for $|\lambda| < 1/\|K\|$ on condition that we consider convergence in $L^2[a, b]$. To establish this we have not used the fact that $K(x, y)$ is a Volterra kernel; it is this fact which enables us to obtain both uniform pointwise convergence and this convergence for $|\lambda| \geqslant 1/\|K\|$.

We shall prove directly that series (47) has this property of convergence over $[0, 1]$.

If

$$H(x) = \int_0^x |K(x, y)|^2\,\mathrm{d}y$$

and

$$G(y) = \int_y^1 |K(x, y)|^2 \, dx$$

by hypothesis H and G are integrable over $[0, 1]$. We shall put

$$H = \int_0^1 H(x) \, dx \quad \text{and} \quad G = \int_0^1 G(y) \, dy$$

We shall denote

$$\int_0^x H(u) \, du$$

by $h(x)$. This is an increasing function whose limit when $x \to 1$ is H.

We shall prove that all the kernels K_n are also Volterra kernels, which will give

$$K_2(x, y) = \int_0^1 K(x, u)K(u, y) \, du$$

Now $K(x, u)K(u, y)$ is non-zero only if $x \geqslant u \geqslant y$. Therefore

$$K_2(x, y) = 0 \qquad\qquad \text{if } y > x$$

$$K_2(x, y) = \int_y^x K(x, u)K(u, y) \, du \quad \text{if } y \leqslant x$$

$K_2(x, y)$ is therefore a Volterra kernel, as also are $K_3(x, y), \ldots, K_n(x, y)$.
If $y \leqslant x$

$$|K_2(x, y)|^2 \leqslant \int_y^x |K(x, u)|^2 \, du \int_y^x |K(u, y)|^2 \, du$$

$$\leqslant H(x)G(y)$$

(from Schwarz's inequality). In the same way

$$|K_3(x, y)|^2 \leqslant \int_y^x |K(x, u)|^2 \, du \int_y^x |K_2(u, y)|^2 \, du$$

$$\leqslant H(x)G(y) \int_y^x H(u) \, du$$

Thus

$$|K_2(x, y)|^2 \leqslant H(x)G(y)$$

$$|K_3(x, y)|^2 \leqslant H(x)G(y)[h(x) - h(y)]$$

and by recurrence

$$|K_n(x, y)|^2 \leqslant H(x)G(y) \frac{[h(x) - h(y)]^{n-2}}{(n-2)!}$$

so that

$$|K^n f(x)|^2 = \left| \int_0^x K_n(x, y) f(y)\, \mathrm{d}y \right|^2$$

$$\leqslant \int_0^x H(x) G(y) \frac{[h(x) - h(y)]^{n-2}}{(n-2)!}\, \mathrm{d}y \cdot \|f\|_2^2$$

$$\leqslant H(x) \frac{[h(x)]^{n-2}}{(n-2)!} \int_0^x G(y)\, \mathrm{d}y \cdot \|f\|_2^2$$

$$\leqslant \frac{M^n}{(n-2)!} \|f\|_2^2$$

denoting by M a number greater than or equal to H, G and $h(1)$. Thus

$$\forall \lambda \quad |\lambda^n K^n f(x)|^2 \leqslant \frac{(\lambda^2 M)^n}{(n-2)!} \|f\|_2^2$$

which is the general term of a series uniformly and absolutely convergent $\forall \lambda$.

Uniqueness of the solution: this is obvious if $|\lambda| < 1/\|K\|$.

In the general case we use Theorem 2.2.7 from the previous section. Let R be the operator defined over $L^2[0, 1]$ by the relation $R\varphi = f + \lambda K\varphi$. φ is a solution of (46) if $\varphi = f + \lambda K\varphi$, i.e. if $\varphi = R\varphi$. It is sufficient to show that this last equation has only one solution, and for that it is sufficient to prove that there exists an integer m and a real number α $(0 < \alpha < 1)$ such that

$$\|R^m \varphi_1 - R^m \varphi_2\| \leqslant \alpha \|\varphi_1 - \varphi_2\|$$

Now

$$R^2 \varphi = f + \lambda Kf + \lambda^2 K^2 \varphi$$
$$R^3 \varphi = f + \lambda Kf + \lambda^2 K^2 f + \lambda^3 K^3 \varphi$$
$$\vdots$$
$$R^n \varphi = f + \lambda Kf + \lambda^2 K^2 f + \ldots + \lambda^{n-1} K^{n-1} f + \lambda^n K^n \varphi$$

and from the above upper bound (slightly modified, because R is not a linear operator)

$$\|R^n \varphi_1 - R^n \varphi_2\|^2 \leqslant \frac{(\lambda M)^n}{(n-2)!} \|\varphi_1 - \varphi_2\|^2$$

which completes the proof.

Remark   If we no longer require the solutions to belong to the Hilbert space $L^2[a, b]$, the preceding reasoning no longer applies and there can be several solutions; in particular there can be a non-zero solution if $f \equiv 0$, which is the case with the equation

$$\varphi(x) = \int_0^x y^{x-y} \varphi(y) \, dy$$

which admits the solution $\varphi(x) = x^{x-1}$.

Example   Let

$$K(x, y) = \begin{cases} \exp(x-y) & \text{for } 0 \leqslant y \leqslant x \leqslant 1 \\ 0 & \text{for } 0 \leqslant x \leqslant y \end{cases}$$

and let

$$\varphi(x) = \lambda \int_0^x \exp(x-y) \varphi(y) \, dy + f(x) \tag{48}$$

We calculate

$$K_2(x, y) = \int_y^x \exp(x-z) \exp(z-y) \, dz = (x-y) \exp(x-y)$$

$$K_3(x, y) = \int_y^x \exp(x-z)(z-y) \exp(z-y) \, dz = \exp(x-y) \frac{(x-y)^2}{2!}$$

and by recurrence

$$K_n(x, y) = \exp(x-y) \frac{(x-y)^{n-1}}{(n-1)!}$$

Then

$$H(x, y, \lambda) = \exp[(x-y)(1+\lambda)]$$

and

$$\varphi(x) = f(x) + \lambda \int_0^x \exp[(x-y)(1+\lambda)] f(y) \, dy$$

Singular kernels

The reasoning we have used which proves, at the same time, the uniqueness and the existence of solutions if $K(x, y)$ is square integrable can be applied if

$$K(x, y) = \frac{H(x, y)}{(x-y)^\alpha}$$

where $0 \leqslant \alpha < \frac{1}{2}$ and $H(x, y)$ is a continuous function.

By considering iterated kernels, we can easily prove the following result.

Theorem 3.2.3

Let

$$\varphi(x) = \lambda \int_0^x K(x, y) \varphi(y) \, dy + f(x) \tag{49}$$

be an equation in which

$$K(x, y) = \frac{H(x, y)}{(x - y)^{\alpha}}$$

We assume that H and f are continuous over $[a, b]$ and $0 \leqslant \alpha < 1$. Then there exists a unique solution of (49).

The proof consists in verifying that there exists n_0 such that $\mathbb{R}^{n_0}$ is a contraction.

Volterra equations of the first kind

Definition 3.2.4
Let $K(x, y)$ be a Volterra kernel over $[a, b] \times [a, b]$. We define a **Volterra equation of the first kind** as an equation of the form

$$\int_a^x K(x, y)\varphi(y)\,dy = f(x) \tag{50}$$

where f is given.

In general these equations are more difficult to solve than those of the second kind unless f and K are sufficiently regular. In this case we are led back to equations of the second kind.

Theorem 3.2.5
Suppose f and K are of class C^1 and $K(x, x) \neq 0$ $\forall x \in [a, b]$. Equation (50) then reduces to an equation of the second kind.

Proof Let φ be a solution of (50). Each member of this equality can be differentiated, so that

$$K(x, x)\varphi(x) + \int_a^x \frac{\partial}{\partial x} K(x, y)\varphi(y)\,dy = \frac{df}{dx}$$

i.e.

$$\varphi(x) = \frac{df/dx}{K(x, x)} - \int_a^x \frac{1}{K(x, x)} \frac{\partial}{\partial x} K(x, y)\varphi(y)\,dy$$

See *Exercises 27 to 29* at the end of the chapter.

Eigenvalues of a Volterra kernel

Proposition 3.2.6
Let $K(x, y)$ be a square integrable Volterra kernel over $[a, b] \times [a, b]$. The associated operator K in $L^2[a, b]$ has no non-zero eigenvalue.

Proof Let $1/\lambda$ be a non-zero eigenvalue of K. There exists φ (non-zero) such that

$$\frac{1}{\lambda}\varphi = K\varphi$$

$$\varphi = \lambda K\varphi = \lambda^2 K^2 \varphi = \ldots = \lambda^n K^n \varphi \quad \forall n \in \mathbb{N}$$

Now we have shown that $|\lambda^n K^n \varphi(x)|$ has limit zero if $n \to \infty$, which is absurd.

Comment The Volterra equation $(I - \lambda K)\varphi = f$ has a unique solution φ for all values of λ (if $\lambda = 0$, $\varphi \equiv f$). This would not be the case if there existed a non-zero eigenvalue, nor will it be the case for Fredholm equations.

We must not forget, however, that $\mathcal{N}(I - \lambda K) = \{0\}$ does not imply that $(I - \lambda K)^{-1}$ exists in the general case.

3.3. Fredholm integral equations

Definition 3.3.1
Let $K(x, y)$ be a square integrable kernel over $[a, b] \times [a, b]$, λ a real number (non-zero) and $f \in L^2[a, b]$. We define a **Fredholm equation of the second kind** as the following equation, where the unknown $\varphi \in L^2[a, b]$:

$$\varphi(x) = \lambda \int_a^b K(x, y)\varphi(y)\, dy + f(x) \tag{51}$$

We denote by K the operator with kernel $K(x, y)$. Equation (51) can then be written $(I - \lambda K)\varphi = f$. As before, if $|\lambda| < 1/\|K\|$ there exists a unique solution

$$\varphi = (I - \lambda K)^{-1} f$$

$$= \lim_{n \to \infty} L^2(f + \lambda Kf + \lambda^2 K^2 f + \ldots + \lambda^n K^n f)$$

The following theorem can be proved.

Theorem 3.3.2
Let

$$H(x, y, \lambda) = \lim_{n \to \infty} \left[K(x, y) + \lambda K_2(x, y) + \ldots + \lambda^{n-1} K_n(x, y) \right]$$

where $K_n(x, y)$ is the kernel of the nth iterate K_n of K.

(a) Let $|\lambda| < 1/\|K\|$; the above series is uniformly convergent over $[a, b]$ and the unique solution of (51) is

$$\varphi(x) = f(x) + \lambda \int_a^b H(x, y, \lambda) f(y)\, dy$$

(b) For every number λ such that $1/\lambda$ is not an eigenvalue of K, there exists a function $\tilde{H}(x, y, \lambda)$ such that

$$\varphi(x) = f(x) + \lambda \int_a^b \tilde{H}(x, y, \lambda) f(y) \, dy$$

is the unique solution of (51) (if $\lambda < 1/\|K\|$, $\tilde{H}(x, y, \lambda) = H(x, y, \lambda)$).

(c) If $1/\lambda$ is an eigenvalue of K there exists either no solution or an infinity of solutions, in accordance with the Fredholm alternative which we shall reformulate.

The numbers λ such that $1/\lambda$ is an eigenvalue of K are often called **characteristic numbers** of K (or of (51)).

THE FREDHOLM ALTERNATIVE

Theorem 3.3.3

Let K^* be the transposed operator of K; this is the integral operator whose kernel is $K(y, x)$ and whose eigenvalues are the conjugates of those of K. Let μ be one of these eigenvalues and $\lambda = 1/\mu$. Equation (51) can be written $(\mu I - K)\varphi = \mu f$.

(a) There is no solution if f is not orthogonal to the eigenfunctions of K^* associated with $\bar{\mu}$.

(b) If, however, f is orthogonal to these functions there is an infinity of solutions.

Proof It is easy to verify that K^* has the form stated since

$$\langle Kf, g \rangle = \int_a^b \left[\int_a^b K(x, y) f(y) \, dy \right] g(x) \, dx$$

$$= \int_a^b \left[\int_a^b K(x, y) g(x) \, dx \right] f(y) \, dy$$

$$= \langle f, K^* g \rangle$$

Since K is compact, K and K^* have conjugate eigenvalues; the rest of the proof is an immediate application of Theorem 2.2.13.

It was in this form that the Fredholm alternative was proved for the first time by Fredholm himself.

Example Let

$$\varphi(x) = f(x) + \lambda \int_0^1 \exp(x - y) \, \varphi(y) \, dy \tag{52}$$

$$K(x, y) = \exp(x - y)$$

$$K_2(x, y) = \int_0^1 \exp(x - z) \exp(z - y) \, dz = K(x, y) \ldots$$

Therefore

$$K_n(x, y) = K(x, y)$$

and

$$H(x, y, \lambda) = \exp(x - y)(1 + \lambda + \lambda^2 + \ldots + \lambda^n + \ldots)$$

$$= \exp(x - y)\frac{1}{1 - \lambda} \quad \text{if } |\lambda| < 1$$

Then

$$\varphi(x) = f(x) + \frac{\lambda}{1 - \lambda}\exp(x)\int_0^1 \exp(-y)f(y)\,dy \qquad (53)$$

We can establish in fact that the relation (53) again defines the solution of (52) for $|\lambda| > 1$; therefore $\tilde{H}(x, y, \lambda) = H(x, y, \lambda)$.

Let us seek the eigenvalues and eigenfunctions of K^* whose kernel is $\exp(y - x)$. Naturally $\mu = 0$ is an eigenvalue since there exist functions such that

$$\int_0^1 \exp(y)\,\varphi(y)\,dy = 0$$

There is actually an infinity of them, which are independent. However, this eigenvalue is not appropriate since we suppose $\lambda < \infty$.

A non-zero number μ is an eigenvalue if there exists $\varphi \neq 0$ such that

$$\int_0^1 \exp(y - x)\,\varphi(y)\,dy = \mu\varphi(x)$$

Therefore $\varphi(x) = k\exp(-x)$, where k is such that

$$k\exp(-x)\int_0^1 \exp(y)\exp(-y)\,dy = \mu k\exp(-x)$$

Thus $\mu = 1$ is the only non-zero eigenvalue of K^*; we find that formula (53) can be extended $\forall \lambda \neq 1$.

We know then that if $\lambda = 1$ solutions exist only if

$$\int_0^1 \exp(-y)f(y)\,dy = 0$$

This can be obtained directly.

For $\lambda = 1$ (52) can be written

$$\varphi(x) = f(x) + \exp(x)\int_0^1 \exp(-y)\,\varphi(y)\,dy$$

For

$$A(\varphi) = \int_0^1 \exp(-y)\,\varphi(y)\,dy$$

if there exists a solution φ, it is such that $\varphi(x) = f(x) + A(\varphi) \exp(x)$ and in its turn $A(\varphi)$ is such that

$$A(\varphi) = \int_0^1 \exp(-y)\, \varphi(y)\, dy$$

$$= \int_0^1 \exp(-y) f(y)\, dy + A(\varphi) \int_0^1 \exp(-y) \exp(y)\, dy$$

$$= A(\varphi) + \int_0^1 \exp(-y) f(y)\, dy$$

If

$$\int_0^1 \exp(-y) f(y)\, dy \neq 0$$

there is no solution. If on the contrary

$$\int_0^1 \exp(-y) f(y)\, dy = 0$$

$$\varphi(x) = f(x) + c \exp(x)$$

We can establish that the solutions depend on d^* parameters, where d^* is the dimension of the eigenspace of K^* associated with the eigenvalue $\mu = 1$ (cf. Theorem 2.2.13).

Case of a symmetric kernel

Suppose that the kernel $K(x, y)$ is square integrable over $[a, b] \times [a, b]$ and in addition is symmetric; K is then a compact self-adjoint operator to which we can apply the results of Theorems 2.3.3 and 2.3.4.

If φ is a solution, from the Hilbert–Schmidt theorem $\varphi - f = K(\lambda\varphi)$ is a uniformly and absolutely convergent series.

Theorem 3.3.4

Let $K(x, y)$ be a symmetric kernel which is square integrable over $[a, b] \times [a, b]$. Let $(\lambda_n, h_n)_{n \in \mathbb{N}}$ be the eigenvalues and eigenfunctions of the operator K associated with this kernel (with $\|h_n\| = 1 \; \forall n$).

If $\lambda \neq \lambda_n^{-1} \; \forall n$, the equation

$$\varphi(x) = f(x) + \lambda \int_a^b K(x, y)\varphi(y)\, dy$$

has as its unique solution

$$\varphi(x) = f(x) + \sum_{n=0}^{\infty} \frac{\lambda \lambda_n}{1 - \lambda \lambda_n} \langle f, h_n \rangle h_n(x)$$

this series being uniformly and absolutely convergent. This formula can also be written

$$\varphi(x) = f(x) + \int_a^b \sum_{n=0}^{\infty} \frac{\lambda \lambda_n}{1 - \lambda \lambda_n} h_n(x) h_n(y) f(y) \, dy$$

If $\lambda \neq \lambda_{n_0}^{-1}$ and if $\langle f, h_{n_0} \rangle = 0$ for every eigenfunction associated with λ_{n_0}, we have the solutions

$$\varphi(x) = f(x) + \sum_{\substack{n=0 \\ n \neq n_0}}^{\infty} \frac{\lambda \lambda_n}{1 - \lambda \lambda_n} \langle f, h_n \rangle h_n(x) + \sum_{n=n_0} c_{n_0} \varphi_{n_0}$$

and the convergence is absolute and uniform.

   Example　Let

$$K(x, y) = \begin{cases} x(y-1) & \text{if } 0 \leqslant x \leqslant y \leqslant 1 \\ y(x-1) & \text{if } y \leqslant x \leqslant 1 \end{cases}$$

Solve

$$\varphi(x) = \lambda \int_0^1 K(x, y) \varphi(y) \, dy + x$$

Let us seek the eigenvalues and eigenfunctions of the operator K, i.e. the numbers λ and functions g such that

$$\int_0^1 K(x, y) g(y) \, dy = \lambda g(x)$$

We have already found these eigenvalues in the preceding section as an example of approximation to eigenvalues (p. 362); they are the numbers

$$\lambda_n = -\frac{1}{(n+1)^2 \pi^2}$$

The eigenfunctions are the solutions of

$$\frac{d^2 g}{dx^2} = -(n+1)^2 \pi^2 g$$

such that $g(0) = 0 = g(1)$; those which are of norm 1 are $h_n(x) = \sqrt{2} [\sin(n+1)\pi x]$ $(n \in \mathbb{N})$.
   Let us examine whether x is orthogonal to the latter:

$$\int_0^1 x \sin(n+1)\pi x \, dx = \left[-\frac{x \cos(n+1)\pi x}{(n+1)\pi} \right]_0^1 + \int_0^1 \frac{\cos(n+1)\pi x}{(n+1)\pi} \, dx$$

$$= \frac{(-1)^n}{(n+1)\pi}$$

Therefore if $\lambda \neq -(n+1)^2\pi^2$

$$\varphi(x) = x + \sum_{p=1}^{\infty} \frac{-\lambda/p^2\pi^2}{1 + \lambda/p^2\pi^2} \langle x, \sqrt{2}\sin p\pi x \rangle \sqrt{2}\sin p\pi x$$

$$= x + \frac{2\lambda}{\pi} \sum_{p=1}^{\infty} \frac{(-1)^p}{p} \frac{\sin(p\pi x)}{\lambda^2 + p^2\pi^2}$$

We can easily verify that this series converges uniformly and absolutely. If $\lambda = -(n+1)^2\pi^2$ the equation has no solution.

See **Exercises 30 to 34** at the end of the chapter.

Remark For degenerate kernels we shall introduce a simpler method, although Theorem 3.2.2 could be applied. Moreover, we shall be able to handle in this way the example below.

DEGENERATE KERNELS

We define a degenerate kernel as a function $K(x, y)$ of the form

$$K(x, y) = \sum_{i=1}^{p} f_i(x)g_i(y)$$

Such a kernel has at most p eigenvalues, which are very easy to determine.
We must find λ and h such that

$$\lambda h(x) = \int_a^b \sum_{i=1}^{p} f_i(x)g_i(y)h(y)\,dy$$

i.e.

$$\lambda h(x) = \sum_{i=1}^{p} \left[\int_a^b g_i(y)h(y)\,dy \right] f_i(x)$$

Then there exist constants (unknown and dependent on h) $c_1, \ldots, c_p$ such that

$$\lambda h(x) = c_1 f_1(x) + c_2 f_2(x) + \ldots + c_p f_p(x)$$

Since

$$c_1 = \int_a^b g_1(x)h(x)\,dx$$

$$\lambda c_1 = c_1 \int_a^b g_1(x)f_1(x)\,dx + c_2 \int_a^b g_1(x)f_2(x)\,dx + \ldots + c_p \int_a^b g_1(x)f_p(x)\,dx$$

Thus we obtain p relations; writing

$$a_{ij} = \int_a^b g_i(x)f_i(x)\,dx$$

we see that $c_1, c_2, \ldots, c_p$ are solutions of the system

$$
\left.
\begin{aligned}
(\lambda - a_{11})c_1 + \quad a_{12}\,c_2 + \ldots + \quad a_{1p}\,c_p &= 0 \\
a_{21}\,c_1 + (\lambda - a_{22})c_2 + \ldots + \quad a_{2p}\,c_p &= 0 \\
\vdots \qquad\qquad\qquad\qquad \\
a_{p1}\,c_1 + \quad a_{p2}\,c_2 + \ldots + (\lambda - a_{pp})c_p &= 0
\end{aligned}
\right\}
\tag{54}
$$

The eigenvalues are then the values of λ which make the determinant of this system zero—it is called the Fredholm determinant. We next calculate the eigenfunctions. We could calculate the solutions directly.

Example   Let $[a, b] = [0, \pi]$.

$$
K(x, y) = \cos(x + y)
$$
$$
= \cos x \cos y - \sin x \sin y
$$

h is an eigenfunction associated with λ if

$$
\lambda h(x) = \cos x \int_0^\pi h(y) \cos y \, dy - \sin x \int_0^\pi h(y) \sin y \, dy
$$

Thus $\lambda h(x) = A \cos x - B \sin x$; λ, A and B must then satisfy

$$
\lambda A = A \int_0^\pi \cos^2 y \, dy - B \int_0^\pi \cos y \sin y \, dy
$$

$$
\lambda B = A \int_0^\pi \sin y \cos y \, dy - B \int_0^\pi \sin^2 y \, dy
$$

This system reduces to

$$
\lambda A = A \frac{\pi}{2} \qquad \lambda B = -B \frac{\pi}{2}
$$

Thus there are two eigenvalues

$$
\lambda_1 = \frac{\pi}{2} \quad \text{with eigenfunction } h_1(x) = \frac{2}{\pi} \cos x
$$

$$
\lambda_2 = -\frac{\pi}{2} \quad \text{with eigenfunction } h_2(x) = \frac{2}{\pi} \sin x
$$

The rest of the solution of the equation

$$
\varphi(x) = f(x) + \lambda \int_a^b K(x, y)\varphi(y) \, dy
$$

is identical with that for the general case with symmetric kernels.

If $|\lambda| \neq 2/\pi$ there is a unique solution:

$$\varphi(x) = f(x) + \frac{2\lambda}{2 - \lambda\pi}\left[\int_0^\pi f(y)\cos y\,\mathrm{d}y\right]\cos x$$

$$- \frac{2\lambda}{2 + \lambda\pi}\left[\int_0^\pi f(y)\sin y\,\mathrm{d}y\right]\sin x$$

For example, let $f(x) = \cos 2x$:

if $|\lambda| \neq \dfrac{2}{\pi}$ $\varphi(x) = \cos 2x + \dfrac{4\lambda}{3\lambda\pi + 6}\sin x$

if $\lambda = -\dfrac{2}{\pi}$ there is no solution for $\displaystyle\int_0^\pi \cos 2x \sin x\,\mathrm{d}x = -\dfrac{2}{3}$

if $\lambda = \dfrac{2}{\pi}$ there is an infinity of solutions of the form

$$\varphi(x) = \cos 2x + \frac{2}{3\pi}\sin x + k\cos x \quad \text{(where } k \text{ is arbitrary)}$$

See *Exercises 35 and 36* at the end of the chapter.

3.4. The Sturm–Liouville problem; special functions

REGULAR PROBLEMS

We can now give a complete answer to the question which we set ourselves at the beginning of this chapter.

Definition 3.4.1 (The regular Sturm–Liouville problem)
We denote by this name the study of solutions of an equation of the type

$$\frac{\mathrm{d}}{\mathrm{d}x}\left[p(x)\frac{\mathrm{d}f}{\mathrm{d}x}(x)\right] + q(x)f(x) + s(x)\lambda f(x) = 0 \tag{55}$$

where p, q and s are three functions continuous over a closed, bounded interval $[a, b]$, where p is of class C^1 and $p(x) > 0$ $\forall x \in [a, b]$ and $s(x) > 0$ $\forall x \in [a, b]$. This equation has the boundary conditions

$$\left.\begin{array}{l} a_0 f(a) + a_1 \dfrac{\mathrm{d}f}{\mathrm{d}x}(a) = 0 \\[2mm] b_0 f(b) + b_1 \dfrac{\mathrm{d}f}{\mathrm{d}x}(b) = 0 \end{array}\right\} \tag{56}$$

The conditions (56) are described as separated; the first involves a, the second b (see Definition 1.3.1).

Since $s(x) > 0$, this problem is the search for eigenvalues and eigenfunctions of the operator

$$\frac{1}{s(x)}\frac{\mathrm{d}}{\mathrm{d}x}\left[p(x)\frac{\mathrm{d}f}{\mathrm{d}x}\right] + q(x)f(x) = Af(x)$$

with the boundary conditions (56).

If λ is not an eigenvalue, $f \equiv 0$ is the only solution.

Proposition 3.4.2

The eigenvalues λ_n of a regular problem are simple.

Proof Let λ be an eigenvalue and φ_1 and φ_2 be two eigenfunctions; they are both solutions of (55) and they satisfy (56). Consequently from Theorem 1.3.2 $p(x)w(\varphi_1, \varphi_2)x$ is constant. This constant is

$$p(a)w(\varphi_1, \varphi_2)(a) = p(a)\begin{vmatrix} \varphi_1(a) & \varphi_2(a) \\ \varphi_1'(a) & \varphi_2'(a) \end{vmatrix} = 0 \quad \text{from (56)}$$

Therefore φ_1 is proportional to φ_2.

We can then prove the following fundamental theorem.

Theorem 3.4.3

Over a closed bounded interval $[a, b]$ let p, q and s be three continuous functions with $p > 0$, $s > 0$ and p of class C^1. Let

$$\frac{\mathrm{d}}{\mathrm{d}x}\left[p(x)\frac{\mathrm{d}f}{\mathrm{d}x}\right] + q(x)f(x) + \lambda s(x)f(x) = 0 \tag{57}$$

and

$$\left.\begin{aligned} a_0 f(a) + a_1 f'(a) &= 0 \\ b_0 f(b) + b_1 f'(b) &= 0 \end{aligned}\right\} \tag{58}$$

The search for numbers λ and non-zero functions f_λ which are solutions of (57) and (58) is a regular Sturm–Liouville problem; the numbers $-\lambda$ and functions f_λ are the eigenvalues and eigenfunctions of the operator defined over the vector space of twice-differentiable functions satisfying (58) by

$$Af = \left[\frac{\mathrm{d}}{\mathrm{d}x}(pf') + qf\right]\frac{1}{s}$$

(a) There exists a denumerable family of real simple eigenvalues of A: $\lambda_0 < \lambda_1 < \ldots < \lambda_n < \ldots$ where

$$\lim_{n \to \infty} \lambda_n = \infty$$

and an eigenfunction φ_n associated with λ_n satisfying

$$\frac{d}{dx}(p\varphi_n') + q\varphi_n + \lambda_n s\varphi_n = 0$$

(b) The sequence $(\varphi_n)_{n \in \mathbb{N}}$ is a basis of orthogonal vectors of $L_s^2[a, b]$. φ_n has exactly n zeros in $]a, b[$, and φ_{n-1} has a zero between two consecutive zeros of φ_n.

(c) If φ is continuous and satisfies (58) and if φ' is piecewise continuous

$$\varphi(x) = \sum_{n=0}^{\infty} \langle \varphi, \varphi_n \rangle_{L_s^2(a, b)} \varphi_n(x)$$

This series converges in $L_s^2(a, b)$ and uniformly and absolutely over $[a, b]$.

Proof Consider the associated homogeneous problem

$$\frac{d}{dx}\left[p(x)\frac{df}{dx} \right] + q(x)f(x) = 0 \tag{59}$$

$$\left. \begin{array}{c} a_0 f(a) + a_1 \dfrac{df}{dx}(a) = 0 \\[2mm] b_0 f(x) + b_1 \dfrac{df}{dx}(b) = 0 \end{array} \right\} \tag{60}$$

Let G be the Green function (generalized).

Suppose first that this problem admits no solution other than zero, and assume that $s \equiv 1$. Let L be the operator

$$f \rightsquigarrow \frac{d}{dx}\left[p\frac{df}{dx} \right] + qf$$

defined over the subspace H of $L^2[a, b]$ of twice-differentiable functions satisfying (60) and let G be the operator associated with the Green function; G is a compact self-adjoint operator from $L^2[a, b]$ into itself and $\mathcal{R}(G) = H$. We know that $Lf = g \Leftrightarrow f = Gg$.

Let λ be a non-zero eigenvalue of G: $\lambda f = Gf$ and therefore

$$Lf = \frac{1}{\lambda}f$$

Let μ be a non-zero eigenvalue of L: $Lf = \mu f$ and therefore

$$\frac{1}{\mu}f = Gf$$

Since the non-zero eigenvalues of L are simple, the same is true for G. Let us now examine the case of zero eigenvalues.

L has no zero eigenvalue since $Lf = 0$ has only one solution, $f \equiv 0$, as the homogeneous problem has only the solution $f \equiv 0$. G has no zero eigenvalue either. In fact Gg, which is the solution of

$$\frac{d}{dx}(pf') + qf = g$$

can be zero only if $g(x) \equiv 0$.

If the homogeneous problem has no non-zero solution, zero is an eigenvalue neither of L nor of G. Therefore the eigenvalues of L are exactly the reciprocals of those of G which are infinite in number since they are simple and tend towards zero, which gives the result.

If the homogeneous problem has a non-zero solution, say f_0, zero is a simple eigenvalue of L and f_0 is an eigenvector.

If we seek solutions orthogonal to f_0, the problem is situated in a subspace of H in which the situation is again that in which zero is not an eigenvalue and the conclusions are therefore those which we have just found.

As the eigenfunctions are the same for L and G (except possibly f_0), the orthogonality and existence properties of a basis result immediately from the corresponding properties for G (established, for example, in Theorem 3.1.3).

The uniform convergence results from the Hilbert–Schmidt theorem when we apply L to f, since then the function f is of the form Gg where $Lf = g$; it is therefore true at least for $f \in H$.

The result on the zeros of eigenvalues is a consequence of a theorem of Sturm which we shall not state exactly.

If $s \not\equiv 1$, since the problem is regular s is bounded above and below and putting

$$Kg(y) = \int_a^b G(x, y)s(y)g(y)\,dy$$

Kg is the solution of

$$\frac{d}{dx}\left[p(x)\frac{df}{dx}\right] + q(x)f(x) + s(x)g(x) = 0$$

K has the same properties as G and all the results hold, by considering the functions $\varphi(x)[s(x)]^{1/2}$ and the basis $\{\varphi_n(x)[s_n(x)]^{1/2}\}_{n \in \mathbb{N}}$.

Examples

(a) We take up again the problem at the beginning of this chapter which is represented by the equation

$$\frac{\partial u}{\partial t}(x, t) = k\frac{\partial^2 u}{\partial x^2}(x, t) \qquad 0 < x < \pi \text{ and } 0 < t \tag{61}$$

and

$$u(0, t) = 0 = u(\pi, t)$$

$$u(x, 0) = f(x) \qquad f(0) = f(\pi) = 0$$

By separation of variables, we are led to the equations

$$\frac{d^2\varphi}{dx^2} + \lambda\varphi(x) = 0 \tag{62}$$

$$\frac{dT}{dt} + \lambda kT(t) = 0 \tag{63}$$

$$\varphi(0) = 0 = \varphi(\pi) \tag{64}$$

We are looking for $u(x, t)$ in the form

$$\sum_n \varphi_n(x)T_n(t)$$

The problem (62), (64) is regular; its eigenvalues are $\lambda_n = (n + 1)^2$, $n \in \mathbb{N}$, and its eigenfunctions are $\varphi_n(x) = \sin nx$ with norm $(\pi/2)^{1/2}$. For $\lambda_n = (n + 1)^2$ the solution of (63) is $T(t) = T(0) \exp[-(n + 1)^2 kt]$. We assume that f and df/dx are continuous and that f satisfies (64); it can then be expanded in the form

$$f(x) = \frac{2}{\pi} \sum_{n=1}^{\infty} \left[\int_0^{\pi} f(u) \sin nu \, du \right] \sin nx$$

Then

$$u(x, t) = \sum_{n=1}^{\infty} \left[\frac{2}{\pi} \int_0^{\pi} f(u) \sin nu \, du \right] \sin nx \exp(-n^2 kt)$$

and the convergence is uniform and absolute over $[0, \pi]$.

(b) Consider throughout the metallic bar discussed on page 307. Suppose always that the end with zero abscissa is kept at zero temperature but that the other end is thermally isolated, i.e. that no heat passes through this end face.

The new boundary conditions are then

$$u(0, t) = 0 = \frac{\partial u}{\partial x}(\pi, t) \qquad t > 0 \text{ and } \pi > x > 0$$

$$u(x, 0) = f(x)$$

We again use Eqn (1) and proceed in the same way as before:

$$\frac{d^2\varphi}{dx^2} + \lambda\varphi(x) = 0 \tag{65}$$

has conditions

$$\varphi(0) = 0 = \frac{d\varphi}{dx}(\pi) \tag{66}$$

This is again a regular problem; its eigenvalues are this time $\lambda_n = \frac{1}{4}(2n - 1)^2$ and its eigenfunctions are

$$\varphi_n(x) = \left(\frac{2}{\pi}\right)^{1/2} \sin\left(\frac{2n - 1}{2}\right)x$$

The solution is

$$u(x, t) = \sum_{n=1}^{\infty} \alpha_n \exp\left[-\frac{1}{4}(2n-1)^2 kt \right] \left(\frac{2}{\pi}\right)^{1/2} \sin\left(\frac{2n-1}{2}\right) x$$

where α_n is the nth coefficient in the expansion of f in terms of the basis

$$\left(\frac{2}{\pi}\right)^{1/2} \sin\left(\frac{2n-1}{2}\right) x$$

$$\alpha_n = \left(\frac{2}{\pi}\right)^{1/2} \int_0^{\pi} f(y) \sin\left(\frac{2n-1}{2} y\right) dy$$

If f and df/dx are continuous, and if $df(\pi)/dx = 0$, the convergence is uniform and absolute.

(c) It is required to find a function $u(x, y)$ defined over the rectangle $0 \leqslant x \leqslant a, 0 \leqslant y \leqslant b$ (Fig. 6.5), such that

$$\frac{\partial^2 u}{\partial x^2} + \frac{\partial^2 u}{\partial y^2} = 0$$

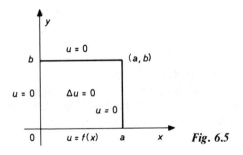

Fig. 6.5

and

$$u(0, y) = 0 = u(a, y) \quad \forall y \in [0, b]$$

$$u(x, 0) = f(x) \quad \forall x \in [0, a]$$

$$u(x, b) = 0$$

Such a problem represents the search for the equilibrium temperature of a homogeneous rectangle which does not exchange heat with its surroundings through its faces and whose edges are maintained at temperatures 0 and $f(x)$ as indicated in Fig. 6.5.

The problem also represents the search for the electrostatic potential of a cylinder with rectangular base whose faces are maintained at potentials 0 and $f(x)$ as indicated, and which has no charge in its interior.

If we suppose that $f(0) = 0 = f(a)$, it is also the study of the deformation of a membrane stretched over the rectangle whose edge $y = 0$ is displaced in accordance with $f(x)$.

In seeking a solution of the form $u(x, y) = X(x)Y(y)$ we are led to the following Sturm–Liouville problem

$$\frac{d^2 X}{dx^2} + \lambda X(x) = 0$$

with $X(0) = X(a) = 0$ and to the equation

$$\frac{d^2 Y}{dy^2} - \lambda Y(y) = 0$$

with the condition $Y(b) = 0$.

The eigenvalues of the Sturm–Liouville problem are

$$\lambda_n = \frac{(n + 1)^2 \pi^2}{a^2}$$

and the eigenfunctions are

$$X_n(x) = C_n \sin \frac{(n + 1)\pi x}{a}$$

The solutions of the equation are

$$Y(y) = K \sinh\left[\frac{n\pi}{a}(b - y)\right]$$

If f and its first derivative are continuous and if $f(0) = 0 = f(a)$, then

$$u(x, y) = \sum_{n=1}^{\infty} \alpha_n \sinh\left[\frac{n\pi}{a}(b - y)\right] \sin\frac{n\pi x}{a}$$

and the limit is absolute and uniform, α_n representing the nth coefficient in the series expansion of f:

$$f(x) = \sum_{n=1}^{\infty} \alpha_n \sinh\frac{n\pi b}{a} \sin\frac{n\pi x}{a}$$

where

$$\alpha_n = \frac{2}{a \sinh(n\pi b/a)} \int_0^a f(t) \sin\frac{n\pi t}{a}\, dt$$

Periodic Sturm–Liouville problems

Definition 3.4.4
The problem

$$\frac{d}{dx}\left[p(x)\frac{df}{dx}\right] + q(x)f(x) + \lambda s(x)f(x) = 0$$

where $p(a) = p(b)$ and where the boundary conditions are $f(a) = f(b)$ and

$$\frac{df}{dx}(a) = \frac{df}{dx}(b)$$

is called a periodic Sturm–Liouville problem.

This problem is not regular since the boundary conditions are not separated. If in addition the other conditions required for a problem to be regular are fulfilled, the only difference is that there can be double eigenvalues.

Example

$$\frac{d^2 f}{dx^2} + \lambda f(x) = 0 \qquad f(-\pi) = f(\pi) \qquad \frac{df}{dx}(-\pi) = \frac{df}{dx}(\pi)$$

This problem has eigenvalues $\lambda_0 = 0$, $\lambda_n = n^2$ $\forall n \in \mathbb{N}$. λ_0 is a simple eigenvalue; 1 is an eigenfunction associated with λ_0. n^2 is a double eigenvalue; two independent eigenfunctions corresponding to this eigenvalue are $\sin nx$ and $\cos nx$.

This system is clearly a basis of $L^2(-\pi, \pi)$.

Remark   We shall see in Exercise 40 how the situation changes in the case of a singular problem—e.g. the existence of a non-denumerable infinity of eigenvalues.

See *Exercises 37 to 45* at the end of the chapter.

EXAMPLES OF SINGULAR PROBLEMS—SPECIAL FUNCTIONS

The following are examples of singular problems in the sense that either $p(x)$ is not strictly positive for all values of x or the interval $[a, b]$ is infinite, or the boundary conditions are different.

We shall not prove any of the results stated, some of which could be proved in a similar way to that which we used for regular problems; other proofs would be more complicated.

The results we shall give concern differential equations which occur in many problems in physics. In particular, we shall indicate the existence of bases of eigenfunctions often called **special functions**. We must note that we are concerned not with arbitrary bases of Hilbert spaces but with bases giving rise to series expansions which are uniformly and absolutely convergent in accordance with the Hilbert–Schmidt theorem, or similar theorems.

Legendre polynomials

Theorem 3.4.5

Consider the following problem: solve, over the interval $]-1, +1[$, the equation

$$\frac{d}{dx}\left[(1-x^2)\frac{df}{dx}\right] + \lambda f(x) = 0 = (1-x^2)\frac{d^2f}{dx^2} - 2x\frac{df}{dx} + \lambda f(x) \qquad (67)$$

with the conditions

$$\lim_{x \to -1} f(x) < \infty \qquad \lim_{x \to +1} f(x) < \infty \qquad (68)$$

The solutions of this problem are the numbers $\lambda_n = n(n+1)$ and the functions

$$P_n(x) = (n + \tfrac{1}{2})^{1/2} \frac{1}{2^n n!} \frac{d^n}{dx^n} (x^2 - 1)^n$$

called **Legendre polynomials**. Thus

$$(1 - x^2)P_n''(x) - 2xP_n'(x) + n(n+1)P_n(x) = 0$$

As in the regular case we thus obtain a basis of $L^2(-1, +1)$ and we can state a theorem on pointwise convergence.

Theorem 3.4.6

The Legendre polynomials form an orthonormal basis of $L^2(-1, +1)$; if $\varphi \in L^2(-1, +1)$

$$a_n = \langle \varphi, P_n \rangle = \int_{-1}^{+1} \varphi(x)P_n(x)\,dx$$

Then

$$\lim_{p \to \infty} \left\| \varphi - \sum_0^p a_n P_n \right\|_2^2 = \lim_{p \to \infty} \int_{-1}^{+1} \left| \varphi - \sum_0^p a_n P_n \right|^2 (x)\,dx = 0$$

If φ and φ' are piecewise continuous

$$\lim_{p \to \infty} \sum_0^p a_n P_n(x) = \frac{1}{2}[\varphi(x^+) + \varphi(x^-)]$$

If φ is continuous the convergence is uniform over every segment $-1 < a \leqslant x \leqslant b < +1$.

Properties   We can verify that

$$P_0(x) = \frac{1}{\sqrt{2}}$$

$$P_1(x) = \left(\frac{3}{2}\right)^{1/2} x$$

$$P_2(x) = \left(\frac{5}{2}\right)^{1/2} \frac{3x^2 - 1}{2}$$

$$P_3(x) = \left(\frac{7}{2}\right)^{1/2} \frac{5x^3 - 3x}{2}$$

$$P_4(x) = \frac{3}{8\sqrt{2}} (35x^4 - 30x^2 + 3)$$

$$P_5(x) = \left(\frac{11}{2}\right)^{1/2} \frac{63x^5 - 70x^3 + 15x}{8} \ldots$$

These polynomials are alternately even and odd; P_n, which is of degree n, has n zeros in $]-1, +1[$, and between two consecutive zeros of P_n there is a zero of P_{n-1}.

We can verify the recurrence relation

$$\frac{n+1}{(2n+3)^{1/2}} P_{n+1}(x) = \frac{2n+1}{(2n+1)^{1/2}} xP_n(x) - \frac{n}{(2n-1)^{1/2}} P_{n-1}(x)$$

In the next chapter we shall study Legendre's equation: if $v > 0$ and $x \in]-1, +1[$, $(1 - x^2)f''(x) - 2xf'(x) + v(v + 1)f(x) = 0$. This equation has polynomial solutions if v is an integer.

As with Fourier series, we can consider expansions in series of even or of odd Legendre polynomials. If f is defined over $[0, 1]$ and we extend it to $[-1, +1]$ in such a way that it becomes an even function, this new function can be expanded in a series of Legendre polynomials in which only polynomials of even degree appear; since the function is equal to f over $[0, 1]$, f can be expanded in a series of even Legendre polynomials. In the same way we can extend f into an odd function over $[-1, +1]$ and hence deduce an expansion in a series of odd Legendre polynomials. Moreover, these two cases correspond to the Sturm–Liouville problems.

Theorem 3.4.7
(i) The Sturm–Liouville problem

$$\frac{d}{dx}\left[(1 - x^2)\frac{df}{dx}\right] + \lambda f(x) = 0 \quad \text{for } x \in [0, 1[\tag{69}$$

$$\left.\begin{array}{l} f'(0) = 0 \\ \lim_{x \to 1} f(x) < \infty \end{array}\right\} \tag{70}$$

has as solutions the numbers $\lambda_n = 2n(2n+1)$ and the polynomials

$$\sqrt{2}\,P_{2n} = (4n+1)^{1/2}\,\frac{1}{2^{2n}(2n)!}\,\frac{d^{2n}}{dx^{2n}}\,(x^2-1)^{2n}$$

The latter form an orthornomal basis of $L^2(0, 1)$, and every function $\varphi \in L^2(0, 1)$ whose derivative is piecewise continuous satisfies

$$\lim_{p \to \infty} \sum_0^p 2P_{2n}(x)\left[\int_0^1 \varphi(y)P_{2n}(y)\,dy\right] = \frac{1}{2}\,[\varphi(x^+) + \varphi(x^-)]$$

(ii) The Sturm–Liouville problem

$$\frac{d}{dx}\left[(1-x^2)\frac{df}{dx}\right] + \lambda f(x) = 0 \quad \text{for } x \in [0, 1[\tag{71}$$

$$\left.\begin{array}{c} f(0) = 0 \\[4pt] \lim_{x \to 1} f(x) < \infty \end{array}\right\} \tag{72}$$

has as solutions the numbers $\lambda_n = (2n+1)(2n+2)$ and the polynomials

$$\sqrt{2}\,P_{2n+1} = (4n+3)^{1/2}\,\frac{1}{2^{2n+1}(2n+1)!}\,\frac{d^{2n+1}}{dx^{2n+1}}\,(x^2-1)^{2n+1}$$

The latter also form an orthonormal basis of $L^2(0, 1)$, and every function $\varphi \in L^2(0, 1)$ whose derivative is piecewise continuous satisfies

$$\lim_{p \to \infty} \sum_0^p 2P_{2n+1}(x)\left[\int_0^1 \varphi(y)P_{2n+1}(y)\,dy\right] = \frac{1}{2}\,[\varphi(x^+) + \varphi(x^-)]$$

In these two cases if φ is continuous the convergence is uniform over every segment $0 \leqslant a \leqslant x \leqslant b < 1$.

Hermite and Laguerre polynomials

Theorem 3.4.8

Consider over $\mathbb{R}$ the equation

$$\frac{d}{dx}\left[\exp(-x^2)\frac{df}{dx}\right] + \lambda \exp(-x^2)\,f(x) = 0 = \frac{d^2f}{dx^2} - 2x\,\frac{df}{dx} + \lambda f$$

with conditions $\exists k > 0,\ k \in \mathbb{N}$:

$$\lim_{x \to \infty} \frac{f(x)}{x^k} < \infty, \qquad \lim_{x \to -\infty} \frac{f(x)}{|x|^k} < \infty$$

The solutions of this problem are the numbers $\lambda_n = 2n$ and the polynomials H_n, called **Hermite polynomials**:

$$H_n(x) = \frac{(-1)^n}{(2^n n!\,\pi^{1/2})^{1/2}}\,\exp(x^2)\,\frac{d^n \exp(-x^2)}{dx^n}$$

H_n has n zeros over $\mathbb{R}$; between two consecutive zeros of H_n there is one zero of H_{n-1}.

The Hermite polynomials form an orthonormal basis of $L^2[\exp(-x^2), \mathbb{R}]$. If $\varphi \in L^2[\exp(-x^2), \mathbb{R}]$

$$a_n = \langle \varphi, H_n \rangle = \int_{-\infty}^{\infty} \varphi(x)H_n(x)\exp(-x^2)\,\mathrm{d}x$$

Then

$$\varphi(x) = \lim_{p \to \infty} \sum_{0}^{p} a_n H_n(x)$$

in $L^2[\exp(-x^2), \mathbb{R}]$. If φ and φ' are piecewise continuous the preceding sequence converges towards $\frac{1}{2}[\varphi(x^+) + \varphi(x^-)]$; if φ is continuous the convergence is uniform over every interval $a \leqslant x \leqslant b$.

We can verify that

$$H_0(x) = \pi^{-1/4} \qquad H_1(x) = \pi^{-1/4}x\sqrt{2}$$

$$H_2(x) = \pi^{-1/4}\frac{2x^2 - 1}{\sqrt{2}} \qquad H_3(x) = \pi^{-1/4}\frac{2x^3 - 3x}{\sqrt{3}}$$

$$H_4(x) = \pi^{-1/4}\frac{4x^4 - 12x^2 - 3}{2\sqrt{6}} \quad \cdots$$

We can verify that $(n+1)^{1/2}H_{n+1}(x) = x\sqrt{2}\,H_n(x) - n^{1/2}H_{n-1}(x)$.

Theorem 3.4.9

Consider over $\mathbb{R}^+$ the equation

$$\frac{\mathrm{d}}{\mathrm{d}x}\left[x\exp(-x)\frac{\mathrm{d}f}{\mathrm{d}x}\right] + \lambda\exp(-x)f(x) = 0 = x\frac{\mathrm{d}^2f}{\mathrm{d}x^2} + (1-x)\frac{\mathrm{d}f}{\mathrm{d}x} + \lambda f(x)$$

with the conditions

f remains bounded when $x \to 0$

for every solution f there exists $k > 0$ such that $\displaystyle\lim_{x \to \infty}\frac{f(x)}{x^k} < \infty$

The solutions of this problem are the numbers $\lambda_n = n$ and the polynomials

$$L_n(x) = \frac{1}{n!}\exp(x)\frac{\mathrm{d}^n}{\mathrm{d}x^n}\left[x^n\exp(-x)\right]$$

called **Laguerre polynomials**. L_n has n zeros over $\mathbb{R}^+$ and between two consecutive zeros of L_n there is exactly one zero of L_{n-1}.

The Laguerre polynomials form an orthonormal basis of $L^2[\exp(-x), \mathbb{R}^+]$. If $\varphi \in L^2[\exp(-x), \mathbb{R}^+]$

$$a_n = \langle \varphi, L_n \rangle = \int_0^{\infty} \varphi(x)L_n(x)\exp(-x)\,\mathrm{d}x$$

Then

$$\varphi(x) = \lim_{p \to \infty} \sum_0^p a_n L_n(x)$$

in $L^2[\exp(-x), \mathbb{R}^+]$. If φ and φ' are piecewise continuous the preceding sequence converges towards $\frac{1}{2}[\varphi(x^+) + \varphi(x^-)]$; if φ is continuous the convergence is uniform over every interval $0 < a \leqslant x \leqslant b$.

We note that

$$L_0(x) = 1 \qquad L_1(x) = 1 - x \qquad L_2(x) = \frac{x^2}{2} - 2x + 1$$

$$L_3(x) = \frac{1}{6}[-x^3 + 9x^2 - 18x + 6] \quad \dots$$

We can verify that $(n + 1)L_{n+1}(x) = (2n + 1 - x)L_n(x) - nL_{n-1}(x)$.

In the next chapter we shall come across the Hermite and Laguerre equations again.

Bessel functions

Before stating the Sturm–Liouville problem in which the Bessel functions appear, we shall recall and complete some results from Chapter 2.

If $v \geqslant 0$ we define the Bessel equation with index v as the following equation over $]0, \infty[$:

$$x^2 \frac{d^2 f}{dx^2} + x \frac{df}{dx} + (x^2 - v^2)f(x) = 0$$

We shall see in the next chapter that this equation possesses a series solution which remains bounded when $x \to 0$; it is called a Bessel function of the first kind of index v and is denoted J_v:

$$J_v(x) = \left(\frac{x}{2}\right)^v \left[\frac{1}{\Gamma(v+1)} - \left(\frac{x}{2}\right)^2 \frac{1}{\Gamma(v+2)} + \left(\frac{x}{2}\right)^4 \frac{1}{2!\,\Gamma(v+3)} + \dots \right.$$

$$\left. + (-1)^p \left(\frac{x}{2}\right)^{2p} \frac{1}{p!\,\Gamma(v+p+1)} + \dots \right]$$

(We recall that

$$\Gamma(t) = \int_0^\infty \exp(-u)\,u^{t-1}\,du$$

and that, $\forall n \in \mathbb{N}$, $\Gamma(n+1) = n!$.) In the next chapter we shall give some properties of the functions J_v. Here we note some of them for the functions J_n $(n \in \mathbb{N})$.

We can verify that

$$xJ_{n+1}(x) = 2nJ_n(x) - xJ_{n-1}(x)$$

$$J_{n+1}(x) = -x^n \frac{d}{dx}[x^{-n}J_n(x)]$$

and

$$x^n J_{n-1}(x) = \frac{d}{dx}[x^n J_n(x)]$$

In particular $J_0' = -J_1$ and

$$\int_0^y xJ_0(x)\,dx = yJ_1(y)$$

J_n has an infinity of zeros over $\mathbb{R}^+$ which form a sequence tending to infinity; the difference between two consecutive zeros has limit π when x tends to infinity. Between two consecutive zeros of J_{n+1} there is exactly one zero of J_n; if $x \to \infty$ J_n is equivalent to

$$\frac{A_n}{x^{1/2}}\sin(x + \vartheta_n)$$

(see Fig. 6.6). In particular

$$J_0(x) = \frac{1}{\pi}\int_0^\pi \cos(x\sin\vartheta)\,d\vartheta \sim \left(\frac{2}{\pi x}\right)^{1/2}\sin\left(x + \frac{\pi}{4}\right)$$

$$J_{1/2} = \left(\frac{2}{\pi x}\right)^{1/2}\sin x$$

The Bessel functions enable us to solve the following Sturm–Liouville problem. Let

$$x^2 f''(x) + xf'(x) + (s^2 x^2 - v^2)f(x) = 0 \quad \text{over }]0, a[\tag{73}$$

We seek the numbers s such that there exists over $]0, a[$ a solution f, not identically zero, satisfying in addition $f(a) = 0$ and

$$\lim_{x \to 0} f(x) < \infty$$

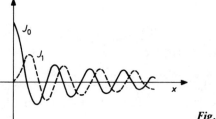

Fig. 6.6

We shall show that $s = 0$ cannot be admitted. If $s = 0$ (73) becomes $x^2 f''(x) + x f'(x) - v^2 f(x) = 0$, a Euler equation whose solutions are $Ax^v + Bx^{-v}$ if $v \neq 0$ (x^r is a solution of $r(r - 1) + r - v^2 = 0$). If $v > 0$, B must vanish if we are to have

$$\lim_{x \to 0} f(x) < \infty$$

but then $Aa^v = 0$ implies that $A = 0$; if $v = 0$ the solutions are $A + B \log x$ which cannot fit unless $A = B = 0$.

Thus there exists a non-zero solution of (73) satisfying the conditions imposed only if $s \neq 0$.

Now let φ be defined over $]0, a/s]$ by $\varphi(z) = f(z/s)$ or $f(x) = \varphi(sx)$ so that $f'(x) = s\varphi'(sx)$ and $f''(x) = s^2 \varphi''(sx)$. Thus

$$s^2 x^2 \varphi''(sx) + sx\varphi'(sx) + (s^2 x^2 - v^2)\varphi(sx) = 0$$

$$\varphi(sa) = 0 \qquad \lim_{x \to 0} \varphi(sx) < \infty$$

At every point $y = sx$, $y^2 \varphi''(y) + y\varphi'(y) + (y^2 - v^2)\varphi(y) = 0$. Then $\varphi(y)$ is proportional to $J_v(y)$: $f(x) = J_v(sx)$, s being determined by the condition $f(a) = 0 = J_v(sa)$. Let $\{b_n\}_{n \geq 1}$ be the sequence of zeros of J_v and s_n the sequence of numbers such that $s_n a = b_n$: the solutions of the problem are thus s_n and $J_v(s_n x)$.

Theorem 3.4.10

Let $v \geq 0$ be fixed, $a > 0$ and $\{b_n\}_{n \geq 1}$ be the sequence of zeros of J_v. Put $s_n = b_n/a$, so that $\forall n \geq 1$ $J_v(s_n a) = 0$. The Sturm–Liouville problem

$$\frac{1}{x} \frac{d}{dx}\left(x \frac{df}{dx}\right) + \left(\lambda - \frac{v^2}{x^2}\right) f(x) = 0 = x^2 f''(x) + xf'(x) + (\lambda x^2 - v^2) f(x)$$

(74)

$$f(a) = 0 \qquad \lim_{x \to 0} f(x) < \infty$$

(75)

has as its solutions

$$\lambda_n = s_n^2 \ (n \geq 1) \quad \text{and} \quad f_n(x) = J_v(s_n x) \ (n \geq 1)$$

If

$$N_{v,n} = \int_0^a J_v^2(s_n x) x \, dx$$

the sequence

$$\left\{ \frac{J_v(s_n x)}{N_{v,n}} \right\}_{n \geq 1}$$

forms an orthonormal basis of $L^2[(0, a), x\,dx]$. Let φ be a function such that

$$\int_0^a \varphi^2(x)x\,dx < \infty$$

and let

$$\alpha_n = \left\langle \varphi, \frac{J_\nu(s_n x)}{N_{\nu,n}} \right\rangle = \int_0^a x\varphi(x)\frac{J_\nu(s_n x)}{N_{\nu,n}}\,dx$$

Then

$$\lim_{p \to \infty} \sum_1^p \alpha_n \frac{J_\nu(s_n x)}{N_{\nu,n}} = \varphi$$

in $L^2[(0, a), x\,dx]$.

If φ and φ' are piecewise continuous over $[0, a]$,

$$\lim_{p \to \infty} \sum_{n=1}^p \frac{J_\nu(s_n x)}{N_{\nu,n}^2} \int_0^a x\varphi(x)J_\nu(s_n x)\,dx = \frac{1}{2}[\varphi(x^+) + \varphi(x^-)]$$

If φ is continuous the convergence is uniform.

We note that $J_\nu(s_n x)$ vanishes at the $n-1$ points

$$x_p = \frac{b_p}{b_n} a$$

for $0 < p \leqslant n - 1$ of $]0, a[$. Between two consecutive zeros of $J_\nu(s_{n+1} x)$ there is a zero of $J_\nu(s_n x)$; in fact we can show that

$$\frac{b_{n+1}}{b_n} < \frac{b_{p+1}}{b_p}$$

if $n > p$ and therefore that

$$\frac{b_p}{b_{n+1}} < \frac{b_p}{b_n} < \frac{b_{p+1}}{b_{n+1}}$$

The pth zero of $J_\nu(s_n x)$ is between the pth and the $(p + 1)$th zeros of $J_\nu(s_{n+1} x)$. It is often useful to know that

$$N_{p,n}^{\;2} = \frac{a^2}{2}[J_{p+1}(s_n a)]^2$$

and in particular that

$$N_{0,n}^{\;2} = \frac{a^2}{2}[J_1(s_n a)]^2$$

Example   Consider a solid cylinder, of radius R, whose points are specified

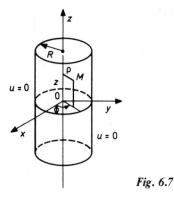

Fig. 6.7

by their cylindrical coordinates Φ, ρ and z where $\Phi \in [0, 2\pi[$ and $p \geqslant 0$ (Fig. 6.7).

We study the temperature in the interior of this cylinder, of which we suppose that the height is infinite (i.e. very large) and the lateral surface is maintained at zero temperature.

We suppose that the initial distribution $u(\rho, 0) = f(\rho)$ is a function of ρ only, so that u is a function only of ρ and t. This temperature $u(\rho, t)$ is a solution of

$$\left. \begin{aligned} \frac{\partial u}{\partial t} &= k\left(\frac{\partial^2 u}{\partial \rho^2} + \frac{1}{\rho}\frac{\partial u}{\partial \rho}\right) \\ u(R, t) &= 0 \\ u(\rho, 0) &= f(\rho) \end{aligned} \right\} \qquad 0 < \rho < R \text{ and } 0 < t$$

We seek a solution by separation of variables, putting $u(\rho, t) = \varphi(\rho)g(t)$. This brings us to the following Sturm–Liouville problem and equation:

$$\rho\frac{d^2\varphi}{d\rho^2} + \frac{d\varphi}{d\rho} - \lambda\rho\varphi(\rho) = 0$$

$$\varphi(R) = 0 \qquad \lim_{\rho \to 0} \varphi(\rho) < \infty \qquad \lim_{\rho \to 0} \frac{d\varphi}{d\rho} < \infty \qquad (76)$$

$$\frac{dg}{dt} - \lambda k g(t) = 0 \qquad (77)$$

Equation (76) can also be written

$$\frac{1}{\rho}\left[\frac{d}{d\rho}\left(\rho\frac{d\varphi}{d\rho}\right)\right] = \lambda\varphi$$

which is a problem of the preceding type for $v = 0$; the eigenvalues are $-s_n^2$ where s_n is such that $J_0(s_n R) = 0$ and the eigenfunctions are $J_0(s_n\rho)$.

Equation (77) has solutions $g(t) = g(0) \exp(-s_n^2 kt)$.

By supposing $f'(\rho)$ to be piecewise continuous (which is physically the usual situation) we can expand $f(\rho)$ in $L^2\{[0, R], x\,dx\}$, and the temperature $u(\rho, t)$ is the uniform and absolute limit of the series

$$u(\rho, t) = \frac{2}{R^2} \sum_{n=1}^{\infty} \frac{J_0(s_n\rho)}{[J_1(s_nR)]^2} \exp(-s_n^2 kt) \int_0^R yf(y)J_0(s_n y)\,dy$$

In the next chapter we shall study the series expansions of the special functions.

See **Exercises 46 and 47** at the end of the chapter.

Exercises

1. Solve the following equation, after finding the Green function or the generalized Green function

$$\frac{d^2f}{dx^2} = g(x) \qquad f(0) = f'(1) = 0$$

2. Repeat Exercise 1 for

$$\frac{d^2f}{dx^2} = g(x) \qquad \begin{cases} f(0) = f'(0) \\ f(1) = f'(1) \end{cases}$$

3. Repeat Exercise 1 for

$$\frac{d^2f}{dx^2} = g(x) \qquad \begin{cases} f(0) = 0 \\ f(1) + f'(1) = 2 \end{cases}$$

4. Repeat Exercise 1 for

$$\frac{d^2f}{dx^2} = g(x) \qquad \begin{cases} f(0) + f(1) = 0 \\ f'(0) + f'(1) = 0 \end{cases}$$

5. Repeat Exercise 1 for

$$\frac{d^2f}{dx^2} = g(x) \qquad f'(0) = 0 = f'(1)$$

6. Repeat Exercise 1 for

$$\frac{d^2f}{dx^2} = g(x) \qquad \begin{cases} f(1) = f(-1) \\ f'(1) = f'(-1) \end{cases}$$

7. Find the Green functions for the following problem. What functions g can be used on the right-hand side?

$$(1 - x^2)\frac{d^2f}{dx^2} - 2x\frac{df}{dx} = 0 \qquad f(0) = 0 = f'(1)$$

8. Repeat Exercise 7 for

$$x\frac{d^2f}{dx^2} + \frac{df}{dx} - \frac{n^2}{x}f = 0 \qquad \begin{cases} f(1) = 0 \\ \lim_{x \to 0} |f(x)| < \infty \end{cases}$$

9. Repeat Exercise 7 for

$$\frac{d}{dx}\left[(1 - x^2)\frac{df}{dx}\right] = 0 \qquad \lim_{x \to \pm 1} |f(x)| < \infty$$

10. Solve the following equation with the conditions indicated:

$$x\frac{d^2f}{dx^2} + \frac{df}{dx} = 0 \quad \text{over } [0, a] \qquad f(a) = 0 \qquad \lim_{x \to 0} f(x) < \infty$$

11. Repeat Exercise 10 for

$$\frac{d^2f}{dx^2} + \frac{df}{dx} = 1 \qquad f(0) = 0 = f(1)$$

12. Repeat Exercise 10 for

$$\frac{d^2f}{dx^2} = \sin x \qquad \begin{cases} f(0) = 0 \\ f(1) + 2f'(1) = 0 \end{cases}$$

13. Repeat Exercise 10 for

$$\frac{d^2f}{dx^2} + 4f = \exp(x) \qquad \begin{cases} f(0) = 0 \\ f'(1) = 0 \end{cases}$$

14. Repeat Exercise 10 for

$$\frac{d^2f}{dx^2} + f = x \qquad f(0) = 0 = f\left(\frac{\pi}{2}\right)$$

15. Repeat Exercise 10 for

$$\frac{d^2f}{dx^2} + \pi^2 f = \cos \pi x \qquad \begin{cases} f(0) = f(1) \\ f'(0) = f'(1) \end{cases}$$

16. The method we have used to seek solutions of

$$\frac{\partial f}{\partial t}(x, t) = \frac{\partial^2 f}{\partial x^2}(x, t) \quad \text{and} \quad \begin{cases} f(x, 0) = g(x) \\ f(0, t) = 0 = f(L, t) \end{cases} \begin{matrix} 0 < x < L \\ 0 < t \end{matrix}$$

is called the **method of separation of variables**.
   (a) Can we apply it to the equation

$$\frac{\partial^2 f}{\partial x^2} + \frac{\partial^2 f}{\partial x \, \partial y} + \frac{\partial^2 f}{\partial y^2} = 0?$$

   (b) Formulate the eigenvalue problem obtained by the method of separation of variables applied to

$$\frac{\partial^2 f}{\partial t^2} = \frac{1}{a^2}\frac{\partial^2 f}{\partial x^2} + F(x)\sin\omega t \quad \text{with } f(0, t) = 0 = f(1, t)$$

(c) Repeat part (b) for

$$x^2\frac{\partial^2 f}{\partial x^2} + y^2\frac{\partial^2 f}{\partial y^2} + ax\frac{\partial f}{\partial x} - by\frac{\partial f}{\partial y} + cf = 0$$

with $f(1, y) = 0 = f(2, y)$ and $f(x, 1) = 0 = f(x, 2)$.

17. (a) Verify by integrating by parts n times that

$$\int_{-1}^{+1} x^p\frac{d^n}{dx^n}[(x^2 - 1)^n]\,dx = 0 \qquad \forall p < n$$

Hence deduce that $\langle P_p, P_q \rangle = 0$ for $p \neq q$ in $L^2(0, 1)$ (P_n stands for the Legendre polynomial of order n). Verify also that

$$\int_{-1}^{1}[P_n(x)]^2\,dx = 1$$

(b) If $v(x) = (x^2 - 1)^n$, prove that

$$(x^2 - 1)v''(x) - 2(n - 1)xv'(x) - 2nv(x) = 0$$

Hence deduce that P_n is a solution of

$$(1 - x^2)\frac{d^2 f}{dx^2} - 2x\frac{df}{dx} + n(n + 1)f(x) = 0$$

18. Let E be the set of sequences of real numbers $U = \{u_p\}_{p \in \mathbb{N}}$ such that all the terms of the sequence U are zero starting from the term of a certain order (depending on U). Prove that E is a vector space.

If $U \in E$, $V \in E$, prove that

$$\langle U, V \rangle = \sum_{p=1}^{\infty} u_p v_p$$

is a scalar product over E.

If $\{U_n\}_{n \in \mathbb{N}}$ is the sequence whose general term is

$$U_n = \left(1, \frac{1}{2}, \frac{1}{3}, \ldots, \frac{1}{n}, 0, 0, \ldots\right)$$

$\{U_n\}_{n \in \mathbb{N}}$ is a sequence in E; prove that it is a Cauchy sequence. Hence deduce that E, with the above scalar product, is not a Hilbert space.

19. Let H be a Hilbert space and let $\{u_n\}_{n \in \mathbb{N}}$ be a sequence of elements of H such that there exists in E an element u such that

$$\lim_{n \to \infty} \|u_n\| = \|u\| \quad \text{and} \quad \lim_{n \to \infty} \langle u_n, u \rangle = \|u\|^2$$

Prove that

$$\lim_{n \to \infty} u_n = u$$

20. Let u and v be two elements in a Hilbert space H such that $|\langle u, v \rangle| = \|u\| \cdot \|v\|$. Prove that u and v are two linearly independent elements of H.

21. Give the Fourier expansions of the following functions.

(a) $f(x) = x$ over $[-\pi, +\pi]$

(b) $f(x) = \begin{cases} 1 & -1 \leqslant x < 0 \\ \frac{1}{2} & x = 0 \\ x & 0 \leqslant x \leqslant 1 \end{cases}$

(c) $f(x) = x^2$ over $[-\pi, +\pi]$

Hence deduce the sum of the series

$$1 - \frac{1}{2^2} + \frac{1}{3^2} - \cdots - \frac{(-1)^n}{n^2} + \cdots$$

(d) $f(x) = ax^2 + bx$ over $[0, 1]$. Hence deduce the sum of the series

$$\sum_{n=1}^{\infty} \frac{1}{n^2}$$

(e) Let $\alpha \notin Z - \{0\}$ and $\cos \alpha x \in [0, \pi]$. Deduce the following results:

$$\frac{\pi}{\sin \pi \alpha} = \frac{1}{\alpha} + \sum_{n=1}^{\infty} (-1)^n \left(\frac{1}{\alpha + n} + \frac{1}{\alpha - n} \right)$$

$$\frac{\pi}{\tan \pi \alpha} = \frac{1}{\alpha} + \sum_{n=1}^{\infty} \left(\frac{1}{\alpha + n} + \frac{1}{\alpha - n} \right)$$

(f) Expand

$$\int_0^x \left(\frac{\pi}{\tan \pi \alpha} - \frac{1}{\alpha} \right) d\alpha$$

in series for $|x| < 1$; deduce, for $|x| < 1$,

$$\log \frac{\sin \pi x}{\pi x} = \sum_1^{\infty} \log \left(1 - \frac{x^2}{n^2} \right)$$

22. Show that

$$\int_0^1 \left(\frac{\sin xy}{y} \right)^2 dy < \infty$$

Prove that the operator A defined over $L^2(0, 1)$ by

$$Af(x) = \int_0^1 \frac{\sin(xy)}{y} f(y) \, dy$$

is a linear operator from $L^2(0, 1)$ into itself. Prove that this operator is not bounded.

23. (a) Let E be a Banach space and A a bounded operator from E into E such that there exists $\alpha < 1$ and n_0 satisfying, $\forall f, g \in E$, $\|A^{n_0}f - A^{n_0}g\| \leqslant \alpha \|f - g\|$. Prove that the equation $Af = f$ has a unique solution in E.

(b)   Let $B \in \mathcal{L}(E, E)$, $g \in E$; consider in E the equation

$$u = \lambda Bu + g \tag{78}$$

where λ is a fixed real number. Writing the equation in the form $u = Au$, prove that (78) has a unique solution if $|\lambda|$ is small enough.

24. Let $K(x, y)$ be a real function continuous over $[0, 1] \times [0, 1]$. With every function f in $L^2(0, 1)$ we associate the function Kf defined by

$$Kf(x) = \int_0^1 K(x, y)f(y) \, dy$$

Show that K defines a bounded linear operator from $L^2(0, 1)$ into itself. Find K^*; under what condition is K self-adjoint?

If $K(x, y) = h(x)g(y)$, describe $\mathcal{R}(K)$ and $\mathcal{K}(K)$.

25. Denote by E_1 the vector space of functions defined over an interval $[a, b]$ and of class C^1. We write

E_2 for the subspace in E_1 of functions such that $f(a) = 0$
E_3 for the subspace in E_2 of functions such that $f(b) = 0$
E_4 for the subspace in E_1 of functions such that $f(b) = kf(a)$ (k fixed)

Consider the operators T_1, T_2, T_3, T_4 defined over E_1, E_2, E_3 and E_4 respectively by

$$\frac{df}{dx} = T_i f \qquad i = 1, 2, 3, 4$$

Determine whether these operators are invertible; if so, determine the inverse operators.

Determine the eigenvalues and eigenfunctions of these operators. If λ is not an eigenvalue of one of the operators, examine whether $(T_i - \lambda I)$ is invertible; if so, calculate this inverse.

26. Denote by F_1 the vector space of functions defined over an interval $[a, b]$ and of class C^2. Denote by F_2 the subspace of functions of F_1 such that $f(a) = 0 = f(b)$ and by F_3 the subspace of functions of F_1 such that $f(a) = \alpha$ and $f'(a) = \beta$ (α and β being two fixed numbers).

Let p_0, p_1 and p_2 be three continuous functions, $p_0(x) > 0$ $\forall x \in [a, b]$. Define three operators by

$$A_i f(x) = p_0(x) \frac{d^2 f}{dx^2} + p_1(x) \frac{df}{dx} + p_2(x) f(x)$$

$Af(x)$ is an element of $C(a, b)$, the space of functions continuous over $[a, b]$.

(a) Consider A_1 as an operator from F_1 into $C(a, b)$. Is A_1 invertible?
(b) Consider A_2 as an operator from F_2 into $C(a, b)$. Is A_2 invertible?
(c) Consider A_3 as an operator from F_3 into $C(a, b)$. Is A_3 invertible?

27. Let $F(x)$ be defined over $[0, A]$ such that $F(x) \neq 0$ $\forall x \in [0, A]$. Let

$$K(x, y) = \begin{cases} F(x)/F(y) & \text{if } 0 \leqslant y \leqslant x \leqslant A \\ 0 & \text{otherwise} \end{cases}$$

Solve the equation

$$\varphi(x) = f(x) + \lambda \int_0^x K(x, y) \varphi(y) \, dy$$

28. Solve over $[0, A]$ the Volterra equations with the following kernels for $y \leqslant x$:

(a) $K(x, y) = x - y$

(b) $K(x, y) = \exp(x - y)$

(c) $K(x, y) = \exp(x^2 - y^2)$

(d) $K(x, y) = \dfrac{\cosh x}{\cosh y}$

29. Solve the following equations:

(a) $\varphi(x) = \exp(x^2) + \displaystyle\int_0^x \exp(x^2 - y^2) \, \varphi(y) \, dy$

(b) $\varphi(x) = \exp(x) + \displaystyle\int_0^x \exp(x - y) \, \varphi(y) \, dy$

(c) $\varphi(x) = \exp(x) \sin x + \displaystyle\int_0^x \frac{2 + \cos x}{2 + \cos y} \, \varphi(y) \, dy$

(d) $\varphi(x) = 1 + x^2 + \displaystyle\int_0^x \frac{1 + x^2}{1 + y^2} \, \varphi(y) \, dy$

30. Solve

$$\varphi(x) = \lambda \int_0^1 \varphi(u) \, du + 1$$

for $\lambda \neq 1$. What happens if $\lambda = 1$?

31. Solve

$$\varphi(x) = \lambda \int_a^b K(x, y)\varphi(y)\,dy + f(x)$$

in the following cases:

(a) $[a, b] = [-1, +1]$ $K(x, y) = x - y$

(b) $[a, b] = [0, 1]$ $K(x, y) = xy$

(c) $[a, b] = [-\pi, +\pi]$ $K(x, y) = x + \sin y$

(d) $[a, b] = [0, 1]$ $K(x, y) = x \exp(y)$

(e) $[a, b] = [0, 1]$ $K(x, y) = \exp(x + y)$

(f) $[a, b] = [-1, +1]$ $K(x, y) = xy$

32. Solve

$$\varphi(x) = \lambda \int_0^1 K(x, y)\varphi(y)\,dy + \cos \pi x$$

where

$$K(x, y) = \begin{cases} (x + 1)y & \text{if } 0 \leqslant x \leqslant y \leqslant 1 \\ (y + 1)x & \text{if } y \leqslant x \leqslant 1 \end{cases}$$

33. Solve

$$\varphi(x) + \lambda \int_0^1 K(x, y)\varphi(y)\,dy = \exp(x)$$

for the kernel

$$K(x, y) = \begin{cases} \sinh x \sinh(y - 1)\dfrac{1}{\sinh 1} & \text{for } 0 \leqslant x \leqslant y \leqslant 1 \\ \sinh y \sinh(x - 1)\dfrac{1}{\sinh 1} & \text{for } y \leqslant x \leqslant 1 \end{cases}$$

34. Solve

$$\varphi(x) + \int_0^\pi K(x, y)\varphi(y)\,dy = \sin x$$

for

$$K(x, y) = \begin{cases} \sin\left(x + \dfrac{\pi}{4}\right)\sin\left(y - \dfrac{\pi}{4}\right) & \text{for } 0 \leqslant x \leqslant y \leqslant \pi \\ \sin\left(y + \dfrac{\pi}{4}\right)\sin\left(x - \dfrac{\pi}{4}\right) & \text{for } y \leqslant x \leqslant \pi \end{cases}$$

35. Solve the following equations:

(a) $\varphi(x) = \int_{-1}^{+1} \exp(\arcsin x)\,\varphi(y)\,dy + \tan x$

(b) $\varphi(x) = \lambda \int_{0}^{2\pi} \sin x \cos y\,\varphi(y)\,dy + \sin x$

(c) $\varphi(x) = \frac{1}{2}\int_{-1}^{+1}\left[x - \frac{1}{2}(3y^2 - 1) + \frac{1}{2}y(3x^2 - 1)\right]\varphi(y)\,dy + 1$

36. Solve, if possible, the following equations:

(a) $\varphi(x) = \lambda \int_{0}^{1} xy^2\varphi(y)\,dy + \alpha x + \beta$

(b) $\varphi(x) = \lambda \int_{0}^{\pi} \cos^2 x\,\varphi(y)\,dy + 1$

(c) $\varphi(x) = \lambda \int_{0}^{2\pi}\left[\frac{1}{\pi}\cos x \cos y + \frac{1}{\pi}\sin 2x \sin 2y\right]\varphi(y)\,dy + \sin x$

37. Solve the following Sturm–Liouville problems:

(a) $\dfrac{d^2f}{dx^2} + \lambda f = 0 \qquad f(0) = 0 = f(\pi)$

(b) $\dfrac{d^2f}{dx^2} + \lambda f = 0 \qquad f(0) = 0 = \dfrac{df}{dx}\left(\dfrac{\pi}{2}\right)$

(c) $\dfrac{d^2f}{dx^2} + \lambda f = 0 \qquad \dfrac{df}{dx}(0) = 0 = \dfrac{df}{dx}(\pi)$

(d) $\dfrac{d^2f}{dx^2} + \lambda f = 0 \qquad f(0) + \dfrac{df}{dx}(0) = 0 = f(1)$

38. Solve the following periodic Sturm–Liouville problems:

(a) $\dfrac{d^2f}{dx^2} + \lambda f = 0 \qquad \begin{cases} f(1) = f(-1) \\ \dfrac{df}{dx}(1) = \dfrac{df}{dx}(-1) \end{cases}$

(b) $\dfrac{d^2f}{dx^2} + \lambda f = 0 \qquad \begin{cases} f(0) = f(2\pi) \\ \dfrac{df}{dx}(0) = \dfrac{df}{dx}(2\pi) \end{cases}$

39. Solve the following Sturm–Liouville problems:

(a) $\dfrac{d}{dx}\left[(2 + x^2)\dfrac{df}{dx}\right] + \lambda f(x) = 0 \qquad\qquad f(1) = f(-1) = 0$

(b) $(1 + x^2)\dfrac{d^2f}{dx^2} + 2(1 + x)\dfrac{df}{dx} + 3\lambda f(x) = 0 \qquad f(0) = 0 = f(1)$

(c) $\dfrac{d}{dx}\left(x^3\dfrac{df}{dx}\right) + \lambda x f(x) = 0 \qquad\qquad f(1) = 0 = f(e)$

40. Let

$$x^2 f''(x) + axf'(x) = \lambda f(x) \tag{79}$$

(a) Solve (79) with $f(1) = f(2) = 0$.

(b) What happens if we require $f(0) = f(1) = 0$? Is the problem still regular? What is the set of eigenvalues?

41. Solve the following problem by separation of variables:

$$\dfrac{\partial u}{\partial t}(x, t) = k\dfrac{\partial^2 u}{\partial x^2}(x, t) \qquad \begin{cases} u(0, t) = 0 \\ u(\pi, t) = a \end{cases}$$

$$u(x, 0) = 0 \quad \text{for } \begin{cases} 0 < x < \pi \\ 0 < t \end{cases}$$

42. Solve the following problem by separation of variables:

$$\dfrac{\partial u}{\partial t}(x, t) = k\dfrac{\partial^2 u}{\partial x^2}(x, t) \qquad \begin{cases} u(0, t) = 0 \\ u(L, t) = 0 \end{cases} \quad 0 < t$$

and

$$\begin{cases} u(x, 0) = A \quad \text{if } 0 < x \leqslant \dfrac{L}{2} \\ u(x, 0) = 0 \quad \text{if } \dfrac{L}{2} < x \leqslant L \end{cases}$$

43. Solve the following problem by separation of variables:

$$\dfrac{\partial u}{\partial t}(x, t) = k\dfrac{\partial^2 u}{\partial x^2}(x, t) \qquad \begin{cases} \dfrac{\partial u}{\partial x}(0, t) = \dfrac{\partial}{\partial x}(L, t) \quad \text{for } 0 < t \\ u(x, 0) = f(x) \qquad\qquad \text{for } 0 < x < L \end{cases}$$

(This concerns the temperature in a bar whose ends are isolated.)

44. Solve the following problem by separation of variables:

$$\dfrac{\partial u}{\partial t}(x, t) = k\dfrac{\partial^2 u}{\partial x^2}(x, t) - hu(x, t) \quad \text{where } h > 0 \text{ is a constant}$$

$$\dfrac{\partial u}{\partial t}(0, t) = \dfrac{\partial u}{\partial t}(L, t) = 0 \qquad\qquad \text{for } 0 < t$$

$$u(x, 0) = f(x) \qquad\qquad \text{for } 0 < x < L$$

This concerns a bar whose ends are isolated but which exchanges heat with the surrounding medium at each point x.

Compare the result with that of Exercise 42.

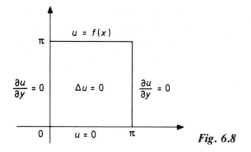

Fig. 6.8

45. Consider a square plate of side π (Fig. 6.8). Solve the following problem by separation of variables:

$$\frac{\partial^2 u}{\partial x^2} + \frac{\partial^2 u}{\partial y^2} = 0 \qquad \frac{\partial u}{\partial y}(0, y) = 0 = \frac{\partial u}{\partial y}(\pi, y)$$

$$u(x, 0) = 0 \quad \text{and} \quad u(x, \pi) = f(x)$$

What happens if $f(x) \equiv a$?

46. Consider a sphere of radius R whose points are specified by their spherical coordinates (Φ, ϑ, r) where $\Phi \in [0, 2\pi]$, $\vartheta \in [0, \pi]$, $r \in [0, R]$ (Fig. 6.9). It is

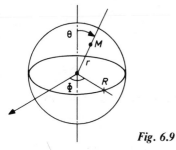

Fig. 6.9

required to solve, in this domain, the equation $\Delta f = 0$ which is written, in spherical coordinates,

$$r \frac{\partial^2}{\partial r^2} [rf(\vartheta, r)] + \frac{1}{\sin \vartheta} \frac{\partial}{\partial \vartheta} \left[\sin \vartheta \frac{\partial f}{\partial \vartheta}(\vartheta, r) \right] = 0$$

(We assume that f does not depend on Φ.) Putting $x = \cos \vartheta$, we seek

$$g(x, r): r \frac{\partial^2}{\partial r^2} [rg(x, r)] + \frac{\partial}{\partial x} \left[(1 - x^2) \frac{\partial g}{\partial x} \right] = 0$$

Suppose that f is sufficiently regular; solve the problem by separation of variables. (For the physical interpretation, see example (c) on page 388.)

47. Consider a circular membrane of radius R whose points are specified by their polar coordinates $\rho \in [0, R]$, $\vartheta \in [-\pi, +\pi]$. Suppose that this membrane is placed in a horizontal plane $z = 0$ and is fixed by its circumference. It is displaced vertically according to a law $g(\rho, \vartheta)$ and its position $f(\rho, \vartheta, t)$ at the instant t is observed (Fig. 6.10).

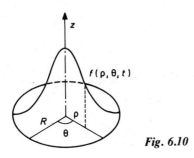

Fig. 6.10

f is a solution of the equation

$$\frac{\partial^2 f}{\partial t^2} = k^2 \left(\frac{\partial^2 f}{\partial \rho^2} + \frac{1}{\rho} \frac{\partial f}{\partial \rho} + \frac{1}{\rho^2} \frac{\partial^2 f}{\partial \vartheta^2} \right)$$

with

$$f(R, \vartheta, t) = 0 \qquad \forall t \geqslant 0$$

$$f(\rho, \vartheta, 0) = g(\rho, \vartheta)$$

$$\frac{\partial f}{\partial t} (\rho, \vartheta, 0) = 0$$

Find, by separation of variables, the solution of this system, supposing g to be sufficiently regular.

Chapter 7

Series expansions

Part 1: Generalities

1.1. Introduction

Let

$$A(x)\frac{d^2f}{dx^2} + B(x)\frac{df}{dx} + C(x)f = 0 \qquad (1)$$

be a linear differential equation of the second order. Suppose that in a neighbourhood of the origin the functions A, B and C can be expanded in series and thus are infinitely differentiable. All the solutions are therefore also infinitely differentiable. It is then natural to look for a solution which can itself be expanded in series; since the equation is linear, we can proceed by substitution. Let

$$f(x) = u_0 + u_1 x + \ldots + u_n x^n + \ldots$$

Then

$$C(x)f(x) = (c_0 + c_1 x + \ldots)(u_0 + u_1 x + \ldots + u_n x^n + \ldots)$$

$$B(x)\frac{df}{dx} = (b_0 + b_1 x + \ldots)[u_1 + 2u_2 x + \ldots + (n+1)u_{n+1}x^n + \ldots]$$

$$A(x)\frac{d^2f}{dx^2} = (a_0 + a_1 x + \ldots)[2u_2 + 3(2u_3 x) + \ldots$$

$$+ (n+2)(n+1)u_{n+2}x^n + \ldots]$$

and necessarily

$$0 = c_0 u_0 + b_0 u_1 + 2a_0 u_2$$

$$0 = (c_1 u_0 + c_0 u_1) + (2b_0 u_2 + b_1 u_1) + [3(2u_3 a_0) + 2u_2 a_1]$$

$$0 = (c_0 u_n + c_1 u_{n-1} + \ldots + c_n u_0) + [(n+1)b_0 u_{n+1} + \ldots + b_n u_1]$$

$$+ [(n+2)(n+1)u_{n+2}a_0 + \ldots + 2a_n u_2]$$

In the simplest case we shall be able to fix u_0 and u_1 arbitrarily, and hence deduce u_2, then $u_3, \ldots$ and u_{n+1} when we know $u_0, u_1, \ldots, u_n$. By choosing $(u_0 = 1, u_1 = 0)$ and then $(u_0 = 0, u_1 = 1)$ we shall obtain two solutions whose Wronskian equals 1 for $x = 0$, so that the two solutions are independent; if f_1 and f_2 are these solutions, all solutions will then be of the form $\lambda_1 f_1 + \lambda_2 f_2$ and may be expanded in series.

However, it can happen that the series expansions of A, B and C are such that the preceding equations cannot be solved in this way; it can also happen that we can find an expansion, i.e. we determine the coefficients $u_0, u_1, \ldots, u_n, \ldots$ step by step, but that the series

$$\sum_{n=0}^{\infty} u_n x^n$$

does not converge for any non-zero value of x.

This is not surprising since an equation as regular as (1) can have solutions which are not bounded if x tends to zero or even are not capable of expansion in series, e.g.

$$x \frac{d^2 f}{dx^2} + \frac{df}{dx} = 0$$

whose solutions are $f(x) = a + b \log x$.

Examples
(a) Let

$$\frac{d^2 f}{dx^2} + xf = 0$$

If there is a solution which can be expanded in series, with the above notation its coefficients must satisfy the relations

$$u_2 = 0 \quad u_0 + (3)(2)u_3 = 0 \quad \ldots \quad (n+3)(n+2)u_{n+3} + u_n = 0$$

We thus obtain two series f_1 and f_2; it is easy to check that they are convergent:

$$f_1(x) = 1 - \frac{x^3}{(3)(2)} + \frac{x^6}{(6)(5)(3)(2)} + \ldots + (-1)^n \frac{x^{3n}}{3n(3n-1)\ldots(6)(5)(3)(2)} + \ldots$$

$$f_2(x) = x - \frac{x^4}{(4)(3)} + \frac{x^7}{(7)(6)(4)(3)} + \ldots$$

$$+ (-1)^n \frac{x^{3n+1}}{(3n+1)3n(3n-2)(3n-3)\ldots(7)(6)(4)(3)} + \ldots$$

Every solution of (1) is of the form $u_0 f_1 + u_1 f_2$.

(b) $$x^2 \frac{d^2 f}{dx^2} + (3x - 1)\frac{df}{dx} + f(x) = 0$$

The coefficients of a series solution must satisfy the relations $u_1 = u_0$, $u_2 = 2u_1, \ldots, (n+1)u_{n+1} = (n+1)^2 u_n$.

Thus $u_n = n! \, u_0$; there is only one series satisfying these conditions

$$f(x) = u_0 \sum_0^\infty n! \, x^n$$

which converges only for $x \equiv 0$.

To simplify the work, we shall confine ourselves to linear equations of the second order. The theorems we shall state can be modified to apply to linear equations of order n. This generalization is simple as far as regular points are concerned (particularly Theorems 2.1.1, 2.1.3 and 2.1.4 in Part 2); it is more intricate and the statements are more complicated for singular points, which we shall study in Part 3.

We recall that if a function of a complex variable is differentiable it is in fact infinitely differentiable and that this is so iff it can be expanded in a **power series**. Moreover, if in a domain containing a disc it is differentiable except at the centre of the disc, we know that it can be expanded in a **Laurent series**. For these reasons, and for others (for example the notion of **analytical continuation**), it appears that the most appropriate framework for our study is the consideration of differential equations in the complex plane.

For convenience we shall first give a short review of functions of complex variables.

1.2. Functions of complex variables

We shall only give a list of the principal concepts and essential properties which we shall be using. We shall denote the complex plane by $\mathbb{C}$.

UNIFORM FUNCTIONS; MANY-VALUED FUNCTIONS

Definition 1.2.1

A function

$$\mathbb{C} \to \mathbb{C}$$

$$z \rightsquigarrow f(z)$$

is **uniform** if for all z in $\mathbb{C}$ the set of values $f(z)$ is a single point. In the converse case it is **many-valued**.

The function $z \rightsquigarrow z^2 = f_1(z)$ which makes the single point $(\rho^2, 2\vartheta + 4k\pi)$ correspond to the point $(\rho, \vartheta + 2k\pi)$ is uniform; the function $z \rightsquigarrow z^{1/n} = f_2(z)$ which makes the points $(\rho^{1/n}, \vartheta/n + 2k\pi/n)$ correspond to the point $(\rho, \vartheta + 2k\pi)$ is many-valued, as also is the function $z \rightsquigarrow \log z = f_3(z)$ which makes the points $\log \rho + i(\vartheta + 2k\pi)$ correspond to the point $(\rho, \vartheta + 2k\pi)$. If we fix k in f_2

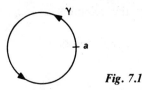

Fig. 7.1

or f_3 we obtain a **branch** of these functions. To study a branch, we can study its variation along a closed curve γ. If a is a point of γ (Fig. 7.1) we shall denote by $f(a)$ the value given by the branch at a, and by $f(a_+)$ this value after we have traversed γ in the positive sense and returned to a. It is clear that, whatever γ may be, $f(a) = f(a_+)$ if the function is uniform, whereas if the function is many-valued the result depends on γ. Let γ be the circle of radius ρ centred at the origin and let f_1, f_2 and f_3 be the above functions with $k = 0$. If $a = (\rho, 0)$, $f_1(a) = f_1(a_+)$ whereas $f_2(a_+) = (\rho^{1/n}, 2\pi/n) \neq f_2(a)$ and $f_3(a_+) = f_3(a) + 2i\pi$. If γ is traversed a second time the values of f_2 and f_3 change again. If γ is a closed curve which does not encircle the origin, $f_1(a) = f_1(a_+)$, $f_2(a) = f_2(a_+)$ and $f_3(a) = f_3(a_+)$.

If we study a many-valued function in the interior of a domain bounded by a closed curve along which the branches do not change, we say that we have defined a **uniform branch** of that function.

We shall not dwell upon these phenomena, but shall merely draw attention to the existence of particular points in the neighbourhood of which we cannot define a uniform branch of a many-valued function.

Note   We recall that a neighbourhood of a point z_0 is a set containing a disc $|z - z_0| < a$ and containing in particular the point z_0 itself.

INTEGRATION

Let γ be a piecewise differentiable function; we define the curvilinear integral

$$\int_\gamma \varphi(x, y)\, dx$$

of a bounded and piecewise continuous function as the limit of

$$\sum_{i=0}^{n} \varphi(x_i, y_i)(x_{i+1} - x_i)$$

where the points $\{M_i = (x_i, y_i)\}$ are points of γ such that

$$\sup_i \text{distance }(M_{i+1}, M_i)$$

tends to zero when n tends to infinity. We can show that this limit exists and does not depend on the subdivision. We define in the same way the integral of a uniform function along a curve γ.

Definition 1.2.2

If γ is a piecewise differentiable curve and f is a continuous function

$$\int_\gamma f(z)\,dz = \lim_{n\to\infty} \sum_{i=0}^n f(z_i)(z_{i+1} - z_i)$$

where z_i is a sequence of points of γ such that

$$\lim_{n\to\infty} \sup_i |z_{i+1} - z_i| = 0$$

If

$$f(z) = P(x, y) + iQ(x, y)$$

$$\int_\gamma f(z)\,dz = \int_\gamma P\,dx - Q\,dy + i \int_\gamma P\,dy + Q\,dx$$

Formally this relation is written

$$\int_\gamma f(z)\,dz = \int_\gamma (P + iQ)(dx + i\,dy)$$

In the same way

$$\int_\gamma f(z)\,dz = \int (P + iQ)(r, \vartheta)\exp(i\vartheta)\,(dr + ir\,d\vartheta)$$

Let a and b be two points in $\mathbb{C}$, and let γ be a curve joining a and b; in general

$$\int_\gamma f(z)\,dz$$

depends on γ.

ANALYTIC FUNCTIONS

Definition 1.2.3

A uniform function is analytic in a neighbourhood of a point z_0 if it is differentiable at that point. We know that it is then of class C^∞.

Theorem 1.2.4

A function is **analytic** in a domain iff it is equal to the sum of its Taylor series:

$$f(z) = f(z_0) + a_0(z - z_0) + a_1(z - z_0)^2 + \ldots + a_n(z - z_0)^n + \ldots$$

where

$$a_n = \frac{1}{n!} f^{(n)}(z_0)$$

The convergence is uniform in every disc $|z - z_0| \leqslant r$ contained in the domain in which f is analytic.

Such series can be differentiated term by term.

If the Taylor expansion in the neighbourhood of z_0 commences with the term $a_p(z - z_0)^p$, we say that z_0 is a zero of order p and $f(z)$ is equal to $(z - z_0)^p[\varphi(z)]$ where φ is non-zero in a neighbourhood of z_0; the zeros of an analytic function are thus isolated.

Theorem 1.2.5 (Cauchy's theorem)

If f is an analytic function in a disc with centre z_0 and if γ is a closed curve contained in this disc

$$\int_\gamma f(z)\,dz = 0$$

Therefore, if a and b are two points of this disc, the integrals of f along all the curves contained in the disc having a and b as end-points are equal.

We can then define a primitive

$$\int_a^z f(u)\,du$$

which is therefore analytic in the disc in question; two primitives

$$\int_{a_1}^z f(u)\,du$$

and

$$\int_{a_2}^z f(u)\,du$$

differ by a constant.

This property can be extended to domains more general than a disc; these are the domains bounded by closed curves and such that any two points of the domain can be joined by a curve situated entirely in the domain and which have no 'holes' (simply connected domains).

Theorem 1.2.6 (Uniqueness theorem)

Let f_1 and f_2 be two functions analytic in a domain D, a a point situated in the interior of D and $\{a_n\}_{n \in \mathbb{N}}$ a sequence of points of D converging to a such that $f_1(a_n) = f_2(a_n)$ $\forall n \in \mathbb{N}$. Then $f_1 \equiv f_2$ in D.

SINGULAR POINTS, LAURENT EXPANSIONS

Theorem 1.2.7

Let f be a function analytic in an annulus $r < |z - z_0| < R$; then, in this annulus, f admits the expansion

$$f(z) = \sum_{-\infty}^{\infty} c_n(z - z_0)^n = \ldots + \frac{c_{-p}}{(z - z_0)^p} + \ldots + \frac{c_{-1}}{z - z_0} + c_0 + c_1(z - z_0) +$$

$$\ldots + c_q(z - z_0)^q + \ldots$$

The coefficients $\{c_p\}_{p \in \mathbb{Z}}$ are unique and the expansion is called the Laurent expansion of f. The convergence is uniform in every closed domain contained within the annulus; it can therefore be differentiated term by term. The series

$$\sum_{n=0}^{\infty} c_n(z - z_0)^n$$

is called the regular part of the expansion. The series

$$\sum_{p=-\infty}^{-1} c_p(z - z_0)^p$$

is called the principal part of the expansion.

Definition 1.2.8

Let f be a function uniform in a domain D and $z_0 \in D$.

(a) z_0 is a **regular point** of f if

$$\lim_{z \to z_0} f(z) = f(z_0)$$

In all other cases z_0 is a **singular** point. z_0 is a **removable singular point** if

$$\lim_{z \to z_0} f(z) = a \quad (|a| < \infty)$$

(b) z_0 is a **pole** of f if

$$\lim_{z \to z_0} |f(z)| = \infty$$

(c) z_0 is an **essential singular point** if

$$\lim_{z \to z_0} f(z)$$

does not exist. A singular point z_0 is called isolated if there exists $\rho > 0$ such that, in the annulus $0 < |z - z_0| < \rho$, f is analytic.

If z_0 is an isolated singular point, f can then be expanded in a Laurent series within an annulus $0 < |z - z_0| < \rho$. Thus z_0 is a removable singular point iff the expansion has no principal part. It is a pole of order p iff the principal part reduces to

$$\frac{c_{-p}}{(z - z_0)^p} + \ldots + \frac{c_{-1}}{z - z_0} \quad (c_{-p} \neq 0)$$

It is an essential singular point iff the principal part has an infinity of non-zero terms.

POINTS AT INFINITY

On putting $\zeta = 1/z$ and $g(\zeta) = f(1/\zeta)$, we can study f in the neighbourhood of infinity by studying g in the neighbourhood of zero; we say that f has a singular point at infinity if g has one at the origin. We define thus a pole or an essential singular point at infinity.

If infinity is an isolated singular point of f there exists a Laurent expansion of f in the annulus $|z| > R$; the nature of this expansion determines in its turn the nature of the singular point at infinity.

Thus, a function which is analytic over the whole plane (including the point at infinity) is necessarily constant.

ANALYTIC CONTINUATION

Definition 1.2.9

Let f be a function analytic in an open domain D_0, and let D_1 be another open domain such that $D_0 \cap D_1 \neq \emptyset$.

We say that f admits an analytic continuation in D_1 if there exists a function g, analytic in D_1, such that $f = g$ in $D_0 \cap D_1$; according to the uniqueness theorem, such a continuation is necessarily unique.

Definition 1.2.10 (Analytic continuation along a curve γ)

Let f_0 be a function analytic in a disc C_0 centred at a_0, and let γ be a curve passing through a_0 (Fig. 7.2). We say that we have obtained an analytic continuation of f along γ if we have found

(a) points $a_1, \ldots, a_n$ of γ and discs $C_1, \ldots, C_n$ centred on these points such that
$\forall i \ C_i \cap C_{i+1} \neq \emptyset$;
(b) functions $f_1, \ldots, f_n$ analytic in $C_1, \ldots, C_n$ respectively such that f_1 is a

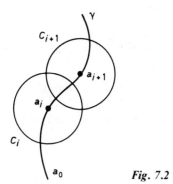

Fig. 7.2

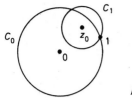

Fig. 7.3

continuation of $f_0, \ldots$ and f_n is a continuation of f_{n-1}. Such a continuation, if it exists, is unique in the following sense: if $b_1, \ldots, b_p$, $D_1, \ldots, D_p$ and $g_1, \ldots, g_p$ is another continuation, if a point z of γ belongs to C_k and D_q then $f_k \equiv g_q$ within $C_k \cap D_q$.

Examples
(a) Let

$$f(z) = 1 + z + z^2 + \ldots + z^n + \ldots$$

and $C_0 = \{|z| < 1\}$. Within C_0, f is analytic and $f = 1/(1-z)$. Let z_0 be an interior point of C_0 other than the origin and C_1 the disc centred at z_0 and passing through the point 1 (Fig. 7.3).

$$\frac{1}{1-z} = \frac{1}{(1-z_0) - (z-z_0)}$$

$$= \frac{1}{1-z_0}\left[1 + \frac{z-z_0}{1-z_0} + \ldots + \frac{(z-z_0)^n}{(1-z_0)^n} + \ldots\right]$$

is defined for $|z - z_0| < |1 - z_0|$ and continues f_0 into C_1. We can define step by step a continuation of f_0 over the whole plane except the point 1.
  (b) Let

$$f_0(z) = \sum_{k=1}^{\infty} z^{2^k} = z^2 + z^4 + z^8 + \ldots$$

analytic in $\{|z| < 1\}$. This function cannot be continued outside this circle. We shall verify this as follows: we shall show that if $z_n = 1^{1/2^n}$ then

$$\lim_{z \to z_n} f_0(z) = \infty$$

so that we cannot construct a continuation in a neighbourhood of z_n, and since every point of $|z| = 1$ is within an arbitrarily small neighbourhood of such a point (provided that n is chosen sufficiently large), continuation is impossible outside $|z| < 1$.
  Let us prove the property stated. If $n = 0$, $1^{1/2^n} = 1$ and

$$\lim_{z \to 1} \sum_{1}^{\infty} z^{2^k} = \infty$$

We note that

$$f_0(z) = z^2 + \sum_{k=1}^{\infty} (z^2)^{2^k} = z^2 + f_0(z^2)$$

Since

$$\lim_{z^2 \to 1} f_0(z^2) = \infty \qquad \lim_{z \to 1^{1/2}} f_0(z) = \infty$$

Similarly, since

$$f_0(z) = z^2 + z^4 + \ldots + z^{2^n} + f_0(z^{2^n})$$

and since

$$\lim_{z^{2^n} \to 1} f_0(z^{2^n}) = \lim_{z \to 1^{1/2^n}} f_0(z^{2^n}) = \infty$$

then

$$\lim_{z \to 1^{1/2^n}} f_0(z) = \infty$$

(c) Let $a_0 = 1$ and $\gamma = \{\exp(i\vartheta), \vartheta \in [0, 2\pi]\}$ (Fig. 7.4). Let $C_0 = |z - 1| < 1$ and

$$f_0(z) = (z - 1) + \frac{(z - 1)^2}{2} + \ldots + \frac{(z - 1)^n}{n} + \ldots$$

f_0 is analytic within C_0. We can continue it along γ by putting $f(z) = \log r + i\vartheta$

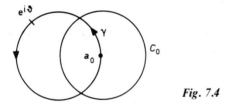

Fig. 7.4

within every circle centred at $\exp(i\vartheta)$ and with radius 1; in this case, in a neighbourhood of $\vartheta = 2\pi$, $f_0(z)$ is continued by $f_0(z) + i\pi$.

ASYMPTOTIC EXPANSION

Definition 1.2.11
If f is a function defined in a domain D containing a curve γ tending to infinity, we say that f admits an asymptotic expansion when z tends to infinity along γ if there exists a sequence $\{a_n\}_{n \in \mathbb{N}}$ such that

$$\lim_{\substack{z \to \infty \\ z \in \gamma}} z^n \left[f(z) - \sum_{k=0}^{n} \frac{a_k}{z^k} \right] = 0 \qquad \forall n \in \mathbb{N}$$

Thus when z is large

$$f(z) - \sum_{k=0}^{n} \frac{a_k}{z^k}$$

is, for each n, infinitesimal in comparison with $1/z^n$.

The series provides for each n an approximation to $f(z)$ which, for fixed n, is closer the larger z is. The series

$$\sum_{k=0}^{\infty} \frac{a_k}{z^k}$$

is called an **asymptotic expansion** of f and we write

$$f \sim \sum_{k=0}^{\infty} \frac{a_k}{z^k}$$

more generally, if

$$\frac{f}{g}(z) \sim \sum_{0}^{\infty} \frac{a_k}{z^k}$$

we write

$$f \sim g \sum_{0}^{\infty} \frac{a_k}{z^k}$$

and we speak again of an asymptotic expansion of f.

It must be noted that it is possible that the series

$$\sum_{k=0}^{\infty} \frac{a_k}{z^k}$$

is not convergent for any value of z.

If it exists, such an expansion is unique on γ, but two different functions can have the same asymptotic expansion. The series which is identically zero is clearly an asymptotic expansion of the function $f \equiv 0$ but it is also an asymptotic expansion along the positive real axis of $\exp(-x)$ since

$$\lim_{x \to \infty} x^n \exp(-x) = 0 \quad \forall n$$

The following properties are easy to prove.

If

$$f \sim \sum_{n=0}^{\infty} \frac{a_n}{z^n}$$

and

$$g \sim \sum_{n=0}^{\infty} \frac{b_n}{z^n}$$

then

$$(f+g)(z) \sim \sum_{n=0}^{\infty} \frac{a_n + b_n}{z^n}$$

and

$$(f \cdot g)(z) \sim \sum_{n=0}^{\infty} \frac{a_0 b_n + \ldots + a_n b_0}{z^n}$$

i.e. the formal expansion of

$$\left(\sum_{k=0}^{\infty} \frac{a_k}{z^k} \right) \left(\sum_{n=0}^{\infty} \frac{b_n}{z^n} \right)$$

Also if

$$f \sim \sum_{n=2}^{\infty} \frac{a_n}{z^n}$$

then

$$\int_{z}^{\infty} f(z) \, dz \sim \sum_{n=2}^{\infty} \frac{a_n}{(n-1)z^{n-1}}$$

where the integral is taken along γ from z to infinity.

In general, we cannot differentiate an asymptotic expansion term by term; e.g. $\exp(-x)[\sin \exp(x)] \sim 0$ over $x > 0$ whereas the derivative $-\exp(-x)\sin \exp(x) + \cos \exp(x)$ has no limit as x tends to ∞.

Example Let

$$f(x) = \int_{-\infty}^{x} \frac{1}{u} \exp(u) \, du$$

be a function defined for $x < 0$ which is called the exponential integral. Integrating by parts

$$f(x) = \left[\frac{1}{u} \exp(u) \right]_{-\infty}^{x} + \int_{-\infty}^{x} \frac{1}{u^2} \exp(u) \, du$$

Then again

$$f(x) = \frac{1}{x} \exp(x) + \left[\frac{1}{u^2} \exp(u) \right]_{-\infty}^{x} + \int_{-\infty}^{x} \frac{2}{u^3} \exp(u) \, du$$

and by recurrence

$$f(x) = \exp(x) \left(\frac{1}{x} + \frac{1!}{x^2} + \frac{2!}{x^3} + \ldots + \frac{p!}{x^{p+1}} \right) + (p+1)! \int_{-\infty}^{x} \frac{\exp(u)}{u^{p+2}} \, du$$

Let

$$x \exp(-x) f(x) = 1 + \frac{1!}{x} + \ldots + \frac{p!}{x^p} + S_p(x)$$

$$S_p(x) = (p+1)! \, x \int_{-\infty}^{x} \frac{\exp(u-x)}{u^{p+2}} \, du$$

$$= \left[(p+1)! \, x \frac{\exp(u-x)}{u^{p+2}} \right]_{-\infty}^{x} + (p+2)! \, x \int_{-\infty}^{x} \frac{\exp(u-x)}{u^{p+3}} \, du$$

so that

$$|S_p(x)| \leqslant (p+1)! \, |x|^{-p-1} + (p+2)! \, |x| \left| \int_{-\infty}^{x} \frac{du}{u^{p+3}} \right|$$

$$\leqslant 2(p+1)! \, |x|^{-(p+1)}$$

Thus

$$x \exp(-x) f(x) \sim \sum_{p=0}^{\infty} \frac{p!}{x^p}$$

this last series being divergent for all non-zero x.

Remark There is of course no difficulty in defining an asymptotic expansion if z tends towards zero instead of tending towards infinity.

See *Exercise 1* at the end of the chapter.

Part 2: Regular points; Fuchs's theorem

2.1. Fundamental theorems

Let

$$\frac{d^2 f}{dz^2} + p(z) \frac{df}{dz} + q(z) f = 0 \qquad (2)$$

Suppose that p and q are analytic in the domain $0 < |z - z_0| < R$ which we shall denote by D. If z_1 is a point of D, we can verify easily that there exists a neighbourhood U of z_1 in which there is a unique solution of (2) such that $f(z_1) = \alpha$ and $f'(z_1) = \beta$. This solution is, of course, analytic in U. The proof is the same as for equations depending on a real variable.

In the same way we can prove that there exists, in the neighbourhood of every point, a fundamental system of solutions f_1 and f_2 as functions of which we can express all the others.

When we wish to extend the domain of definition of the solutions we are led to examine whether the solutions are uniform functions.

If γ is a curve contained in D and f a solution, every point of γ is the centre of a disc, at every point of which the Cauchy problem has a unique solution. We can therefore construct an analytic continuation of f along the whole of γ; since p and q are uniform, the function thus obtained is again a solution of (2).

We shall prove the following result.

Theorem 2.1.1

Let

$$\frac{\mathrm{d}^2 f}{\mathrm{d}z^2} + p(z)\frac{\mathrm{d}f}{\mathrm{d}z} + q(z)f = 0 \tag{3}$$

be an equation in $\mathbb{C}$ where p and q are analytic in the punctured disc $0 < |z - z_0| < R$ which we shall denote by D. Then there exists a solution of (3) of the form $(z - z_0)^r \varphi(z)$ where r is a complex constant and φ is analytic (and therefore uniform) in D.

Proof Let z be a point of D and γ the circle with centre z_0 passing through z (Fig. 7.5). There exist in the neighbourhood of z two independent solutions f_1 and f_2. Traversing γ starting from z in the positive direction, we can define an

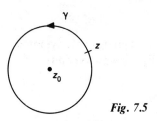

Fig. 7.5

analytic continuation of f_1 and f_2, and when we return to the point z $f_1(z)$ has become $F_1(z)$ and $f_2(z)$ has become $F_2(z)$. F_1 and F_2 are again solutions, as we have remarked. They are independent, for if they were not there would exist two constants λ and μ, not both zero, such that $\lambda F_1(z) + \mu F_2(z) = 0$; by traversing γ in the negative direction we could deduce from this that $\lambda f_1(z) + \mu f_2(z) = 0$ which would contradict the independence of f_1 and f_2.

Since F_1 and F_2 are solutions of (3), there exist constants λ_1, μ_1 and λ_2, μ_2 such that

$$F_1(z) = \lambda_1 f_1(z) + \lambda_2 f_2(z)$$

$$F_2(z) = \mu_1 f_1(z) + \mu_2 f_2(z)$$

or if we denote by M the matrix

$$M = \begin{pmatrix} \lambda_1 & \lambda_2 \\ \mu_1 & \mu_2 \end{pmatrix}$$

$$\begin{pmatrix} F_1 \\ F_2 \end{pmatrix} = M \begin{pmatrix} f_1 \\ f_2 \end{pmatrix}$$

Since F_1 and F_2 are independent, $\det M \neq 0$.

Suppose that there exists a solution having the stated property. If this is $(z - z_0)^r \varphi(z)$, consider the analytic continuation of this solution along γ. After one revolution $\varphi(z)$ takes the same value again whereas $(z - z_0)^r = \exp\{r(\log|z - z_0| + i\vartheta)\}$ becomes $\exp(2i\pi r)\exp(r\log|z - z_0|)$ and is thus multiplied by the constant $K = \exp(2i\pi r)$ (r is, *a priori*, complex).

The proof of this theorem thus reduces to that of the existence of a solution whose value at z, after analytic continuation along γ, is equal to its initial value multiplied by a constant K which we write $\exp(2i\pi r)$.

Let f be a solution

$$f = (f_1, f_2)\binom{a_1}{a_2}$$

After analytic continuation along γ, f has become

$$F = (F_1, F_2)\binom{a_1}{a_2} = (f_1, f_2)\tilde{M}\binom{a_1}{a_2}$$

For $F(z)$ to equal $Kf(z)$ it is necessary and sufficient that

$$K\binom{a_1}{a_2} = \tilde{M}\binom{a_1}{a_2}$$

i.e. that K be an eigenvalue of $\tilde{M}$ and

$$\binom{a_1}{a_2}$$

a corresponding eigenvector.

Since $\tilde{M}$ is invertible there exists at least one non-zero eigenvalue of $\tilde{M}$ which we can write $\exp(2i\pi r)$ and, if $f = a_1 f_1 + a_2 f_2$, $f(z)$ becomes, after continuation along γ, $\exp(2i\pi r) f(z)$. By putting $\varphi(z) = (z - z_0)^{-r} f(z)$, we obtain a function analytic in the neighbourhood of z which, by analytic continuation along γ, becomes $(z - z_0)^{-r} \exp(-2i\pi r) \exp(2i\pi r) f(z)$; it is therefore uniform in D. If there exist two distinct eigenvalues of M we obtain thus a fundamental system of solutions. In all cases we can find a second solution by reducing the order, putting

$$f(z) = g(z)[(z - z_0)^r \varphi(z)]$$

We shall see what forms this new solution can take.

This result is of course, only a statement of existence, since M is not known; it is none the less essential, as much for understanding as for proving Fuchs's theorem.

Since the function $(z - z_0)^r$ has behaviour in the neighbourhood of z_0 which can be quite complex, we are led to the following definitions.

Definition 2.1.2

A point z_0 is called an **ordinary point** of (3) if p and q are analytic in a neighbourhood (including z_0) of that point; it is called singular in the converse case.

A singular point is called **regular** if, for every solution f of (3), there exists a real positive number r_0 such that

$$\lim_{z \to z_0} (z - z_0)^{r_0} f(z) = 0$$

(We assume that $\arg(z - z_0) \in [0, 2\pi[$.)

For the solution $(z - z_0)^r \varphi(z)$ this amounts to assuming that the principal part of the Laurent expansion of φ contains only a finite number of terms:

$$\varphi(z) = \frac{a_{-p}}{(z - z_0)^p} + \ldots + a_0 + a_1 z + \ldots$$

Putting $s = r - p$, we establish that Theorem 2.1.1 affirms that for a regular point there exists a solution of the form $f(z) = (z - z_0)^s \varphi_0(z)$ where φ_0 is analytic for $|z - z_0| < R$; we must then be able, by identifying coefficients, to find s and the coefficients in the expansion of φ_0.

It remains to find under what conditions a point is regular and to give an expression for a second solution independent of the first; this is the object of the following theorems.

By replacing z by $z - z_0$, we can assume that $z_0 = 0$; we shall do this to simplify the writing.

Theorem 2.1.3 (Fuchs's theorem)
The origin $z = 0$ is a regular singular point of the equation

$$\frac{d^2 f}{dz^2} + p(z) \frac{df}{dz} + q(z) f = 0 \tag{4}$$

iff $P(z) = z p(z)$ and $Q(z) = z^2 q(z)$ are analytic in a disc $\{|z| < R\}$. This is equivalent to stating that p and q have within the annulus $0 < |z| < R$ a Laurent expansion of the form

$$p(z) = \frac{u_{-1}}{z} + u_0 + u_1 z + \ldots$$

$$q(z) = \frac{v_{-2}}{z^2} + \frac{v_{-1}}{z} + v_0 + v_1 z + \ldots$$

Theorem 2.1.4
Let

$$z^2 \frac{d^2 f}{dz^2} + z P(z) \frac{df}{dz} + Q(z) f = 0 \tag{5}$$

where P and Q are analytic within the disc $|z| < R$. Let

$$s(s-1) + P(0)s + Q(0) = 0 \tag{6}$$

which is called the **indicial equation**.

Let s_1 be the root with greatest real part.

(a) If (6) has two distinct roots s_1 and s_2 whose difference is not an integer, there exist two independent solutions of (5) of the form $f_1(z) = z^{s_1}\varphi(s)$ and $f_2(z) = z^{s_2}\psi(z)$ where

$$\varphi(z) = 1 + a_1 z + a_2 z^2 + \dots$$

$$\psi(z) = 1 + b_1 z + b_2 z^2 + \dots$$

φ and ψ being analytic in $\{|z| < R\}$ ($\varphi(0) \neq 0$ and $\psi(0) \neq 0$).

(b) If $s_1 = s_2 + n$, $n \in \mathbb{N}$, f_1 is a solution of (5) and there exists a second solution f_2, independent of f_1, of the form

$$f_2(z) = \lambda f_1(z) \log z + z^{s_2}\psi(z)$$

where $\psi(z) = c_0 + c_1 z + \dots$ is analytic in $\{|z| < R\}$ (λ, c_0 and some coefficients c_k can be zero).

(c) If s_1 is a double root of (6), f_1 is a solution of (5) and there exists a solution f_2, independent of f_1, of the form

$$f_2(z) = f_1(z) \log z + z^{s_1+1}\psi(z)$$

where $\psi(z) = d_0 + d_1 z + \dots$ is analytic in $\{|z| < R\}$ (same remark as for (b)).

Proof

(a) *The necessary condition in Theorem 2.1.3*: we know that there exists a solution g_1 of the form $z^s\varphi_0(z)$ where φ_0 is analytic for $|z| < R_0$ (cf. p. 426). We can assume that s is chosen such that $\varphi_0(0) \neq 0$, so that $\varphi_0(z) \neq 0$ for $|z| < R$.

We can find g_2 independent of g_1 by putting $g_2(z) = g_1(z)h(z)$; the function h, which is to be determined, is analytic for $0 < |z| < R$ and its derivative l is a solution of

$$\frac{dl}{dz} + \left[p(z) + 2\frac{g_1'}{g_1}(z) \right] l(z) = 0 \tag{7}$$

Let γ be the circle with centre O passing through a given point z of $0 < |z| < R$. Continuing l along γ we obtain, on returning to z, a function L which is also a solution of (7) since p and g_1'/g_1 are uniform functions. L is therefore proportional to l and there exists a non-zero constant λ such that $L(z) = \lambda l(z)$. If we put $\lambda = \exp(2i\pi r_1)$, we have $l(z) = z^{r_1}\psi(z)$, since ψ is analytic in $0 < |z| < R$. Thus

$$l(z) = \sum_{-\infty}^{+\infty} a_k z^{r_1+k}$$

and h, which is a primitive of l, is of the form

$$h(z) = a \log z + z^{r_1}\eta(z)$$

Since the origin is a regular point, the Laurent expansion of η has only a finite number of terms and thus so also has that of ψ. By changing r_1 we can then assume that ψ and η are analytic for $|z| < R$ and $\psi(z) \neq 0$ in this set:

$$g_2(z) = g_1(z)[a \log z + z^{r_1}\eta(z)]$$

We then obtain the necessary condition by calculating p and q as functions of g_1 and g_2. Thus

$$p(z) = \frac{g_1''g_2 - g_2''g_1}{g_1g_2' - g_2g_1'}(z)$$

$$= -\frac{h''}{h'}(z) - 2\frac{g_1'}{g_1}(z)$$

$$= -\frac{l'}{l}(z) - 2\frac{g_1'}{g_1}(z)$$

$$= -\frac{r_1}{z} - \frac{\psi'}{\psi}(z) - 2\left(\frac{s}{z} + \frac{\varphi'}{\varphi}\right)$$

$$= -\frac{\psi'}{\psi}(z) - 2\frac{\varphi'}{\varphi}(z) - \frac{r_1 + 2s}{z}$$

which has at most a pole of order one at the origin.

The result for q could be proved in the same way.

(b) *The sufficient condition in Theorem 2.1.3 and in Theorem 2.1.4*: we know already the existence of the solution f_1 and the form of a solution f_2 independent of f_1, but since we do not know M it remains to identify the solutions. This will prove Theorem 2.1.4 at the same time and the converse of the theorem preceding it.

We shall show how we can in practice identify these solutions. Let

$$P(z) = \sum_{n=0}^{\infty} p_n z^n \quad \text{and} \quad Q(z) = \sum_{n=0}^{\infty} q_n z^n$$

We know that there exists a solution f_1 of the form

$$f_1(z) = \sum_{n=0}^{\infty} a_n z^{n+s}$$

so that

$$z^2 f_1''(z) = \sum_{n=0}^{\infty} (n+s)(n+s-1)a_n z^{n+s}$$

$$zP(z)f_1'(z) = \left[\sum_{n=0}^{\infty} (n+s)a_n z^{n+s}\right]\left(\sum_{n=0}^{\infty} p_n z^n\right)$$

$$Q(z)f_1(z) = \left(\sum_{n=0}^{\infty} a_n z^{n+s}\right)\left(\sum_{n=0}^{\infty} q_n z^n\right)$$

The coefficient of z^s is thus $a_0[s(s-1) + p_0 s + q_0]$; since a_0 is not zero, s is one of the roots of the indicial equation (6).

Let us denote by s_1 the root of (6) with largest real part and verify that we can find successively $a_1, \ldots, a_n$ by equating to zero the coefficients of $z^{s+1}, z^{s+2}, \ldots$. We shall also verify the convergence of the series obtained within the disc $\{|z| < R\}$.

Suppose that $s_1 \neq s_2$ and $s_1 - s_2$ is not an integer. Let $E(s) = s(s-1) + p_0 s + q_0$. The coefficient of z^{n+s} is

$$E(s+n)a_n + \sum_{k=0}^{n-1} [p_{n-k}(k+s) + q_{n-k}]a_k$$

which determines a_n as a function of $a_0, \ldots, a_{n-k}$ on condition that $E(s+n) \neq 0$.

Now, since s_1 is the root with largest real part, $E(s_1 + n)$ is non-zero for every integer $n > 0$; since $s_1 - s_2$ is not an integer $E(s_2 + n) \neq 0 \;\forall n > 0$. Thus we can find $f_1(z) = z^{s_1}\varphi(z)$ and $f_2(z) = z^{s_2}\psi(z)$.

We shall show that φ is convergent for $|z| < R$. Suppose that $a_0 = 1$. We know that, if $|z| \leq r < R$, $P(z)$ and $Q(z)$ converge and there exists M $(1 < M < \infty)$ such that, $\forall n, |p_n| \leq M/r^n$ and $|q_n| \leq M/r^n$. We shall assume also that $|s_1 p_n + q_n| \leq M/r^n$. The equation enabling us to calculate a_1 is $0 = a_1 E(s+1) + p_1 s_1 + q_1$. Then

$$|a_1||E(s_1 + 1)| = |s_1 p_1 + q_1|$$

Now

$$E(s_1 + 1) = E(s_1) + 2s_1 + p_0$$
$$= 2s_1 + p_0$$
$$= 2s_1 - s_1 - s_2 + 1$$
$$= 1 + s_1 - s_2$$

Thus

$$|a_1| < \frac{M}{r}$$

Suppose that $|a_k| \leq M^k/r^k$ for $1 \leq k \leq n-1$. Then since $E(s_1 + n) = n(n + s_1 - s_2)$

$$|a_n| \leq \frac{1}{n(n + s_1 - s_2)} \left(\sum_{k=1}^{n-1} |a_k| |s_1 p_{n-k} + q_{n-k}| + |s_1 p_n + q_n| + \sum_{k=1}^{n-1} k|p_{n-k}||a_k| \right)$$

$$\leq \frac{1}{n(n + s_1 - s_2)} \left(\sum_{k=1}^{n-1} \frac{M^k}{r^k} \frac{M}{r^{n-k}} + \frac{M}{r^n} + \sum_{k=1}^{n-1} k \frac{M}{r^{n-k}} \frac{M^k}{r^k} \right)$$

and as $M > 1$

$$|a_n| \leqslant \frac{1}{n(n + s_1 - s_2)} n \frac{M^n}{r^n} + \frac{M^n}{r^n} \frac{n(n-1)}{2}$$

$$\leqslant \frac{n+1}{2n} \frac{M^n}{r^n}$$

$$< \frac{M^n}{r^n}$$

The series $\varphi(z)$ is therefore convergent for $|z| \leqslant r < R$.

We could prove in the same way the convergence of ψ and of the various series which appear in the statement.

Remarks

(i) 0 *an ordinary point*: in this case, the indicial equation is $s(s-1) = 0$. We can apply the preceding result and find a fundamental system of solutions $f_1(z) = 1 + a_1 z + a_2 z^2 + \dots$ and $f_2(z) = z + b_2 z^2 + b_3 z^3 + \dots$ ($\lambda = 0$). But we can establish directly that the Cauchy problem has a unique solution for the point zero which is analytic.

(ii) *Real solutions*: if the coefficients q_n and p_n are real for all n, $s_1 = a + ib$ and $s_2 = a - ib$ so that if for x real

$$f_1(x) = x^{a+ib} \sum_{n=0}^{\infty} a_n x^n \qquad f_2(x) = x^{a-ib} \sum_{n=0}^{\infty} \bar{a}_n x^n$$

both $f_1 + f_2$ and $(1/i)(f_1 - f_2)$ are independent real solutions.

(iii) *The point at infinity*: by making the change of variable $z = 1/t$, we define regularity of the point at infinity. If the origin is a regular point of the transformed equation, infinity is a regular point of the initial equation and we can find a solution of the form $(1/z)^q \varphi(1/z)$ where $\varphi(1/z)$ is analytic for $|z| > R$ and a second solution whose form depends on the indicial equation.

We can prove that infinity is a regular point iff, for $|z| > R$, p and q have expansions of the form

$$p(z) = \frac{p_1}{z} + \frac{p_2}{z^2} + \dots \qquad q(z) = \frac{q_2}{z^2} + \frac{q_3}{z^3} + \dots$$

See *Exercises 2 to 5* at the end of the chapter.

2.2. Examples

LEGENDRE'S EQUATION (see Chapter 6, p. 391)

Let us study the following equation, called Legendre's equation:

$$(1 - z^2)\frac{d^2 f}{dz^2} - 2z \frac{df}{dz} + v(v+1)f(z) = 0 \qquad v > 0 \tag{8}$$

Suppose first that $|z| < 1$. In this domain (8) can be written in the form

$$\frac{d^2 f}{dz^2} - \frac{2z}{1-z^2} \frac{df}{dz} + \frac{v(v+1)}{1-z^2} f(z) = 0$$

Let

$$p(z) = -2z \sum_{k=0}^{\infty} z^{2k} \qquad q(z) = v(v+1) \sum_{k=0}^{\infty} z^{2k}$$

We can then apply Theorem 2.1.4 and all points of this domain are ordinary points; all the solutions are therefore analytic. Putting

$$f(z) = \sum_{n=0}^{\infty} a_n z^n$$

and substituting this expression in (8), we obtain

$$(1-z^2) \sum_{n=0}^{\infty} n(n-1)a_n z^{n-2} - 2z \sum_{n=0}^{\infty} na_n z^{n-1} + v(v+1) \sum_{n=0}^{\infty} a_n z^n = 0$$

Equating the coefficients of $z^0, z^1, \ldots, z^n$ to zero we obtain successively

$$2a_2 + v(v+1)a_0 = 0$$
$$(3)2a_3 + (v+2)(v-1)a_1 = 0$$
$$\vdots$$
$$(n+2)(n+1)a_{n+2} + (v+n+1)(v-n)a_n = 0$$

Thus

$$a_{2p} = \frac{(-1)^p}{(2p)!} a_0 v(v-2) \ldots (v-2p+2)(v+1)(v+3) \ldots (v+2p-1)$$

$$a_{2p+1} = \frac{(-1)^p}{(2p+1)!} a_1 (v-2)(v-3) \ldots (v-2p+1)(v+2)(v+4) \ldots (v+2p)$$

We thus obtain two independent solutions:

$$f_1(z) = 1 - \frac{v(v+1)}{2!} z^2 + \frac{v(v-2)(v+1)(v+3)}{4!} z^4 + \ldots$$

$$f_2(z) = z - \frac{(v-1)(v+2)}{3!} z^3 + \frac{(v-1)(v-3)(v+2)(v+4)}{5!} z^5 + \ldots$$

It is clear that, if v is an integer, f_1 or f_2 is a polynomial. We meet again the Legendre polynomials which we discussed in Chapter 6, to which the reader is referred to study the principal properties of the polynomials. Some proofs are proposed as an exercise.

Behaviour at infinity: Legendre functions of the second kind

Infinity is a singular point of Legendre's equation. To study it, we put $u = 1/z$

and $f(z) = f(1/u) = g(u)$. Then

$$f'(z) = f'\left(\frac{1}{u}\right) = -u^2 g'(u)$$

$$f''(z) = f''\left(\frac{1}{u}\right) = u^4 g''(u) + 2u^3 g'(u)$$

Equation (8) becomes for g

$$u^2 g''(u) + \frac{2u^3}{u^2 - 1} g'(u) + \frac{v(v+1)}{u^2 - 1} g(u) = 0 \tag{9}$$

Let

$$P(u) = -2u^2(1 + u^2 + u^4 + \ldots)$$

$$Q(u) = -v(v+1)(1 + u^2 + u^4 + \ldots)$$

These series converge for $|u| < 1$ and $u = 0$ is a regular singular point of (9). Therefore infinity is a regular singular point of Legendre's equation.

The indicial equation is $s(s-1) - (v+1)v = 0$ whose roots are $s_1 = v+1$ and $s_2 = -v$. For $|u| < 1$ there therefore exists a solution of (9) of the form $g_1(u) = u^{v+1}[b_0 + b_1 u + b_2 u^2 + \ldots]$ and hence, for $|z| > 1$ there exists a solution of (8) of the form

$$f_1(z) = \frac{b_0}{z^{v+1}} + \frac{b_1}{z^{v+2}} + \ldots + \frac{b_n}{z^{v+n}} + \ldots$$

We note that if $v = n$ the solution f_2 corresponding to s_2 is not independent of f_1.

Definition 2.2.1

If n is an integer the function f_1 is called a **Legendre function of the second kind**. We denote it by Q_n and referring back to (8) we find that

$$Q_n(z) = \frac{2^n(n!)^2}{(2n+1)! \, z^{n+1}} \left[1 + \frac{(n+1)(n+2)}{2(2n+3)} \frac{1}{z^2} \right.$$

$$\left. + \frac{(n+1)(n+2)(n+3)(n+4)}{(2)4(2n+3)(2n+5)} \frac{1}{z^4} + \ldots \right]$$

If P_n is the Legendre polynomial with index n we can show that P_n and Q_n are independent and form a fundamental system of solutions.

BESSEL'S EQUATION (see Chapter 6, p. 395)

First we recall three of the commonest forms in which Bessel's equation is met (see in particular page 76):

$$z^2 \frac{d^2f}{dz^2} + z \frac{df}{dz} + (z^2 - v^2)f(z) = 0 \qquad v \geqslant 0 \tag{10}$$

For $z = x$, a positive real number, by putting $f(x) = x^{1/2}g(x)$ we obtain

$$\frac{d^2g}{dx^2} + \left(1 + \frac{1 - 4v^2}{4x^2}\right)g(x) = 0 \tag{11}$$

Finally, for all $x > 0$, putting $f(x) = x^v h(x)$ we find

$$x \frac{d^2h}{dx^2} + (2v + 1)\frac{dh}{dx} + xh = 0 \tag{12}$$

We shall study Bessel's equation in the form (10). The singular points are $z = 0$ and $z = \infty$. We can verify that $z = \infty$ is not a regular point (we shall see later (p. 436) how to study this) whereas $z = 0$ is a regular point, from Theorem 2.1.4. The indicial equation is $s^2 - v^2 = 0$, whose roots are $s_1 = v$ and $s_2 = -v$.

Thus there exists a solution of (10) of the form $f_1(z) = z^v(1 + a_1 z + a_2 z^2 + \ldots)$ whose coefficient can be obtained by substituting this expression in (10). We find

$$a_{2p} = \frac{(-1)^p}{2^{2p}p! \,(v + 1)(v + 2)\ldots(v + p)}$$

and

$$a_{2p+1} = 0$$

Naturally we obtain a solution also by putting

$$J_v(z) = f_1(z)\frac{1}{2^v}\frac{1}{\Gamma(v + 1)}$$

We recall that

$$\Gamma(v) = \int_0^\infty x^{v-1}\exp(-x)\,dx$$

We can verify by integrating by parts that $\Gamma(v + 1) = v\Gamma(v)$ and that $\Gamma(n + 1) = n! \;\forall n \geqslant 0,\, n \in \mathbb{N};\, \Gamma(x) \neq 0\;\forall x$ and $\Gamma(-p) = \infty\;\forall p \in \mathbb{N}$. We can thus write the expansion of $J_v(z)$.

Definition 2.2.2

We define a **Bessel function of the first kind** with index v, denoted $J_v(z)$, as the following solution of (10):

$$J_v(z) = \left(\frac{z}{2}\right)^v\left[\frac{1}{\Gamma(v + 1)} - \left(\frac{z}{2}\right)^2\frac{1}{\Gamma(v + 2)} + \left(\frac{z}{2}\right)^4\frac{1}{\Gamma(v + 3)2!} + \ldots\right.$$

$$\left. + (-1)^p\left(\frac{z}{2}\right)^{2p}\frac{1}{p!\,\Gamma(v + p + 1)} + \ldots\right]$$

This series is convergent for all z.

In accordance with Theorem 2.1.4, if $2v \neq n, n \in \mathbb{N}, s_1$ and s_2 are different and $s_1 - s_2$ is not an integer. Thus there exists a second solution f_2 of (10), independent of J_v, of the form

$$f_2(z) = z^{-v}(1 + b_1 z + b_2 z^2 + \ldots)$$

We obtain its coefficients in the same way and we denote by $J_v(z)$ the following function proportional to f_2:

$$J_{-v}(z) = \left(\frac{z}{2}\right)^{-v} \sum_{p=0}^{\infty} (-1)^p \left(\frac{z}{2}\right)^{2p} \frac{1}{p! \, \Gamma(p - v + 1)}$$

We can easily verify that J_v and J_{-v} are again independent if v (not an integer) is equal to $n/2$. Thus we have proved the following theorem.

Proposition 2.2.3

If v is not an integer, Bessel's equation (10) has a fundamental system of solutions formed from the functions J_v and J_{-v}.

J_v is the only solution of (10) which remains bounded when z tends to zero.

Case where $v = n$

In this case $J_n(z)$ still has the same expression but as $\Gamma(p - n + 1) = \infty$ if $p \leqslant n - 1$

$$J_{-n}(z) = \sum_{p=0}^{\infty} \frac{(-1)^{n+p}}{p! \, (n + p)!} \left(\frac{z}{2}\right)^{n+2p}$$

$$= (-1)^n J_n(z)$$

J_{-n} is thus not independent of J_n and it is necessary to find a second solution. Following the general method we can put $N_n(z) = \varphi(z) J_n(z)$. We find for $\varphi(z)$ an expression of the form $\varphi(z) = \log z + \psi(z)$ where ψ is a Laurent series.

Bessel functions of the second kind

Another method of finding a function independent of J_n consists in making use of a fundamental system (J_v, Y_v) for $v \neq n$ and verifying that if v tends to n this system remains a fundamental system.

Proposition 2.2.4

If $v \notin \mathbb{N}$ we define a **Bessel function of the second kind with index v** as the quantity

$$Y_v(z) = \frac{J_v(z) \cos \pi v - J_{-v}(z)}{\sin \pi v}$$

J_v and Y_v form a fundamental system of solutions of (10).

If $v \to n$, Y_v has a limit

$$N_n(z) = \lim_{v \to n} Y_v(z)$$

J_n and N_n form a fundamental system of solutions of (10).

J_n is the only solution of (10) which remains bounded when z tends to zero.

PROPERTIES OF BESSEL FUNCTIONS

We shall indicate only some of the fundamental properties. The proofs of these are for the most part proposed as exercises.

Recurrence relations

$$J_{v+1}(z) + J_{v-1}(z) = \frac{2v}{z} J_v(z) \qquad z \neq 0$$

$$J_{v+1}(z) - J_{v-1}(z) = -2 \frac{dJ_v}{dz}(z)$$

Zeros of Bessel functions

For x real we study the set of x such that $J_v(x) = 0$. We can verify that $J_0(x)$ has a zero in every interval of length π; hence we deduce that $J_n(x)$ has an infinity of zeros for every n, which can be proved also for N_n.

Integral representation

We can show that $\forall n \in \mathbb{N}$ for all real x

$$J_n(x) = \frac{1}{2\pi} \int_{-\pi}^{\pi} \exp(ix \sin \vartheta) \exp(-in\vartheta) \, d\vartheta = \frac{1}{\pi} \int_0^{\pi} \cos(x \sin \vartheta - n\vartheta) \, d\vartheta$$

Hence we deduce that $J_n(x)$ is the nth coefficient in the Fourier expansion of the periodic function $\exp(ix \sin \vartheta)$:

$$\exp(ix \sin \vartheta) = \sum_{p=-1}^{+\infty} J_p(x) \exp(ip\vartheta)$$

which enables us to establish, for example, the formulae

$$J_{2n}(x) = \frac{1}{\pi} \int_0^{\pi} \cos(x \sin \vartheta) \cos 2n\vartheta \, d\vartheta$$

and

$$J_{2n+1}(x) = \frac{1}{\pi} \int_0^{\pi} \sin(x \sin \vartheta) \sin(2n+1)\vartheta \, d\vartheta$$

Behaviour at infinity

We can show that when x (real) tends to $+\infty$, $J_n(x)$ is equivalent to

$$\frac{A_n}{x^{1/2}} \sin(x + \Theta_n)$$

where A_n and Θ_n are constants different for each index n.

The properties indicated in Chapter 6 are also very important; we refer the reader to them.

See *Exercises 6 to 20* at the end of the chapter.

Part 3: Non-regular points; asymptotic expansions

3.1. Non-regular points

Let

$$\frac{d^2f}{dz^2} + p(z)\frac{df}{dz} + q(z)f(z) = 0 \tag{13}$$

In Part 2 we showed how we can use convergent series to find solutions in the neighbourhood of regular singular points.

These points are characterized by the fact that the Laurent expansions of p and q are respectively expressions of the form

$$p(z) = \frac{a_{-1}}{z} + a_0 + a_1 z + \ldots$$

$$q(z) = \frac{b_{-2}}{z^2} + \frac{b_{-1}}{z} + b_0 + b_1 z + \ldots$$

We shall now study singular points for which these Laurent expansions can contain other terms. We shall then be led to use asymptotic expansions, and therefore divergent series; for this purpose we shall introduce the notion of a formal series.

Definition 3.1.1

Let

$$\frac{d^2f}{dz^2} + p(z)\frac{df}{dz} + q(z)f = 0$$

We assume that p and q are analytic in the set $0 < |z| < R$ and that their Laurent expansions are

$$p(z) = \frac{a_{-k-1}}{z^{k+1}} + \frac{a_{-k}}{z^k} + \ldots + \frac{a_{-1}}{z} + a_0 + a_1 z + \ldots$$

$$q(z) = \frac{b_{-2(k+1)}}{z^{2(k+1)}} + \frac{b_{-2k-1}}{z^{2k+1}} + \ldots + \frac{b_{-2}}{z^2} + \frac{b_{-1}}{z} + b_0 + b_1 z + \ldots$$

where $k \geqslant 0$ and at least one of the coefficients a_{-k-1} and $b_{-2(k+1)}$ is non-zero.

We say then that the origin is a singular point of rank k. If $k = 0$ it is a regular singular point or a singular point of the first kind; if $k > 0$ it is a non-regular singular point of rank k or a singular point of the second kind of rank k.

We can of course define the rank of a singularity at z_0 by putting $y = z - z_0$.

Singularity at infinity

By putting $z = 1/u$ and $f(z) = f(1/u) = g(u)$ we define the rank of a singularity at infinity as that of the singularity at zero of the transformed equation, g being a solution of the equation

$$\frac{d^2g}{du^2} - \frac{1}{u^2}\left[p\left(\frac{1}{u}\right) - 2u\right]\frac{dg}{du} + \frac{1}{u^4}q\left(\frac{1}{u}\right)g(u) = 0 \tag{14}$$

We obtain the following result.

Proposition 3.1.2
Infinity is a singular point of order k of the equation (13) iff, for $|z| > R$, p and q have Laurent expansions of the form

$$p(z) = a_{k-1}z^{k-1} + a_{k-2}z^{k-2} + \ldots + a_0 + \frac{a_{-1}}{z} + \ldots$$

$$q(z) = b_{2k-2}z^{2k-2} + b_{2k-3}z^{2k-3} + \ldots + b_0 + \frac{b_{-1}}{z} + \ldots$$

where, naturally, at least one of the coefficients a_{k-1} and b_{2k-2} is non-zero. Infinity is a singular point of order 1 if, for $|z| > R$,

$$p(z) = a_0 + \frac{a_{-1}}{z} + \ldots$$

and

$$q(z) = b_0 + \frac{b_{-1}}{z} + \ldots$$

and one of the numbers a_0 and b_0 is non-zero.

Example Infinity is a non-regular singular point of rank 1 of the equation

$$\frac{d^2f}{dz^2} + \left(1 - \frac{a}{z^2}\right)f(z) = 0$$

or of Bessel's equation with index $v \neq 0$.

3.2. Asymptotic expansions of solutions

FORMAL SERIES; FORMAL SOLUTIONS

Definition 3.2.1
We define a formal series as an expression of the form

$$\sum_{n=-\infty}^{+\infty} a_n z^n$$

which we consider independently of any notion of convergence.

This is in fact a sequence of coefficients a_n. Two formal series

$$\sum_{-\infty}^{+\infty} a_n z^n \quad \text{and} \quad \sum_{-\infty}^{+\infty} b_n z^n$$

are identical if the sequences $\{a_n\}_{n\in\mathbb{N}}$ and $\{b_n\}_{n\in\mathbb{N}}$ are identical.

We define the **sum**, the **product**, the **composite** or the **derivative** of such series as if we were dealing with convergent series. Thus the sum of

$$\sum_{-\infty}^{+\infty} a_n z^n \quad \text{and} \quad \sum_{-\infty}^{+\infty} b_n z^n$$

is the formal series

$$\sum_{-\infty}^{+\infty} (a_n + b_n) z^n$$

The derivative of

$$\sum_{-\infty}^{+\infty} a_n z^n$$

is

$$\sum_{-\infty}^{+\infty} (n+1) a_{n+1} z^n$$

and so on.

Example   The series

$$\sum_{n=0}^{\infty} n! \, z^n$$

which converges only for $z = 0$, considered as a formal series has derivative

$$\sum_{n=0}^{\infty} (n+1)(n+1)! \, z^n$$

and second derivative

$$\sum_{n=0}^{\infty} (n+2)(n+1)(n+2)! \, z^n$$

these two series converging only for $z = 0$.

The asymptotic expansions which we considered in Part 1 are formal series.

Definition 3.2.2
We define a **formal solution** as an expression f in which a formal series

occurs and which is such that

$$\frac{d^2f}{dz^2} + p(z)\frac{df}{dz} + q(z)f(z) = 0$$

the symbol d/dz denoting naturally, for a formal series, the formal derivative.

Example The formal series

$$\sum_0^\infty n!\, z^n$$

is a formal solution of the equation

$$z^2\frac{d^2f}{dz^2} + (3z - 1)\frac{df}{dz} + f(z) = 0$$

(cf. p. 413).

To the end of this chapter, we shall confine ourselves to equations over $\mathbb{R}$ (most of the results are true in $\mathbb{C}$, but the statements are more complicated). The essential result is the following.

Under certain conditions, it can be shown that there exist formal solutions in the neighbourhood of a singular point and that each formal solution is the asymptotic expansion in the neighbourhood of the singular point of an actual solution of the equation.

Mostly, we consider expansions in the neighbourhood of infinity. We shall therefore first state the results for a singular point of the second kind at infinity; then we shall study non-regular singular points at the origin.

We can re-write Eqn (13) in the form

$$\frac{d^2f}{dx^2} + x^{k-1}P(x)\frac{df}{dx} + x^{2(k-1)}Q(x)f(x) = 0 \tag{15}$$

We assume $k \geqslant 1$ and P and Q analytic for $|x| \geqslant R$. Infinity is therefore a singular point of rank k of (15) (Definition 3.1.1).

If we put

$$u(x) = f(x) \quad \text{and} \quad v(x) = x^{-k+1}\frac{df}{dx}$$

this equation is equivalent to the first order system

$$\frac{d}{dx}\begin{pmatrix} u \\ v \end{pmatrix} = x^{k-1}A(x)\begin{pmatrix} u \\ v \end{pmatrix}$$

where

$$A(x) = \begin{bmatrix} 0 & 1 \\ -Q(x) & -(k-1)x^{-k} - P(x) \end{bmatrix}$$

As

$$Q(x) = q_0 + \frac{q_1}{x} + \frac{q_2}{x^2} + \dots$$

and

$$P(x) = p_0 + \frac{p_1}{x} + \frac{p_2}{x^2} + \dots$$

$A(x)$ can be expanded in series in the form

$$A(x) = A_0 + \frac{A_1}{x} + \frac{A_2}{x^2} + \dots$$

where

$$A_0 = \begin{pmatrix} 0 & 1 \\ -q_0 & -p_0 \end{pmatrix}$$

The characteristic equation of A_0 is $\lambda^2 + p_0 \lambda + q_0 = 0$.
Thus we study in fact

$$\frac{dV}{dx} = x^{k-1} A(x) V$$

under the assumption that

$$A(x) = A_0 + \frac{A_1}{x} + \dots$$

We can prove the following theorem.

Theorem 3.2.3
Let

$$\frac{d^2 f}{dx^2} + x^{k-1} P(x) + x^{2(k-1)} Q(x) f(x) = 0$$

where

$$P(x) = p_0 + \frac{p_1}{x} + \dots \quad \text{and} \quad Q(x) = q_0 + \frac{q_1}{x} + \dots$$

are analytic in the neighbourhood of infinity ($x \in \mathbb{R}$). Let λ_1 and λ_2 be the roots of the equation $\lambda^2 + p_0 \lambda + q_0 = 0$. Suppose that $\lambda_1 \neq \lambda_2$. Then there exist two formal solutions of the form

$$\varphi_1(x) = x^{\alpha}\left(a_0 + \frac{a_1}{x} + \dots\right) \exp\left(\lambda_1 \frac{x^k}{k} + \mu_{k-1} x^{k-1} + \dots + \mu_1 x\right)$$

$$\varphi_2(x) = x^{\beta}\left(b_0 + \frac{b_1}{x} + \dots\right) \exp\left(\lambda_2 \frac{x^k}{k} + \nu_{k-1} x^{k-1} + \dots + \nu_1 x\right)$$

where α, β, μ_i and v_i $(1 \leqslant i \leqslant k - 1)$ are complex constants and

$$\sum_{p=0}^{\infty} \frac{a_p}{x^p} \quad \text{and} \quad \sum_{q=0}^{\infty} \frac{b_q}{x^q}$$

are formal series.

Moreover, for every formal solution φ there exists an actual solution f of which φ is the asymptotic expansion in the neighbourhood of infinity.

If $k = 1$

$$\varphi_1(x) = \exp(\lambda_1 x) \, x^\alpha \left(1 + \frac{a_1}{x} + \ldots \right)$$

$$\varphi_2(x) = \exp(\lambda_2 x) \, x^\beta \left(1 + \frac{b_1}{x} + \ldots \right)$$

The corresponding actual solutions are independent.

In the case of a double root there exist formal solutions whose expressions in general contain logarithms. These formal solutions, in a sense to be made more precise, are the asymptotic expansions of actual solutions.

The practical method consists, as in the regular case, of identifying the various coefficients.

Example   Let

$$\frac{d^2 f}{dx^2} + \left(1 - \frac{1}{4x^2} \right) f(x) = 0 \quad \text{for } x > 0 \tag{16}$$

This is a form of Bessel's equation with index $1/\sqrt{2}$ (cf. p. 433), of which one solution is $x^{-1/2} J_{1/\sqrt{2}}(x)$, for which we shall therefore find an asymptotic expansion.

Here

$$P(x) = 0 \qquad Q(x) = 1 - \frac{1}{4x^2}$$

The characteristic equation of A_0 is therefore $\lambda^2 + 1 = 0$ with roots $\lambda_1 = i$ and $\lambda_2 = -i$. Thus we seek a formal solution of the form

$$\varphi_1(x) = \exp(ix) \, x^\alpha \left(1 + \frac{a_1}{x} + \frac{a_2}{x^2} + \ldots \right)$$

By identification we find in the first place that $\alpha = 0$, as the coefficient of $x^{\alpha-1}$ in

$$\frac{d^2 \varphi_1}{dx^2} + \left(1 - \frac{1}{4x^2} \right) \varphi_1(x)$$

is $i\alpha$. We then find in succession $a_1, a_2, \ldots, a_n$ and

$$\varphi_1(x) = \exp(ix)\left\{1 + \frac{i}{2}\frac{1}{4x} + \left(\frac{i}{2}\right)^2 \frac{1}{(4x)^2}\frac{1-(4)(2)}{2!} + \ldots\right.$$

$$\left. + \left(\frac{i}{2}\right)^n \frac{1}{4^n x^n} \frac{[1-4n(n-1)][1-4(n-1)(n-2)]\ldots[1-(4)(2)]}{n!} + \ldots\right\}$$

We see that the series $\exp(-ix)\varphi_1(x)$ is divergent for all x; there exists an actual solution f_1 such that $f_1 \sim \varphi_1$ and a second solution $f_2 \sim \bar{\varphi}_1$.

Hence we deduce the asymptotic behaviour of $J_{1,\sqrt{2}}$ and in general the following result.

Proposition 3.2.4

$$J_\nu(x) \sim \frac{1}{(2\pi)^{1/2}} \exp\left(-\frac{i\pi}{4}\frac{2+\nu}{\nu}\right)\frac{\exp(ix)}{x^{1/2}}\left\{1 + \frac{i}{2}\frac{4\nu^2-1}{4x} + \ldots\right.$$

$$\left. + \left(\frac{i}{2}\right)^n \frac{(4\nu^2-1)(4\nu^2-3)\ldots[4\nu^2-(2n-1)^2]}{(4x)^n} + \ldots\right\}$$

Thus we find the behaviour at infinity:

$$J_\nu(x) \sim \frac{A_\nu}{x^{1/2}}\sin(x+\vartheta_\nu)$$

(see p. 435).

SINGULAR POINTS OF THE SECOND KIND AT THE ORIGIN

To study such a point it is sufficient to put $u = 1/x$ and to study the point at infinity for the new equation.

Example   Take the equation

$$u^2\frac{d^2f}{du^2} + (3u-1)\frac{df}{du} + f(u) = 0 \tag{17}$$

(cf. p. 439). We know that infinity is a regular singular point. Let us examine the origin. By the transformation $u = 1/x$ and $f(u) = f(1/x) = g(x)$, (17) becomes the equation

$$\frac{d^2g}{dx^2} + \left(1-\frac{1}{x}\right)\frac{dg}{dx} + \frac{1}{x^2}g(x) = 0 \tag{18}$$

For this equation infinity is a singular point of rank 1 such that

$$A(x) = \begin{pmatrix} 0 & 1 \\ -\dfrac{1}{x^2} & -1 + \dfrac{1}{x} \end{pmatrix}$$

and the characteristic equation of A_0 is $\lambda(\lambda + 1) = 0$. There exists therefore a formal solution of (18) of the form

$$x^\alpha \left(1 + \frac{a_1}{x} + \frac{a_2}{x^2} + \dots \right)$$

i.e. a solution of (17) of the form

$$\frac{1}{u^\alpha} (1 + a_1 u + a_2 u^2 + \dots)$$

Therefore there exists an actual solution such that

$$\frac{1}{u^n} \lim_{u \to 0} \left[f(u)u^\alpha - (1 + a_1 u + a_2 u^2 + \dots + a_n u^n) \right] = 0 \qquad \forall n \in \mathbb{N}$$

i.e. $f(u) \sim u^{-\alpha}(1 + a_1 u + \dots + a_n u^n + \dots)$ in the neighbourhood of zero.
We shall study this equation in detail as an exercise.

See *Exercises 21 to 24* at the end of the chapter.

Part 4: Series expansions in terms of a parameter

When we study an equation $dx/dt = f(t, x, \lambda)$ where λ is a parameter, we are often led to use series expansions with respect to λ; we shall see examples of this in the next chapter. This part, which is very short, is intended to expound two results both due to Poincaré, as is also the essence of Part 3. These results concern expansions with respect to λ in the neighbourhood of zero and of infinity.

4.1. Series expansions for λ near zero

The following theorem is a consequence of the theorem of differentiability of solutions, a theorem which it generalizes (cf. Chapter 1, p. 50, Theorem 4.1.2).

Theorem 4.1.1 (Poincaré's theorem)

Let $dx/dt = f(t, x, \lambda)$ be an equation in which $x \in \mathbb{R}^n$, $t \in \mathbb{R}$ and $\lambda \in \mathbb{R}$. We assume that

(a) $dx/dt = f(t, x, 0)$ has a solution $\varphi(t)$ with initial value y_0 defined for $|t - t_0| \leqslant T$ and

(b) for $|t - t_0| \leqslant T$, $\|x - \varphi(t)\| < R_0$, $\|\lambda\| \leqslant A$, $f(t, x, \lambda)$ can be expanded in series in terms of these three variables.

Then there exists R such that, if $\|x_0 - y_0\| \leqslant R$ and $\|\lambda\| \leqslant R$, the solution $\varphi(t, t_0, x_0, \lambda)$ of $dx/dt = f(t, x, \lambda)$, such that $x(t_0) = x_0$, can be expanded in series in terms of $(x_0 - y_0)$ and λ over the segment $\|t - t_0\| \leqslant T$.

This theorem is also true if $x \in \mathbb{C}^n$ and $\lambda \in \mathbb{C}^k$; we shall see the use of this in the next chapter.

4.2. Equations containing a large parameter

In Part 3 we studied equations which, in matrix form (p. 440), could be written

$$\frac{dV}{dx} = x^{k-1} A(x) V$$

In particular we studied these equations for x large and $A(x)$ capable of series expansion in the form

$$A(x) = A_0 + \ldots + \frac{A_p}{x^p} + \ldots$$

This enabled us to find asymptotic expansions of the solutions.
In the same way we shall study the equations

$$\frac{dV}{dx} = \lambda^r A(x, \lambda) V$$

for λ large, assuming that

$$A(x, \lambda) = A_0(x) + \frac{A_1(x)}{\lambda} + \ldots + \frac{A_p(x)}{\lambda^p} + \ldots$$

We shall assume that for all x in an interval the eigenvalues of A_0 are distinct, which rules out the delicate (and therefore interesting) case in which the eigenvalues can be equal if $x = x_0$ but otherwise are different (we say that this is an example of a Stokes phenomenon and that x_0 is a transition point).
We shall use formal series in λ

$$\sum_{p=0}^{\infty} X_p(x) \lambda^{-p}$$

whose derivatives are the formal series in λ

$$\sum_{p=0}^{\infty} X_p'(x) \lambda^{-p}$$

(These are derivatives with respect to x.)

By using methods similar to those in Part 3, we can prove the following theorem.

Theorem 4.2.1
Let the following be an equation over the interval $a \leqslant x \leqslant b$:

$$\frac{d^2 f}{dx^2} + \lambda p(x, \lambda) \frac{df}{dx} + \lambda^2 q(x, \lambda) f(x) = 0 \qquad (19)$$

Suppose that $\lambda \in D \subset \mathbb{C}$ where D contains a half-angle, e.g. the half-line $\lambda \geqslant \lambda_0$. We assume that

$$p(x, \lambda) = p_0(x) + \frac{p_1(x)}{\lambda} + \frac{p_2(x)}{\lambda^2} + \dots$$

$$q(x, \lambda) = q_0(x) + \frac{q_1(x)}{\lambda} + \frac{q_2(x)}{\lambda^2} + \dots$$

is D.

Suppose that for all i and all j, $p_i(x)$ and $q_j(x)$ are C^∞ in D and that the series $p(x, \lambda)$ and $q(x, \lambda)$ converge for $\lambda \geqslant \lambda_0$ uniformly in $x \in [a, b]$.

If $r_1(x)$ and $r_2(x)$ are the eigenvalues, assumed to be distinct, of the matrix

$$M_0 = \begin{pmatrix} 0 & 1 \\ -q_0(x) & -p_0(x) \end{pmatrix}$$

and $s_1(x)$ and $s_2(x)$ are primitives of $r_1(x)$ and $r_2(x)$, then there exist formal solutions of (19) of the form

$$\varphi_1(x, \lambda) = \left[\sum_{k=0}^{\infty} a_k(x) \frac{1}{\lambda^k} \right] \exp[\lambda s_1(x)]$$

$$\varphi_2(x, \lambda) = \left[\sum_{k=0}^{\infty} b_k(x) \frac{1}{\lambda^k} \right] \exp[\lambda s_2(x)]$$

Moreover, for each formal solution $\varphi(x, \lambda)$ there exists a family of real solutions $\{f_n(x, \lambda)\}_{n \in \mathbb{N}}$ such that for sufficiently large λ and $\forall x \in [a, b]$

$$\left| f_n(x, \lambda) - \left[\sum_{0}^{n} a_k(x) \frac{1}{\lambda^k} \right] \exp[\lambda^s{}_i(t)] \right| \leqslant \frac{A_n}{|\lambda|^{n+1}} \qquad i = 1, 2$$

Example Let

$$\frac{d^2 f}{dx^2} + [\lambda^2 + g(x)] f = 0 \quad \text{with } g(x) = u_1 x + u_2 x^2 + \dots + u_n x^n + \dots$$

be a convergent series for all x.

$$M_0 = \begin{pmatrix} 0 & 1 \\ \lambda^2 & 0 \end{pmatrix}$$

has eigenvalues $r = \pm i\lambda$ with primitives $\pm i\lambda x$. Therefore there exists a formal solution

$$\varphi_1(x, \lambda) = \exp(i\lambda x)\left[a_0(x) + \frac{a_1(x)}{\lambda} + \ldots \right]$$

with derivative

$$\varphi_1'(x, \lambda) = i\lambda \exp(i\lambda x)\left[a_0(x) + \frac{a_1(x)}{\lambda} + \ldots \right] + \exp(i\lambda x)[a_0'(x) + \ldots]$$

and with second derivative

$$\varphi_1''(x, \lambda) = -\lambda^2 \exp(i\lambda x)[a_0(x) + \ldots] + 2i\lambda \exp(i\lambda x)[a_0'(x) + \ldots]$$
$$+ \exp(i\lambda x)[a_0''(x) + \ldots]$$

By choosing $a_0(x) \equiv 1$ we find step by step the functions $a_n(x)$ by equating the coefficients of λ^p to zero in

$$\left[a_0''(x) + \frac{a_1''(x)}{\lambda} + \ldots \right] + 2i\lambda[a_0'(x) + \ldots] + g(x)[a_0(x) + \ldots]$$

Thus

$$2ia_1'(x) + g(x) = 0$$
$$\vdots$$
$$a_{n-1}''(x) + 2ia_n'(x) + g(x)a_{n-1}(x) = 0$$

Remark   As in the case of non-regular singular points, there are methods of studying the case of double eigenvalues.

See *Exercises 25 and 26* at the end of the chapter.

Exercises

1. Find an asymptotic expansion along $x > 0$ of the following functions:

(a) $\int_x^\infty \frac{1}{u} \exp(x - u)\, dx$

(b) $\int_x^\infty \exp(x^2 - u^2)\, du$

2. Find by expansion in series the solution of the following Cauchy problem and state precisely the domain of convergence:

$$\frac{d^2 f}{dz^2} + z\frac{df}{dz} + f = 0 \qquad f(0) = 1 \qquad \frac{df}{dz}(0) = 2$$

3. For real x find in the form of series (whose convergence conditions are to be given exactly) the solutions of

(a) $(x^2 + 4)\dfrac{d^2f}{dx^2} + 2x\dfrac{df}{dx} - 12f(x) = 0$

(b) $\dfrac{d^2f}{dx^2} - xf(x) = 0$

(c) $\dfrac{d^2f}{dx^2} - f(x) = \sin x$

(d) $x(1 - x)\dfrac{d^2f}{dx^2} + 2(1 - x)\dfrac{df}{dx} + 2f(x) = 0$

4. Study, for real x, the equation

$$4x^4\frac{d^2f}{dx^2} + 4x^3\frac{df}{dx} + (1 - x^2)f(x) = 0$$

Prove that the point at infinity is a regular singular point; calculate the solutions by expansion in series in the neighbourhood of infinity.

Is the origin a regular point?

5. Repeat Exercise 4 for the equation

$$x^2\frac{d^2f}{dx^2} + (3x - 1)\frac{df}{dx} + f(x) = 0$$

Legendre polynomials

6. Prove the Olinde–Rodrigue formula

$$P_n(x) = \frac{1}{2^n n!}\frac{d^n}{dx^n}(x^2 - 1)^n$$

7. Prove directly that the Legendre polynomials form an orthonormal family of $L^2(-1, 1)$.

Gauss's equation

Let

$$z(1 - z)\frac{d^2f}{dz^2} + [\gamma - (\alpha + \beta + 1)z]\frac{df}{dz} - \alpha\beta f(z) = 0 \qquad (20)$$

8. Indicate the singular points of (20) and prove that there exists a solution, analytic in the neighbourhood of zero, which we shall write $F(\alpha, \beta, \gamma, x)$ and which is called a hypergeometric function. Does this give us known results?

(For Exercises 9, 10 and 11 we shall assume that γ, $\gamma - \alpha - \beta$ and $\alpha - \beta$ are not integers.)

9. Find a second solution in the neighbourhood of zero.

10. Repeat Exercise 9 for the point $z = 1$.

11. Repeat Exercise 9 for the point at infinity.

Laguerre's equation (Chapter 6, p. 394); Hermite's equation (Chapter 6, p. 393)

12. Study series expansion of solutions of the Laguerre equation

$$z \frac{d^2 f}{dz^2} + (1 - z) \frac{df}{dz} + \lambda f(z) = 0$$

13. Repeat Exercise 12 for the Hermite equation

$$\frac{d^2 f}{dz^2} - 2z \frac{df}{dz} + \lambda f(z) = 0$$

Bessel functions

14. Prove that for $x > 0$

$$J_{1/2}(x) = \sin x \left(\frac{2}{\pi x} \right)^{1/2} \qquad J_{-1/2}(x) = \cos x \left(\frac{2}{\pi x} \right)^{1/2}$$

15. Prove the recurrence relations

$$J_{\nu+1}(z) + J_{\nu-1}(z) = \frac{2\nu}{z} J_\nu(z)$$

$$J_{\nu+1}(z) - J_{\nu-1}(z) = -2 \frac{dJ_\nu}{dz}(z)$$

16. Prove that

$$\frac{d}{dz} \left(\frac{J_\lambda}{z^\lambda} \right) = -\frac{J_{\lambda+1}}{z^\lambda} \qquad \frac{d}{dz} (z^\lambda J_\lambda) = z^\lambda J_{\lambda-1}$$

Calculate $J_{3/2}(x)$ for real x.

17. Prove that if n is an integer

$$J_{n+1}(z) = -z^n \frac{d}{dz} \left(\frac{J_n}{z^n} \right)$$

Hence deduce that, over $\mathbb{R}^+$, between two consecutive zeros of J_n there exists at least one zero of J_{n+1}. By using the results on the zeros of J_0 describe the set of zeros of J_n.

18. By using the integral representation of J_n prove that

$$J_n(x + y) = \sum_{p = -\infty}^{+\infty} J_p(x) J_{n-p}(y)$$

19. A pendulum consists of a unit mass at the end of a string of variable length $l = gt$. Let ϑ be the angle of the string to the vertical. Small oscillations of the pendulum are solutions of the equation

$$t\frac{d^2\vartheta}{dt^2} + 2\frac{d\vartheta}{dt} + \vartheta = 0 \tag{21}$$

Solve this equation by putting $\varphi(t) = t\vartheta(t)$.

20. Let x be real and positive (cf. p. 435).
  (a) Let $g_n(x) = x^{1/2}J_n(x)$. Of what equation is g a solution? Prove that we can write $g_n(x) = \rho_n(x)\sin\vartheta_n(x)$ and $g_n'(x) = \rho_n(x)\cos\vartheta_n(x)$. Find the equations satisfied by ρ_n and ϑ_n.
  (b) Prove that

$$\lim_{x \to +\infty} \rho_n(x) = A_n < \infty$$

  (c) Let $\psi_n(x) = \vartheta_n(x) - 1$. Find an equation satisfied by ϑ_n. By studying $\psi_n(x) - \psi_n(0)$ prove that

$$\lim_{x \to \infty} \psi_n(x) = B_n$$

Hence deduce an equivalent of g_n when $x \to \infty$.

21. Study the following equation in the neighbourhood of infinity:

$$\frac{d^2f}{dx^2} + \left(1 + \frac{1}{x} + \frac{1}{x^2}\right)f(x) = 0$$

22. Let

$$\frac{d^2f}{dx^2} + \left(1 - \frac{1}{4x^2}\right)f(x) = 0 \tag{22}$$

and let

$$\varphi(x) = \exp(ix) - \frac{1}{4}\int_x^\infty \sin(x-y)\,\varphi(y)\frac{1}{y^2}\,dy \qquad (x > 1) \tag{23}$$

be the integral equation.
  (a) By defining by recurrence the sequence φ_n

$$\varphi_1(x) = \exp(ix) \qquad \varphi_{n+1}(x) = \exp(ix) - \frac{1}{4}\int_x^\infty \sin(x-y)\,\varphi_n(y)\frac{1}{y^2}\,dy$$

prove that (23) has a solution φ such that $|\varphi(x)| \leqslant \exp(\frac{1}{4})$ ($\forall x > 0$).
  (b) Hence deduce that φ, a solution of (23), is also a solution of (22); could this have been foreseen, given the form of formula (23)?
  (c) Prove then that φ has as its asymptotic expansion that of the function $x^{-1/2}J_{1/\sqrt{2}}(x)$.

23. Let

$$x\frac{d^2f}{dx^2}+\frac{df}{dx}+f=0 \qquad (x>0) \tag{24}$$

(a) What is the nature of the point at infinity? Prove that the matrix A_0 has a double eigenvalue.

(b) Putting $x=y^2$, find an equation equivalent to (24) and hence find a formal solution by asymptotic expansion in the neighbourhood of infinity.

(c) Hence deduce a formal solution of (24) in the neighbourhood of infinity. It can be assumed that there exists an actual solution which has this formal solution as its asymptotic expansion.

24. Find in the neighbourhood of zero asymptotic expansions of two independent solutions of the equation, studied on page 439.

$$u^2\frac{d^2f}{du^2}+(3u-1)\frac{df}{du}+f=0 \tag{25}$$

(Solution: we find again the formal solution

$$\varphi_1(u)=\sum_{n=1}^{\infty}n!u^n$$

indicated on page 439 and the actual solution $\varphi_2(u)=(1/u)\exp(-1/u)$.)

25. Study the example on page 445 for $g(x)=x$.

26. State a theorem relating to the solutions of

$$\varepsilon^2\frac{d^2f}{dx^2}+\varepsilon p(x)\frac{df}{dx}+q(x)f=0$$

in the neighbourhood of $\varepsilon=0$.

Chapter 8

Periodic equations and solutions

This chapter is devoted to the study of differential equations

$$\frac{dx}{dt} = f(t, x) \tag{1}$$

where the function f is periodic in t, i.e. $f(t + T) = f(t) \, \forall t$; this of course includes the autonomous case in which $f(t, x)$ does not depend on t.

The study of these equations is of considerable theoretical interest but also corresponds to important practical problems in every situation involving the phenomena of vibration, resonance or the existence of limit cycles.

In the first place, we might enquire whether the periodicity of the right-hand side has implications for the properties of the set of solutions of (1). In the general case, we shall describe only some results, limited but significant, connecting the existence of bounded or stable solutions with the existence of periodic solutions (Theorems 2.1.6, 2.1.7 and 2.1.8). In the case of linear systems with periodic coefficients we shall see that there exists a very intimate relation between linear equations with periodic coefficients and linear equations with constant coefficients, so that the first present global properties characteristic of linear systems, similar to those of the second. Of course if there exists a periodic solution $\varphi(t, t_0, x_0)$ of a homogeneous linear equation, there exists an infinite number of them: $\varphi(t, t_0, \lambda x_0)$.

A second problem is that of the detection of the existence and possibly of the uniqueness of periodic solutions within certain domains and under certain conditions. This can be a question of solutions having the same period as the equation, if it has one, or a multiple of it, or a number close to it, or possibly a different period, as we shall see.

We shall indicate several results of this type and especially with regard to dynamic systems of dimension 2 which are special for the following reason: every closed orbit determines in $\mathbb{R}^2$ two distinct regions and the uniqueness of the solutions implies that a solution whose initial value is in one of these parts remains permanently within it.

We shall indicate finally methods of approximation for periodic solutions. There exist in the literature very many methods, some based on theorems of approximation to the fixed point of a contracting mapping, others, more heuristic, based on methods of expansion in series with respect to 'small' parameters or with respect to certain functions.

These last are in current use in applications but their validity is not always well established. In fact all these methods can be linked with one and the same idea and we have chosen to follow the exposition which J. K. Hale has given of the method perfected by himself and Cesari. This rigorous method is of quite general application; we have given examples of it and we have linked it with certain other classical methods.

Part 1 is devoted to the examination of homogeneous linear systems. We shall first give an account of Floquet's theory and its consequences, in particular with regard to the study of an equation in the neighbourhood of a periodic solution of Eqn (1) by means of the linear variational equation. One section, in particular, is devoted to equations of the second order and to parametric resonance.

In Part 2 we study perturbations of linear equations. With regard to the search for solutions of period T, the situation is different according to whether the unperturbed equation possesses a non-zero solution of period T—the critical case—or does not possess such a solution—the non-critical case. This fact is an aspect of the Fredholm alternative which forms the subject of the first section. In the second section we concern ourselves with the non-critical case.

Sections 2.3 and 2.4 are devoted to the critical case of which the study, following Hale's method, leads to the bifurcation equations (also called the determinant equations). We show how to approximate to the solutions and verify that these approximations are in fact ε-approximate solutions (cf. Chapter 1). Examples are studied at length. Section 2.5 recalls the methods of expansion in series and links them with methods previously expounded.

In Part 3 the principles of the method of centring, or the method of the mean, introduced by Krylov and Bogolyubov, are explained. This method, which is very convenient for obtaining the first-order approximation, is of prime importance because it can be extended to very general situations.

Some of the elements of Poincaré's theory of dynamic systems in $\mathbb{R}^2$ are given in Part 4.

Despite its length this chapter is only an introduction; it has been necessary to omit a certain number of methods. We feel that the exposition which we give is a preparation for the study of these methods, since it presents certain fundamental ideas.

Part 1: Homogeneous linear equations with periodic coefficients

We shall study equations with periodic coefficients of the form

$$\frac{\mathrm{d}x}{\mathrm{d}t} = A(t)x \tag{2}$$

where

$$A: \begin{cases} \mathbb{R} \to \mathscr{L}(\mathbb{R}^d, \mathbb{R}^d) \\ t \rightsquigarrow A(t) \end{cases}$$

is a mapping from $\mathbb{R}$ into the space of continuous linear mappings from $\mathbb{R}^d$ into $\mathbb{R}^d$, so that there exists a unique solution $\varphi(t, t_0, x_0)$ to the Cauchy problem relative to (t_0, x_0) and so that the latter is defined $\forall t \in \mathbb{R}$. We shall suppose that $A(t + T) = A(t)$ $\forall t \in \mathbb{R}$.

These equations have a very close connection with linear equations with constant coefficients; there exists a linear invertible mapping, dependent on t, which transforms every solution of $\mathrm{d}x/\mathrm{d}t = A(t)x$ into a solution of an equation with constant coefficients.

Although we do not know this transformation in general, this result, due to Floquet, has very important consequences relative to the behaviour of the solutions, particularly with regard to their stability and the existence of periodic solutions.

In this connection we recall that in spite of this relationship between equations with periodic coefficients and with constant coefficients, the nature of the eigenvalues of an equation with periodic coefficients gives no information about the behaviour of the solutions (cf. Chapter 3, after Theorem 2.1.2).

Before studying the resolvent of (2), we note that the linear character of this equation implies that if $\varphi(t, t_0, x_0)$ is periodic $\varphi(t, t_0, \lambda x_0)$ is periodic also and that in dimension 2 there can be no limit cycle but there is a centre. In what follows we shall not be interested in the solution $X(t) \equiv 0$ which is of period T $\forall T$.

1.1. Periodic solutions

We shall investigate solutions of period T or $2T$ of Eqn (2).

Let Φ be the resolvent of (2) and x_0 the value for $t = 0$ of a solution of period T; then necessarily $\Phi(T, 0)x_0 = x_0$.

We shall establish that this condition is sufficient, in virtue of the form imposed on Φ by the periodicity of A.

Proposition 1.1.1

$$\forall n \in \mathbb{N}, \forall t \qquad \Phi(t + nT, 0) = \Phi(t, 0)\Phi(nT, 0)$$
$$\forall t, \forall s \qquad \Phi(t + T, s + T) = \Phi(t, s)$$

Proof This results from the uniqueness of the solutions to the Cauchy problem for the equation

$$\frac{\mathrm{d}M}{\mathrm{d}t} = A(t)M \tag{3}$$

where the unknown $M \in \mathscr{L}(\mathbb{R}^d, \mathbb{R}^d)$.

In fact $\tilde{\Phi}(t) = \Phi(t + nT, 0)$ is the solution of (3) such that

$$\tilde{\Phi}(0) = \Phi(nT, 0)$$

It is thus equal to $\Phi(t, 0)\Phi(nT, 0)$. In particular

$$\Phi(nT, 0) = \{\Phi(T, 0)\}^n$$

Moreover

$$\Phi(t + T, s + T) = \Phi(t + T, 0)\Phi(0, s + T)$$

$$= \Phi(t, 0)\Phi(T, 0)\Phi(0, T)\Phi(0, s) = \Phi(t, s)$$

For there to be a periodic solution, it is necessary that 1 should be an eigenvalue of $\Phi(T, 0)$. Then if x_0 is an eigenvector associated with this eigenvalue

$$\varphi(T + t, 0, x_0) = \Phi(T + t, 0)x_0$$

$$= \Phi(t, 0)\Phi(T, 0)x_0$$

$$= \Phi(t, 0)x_0$$

$$= \varphi(t, 0, x_0)$$

and the condition $\Phi(T, 0)x_0 = x_0$ is therefore sufficient.

We note that if x_1 is associated with the possible eigenvalue -1

$$\Phi(2T + t, 0)x_1 = \Phi(t, 0)[\Phi(T, 0)]^2 x_1$$

$$= \Phi(t, 0)x_1$$

and $\varphi(t, 0, x_1)$ has period $2T$.

We have thus proved the following result.

Theorem 1.1.2

Let

$$\frac{\mathrm{d}x}{\mathrm{d}t} = A(t)x \tag{4}$$

where $x \in \mathbb{R}^d$, and let $A(t) \in \mathscr{L}(\mathbb{R}^d, \mathbb{R}^d)$ be continuous with period T. There exists a solution $\varphi(t, 0, x_0)$ of (4) of period T (or of period $2T$) iff 1 (or -1) is an eigenvalue of $\Phi(T, 0)$. The value $\varphi(t, 0, x_0)$ for $t = 0$ is an eigenvector associated with this eigenvalue.

Remark We note in passing that an equation with constant coefficients

$$\frac{dx}{dt} = Ax \tag{5}$$

can have a solution of period T only if 1 is an eigenvalue of $\exp(AT)$, and therefore if there exists an eigenvalue of A of the form $2i\pi n/T$ for an integer n.

CHARACTERISTIC MULTIPLIERS

The following proposition indicates a general property of the eigenvalues of $\Phi(T, 0)$.

Theorem 1.1.3

The eigenvalues of $\Phi(T, 0)$ are termed characteristic multipliers of $A(t)$. Let λ be a characteristic multiplier and x_0 a corresponding eigenvector. If $g(t)$ is the solution of $dx/dt = A(t)x$ such that $g(0) = x_0$, then

$$g(t + nT) = \lambda^n g(t)$$

Thus, there exist solutions of period T or $2T$ iff there exist characteristic multipliers equal to 1 or -1, respectively.

Proof The proof is obvious:

$$g(t + nT) = \Phi(t + nT, 0)x_0$$
$$= \Phi(t, 0)[\Phi(T, 0)]^n x_0$$
$$= \lambda^n g(t)$$

Thus, at every period T the solution is multiplied by λ.

Remarks
(a) g is a real solution only if λ is real. If λ is complex, the real part of g is also a solution but it is not multiplied by a constant at each period T.
(b) We note that, according to Proposition 1.1.8 in Chapter 2, the product of the multipliers is

$$\exp\left[\int_0^T \operatorname{tr} A(s)\,ds\right]$$

EXPANSION IN SERIES OF THE RESOLVENT OF $A(t)$

Since we do not know $\Phi(T, 0)$ we are often led to seek an approximation which will enable us to obtain information about its eigenvalues. This is the case in particular if $A(t)$ can be expanded in series in terms of a parameter ε. Let

$$A(t) = A_0 + \varepsilon A_1(t) + \varepsilon^2 A_2(t) + \ldots + \varepsilon^n A_n(t) + \ldots$$

where $A_p(t + T) = A_p(t)$ for all $p \in \mathbb{N}$ and all t in $\mathbb{R}$, the expansion being uniformly convergent for $\varepsilon \leqslant \varepsilon_0$ and $t \leqslant T$ (and therefore for all t).

Then we know, from Poincaré's theorem (Chapter 7, Theorem 4.1.1), that there exists $\varepsilon_0(T)$ such that, for $\varepsilon \leqslant \varepsilon_0(T)$, $\varphi(t, t_0, x_0)$ can be expanded in series in terms of ε, uniformly in t. The resolvent can therefore be expanded in series, and we shall indicate how this can be done.

If

$$\Phi_\varepsilon(t, 0) = \Phi_0(t) + \varepsilon \Phi_1(t) + \varepsilon^2 \Phi_2(t) + \ldots + \varepsilon^n \Phi_n(t) + \ldots \quad \forall t \in \mathbb{R}$$

$\Phi_0(t, s)$ is the resolvent of A_0 and therefore $\Phi_0(t) = \exp(A_0 t)$. Given the uniform convergence, we can differentiate the above series term by term

$$\frac{\partial}{\partial t} \Phi_\varepsilon(t, 0) = A_0 \exp(A_0 t) + \varepsilon \frac{d\Phi_1}{dt} + \varepsilon^2 \frac{d\Phi_2}{dt} + \ldots + \varepsilon^n \frac{d\Phi_n}{dt} + \ldots$$

Moreover

$$\frac{\partial}{\partial t} \Phi_\varepsilon(t, 0) = A(t) \Phi_\varepsilon(t, 0)$$

$$= [A_0 + \varepsilon A_1(t) + \ldots + \varepsilon^p A_p(t) + \ldots][\exp(A_0 t) + \varepsilon \Phi_1(t) + \ldots]$$

By identifying the coefficients of ε in the two preceding expansions, we find

$$\frac{d\Phi_1}{dt} = A_0 \Phi_1(t) + A_1 \exp(A_0 t)$$

$$\frac{d\Phi_2}{dt} = A_0 \Phi_2(t) + A_1 \Phi_1(t) + A_2 \exp(A_0 t)$$

$$\vdots$$

$$\frac{d\Phi_n}{dt} = A_0 \Phi_n(t) + A_1 \Phi_{n-1}(t) + \ldots + A_{n-1} \Phi_1(t) + A_n \exp(A_0 t)$$

Since $\Phi_\varepsilon(0, 0) = I$ and since $\Phi_0(0) = I$, then $\Phi_n(0) = 0$ $\forall n \geqslant 1$ and

$$\Phi_1(t) = \int_0^t \exp[A_0(t - s)] A_1(s) \exp(A_0 s) \, ds$$

$$\Phi_n(t) = \int_0^t \exp[A_0(t - s)][A_1(s)\Phi_{n-1}(s) + \ldots + A_n(s)\exp(A_0 s)] \, ds \quad \forall n \in \mathbb{N}$$

which enables us to calculate the different coefficients by recurrence.

We often encounter equations of the form

$$\frac{dx}{dt} = [A + \varepsilon P(t)]x \quad \text{where } P(t + T) = P(t) \quad \forall t \in \mathbb{R}$$

We can apply the above procedure to calculate the expansion of $\Phi_\varepsilon(t, 0)$. In this case we find

$$\Phi_1(t) = \int_0^t \exp[A(t - s)] P(s) \exp(As) \, ds \ldots$$

and

$$\Phi_n(t) = \int_0^t \exp[A(t-s)]\, P(s)\Phi_{n-1}(s)\,ds$$

See *Exercise 1* at the end of the chapter.

1.2. Floquet's theory and applications: stability and linearization

We shall establish a link between equations with periodic coefficients and those with constant coefficients. We recall that in Chapter 2 we indicated a formula (due to Dunford) which enables us to associate with every invertible matrix M matrices L such that $\exp(L) = M$. We remarked that those matrices are in general complex. If M is real $M^2 = M \cdot \bar{M} = \exp(L + \bar{L})$ and we can choose P (real) such that $M^2 = \exp(P)$.

Since $\Phi(2T, 0) = [\Phi(T, 0)]^2$, there exists B (real) such that $\Phi(2T, 0) = \exp(2BT)$.

Theorem 1.2.1 (Floquet's theorem)
Let

$$\frac{dx}{dt} = A(t)x \tag{6}$$

where $x \in \mathbb{R}^d$ and $A(t) \in \mathscr{L}(\mathbb{R}^d, \mathbb{R}^d)$ is continuous in t and of period T. Let B (real) be such that $\Phi(2T, 0) = \exp(2BT)$ and $S(t) = \exp(Bt)\Phi(0, t)$. $S(t)$ is invertible and S and S^{-1} are continuous and of period $2T$. Let $y(t) = S(t)x(t)$; y is a solution of the equation

$$\frac{dy}{dt} = By \tag{7}$$

Proof We prove first that y is a solution of (7):

$$\frac{dy}{dt} = B\exp(Bt)\Phi(0, t)x(t) - \exp(Bt)\Phi(0, t)A(t)x(t)$$

$$+ \exp(Bt)\Phi(0, t)\frac{dx}{dt}$$

$$= By(t)$$

It is clear that $S^{-1}(t) = \Phi(t, 0)\exp(-Bt)$ is continuous and of period $2T$ since

$$S^{-1}(t + 2T) = \Phi(t, 0)\Phi(2T, 0)\exp(-2BT)\exp(-Bt) = S^{-1}(t)$$

This theorem enables us to describe the vector space of solutions of (6), but it is particularly important because it enables us to affirm that the coefficients of

$S(t)$ and $S^{-1}(t)$ are bounded and therefore that the solutions x of (6) and y of (7) are of the same nature from the point of view of stability: it is the eigenvalues of B which determine their behaviour.

If ρ is one of these eigenvalues, $\exp(2\rho T)$ is then an eigenvalue of $\Phi(2T, 0) = [\Phi(T, 0)]^2$; this is therefore the square of a multiplier λ and $\exp(\rho T) = \lambda$. If C is such that $\Phi(T, 0) = \exp(CT)$, C is not uniquely defined and not necessarily real (which is why we introduced B); only the real part of its eigenvalues is completely determined. The numbers ρ are equal to the eigenvalues of C, to within $2i\pi n/T$.

Definition 1.2.2
If ρ is an eigenvalue of B, then $\lambda = \exp(\rho T)$ is a characteristic multiplier. If C (not necessarily real) is such that $\Phi(T, 0) = \exp(CT)$, the eigenvalues of C are equal to those of B to within $2i\pi n/T$; we call them **characteristic exponents**. Their multiplicity and their index are the same as those of the eigenvalues of B.

This last result is assured by the Dunford formula already quoted:

$$f(M) = \sum_{i=1}^{r} \sum_{n=0}^{v(\lambda_i)-1} \frac{(M - \lambda_i I)^n}{n!} \frac{d^n f}{dz^n} (\lambda_i) P(\lambda_i)$$

We can then state the following theorem.

Corollary 1.2.3 (Stability)
Let us consider the equation

$$\frac{dx}{dt} = A(t)x \qquad (8)$$

under the preceding conditions.

(a) For (8) to be a uniformly stable system it is necessary and sufficient that all its characteristic multipliers should have modulus less than or equal to 1 and that those of modulus 1 should correspond to non-oscillatory characteristic exponents.

(b) For (8) to be uniformly asymptotically stable it is necessary and sufficient that all the characteristic multipliers should have modulus strictly less than 1, i.e. the characteristic exponents should have strictly negative real parts.

(c) In all other cases the system is unstable.

See *Exercise 2* at the end of the chapter.

Linearization
In Chapter 4 we showed that under certain conditions the stability of a solution $\varphi(t, t_0, x_0)$ of an equation $dx/dt = f(t, x)$ was linked to that of the solution $y(t) \equiv 0$ of the variational linear equation relative to φ. Hence we deduced a corollary in which this equation has constant coefficients (Chapter 4, Theorems 1.2.1 and 1.2.2).

In the same way we can use the above result to discuss the stability of a periodic solution and of an equation whose right-hand term is periodic. We recall, firstly, Proposition 1.1.2 in Chapter 4: if $\varphi(t, t_0, x_0)$ is a solution of

$$\frac{\mathrm{d}x}{\mathrm{d}t} = f(t, x) \tag{9}$$

and if $y(t) = x(t) - \varphi(t, t_0, x_0)$, y is a solution of

$$\frac{\mathrm{d}y}{\mathrm{d}t} = f[t, y + \varphi(t, t_0, x_0)] - f[t, \varphi(t, t_0, x_0)] \tag{10}$$

and $\varphi(t, t_0, x_0)$ is stable (uniformly, asymptotically or asymptotically uniformly) iff $y \equiv 0$ is a stable equilibrium of (10) (uniformly, asymptotically or asymptotically uniformly respectively).

To study (10) we can write it in the form

$$\frac{\mathrm{d}y}{\mathrm{d}t} = \frac{\partial f}{\partial x}[t, \varphi(t, t_0, x_0)]y + g(t, y)$$

and compare it with the equation

$$\frac{\mathrm{d}y}{\mathrm{d}t} = \frac{\partial f}{\partial x}[t, \varphi(t, t_0, x_0)]y$$

by using Theorems 1.2.1 or 1.2.2; if g satisfies the condition

$$\forall \varepsilon > 0 \quad \exists \eta > 0: \|y\| \leqslant \eta \Rightarrow \|g(t, y)\| \leqslant \varepsilon \|y\| \quad \forall t$$

then the solution $y \equiv 0$ of $\mathrm{d}y/\mathrm{d}t = My + g(t, y)$ is asymptotically stable when M is, and unstable if M is unstable.

Consider then $\mathrm{d}x/\mathrm{d}t = A(t)x + f(t, x)$ where A has period T and f satisfies the condition

$$\forall \varepsilon_1 > 0 \quad \exists \eta_1 > 0: \|x\| \leqslant \eta_1 \Rightarrow \|f(t, x)\| \leqslant \varepsilon_1 \|x\|$$

If $S(t)$ is the transformation $\exp(Bt)\Phi(0, t)$ and $y(t) = S(t)x(t)$, then

$$\frac{\mathrm{d}y}{\mathrm{d}t} = By(t) + S(t)f[t, S^{-1}(t)y(t)] = By(t) + g[t, y(t)]$$

Since S and S^{-1} are bounded, we put

$$\alpha = \sup_{0 \leqslant t \leqslant 2T} \|S(t)\| \quad \text{and} \quad \beta = \sup_{0 \leqslant t \leqslant 2T} \|S^{-1}(t)\| \quad 0 < \alpha < \infty, 0 < \beta < \infty$$

and

$$\|g[t, y(t)]\| \leqslant \alpha \|f\{t, S^{-1}[y(t)]\}\|$$

If

$$\|y(t)\| \leqslant \frac{\eta_1}{\beta}$$

then

$$\|S^{-1}y(t)\| \leqslant \eta_1$$

Therefore

$$\|g[t, y(t)]\| \leqslant \alpha\varepsilon_1 \|S^{-1}[y(t)]\| \leqslant \alpha \cdot \beta\varepsilon_1 \|y\|$$

We can thus apply Theorem 1.2.2 to $dy/dt = By$ and $dy/dt = By + g(t, y)$.

Now a solution of $dx/dt = A(t)x + f(t, x)$ has the same nature as the corresponding solution of $dy/dt = By + g(t, y)$. Thus we obtain the following results:

Theorem 1.2.4
Let

$$\frac{dx}{dt} = A(t)x + f(t, x) \tag{11}$$

where A is continuous and periodic and f satisfies the condition

$$\forall \varepsilon > 0 \quad \exists \eta > 0 \colon \|x\| \leqslant \eta \Rightarrow \|f(t, x)\| \leqslant \varepsilon \cdot \|x\|$$

Let

$$\frac{dx}{dt} = A(t)x \tag{12}$$

(a) If (12) is uniformly asymptotically stable, so also is the solution $x \equiv 0$ of (11).

(b) If (12) is unstable the solution $x \equiv 0$ of (11) is also unstable.

Theorem 1.2.5 (Linearization for periodic equations)
Let

$$\frac{dx}{dt} = f(t, x) \tag{13}$$

where f (of class C^2) is such that $f(t + T, x) = f(t, x)$ for all t in $\mathbb{R}$ and x in $\mathbb{R}^d$. Suppose that there is a periodic solution of (13), and let

$$\frac{dz}{dt} = \frac{\partial f}{\partial x}[t, \varphi(t)]z \tag{14}$$

be the linear variational equation of (13) relative to φ. Then

(a) If (14) is uniformly asymptotically stable (if all the characteristic multipliers are of modulus less than 1), φ is a uniformly asymptotically stable solution of (13).

(b) If there exists a multiplier of modulus greater than 1, φ is an unstable solution of (13).

We note that for an autonomous system, when $\varphi(t)$ is a solution of period T of $dx/dt = F(x)$, $d\varphi/dt$ is a periodic solution of the linear variational equation relative to φ since

$$\frac{d}{dt}\left(\frac{d\varphi}{dt}\right) = DF[\varphi(t)]\frac{d\varphi}{dt}$$

There exists therefore a characteristic multiplier of $DF[\varphi(t)]$ of modulus 1.

We note also that in this case, if $t \leadsto \varphi(t)$ is a solution, for all ϑ $t \leadsto \varphi(t + \vartheta)$ is also a solution and that the orbits are identical, so that we are more interested in the stability of this orbit. To specify this situation precisely we introduce the notion of orbital stability.

Definition 1.2.6 (Orbital stability)

Let $dx/dt = f(t, x)$ be an equation in $\mathbb{R}^d$ where f is of class C^1 and of period T. Let φ be a solution (not a constant) of period S of this equation, and let γ be the orbit of φ.

We shall denote by $d(x_0, \gamma)$ the distance from a point x_0 to the closed curve γ.

(a) φ is called **orbitally stable** if

$$\forall \varepsilon > 0 \quad \exists \eta > 0: d[x(t_0), \gamma] < \eta \Rightarrow d\{\varphi[t, t_0, x(t_0)], \gamma\} < \varepsilon \quad \forall t \geqslant t_0$$

(b) If φ is orbitally stable and in addition if there exists $\eta_1 > 0$ such that

$$d[x(t_0), \gamma] < \eta_1 \Rightarrow \lim_{t \to \infty} d\{\varphi[t, t_0, x(t_0)], \gamma\} = 0$$

φ will be called **asymptotically orbitally stable**.

(c) If φ is asymptotically orbitally stable and if for every solution such that $d[x(t_0), \gamma] < \eta_1$ there exists a constant ϑ such that

$$\lim_{t \to \infty} \{\varphi[t, t_0, x(t_0)] - \varphi(t + \vartheta)\} = 0$$

we say that φ is **asymptotically orbitally stable with asymptotic phase**.

It is clear that even in the third case φ is unstable in the ordinary sense. The following two theorems due to Lyapunov can be proved.

Theorem 1.2.7
Let

$$F:\begin{cases} \mathbb{R}^d \to \mathbb{R}^d \\ x \leadsto F(x) \end{cases}$$

be a mapping of class C^1 and let

$$\frac{dx}{dt} = F(x) \tag{15}$$

Let φ be a solution of (15), not a constant, of period T. Let

$$\frac{dz}{dt} = DF[\varphi(t)]z \tag{16}$$

be the linear variational equation relative to φ. Suppose that the $n-1$ multipliers of $DF[\varphi(t)]$ which do not correspond to φ are of modulus strictly less than 1. Then φ is asymptotically orbitally stable with asymptotic phase.

Theorem 1.2.8

Let

$$\frac{dx}{dt} = F(x) + G(x, \varepsilon) \tag{17}$$

where DF and $\partial G(x, \varepsilon)/\partial x$ are continuous. Suppose that $G(x, 0) = 0$ and that $dx/dt = F(x)$ has a periodic solution φ, not constant, of period T and of orbit Γ.

If the $n - 1$ characteristic multipliers of $DF[\varphi(t)]$ which do not correspond to φ are different from 1, there exists $\varepsilon_0 \geqslant 0$ and a neighbourhood G of Γ in which, for every $\varepsilon \leqslant \varepsilon_0$, there is a periodic solution $\psi(t, \varepsilon)$ of (17), of period $T(\varepsilon)$, continuous in (ε, t) and such that $\psi(t, 0) = \varphi(t)$ and $T(0) = T$. If these $n - 1$ multipliers are not of the form $\exp(in\pi)$, $n \in \mathbb{N}$, $\psi(t, \varepsilon)$ is the only periodic solution of (17) in G. If their modulus is strictly less than 1, φ is asymptotically orbitally stable with asymptotic phase; if one of them is greater than 1, φ is unstable.

See *Exercises 3 and 4* at the end of the chapter.

1.3. Equations of the second order; parametric resonance

In this section we shall study equations of the form

$$\frac{d^2x}{dt^2} + f(t)x = 0 \tag{18}$$

where $x \in \mathbb{R}$ and f is a real function continuous over all $\mathbb{R}$ and of period T.

We often meet these equations under the name of **Hill equations**. We note that if

$$\frac{d^2x}{dt^2} + g(t)\frac{dx}{dt} + h(t)x = 0 \tag{19}$$

$$x(t) = u(t)\exp\left[-\frac{1}{2}\int_0^t g(s)\, ds\right]$$

and

$$f = h - \frac{g^2}{4} - \frac{g'^2}{2}$$

Equation (19) is equivalent to

$$\frac{d^2u}{dt^2} + f(t)u = 0$$

but in general x and u do not have the same behaviour with regard to stability or periodicity.

Equation (18) is equivalent to the system

$$\frac{dX}{dt} = \begin{bmatrix} 0 & 1 \\ -f(t) & 0 \end{bmatrix} X$$

We shall denote by $A(t)$ the matrix

$$A(t) = \begin{bmatrix} 0 & 1 \\ -f(t) & 0 \end{bmatrix}$$

The first important statement is that $\operatorname{tr} A(t) = 0$ and therefore that the volume is constant in the phase space (Chapter 2, Proposition 1.1.8 and Theorem 1.1.9). Such a system is called **conservative**.

In the present case we are concerned with the surface of the parallelogram defined by two solutions whose initial values are two independent vectors.

From this fact there result two important consequences.

(a) The solution $X(t) \equiv 0$ cannot be asymptotically stable; therefore no solution can be asymptotically stable.

(b) Let Φ be the resolvent of Eqn (18):

$$\det \Phi(t, s) = \exp \int_s^t [\operatorname{tr} A(u)] \, du = 1 \tag{20}$$

Therefore $\det \Phi(T, 0) = 1$, which establishes the following proposition.

Proposition 1.3.1

The characteristic multipliers of (20) are the roots of the equation $\lambda^2 - \operatorname{tr}[\Phi(T, 0)]\lambda + 1 = 0$.

Consequently, if $|\operatorname{tr}[\Phi(T, 0)]| > 2$ there exist two distinct eigenvalues with product 1; therefore there is a multiplier with modulus greater than 1 and the system is unstable. However, if $|\operatorname{tr}[\Phi(T, 0)]| < 2$ the multipliers are conjugate complex numbers with product 1 and therefore with modulus 1; since they are distinct the roots are simple and the system is stable.

We have thus proved the following result which is basic in the study of the systems in question.

Theorem 1.3.2
Consider the equation

$$\frac{d^2x}{dt^2} + f(t)x = 0$$

in $\mathbb{R}$, where f is continuous and of period T. Let Φ be the resolvent.

(a) If $|\mathrm{tr}[\Phi(T, 0)]| < 2$, the system is uniformly stable and all its solutions are bounded.

(b) If $|\mathrm{tr}[\Phi(T, 0)]| > 2$, the system is unstable and there are unbounded solutions.

(c) If $|\mathrm{tr}[\Phi(T, 0)]| = 2$, there exists a solution of period T or $2T$.

In no case can the system be asymptotically stable.

The study of the stability of Hill equations amounts to the study of the roots of the above equation. There are a considerable number of criteria and of results. We shall first give a result which is simple and easily proved.

Theorem 1.3.3
Take the conditions of Theorem 1.3.2 and assume that f is not identically zero.

(a) If $\forall t \in \mathbb{R} \; f(t) \leqslant 0$, then (20) is unstable.

(b) If

$$0 \leqslant \int_0^T f(t)\,dt$$

and

$$\int_0^T |f(t)|\,dt \leqslant \frac{4}{T}$$

Equation (20) is stable and all its solutions are bounded.

Proof

(a) Let φ be the solution of $d^2x/dt^2 = -f(t)x$ such that $\varphi(0) = 1$ and $\varphi'(0) = 1$. If $f(t) \leqslant 0$ for t near zero, $\varphi''(t) \geqslant 0$. Therefore φ' is increasing, as is φ, and $\forall t \; \varphi'(t) \geqslant 1$. Hence

$$\lim_{t \to \infty} \varphi(t) = +\infty$$

(b) We shall show that under the conditions indicated there cannot be a real characteristic multiplier.

In fact if λ were a real multiplier and x_0 a corresponding eigenvector, then $\varphi(t) = \Phi(t, 0)x_0$ would be a real solution of (20) such that

$$\varphi(t + T) = \lambda\varphi(t): \quad \varphi(T) = \lambda\varphi(0) \qquad \varphi'(T) = \lambda\varphi'(0)$$

(i) If $\forall t \; \varphi(t) \neq 0$ then

$$\frac{\varphi'(0)}{\varphi(0)} = \frac{\varphi'(T)}{\varphi(T)}$$

and

$$0 < \int_0^T \left(\frac{\varphi'}{\varphi}\right)^2 (s)\, ds = -\left[\varphi' \times \frac{1}{\varphi}\right]_0^T + \int_0^T \frac{\varphi''}{\varphi} (s)\, ds$$

$$= \int_0^T -f(s)\, ds$$

which is absurd since

$$0 \leqslant \int_0^T f(s)\, ds$$

(ii) Suppose then that $\varphi(t)$ vanishes. In this case it vanishes an infinite number of times and two consecutive zeros are separated at most by T since $\varphi(t + T) = \lambda\varphi(t)$. Let a and b be two such zeros; for simplicity suppose $0 \leqslant a < b \leqslant T$ and, for example, $\varphi(t) > 0$ for $a < t < b$. Let

$$M = \sup_{a \leqslant t \leqslant b} |\varphi(t)|$$

Then

$$\frac{4}{T} \geqslant \int_0^T |f(s)|\, ds$$

$$\geqslant \int_a^b \left|\frac{\varphi''}{\varphi}\right| (s)\Big|\, ds$$

$$> \frac{1}{M} \int_a^b |\varphi''(s)|\, ds$$

$$\geqslant \frac{1}{M} \left|\int_\alpha^\beta \varphi''(s)\, ds\right|$$

for every pair α, β such that $0 \leqslant \alpha < \beta \leqslant b$.

Thus for every pair α, β such that $a \leqslant \alpha < \beta \leqslant b$, $4M/T > |\varphi'(\beta) - \varphi'(\alpha)|$. Let c be a point of $]a, b[$ at which $M = \varphi(c)$ and

$$\psi(t) = \varphi(t) + \frac{M}{a - c}(t - c)$$

Since $\psi(a) = M = \psi(c)$ there exists $\alpha : a < \alpha < c$ such that $\psi'(\alpha) = 0$ and therefore $\varphi'(\alpha) = M/(c - a)$. In the same way there exists $\beta : c < \beta < b$ such that

$\varphi'(\beta) = M/(c - b)$. Then

$$\frac{4}{T} > \frac{1}{c-a} + \frac{1}{b-c}$$

$$= \frac{b-a}{(c-a)(b-c)}$$

Now

$$(b-a)^2 - 4(b-c)(c-a) = (b-c+c-a)^2 - 4(b-c)(c-a)$$

$$= [(b-c)-c-a)]^2$$

$$\geqslant 0$$

Therefore

$$\frac{T(b-a)}{(b-c)(c-a)} \geqslant \frac{(b-a)^2}{(b-c)(c-a)} \geqslant 4$$

since $T \geqslant b - a$, which contradicts the preceding relation. It is therefore absurd to suppose that there exists a real characteristic multiplier.

There exist numerous results of the same type, more or less elaborate. More generally, if we are interested in equations of the form

$$\frac{d^2 x}{dt^2} + f(t, \alpha, \beta)x = 0 \tag{21}$$

we are led to examine the values of α and β such that (21) is stable or unstable and thus to make the domains of stability obvious.

We shall meet an example of this in a simple case.

Parametric resonance for the equation $d^2 x/dt^2 + \omega^2[1 + \varepsilon a(t)]x = 0$ *where a is continuous and of period* T

In matrix form this equation may be written

$$\frac{dX}{dt} = \begin{bmatrix} 0 & 1 \\ -\omega^2[1 + \varepsilon a(t)] & 0 \end{bmatrix} X$$

We shall denote its resolvent by $\Phi_\varepsilon(t, s)$ and shall study, for ε small, tr $\Phi_\varepsilon(T, 0)$.

We shall call the system

$$\frac{d^2 x}{dt^2} + \omega^2 x = 0$$

the unperturbed system. Its solutions are of the form

$$x(t) = x(0) \cos \omega t + \frac{x'(0)}{\omega} \sin \omega t$$

Their period is $2\pi/\omega$ and their frequency $\omega/2\pi$, which is called the natural frequency of the system.

For $\varepsilon = 0$ the system is stable. If Φ_0 is its resolvent

$$\Phi_0(T, 0) = \begin{pmatrix} \cos \omega T & \dfrac{1}{\omega} \sin \omega T \\ -\omega \sin \omega T & \cos \omega T \end{pmatrix}$$

and

$$\operatorname{tr} \Phi_0(T, 0) = 2 \cos \omega T$$

Let $v/2\pi = 1/T$, the frequency of variations in a. Suppose that

$$\frac{v}{2\pi} \neq \frac{\omega}{2\pi} \times \frac{2}{n}$$

i.e. $\omega \neq n\pi/T \ \forall n \in \mathbb{N}$. Then for $\varepsilon \leqslant \varepsilon_0 \ |\operatorname{tr} \Phi_\varepsilon(T, 0)| < 2$ (since $\omega T \neq n\pi$); the system is stable and all the solutions are bounded.

However, if $v = 2\omega/n$ there exist, in general, values of ε, as close as we wish to zero, such that $|\operatorname{tr} \Phi_\varepsilon(T, 0)| > 2$. For these values the system is unstable and thus there exist solutions which are not bounded; we say that for these values we have **parametric resonance**. If the frequency of variation of a is twice the natural frequency the phenomenon is particularly acute.

It is customary to represent the domains of stability and of instability of the equation in the (ε, ω) plane. We obtain in general domains having the form indicated in Fig. 8.1.

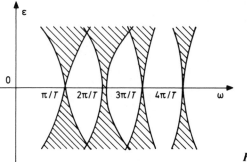

Fig. 8.1

The exact determination of these zones is almost always very intricate. However, we can often find a good approximation to them for small ε by calculating the various terms in the expansion of $\Phi_\varepsilon(t, 0)$ in powers of ε.

We recall that if (ω, ε) is a point on one of the boundary curves of these zones, there exists for the corresponding equation a solution of period $2\pi/\omega$. The study of the boundaries of the stability zones is also that of the existence of periodic solutions.

Example (Mathieu's equation)   We are concerned with the equation

$$\frac{d^2x}{dt^2} + (a + \varepsilon \cos t)x = 0 \tag{22}$$

We shall carry out a complete study only for $a = \omega^2 > 0$. The equation is then equivalent to $dX/dt = AX + \varepsilon P(t)X$ where

$$A = \begin{pmatrix} 0 & 1 \\ -\omega^2 & 0 \end{pmatrix} \quad \text{and} \quad P(t) = \begin{pmatrix} 0 & 0 \\ -\cos t & 0 \end{pmatrix}$$

We shall calculate an approximation to $\Phi_\varepsilon(t, 0)$ by the method of Section 1.1. Thus let

$$\Phi_\varepsilon(t, 0) = \exp(At) + \varepsilon \Phi_1(t) + \varepsilon^2 \Phi_2(t) + \dots$$

We shall calculate $\exp(A2\pi) + \varepsilon \Phi_1(2\pi) + \varepsilon^2 \Phi_2(2\pi)$, the beginning of the expansion of $\Phi_\varepsilon(2\pi, 0)$, and then an approximation to the second order in ε of $\text{tr}\,\Phi_\varepsilon(2\pi, 0)$. We know that

$$\exp[A(t - s)] = \begin{pmatrix} \cos \omega(t - s) & \dfrac{1}{\omega} \sin \omega(t - s) \\ -\omega \sin \omega(t - s) & \cos \omega(t - s) \end{pmatrix}$$

and

$$\text{tr}\,\exp(A2\pi) = 2 \cos 2\omega\pi$$

Let us calculate

$$\Phi_1(t) = \int_0^t \exp[A(t - s)]\,P(s)\,\exp(As)\,ds$$

$$= \int_0^t \begin{pmatrix} -\dfrac{1}{\omega} \sin \omega(t - s) \cos \omega s \cos s & -\dfrac{1}{\omega^2} \sin \omega(t - s) \sin \omega s \cos s \\ -\cos \omega(t - s) \cos \omega s \cos s & -\dfrac{1}{\omega} \cos \omega(t - s) \sin \omega s \cos s \end{pmatrix} ds$$

Thus

$$\text{tr}\,\Phi_1(2\pi) = -\frac{1}{\omega}\int_0^{2\pi} \sin 2\pi\omega \cos s\,ds = 0$$

for every value of ω.

$$\Phi_2(t) = \int_0^t \exp[A(t - s)]\,P(s)\Phi_1(s)\,ds$$

We wish to calculate only the trace of $\Phi_2(2\pi)$; for this it is sufficient to calculate two terms. Writing

$$\Phi_1(s) = \begin{bmatrix} \alpha(s) & \beta(s) \\ \gamma(s) & \delta(s) \end{bmatrix}$$

it is sufficient to calculate the expression

$$\operatorname{tr} \Phi_2(2\pi) = -\frac{1}{\omega} \int_0^{2\pi} [\sin \omega(2\pi - s)\alpha(s) \cos s$$

$$+ \omega\beta(s) \cos \omega(2\pi - s) \cos s] \, ds$$

Now we find, on integrating the expression $\Phi_1(t)$ above, that

$$\alpha(s) = -\frac{1}{4\omega}\left[2 \sin s \sin \omega s + \frac{1}{2\omega + 1} \cos(\omega + 1)s \right.$$

$$\left. + \frac{1}{2\omega - 1} \cos(\omega - 1)s \right] + \frac{1}{4\omega^2 - 1} \cos \omega s$$

$$\omega\beta(s) = -\frac{1}{4\omega}\left[-2 \sin s \cos \omega s + \frac{1}{2\omega + 1} \sin(\omega + 1)s \right.$$

$$\left. + \frac{1}{2\omega - 1} \sin(\omega - 1)s \right] - \frac{1}{4\omega^2 - 1} \sin \omega s$$

Substituting these expressions we find

$$\operatorname{tr} \Phi_2(2\pi) = \frac{1}{4\omega^2} \int_0^{2\pi} 2 \sin s \cos s[\sin \omega s \sin \omega(2\pi - s)$$

$$- \cos \omega s \cos \omega(2\pi - s)] \, ds$$

$$- \frac{1}{\omega}\frac{1}{4\omega^2 - 1} \int_0^{2\pi} \cos s[\cos \omega s \sin \omega(2\pi - s)$$

$$- \sin \omega s \cos \omega(2\pi - s)] \, ds$$

$$+ \frac{1}{4\omega^2}\frac{1}{2\omega + 1} \int_0^{2\pi} \cos s[\cos(\omega + 1)s \sin \omega(2\pi - s)$$

$$+ \sin(\omega + 1)s \cos \omega(2\pi - s)] \, ds$$

$$+ \frac{1}{4\omega^2}\frac{1}{2\omega - 1} \int_0^{2\pi} \cos s \cos(\omega - 1)s \sin \omega(2\pi - s)$$

$$+ \sin(\omega - 1)s \cos \omega(2\pi - s) \, ds$$

It is clear that the first integral is zero. We can prove, by expanding its terms, that the second is also zero. The third is equal to

$$\frac{2\pi \sin 2\pi\omega}{8\omega^2(2\omega + 1)}$$

and the fourth to

$$\frac{2\pi \sin 2\pi\omega}{8\omega^2(2\omega - 1)}$$

Thus

$$\operatorname{tr} \Phi(2\pi) = 2 \cos(2\pi\omega) + \frac{\pi\varepsilon^2}{\omega(4\omega^2 - 1)} \sin(2\omega\pi)$$

(a) First let $\omega \approx 0$: $\operatorname{tr} \Phi(2\pi) \sim 2(1 - 2\omega^2\pi^2) - \pi\varepsilon^2 2\pi(1 + k\omega^2) < 2$. Therefore in the neighbourhood of zero the system is stable (for $a = \omega^2 > 0$).

(b) Let $\omega = \frac{1}{2} + \alpha$: $\operatorname{tr} \Phi(2\pi) \sim - 2(1 - 2\alpha^2\pi^2) - \pi^2\varepsilon^2$. Thus $\operatorname{tr} \Phi(2\pi) = - 2$ iff $4\alpha^2\pi^2 = \pi^2\varepsilon^2$, or $\alpha = \pm\varepsilon/2$ or $\omega = \frac{1}{2} \pm \varepsilon/2$ and $a = \omega^2 = \frac{1}{4} \pm \varepsilon/2$.

(c) Let $\omega = 1 + \beta$: $\operatorname{tr} \Phi(2\pi) \sim 2 - 4\beta^2\pi^2 + 2\pi^2(\varepsilon^2/3)\beta$. Then $\operatorname{tr} \Phi(2\pi) = 2$ if

$$\beta^2 - \beta\frac{\varepsilon^2}{6} = 0$$

which implies that $\beta = 0$ or $\beta = \varepsilon^2/6$. $\beta = 0$ is the first-order approximation to the curve $\beta(\varepsilon)$, whereas for $\beta = \varepsilon^2/6$ the expression for $\operatorname{tr} \Phi(2\pi)$ which we have calculated involves ε^4. To obtain the second-order approximation to $\beta(\varepsilon)$ we must therefore calculate the term in ε^4 in the expression of $\Phi_\varepsilon(s, 0)$. We should then find

$$\beta^2 - \beta\frac{\varepsilon^2}{6} + \frac{5\varepsilon^2}{(24)^2}$$

or

$$\omega^2 = 1 - \frac{\varepsilon^2}{12} \quad \text{and} \quad \omega^2 = 1 + \frac{5\varepsilon^2}{12}$$

In the same way we could study the case where $a = - \omega^2$ is negative. We would find $a = - \varepsilon^2/2$ in the neighbourhood of $\omega = 0$.

In Fig. 8.2 the instability zones are hatched. The actual form is effectively that shown in Fig. 8.3.

See *Exercises 5 to 8* at the end of the chapter.

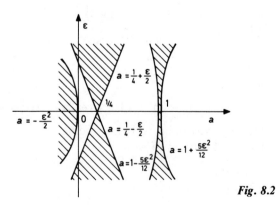

Fig. 8.2

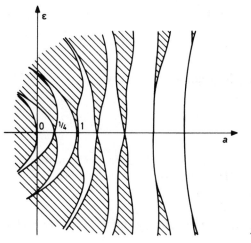

Fig. 8.3

Part 2: Perturbations of periodic linear systems

2.1. Linear equations with non-zero right-hand side

THE FREDHOLM ALTERNATIVE

Notation   If $\mathscr{C}(T)$ is the vector space of continuous functions of period T with values in $\mathbb{R}^d$, for every function φ of $\mathscr{C}(T)$ we shall write

$$\|\varphi\|_T = \sup_{0 \leqslant t \leqslant T} |\varphi(t)|$$

the norm of φ for uniform convergence. We know that $\mathscr{C}(T)$ is a complete space for this norm (Chapter 1, Theorem 3.1.4).

Theorem 2.1.1
Let

$$A: \begin{cases} \mathbb{R} \to \mathscr{L}(\mathbb{R}^d, \mathbb{R}^d) \\ t \rightsquigarrow A(t) \end{cases}$$

and

$$f: \begin{cases} \mathbb{R} \to \mathbb{R}^d \\ t \rightsquigarrow f(t) \end{cases}$$

be two continuous mappings of period T. Let

$$\frac{\mathrm{d}x}{\mathrm{d}t} = A(t)x + f(t) \tag{23}$$

be a linear equation with a non-zero right-hand side and

$$\frac{\mathrm{d}x}{\mathrm{d}t} = A(t)x \tag{24}$$

be the associated homogeneous equation.

If (24) has no non-zero solution of period T, (23) has, for each function f, a unique solution of period T. If Hf is this solution, then there exists K (independent of f) such that $\|Hf\|_T \leqslant K\|f\|_T$.

Proof　Let Φ be the resolvent of A and let y_0 be the initial value of a possible periodic solution:

$$y(t) = \Phi(t, 0)y_0 + \int_0^t \Phi(t, s)f(s)\,\mathrm{d}s$$

and therefore

$$y_0 = \Phi(T, 0)y_0 + \int_0^T \Phi(T, s)f(s)\,\mathrm{d}s$$

By assumption, 1 is not an eigenvalue of $\Phi(T, 0)$ nor of its inverse $\Phi(0, T)$; therefore $\Phi(0, T) - I$ is invertible and the preceding relation can be written

$$y_0 = [\Phi(0, T) - I]^{-1} \int_0^T \Phi(0, s)f(s)\,\mathrm{d}s$$

$$= \Phi(T, 0)\left[y_0 + \int_0^T \Phi(0, s)\,\mathrm{d}s \right]$$

We shall prove that solution with initial value y_0 is periodic and has the upper bound indicated. Writing Hf for this solution

$$Hf(t) = \Phi(t, 0)[\Phi(0, T) - I]^{-1} \int_0^T \Phi(0, s)f(s)\,\mathrm{d}s + \int_0^t \Phi(t, u)f(u)\,\mathrm{d}u$$

$$= \Phi(t, 0)[\Phi(0, T) - I]^{-1} \int_0^T \Phi(0, s)f(s)\,\mathrm{d}s$$

$$+ \Phi(t, 0)[\Phi(0, T) - I]^{-1}[\Phi(0, T) - I] \int_0^t \Phi(0, u)f(u)\,\mathrm{d}u$$

$$= \Phi(t, 0)[\Phi(0, T) - I]^{-1} \int_t^T \Phi(0, s)f(s)\,\mathrm{d}s$$

$$+ \Phi(t, 0)[\Phi(0, T) - I]^{-1} \int_0^t \Phi(0, T)\Phi(0, u)f(u)\,\mathrm{d}u$$

Now $\Phi(0, T)\Phi(0, u) = \Phi(0, T + u)$ and $f(u) = f(u + T)$. Therefore

$$\int_0^t \Phi(0, T)\Phi(0, u)f(u)\,du = \int_T^{T+t} \Phi(0, s)f(s)\,ds$$

Thus

$$Hf(t) = \int_t^{t+T} \Phi(t, 0)[\Phi(0, T) - I]^{-1}\Phi(0, s)f(s)\,ds$$

We know that $\Phi(u + T, 0) = \Phi(u, 0)\Phi(T, 0)$. Therefore $\Phi(u, 0) = \Phi(T + u, 0)\Phi(0, T)$ and

$$Hf(t) = \int_t^{t+T} \Phi(t + T, 0)\Phi(0, T)[\Phi(0, T) - I]^{-1}$$

$$\times \Phi(T, 0)\Phi(0, s + T)f(s + T)\,ds$$

Now for every matrix M such that $M - I$ is invertible,

$$M[M - I]^{-1} = [M - I]^{-1}M$$

(In fact $M[M - I]^{-1} = [M - I + I][M - I]^{-1} = I + [M - I]^{-1}$ and $[M - I]^{-1}M = [M - I]^{-1}[M - I + I] = I + [M - I]^{-1}$.) Therefore

$$\Phi(0, T)[\Phi(0, T) - I]^{-1}\Phi(T, 0) = [\Phi(0, T) - I]^{-1}$$

and

$$Hf(t) = \int_t^{t+T} \Phi(t + T, 0)[\Phi(0, T) - I]^{-1}\Phi(0, s + T)f(s + T)\,ds$$

$$= \int_{t+T}^{t+2T} \Phi(t + T, 0)[\Phi(0, T) - I]^{-1}\Phi(0, s)f(s)\,ds$$

$$= Hf(t + T)$$

Moreover if

$$K_1 = \sup_{0 \leqslant t \leqslant T} \|\Phi(t, 0)\|$$

$$K_2 = \sup_{0 \leqslant t \leqslant T} \sup_{t \leqslant s \leqslant t+T} \|\Phi(0, s)\|$$

and $K = K_1 K_2 \|[\Phi(0, T) - I]^{-1}\|$, $K < \infty$ and $\|Hf\|_T \leqslant K\|f\|_T$.

Since the difference between two solutions of (23) is a solution of (24), the uniqueness is obvious.

A more complex case is that in which there exists a non-zero periodic solution of (24). This is the case which is described by the Fredholm alternative.

The adjoint equation

We recall the definition in Chapter 3.

Definition 2.1.2

Let

$$\frac{dx}{dt} = A(t)x \qquad (25)$$

be a linear equation; the equation

$$\frac{dy}{dt} = -\tilde{A}(t)y \qquad (26)$$

is called the adjoint equation of (25).

Proposition 2.1.3

The resolvent of (26) is $\psi(t, s) = \tilde{\Phi}(s, t)$. There are as many independent periodic solutions of (25) as of (26). If x is a solution of (25) and y a solution of (26), the scalar product $\langle x(t), y(t) \rangle$ is constant.

Proof We know that

$$\frac{\partial}{\partial t} \Phi(s, t) = -\Phi(s, t)A(t)$$

Therefore

$$\frac{\partial}{\partial t} \tilde{\Phi}(s, t) = -\tilde{A}(t)\tilde{\Phi}(s, t)$$

Since $\Phi(t, t) = I = \psi(t, t)$ this proves the first point. The solution of (26) with initial value y_0 is therefore $\tilde{\Phi}(0, t)y_0$. The initial value of a periodic solution of (26) is a vector y_0 such that

$$[\tilde{\Phi}(0, T) - I]y_0 = 0 = \tilde{y}_0[\Phi(0, T) - I]$$

Now $\Phi(0, T) = \Phi(T, 0)^{-1}$. Therefore

$$\tilde{y}_0[\Phi(0, T) - I] = 0 \Rightarrow \tilde{y}_0[\Phi(T, 0)^{-1} - I] = 0$$

which is equivalent to $0 = \tilde{y}_0[\Phi(T, 0) - I]$, an equation which has as many solutions as $[\Phi(T, 0) - I]x_0 = 0$ (this number of independent solutions depends only on the rank of $\Phi(T, 0) - I$). Thus the second point is proved.

Finally

$$\frac{d}{dt} \langle x(t), y(t) \rangle = \langle A(t)x(t), y(t) \rangle + \langle x(t), -\tilde{A}(t)y(t) \rangle = 0$$

The principal theorem is as follows.

Theorem 2.1.4 (The Fredholm alternative)
Let

$$\frac{dx}{dt} = A(t)x + b(t) \tag{27}$$

be a linear equation in $\mathbb{R}^d$, where A and f are continuous and periodic with period T. Suppose that the homogeneous equation

$$\frac{dx}{dt} = A(t)x \tag{28}$$

has p independent solutions of period T. The adjoint equation

$$\frac{dy}{dt} = -\tilde{A}(t)y \tag{29}$$

then also has p independent solutions of period T, which we shall denote by $y_1, y_2, \ldots, y_p$. Then

(a) if $\forall k \leqslant p$

$$\int_0^T \langle y_k(t), b(t) \rangle \, dt = 0$$

there exist p independent solutions of (27) of period T and
(b) if this condition is not fulfilled (27) has no solution of period T.

Remark Before giving the proof, we should note the analogy with Theorem 2.2.13 in Chapter 6.

Let x be a differentiable and periodic function and A be the operator

$$Ax = \frac{dx}{dt} - A(t)x$$

Let us consider the scalar product in $L^2[(0, T), dx]$:

$$\langle Ax, y \rangle = \int_0^T \left\langle \frac{dx}{dt} - A(t)x, y \right\rangle dt$$

$$= [x, y]_0^T - \int_0^T \left\langle x, \frac{dy}{dt} \right\rangle dt - \int_0^T \langle x, \tilde{A}(t)y \rangle \, dt$$

$$= \langle x, \tilde{A}y \rangle$$

putting

$$\tilde{A}y = -\frac{dy}{dt} - \tilde{A}(t)y$$

We must note carefully that the set of differentiable periodic functions is not a complete subspace of $L^2[(0, T), dx]$, so that we cannot directly apply the result quoted. It nevertheless remains true that we are concerned with the same phenomenon. Moreover we shall give a direct proof of this.

Proof of Theorem 2.1.4 x_0 is the initial value of a periodic solution of (27) iff

$$[\Phi(0, T) - I]x_0 = \int_0^T \Phi(0, s)b(s)\, ds$$

$y_k(0)$ is the initial value of a solution y_k iff

$$[\tilde{\Phi}(0, T) - I]y_k(0) = 0$$

Let us denote by A, $\Phi(0, T) - I$, and by $\mathcal{R}(A)$ and $\mathcal{K}(\tilde{A})$ the image of A and the kernel of $\tilde{A}$. We know that $\mathcal{R}(A) \oplus \mathcal{K}(\tilde{A}) = \mathbb{R}^d$ (this is 'the Fredholm alternative'). ($A \in \mathcal{L}(\mathbb{R}^d, \mathbb{R}^d)$ and $\mathbb{R}^d$ is complete.)

By definition there exists x_0 such that

$$Ax_0 = \int_0^T \Phi(0, s)b(s)\, ds$$

iff

$$\int_0^T \Phi(0, s)b(s)\, ds \in \mathcal{R}(A)$$

Moreover $y_1(0), y_2(0), \ldots, y_p(0)$ is a basis of $\mathcal{K}(\tilde{A})$; there exists therefore a periodic solution of (27) iff

$$\left\langle \int_0^T \Phi(0, s)b(s)\, ds, \, y_k(0) \right\rangle = 0 \qquad \forall k \leqslant p$$

i.e. if

$$\forall k \leqslant p \quad \int_0^T \langle \Phi(0, s)b(s), y_k(0) \rangle\, ds = \int_0^T \langle b(s), \tilde{\Phi}(0, s)y_k(0) \rangle\, ds$$

$$= \int_0^T \langle b(s), y_k(s) \rangle\, ds = 0$$

If these relations are satisfied, the set of vectors V such that

$$AV = \int_0^T \Phi(0, s)b(s)\, ds$$

is of the form $V_0 + \mathcal{K}(A)$ where V_0 is one of these vectors; hence there exist p of them which are independent and which are the initial values of p independent periodic solutions of (27).

Examples

(a)

$$\frac{d^2x}{dt^2} + \omega^2 x = f(t)$$

where f is of period $T = 2\pi/\omega$ has solutions of period T iff

$$\int_0^{2\pi/\omega} f(u) \sin \omega u\, du = 0 = \int_0^{2\pi/\omega} f(u) \cos \omega u\, du$$

because

$$\tilde{A} = \begin{pmatrix} 0 & \omega^2 \\ -1 & 0 \end{pmatrix}$$

and a basis of solutions of

$$\frac{dy}{dt} = -\tilde{A}(t)y$$

is

$$\begin{pmatrix} \omega \sin \omega t \\ \cos \omega t \end{pmatrix}, \begin{pmatrix} \omega \cos \omega t \\ -\sin \omega t \end{pmatrix}$$

and for

$$b(t) = \begin{pmatrix} 0 \\ f(t) \end{pmatrix}$$

the conditions of orthogonality are those indicated.

(b)

$$\frac{d^2x}{dt^2} = f(t)$$

where f is of period T, has solutions of period T iff

$$\int_0^T f(u)\, du = 0$$

because

$$\tilde{A} = \begin{pmatrix} 0 & 0 \\ -1 & 0 \end{pmatrix}$$

and the adjoint equation has

$$\begin{pmatrix} 0 \\ a \end{pmatrix}$$

for periodic solutions.

This example is fundamental and we shall often come across this condition later in the text.

BOUNDED SOLUTIONS AND PERIODIC SOLUTIONS

We have already remarked that for linear equations there is a link between the existence of bounded solutions and that of stable solutions (Chapter 3). We shall establish that in the same way there is a link between the existence of bounded solutions and that of periodic solutions for linear systems.

Theorem 2.1.5
Let

$$\frac{dx}{dt} = A(t)x + f(t) \tag{30}$$

where

$$A: \begin{cases} \mathbb{R} \to \mathscr{L}(\mathbb{R}^d, \mathbb{R}^d) \\ t \rightsquigarrow f(t) \end{cases} \quad \text{and} \quad f: \begin{cases} \mathbb{R} \to \mathbb{R}^d \\ t \rightsquigarrow f(t) \end{cases}$$

are continuous functions of period T. If there exists a bounded solution of (30) there exists also a solution of period T.

Proof   If $f \equiv 0$ the result is obvious ($x \equiv 0$). Let Φ be the resolvent of (30); the solution with initial value x_0 is

$$\varphi(t, 0, x_0) = \Phi(t, 0)x_0 + \int_0^t \Phi(t, s)f(s)\,ds$$

so that

$$\varphi(T, 0, x_0) = \Phi(T, 0)x_0 + \int_0^T \Phi(T, s)f(s)\,ds$$

If the homogeneous equation has no solution of period T other than the identically zero solution, we know that there exists a solution of (30) of period T.

Consider the opposite case, i.e. that in which $\{\Phi(T, 0) - I\}$ is not invertible. We know from the Fredholm alternative that there then exists $y \neq 0$ in $\mathbb{R}^d$ having the following two properties:

(a) y is the initial value of a non-zero periodic solution of the adjoint equation, i.e. $\tilde{y}\{\Phi(T, 0) - I\} = 0$.

(b)
$$\left\langle y, \int_0^T \Phi(t, s) f(s)\, ds \right\rangle \neq 0$$

Suppose now that there does not exist a solution of (30) of period T. Let φ be the bounded solution of (30) and φ_0 its initial value:

$$\varphi(T) = \Phi(T, 0)\varphi_0 + \int_0^T \Phi(T, s) f(s)\, ds$$

$$\varphi(2T) = [\Phi(T, 0)]^2 \varphi_0 + \Phi[T, 0] \int_0^T \Phi(T, s) f(s)\, ds + \int_T^{2T} \Phi(T, s) f(s)\, ds$$

Now

$$\int_T^{2T} \Phi(2T, s) f(s)\, ds = \int_0^T \Phi(T + T, u + T) f(u + T)\, du$$

$$= \int_0^T \Phi(T, s) f(s)\, ds$$

and thus, step by step

$$\varphi(nT) = [\Phi(T, 0)]^n \varphi_0 + [\Phi(T, 0)^{n-1} + \Phi(T, 0)^{n-2} + \ldots$$

$$+ \Phi(T, 0) + I] \int_0^T \Phi(T, s) f(s)\, ds$$

Then

$$\langle \varphi(nT), y \rangle = \tilde{y}\Phi(T, 0)^n \varphi_0 + \sum_{p=1}^n \tilde{y}\Phi(T, 0)^{n-p} \int_0^T \Phi(T, s) f(s)\, ds$$

$$= \langle y, \varphi_0 \rangle + n \left\langle y, \int_0^T \Phi(T, s) f(s)\, ds \right\rangle$$

(from (a)) but from relation (b)

$$\lim_{n \to \infty} |\langle \varphi(nT), y \rangle| = +\infty$$

which is absurd if φ is bounded.

Generalization to the non-linear case: case of $\mathbb{R}$ and $\mathbb{R}^2$

The link between the existence of bounded and of periodic solutions exists also for much more general equations in $\mathbb{R}$ and $\mathbb{R}^2$; this results from the structure of these spaces.

Theorem 2.1.6
Let

$$\frac{dx}{dt} = f(t, x) \tag{31}$$

be a differential equation in $\mathbb{R}$ where f is continuous and of period T. Suppose there exists a solution φ, defined and bounded for $t \geqslant 0$. Then there exists a solution of period T.

Proof If $\varphi(T) = \varphi(0)$ the theorem is proved. Suppose, for example, that $\varphi(T) > \varphi(0)$. We shall prove then that $\varphi(2T) > \varphi(T)$.

Let

$$\psi(t) = \varphi(t + T)$$

$$\frac{d\psi}{dt}(t) = \frac{d\varphi}{dt}(t + T) = f[t + T, \varphi(t + T)] = f[t, \psi(t)]$$

$\psi(0) = \varphi(T) > \varphi(0)$; if $\psi(T) \leqslant \varphi(T)$ there exists a pair (u, x_0) such that $\psi(u) = x_0 = \varphi(u)$ (Fig. 8.4) and the Cauchy problem relative to (u, x_0) does not have a

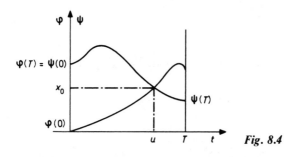

Fig. 8.4

unique solution. Thus

$$\varphi(T) > \varphi(0) \Rightarrow \psi(T) = \varphi(2T) > \varphi(T)$$

and

$$\varphi(t + T) > \varphi(t) \qquad \forall t \in [0, T]$$

We can establish by recurrence that $\forall n \in \mathbb{N}$, $\varphi(t + nT) > \varphi[t + (n-1)T]$, and since φ is bounded above

$$\lim_{n \to \infty} \varphi(t + nT) = g(t) < \infty$$

In particular

$$\lim_{n \to \infty} \varphi(nT) = g(0)$$

and since $\varphi(nT + T) = \varphi[(n+1)T]$

$$g(0) = \lim_{n \to \infty} \varphi[(n+1)T] = g(T)$$

Thus, there exists a function g of period T such that

$$\lim_{n \to \infty} \varphi(t + nT) = g(t)$$

We can show that g is a solution of (31) at least in the sense that g is continuous and satisfies

$$g(t) - g(0) - \int_0^t f[s, g(s)]\, ds = 0$$

Let $g_n(t) = \varphi(t + nT)$ for $n \in \mathbb{N}$; then

$$\left| g(t) - g(0) - \int_0^t f[s, g(s)]\, ds \right|$$

$$\leqslant |g(t) - g_n(t)| + \left| g_n(t) - g(0) - \int_0^t f[s, g_n(s)]\, ds \right|$$

$$+ \left| \int_0^t f[s, g(s)] - f[s, g_n(s)]\, ds \right|$$

From this relation, if g_n converges uniformly to g, g is a solution. Now

$$\forall n \in \mathbb{N} \qquad \sup_{t \leqslant T} |g_n(t)| \leqslant A$$

and $\forall \varepsilon > 0$, $\exists \eta > 0$ such that

$$|t_2 - t_1| < \eta \Rightarrow |g_n(t_1) - g_n(t_2)| < \varepsilon \qquad \forall n \in \mathbb{N}$$

We say that the sequence $\{g_n\}_{n \in \mathbb{N}}$ is uniformly bounded and equicontinuous. A classical theorem of Ascoli states that it then contains a subsequence which converges uniformly to a function which is continuous. This limit can only be g and, since the convergence of $g_n(t)$ to $g(t)$ is monotone, the sequence $\{g_n\}_{n \in \mathbb{N}}$ itself converges uniformly to g.

In the case of $\mathbb{R}^2$ we still have a link of the same type; this time the reason is the following. In $\mathbb{R}^2$ every continuous mapping A, with continuous inverse, from a connected open set G into itself, which conserves orientation, and for which there exists a sequence $x_0, x_1 = Ax_0, \ldots, x_n = A^n x_0, \ldots$ converging in G, possesses a fixed point in G (Browder's theorem).

Theorem 2.1.7
Let

$$\frac{dx}{dt} = f(t, x) \tag{32}$$

be an equation in $\mathbb{R}^2$ where f is continuous and of period T. Suppose that all solutions of (32) exist for $t \geqslant 0$ and that one of them is bounded; there then exists a solution of period T.

Proof Let A be the mapping defined in $\mathbb{R}^2$ by $x \rightarrow Ax = \varphi(T, 0, x)$. By assumption A is well defined. Let x_0 be the initial value of the bounded solution. $\{A^n x_0\}_{n \in \mathbb{N}} = \{\varphi(nT, 0, x_0)\}_{n \in \mathbb{N}}$ is a bounded sequence, and therefore we can extract from it a convergent subsequence so that A has a fixed point y; the solution with initial value y is periodic.

This theorem admits numerous corollaries corresponding to situations in which we can establish the existence of bounded solutions, particularly by the use of auxiliary functions as in Lyapunov's method. The following is an example.

Theorem 2.1.8
In $\mathbb{R}$, let

$$\frac{d^2 x}{dt^2} + f(x)\frac{dx}{dt} + g(x) = h(t)$$

Suppose that

(a) f is continuous and

$$\lim_{x \to \pm\infty} \int_0^x f(u)\,du = +\infty \text{ or } -\infty$$

(b) g is of class C^1 and $xg(x) \geq 0 \ \forall x \in \mathbb{R}$ and
(c) h is continuous of period T and

$$\int_0^T h(u)\,du = 0$$

Then there exists a solution of period T.

Example   The 'forced' Van der Pol equation

$$\frac{d^2 x}{dt^2} + \lambda(1 + x^2)\frac{dx}{dt} + x = \cos \omega t$$

has a solution of period $T = 2\pi/\omega$.

2.2. Perturbation of linear systems: non-critical case

Definition 2.2.1
Consider the equation

$$\frac{dx}{dt} = A(t)x + g(t, x, \varepsilon) \tag{33}$$

where A and g are continuous of period T, $A(t) \in \mathscr{L}(\mathbb{R}^d, \mathbb{R}^d)$ and $g(t, x, \varepsilon) \in \mathbb{R}^d$. We assume that $g(t, 0, 0) = 0$.

We say that (33) is a perturbation of

$$\frac{dx}{dt} = A(t)x \tag{34}$$

This perturbation is called non-critical if (34) has no solution of period T other than the identically zero solution. If g is sufficiently regular, such a system behaves like a linear system with non-zero right-hand side as the following theorem shows.

Theorem 2.2.2

Let

$$\frac{dx}{dt} = A(t)x + g(t, x, \varepsilon) \tag{35}$$

where $A(t) \in \mathcal{L}(\mathbb{R}^d, \mathbb{R}^d)$ and $g(t, x, \varepsilon) \in \mathbb{R}^d$ are continuous and satisfy

$$A(t, T) = A(t)$$

$$g(t + T, x, \varepsilon) = g(t, x, \varepsilon) \qquad \forall t$$

Suppose that g satisfies the following condition.

$$\left.\begin{array}{l} \exists \varepsilon_0 > 0, \exists R > 0 \text{ and a continuous function } k \text{ such that} \\[4pt] \forall t, \forall \varepsilon \leqslant \varepsilon_0 \\[4pt] \begin{array}{l} \forall x_1 \colon \|x_1\| \leqslant r < R \\ \forall x_2 \colon \|x_2\| \leqslant r < R \end{array} \Rightarrow \|g(t, x_1, \varepsilon) - g(t, x_2, \varepsilon)\| \leqslant k(\varepsilon, r)\|x_1 - x_2\| \end{array}\right\} \tag{36}$$

and

k is a non-decreasing function of r and of ε such that $k(0, 0) = 0$
Then $\exists \rho, \varepsilon_1 \colon 0 < \rho < R$ and $0 < \varepsilon_1 < \varepsilon_0$ such that $\forall \varepsilon \leqslant \varepsilon_1$ there exists in the set $\|x\| \leqslant \rho$ a unique solution $\varphi(t, \varepsilon)$ of (35) having period T and $\varphi(t, 0) \equiv 0$.

Remark If $g(t, x, \varepsilon)$ and $(\partial g/\partial x)(t, x, \varepsilon)$ have limit 0 uniformly in t when x and ε tend to 0, the condition (36) is certainly satisfied.

Proof We shall construct, over the set of continuous functions of period T, a contracting operator which will enable us to find the periodic solution by successive approximation.
We use the notation at the beginning of Section 2.1.
A mapping H from $\mathscr{C}(T)$ into itself is a contraction if

$$\exists K < 1 \colon \|H(\psi_1) - H(\psi_2)\|_T \leqslant K\|\psi_1 - \psi_2\|_T \qquad \forall \psi_1 \text{ and } \psi_2 \in \mathscr{C}(T)$$

(Chapter 6, Definition 2.2.6). We know that there then exists a unique fixed point φ of H such that $\varphi = H(\varphi)$ and that $\forall f_0 \in \mathscr{C}(T)$ the sequence f_0, $f_1 = H(f_0)$, $f_2 = H(f_1)$, ..., $f_n = H(f_{n-1})$, ... converges, for the uniform convergence norm, to φ:

$$\lim_{n \to \infty} \|f_n - \varphi\|_T = 0$$

Consider then the set $\{f \in \mathscr{C}(T): \|f\|_T \leqslant \sigma\} = B_\sigma$. We define H over B_σ in the following way. If Φ is the resolvent of A

$$Hf(t) = \int_t^{t+T} \{\Phi(t,\, 0)[\Phi(0,\, T) - I]^{-1} \Phi(0,\, s) g[s,\, f(s),\, \varepsilon]\, ds\}$$

If $g[s, f(s), \varepsilon] = h(s)$ this formula is identical with that which, according to Theorem 2.1.1, gives the periodic solution of a linear equation with non-zero right-hand side. We again use this method which, for an equation which is almost linear (since ε is small), consists in behaving as if it were linear and then iterating the operation, hoping that the sequence thus obtained converges.

In the present case, since f and g are periodic, Hf is periodic. We first determine ρ such that $f \in B_\rho \Rightarrow Hf \in B_\rho$.

$$\forall x: \|x\| \leqslant \sigma \quad \|g(s, x, \varepsilon)\| \leqslant \|g(s, x, \varepsilon) - g(s, 0, \varepsilon)\| + \|g(s, 0, \varepsilon)\|$$

$$\leqslant k(\varepsilon, \sigma)\|x\| + \eta(\varepsilon)$$

where

$$\lim_{\varepsilon \to 0} \eta(\varepsilon) = 0$$

Since $g(s, 0, 0) = 0$ and g is continuous and periodic

$$\forall f \in B_\sigma \qquad \sup_{t \leqslant s \leqslant t+T} \|g[s, f(s), \varepsilon]\| \leqslant k(\varepsilon, \sigma)\|f\|_T + \eta(\varepsilon)$$

By choosing the same upper bound as in Theorem 2.1.1 we find

$$\|Hf\|_T \leqslant K[k(\varepsilon, \sigma)\|f\|_T + \eta(\varepsilon)]$$

From the continuity of k and η we can choose ε_1 and ρ such that

$$\forall \varepsilon < \varepsilon_1,\ \forall \sigma \leqslant \rho \qquad k(\varepsilon, \sigma) < \frac{1}{k} \quad \text{and} \quad \eta < \left[\frac{1}{K} - k(\varepsilon, \sigma)\right]\rho$$

so that

$$\forall f \in B_\rho \qquad \|Hf\|_T \leqslant K\left[k(\varepsilon, \sigma)\|f\|_T + \frac{1 - Kk(\varepsilon, \sigma)}{K}\rho\right] \leqslant \rho$$

H is certainly a contraction since $\forall \varepsilon \leqslant \varepsilon_1,\ \forall \sigma \leqslant \rho,\ \forall \psi_1, \psi_2 \in B_\rho$,

$$\|H(\psi_1) - H(\psi_2)\|_T \leqslant Kk(\varepsilon, \sigma)\|\psi_1 - \psi_2\|_T < \lambda\|\psi_1 - \psi_2\|_T \quad \text{where } \lambda < 1$$

Then let $\varphi \in B_\sigma$ be the fixed point of H; we can prove that φ is a solution of

(35)

$$\frac{d\varphi}{dt} = \frac{d[H\varphi](t)}{dt}$$

$$= \Phi(t, 0)[\Phi(0, T) - I]^{-1}\Phi(0, t + T)g[t + T, f(t + T), \varepsilon]$$
$$- \Phi(t, 0)[\Phi(0, T) - I]^{-1}\Phi(0, t)g[t, f(t), \varepsilon] + A(t)H\varphi(t)$$

Since $\Phi(0, t + T) = \Phi(0, T)\Phi(0, t)$ and since f and g are periodic

$$\frac{d\varphi}{dt} = \Phi(t, 0)[\Phi(0, T) - I]^{-1}[\Phi(0, T) - I]\Phi(0, t)g[t, f(t), \varepsilon] + A(t)\varphi(t)$$

$$= A(t)\varphi(t) + g[t, f(t), \varepsilon]$$

We often construct an approximation to $\varphi(t, \varepsilon)$ by using the sequence

$$f_0(t, \varepsilon) \equiv 0$$

$$f_1(t, \varepsilon) = Hf_0(t, \varepsilon) = \int_t^{t+T} \{\Phi(t, 0)[\Phi(0, T) - I]^{-1}\Phi(0, s)\}g(s, 0, \varepsilon)\,ds$$

$$f_2(t, \varepsilon) = Hf_1(t, \varepsilon)\ldots$$

This sequence naturally depends on ε. If $\varepsilon = 0$, since $g(s, 0, 0) = 0$, $f_1(t, 0) \equiv 0$ as well as $f_n(t, 0)\ \forall n \in \mathbb{N}$; therefore $\varphi(t, 0) \equiv 0$, which completes the proof.

Complement
The following theorems will be proved in Exercises 8 and 10.

Theorem 2.2.3
Let $B \in \mathscr{L}(\mathbb{R}^d, \mathbb{R}^d)$, independent of t and invertible. Let $g(t, x, \varepsilon)$ have the property (36) and let $h(t) \in \mathbb{R}^d$ be a function of period T, continuous and such that

$$\int_0^T h(u)\,du = 0$$

Then there exist ε_1 and ρ, $0 < \rho < R$, $0 < \varepsilon_1 \leqslant \varepsilon_0$, such that $\forall \varepsilon \leqslant \varepsilon_1$

$$\frac{dx}{dt} = \varepsilon[Bx + g(t, x, \varepsilon) + h(t)]$$

has, in the set $\|x\| \leqslant \rho$, a unique solution $\varphi(t, \varepsilon)$ of period T. Moreover $\varphi(t, 0) \equiv 0$.

Theorem 2.2.4
Let B and g satisfy the same conditions as above. Suppose that for every eigenvalue λ of B, Re $\lambda \neq 0$. Then there exist ε_1 and ρ, $0 < \rho < R$, $0 < \varepsilon_1 \leqslant \varepsilon_0$, such that $\forall \varepsilon < \varepsilon_1$ the

equation

$$\varepsilon \frac{dx}{dt} = Bx + g(t, x, \varepsilon)$$

has, in the set $\|x\| \leqslant \rho$, a unique solution $\varphi(t, \varepsilon)$ of period T. Moreover $\varphi(t, 0) \equiv 0$.

Example (*Duffing's equation with forced oscillation*) We shall study solutions of period $2\pi/\omega$ of the equation

$$\frac{d^2x}{dt^2} + \omega_0{}^2 x + \varepsilon x^3 = A \sin \omega t \tag{37}$$

Suppose that $\varepsilon \ll \omega^2 - \omega_0{}^2$ and ω_0/ω is not an integer. For $\varepsilon = 0$

$$\frac{A}{\omega^2 - \omega_0{}^2} \sin \omega t$$

is a solution. We reduce this to the above situation by putting

$$x = \frac{A}{\omega^2 - \omega_0{}^2} \sin \omega t + y$$

where y is a solution of

$$\frac{d^2y}{dt^2} + \omega_0{}^2 y + \varepsilon \left(\frac{A}{\omega^2 - \omega_0{}^2} \sin \omega t + y \right)^3 = 0 \tag{38}$$

Equation (38) is not critical.

To find a periodic solution we shall use the approximation indicated in the proof of Theorem 2.2.2. If $f \equiv 0$, we shall calculate $f_1(t) = [Hf_0](t)$. The resolvent of the linear system is

$$\Phi(t, s) = \begin{pmatrix} \cos \omega_0(t - s) & \dfrac{1}{\omega_0} \sin \omega_0(t - s) \\ -\omega_0 \sin \omega_0(t - s) & \cos \omega_0(t - s) \end{pmatrix}$$

Then

$$Hf_0(t) = \int_t^{(2\pi/\omega) + t} \Phi(t, 0) \left[\Phi\left(0, \frac{2\pi}{\omega}\right) - I \right]^{-1} \Phi(0, s) \begin{pmatrix} 0 \\ \varepsilon B \sin^3 \omega s \end{pmatrix} ds$$

where we have put $B = A^3/(\omega^2 - \omega_0{}^2)^3$. If $\alpha = \pi\omega_0/\omega$ then

$$Hf_0(t) = \int_t^{(2\pi/\omega) + t} \begin{pmatrix} \cos \omega_0 t & \dfrac{1}{\omega_0} \sin \omega_0 t \\ -\omega_0 \sin \omega_0 t & \cos \omega_0 t \end{pmatrix} \begin{pmatrix} -\dfrac{1}{2} & \dfrac{1}{2\omega_0 \tan \alpha} \\ -\dfrac{\omega_0}{2 \tan \alpha} & -\dfrac{1}{2} \end{pmatrix}$$

$$\times \begin{pmatrix} \cos \omega_0 s & -\dfrac{1}{\omega_0} \sin \omega_0 s \\ \omega_0 \sin \omega_0 s & \cos \omega_0 s \end{pmatrix} \begin{pmatrix} 0 \\ \varepsilon B \sin^3 \omega s \end{pmatrix} ds$$

We calculate only the first component, denoted by $y_1(t)$:

$$y_1(t) = \frac{\varepsilon B}{2\omega_0 \sin \alpha} \int_t^{(2\pi/\omega) + t} [\cos \omega_0 t \cos(\omega_0 s - \alpha)$$

$$+ \sin \omega_0 t \sin(\omega_0 s - \alpha)] \sin^3 \omega s \, ds$$

Now $\sin^3 \omega s = \frac{3}{4} \sin \omega s - \frac{1}{4} \sin 3\omega s$ and therefore

$$y_1(t) = \frac{3\varepsilon B}{16\omega_0 \sin \alpha} \int_t^{(2\pi/\omega) + t} [\sin(\omega s - \omega_0 s + \omega_0 t + \alpha)$$

$$+ \sin(\omega s + \omega_0 s - \omega_0 t - \alpha)] \, ds$$

$$- \frac{\varepsilon B}{16\omega_0 \sin \alpha} \int_t^{(2\pi/\omega) + t} [\sin(3\omega s - \omega_0 s + \omega_0 t + \alpha)$$

$$+ \sin(3\omega s + \omega_0 s - \omega_0 t - \alpha)] \, ds$$

Then

$$y_1(t) = \frac{3\varepsilon B}{4(\omega_0^2 - \omega^2)} \sin \omega t - \frac{\varepsilon B}{4} \frac{\sin 3\omega t}{\omega_0^2 - 9\omega^2}$$

and therefore

$$x(t) = \left[\frac{A}{\omega^2 - \omega_0^2} - \frac{3\varepsilon}{4} \frac{A^3}{(\omega^2 - \omega_0^2)^4} \right] \sin \omega t$$

$$- \frac{\varepsilon A^3}{4(\omega^2 - \omega_0^2)^3(\omega_0^2 - 9\omega^2)} \sin 3\omega t$$

It is clear that if $\omega = \omega_0$ or $\omega = \omega_0/3$ the system behaves in a particular way. Moreover, this is a critical system which we shall study further.

We often call the first term the first harmonic and it is customary to represent its amplitude

$$\frac{A}{\omega^2 - \omega_0^2} - \frac{3\varepsilon}{4} \frac{A^3}{(\omega^2 - \omega_0^2)^4}$$

as a function of the amplitude A of the input for each value of ω, or as a function of ω for each value of A.

Figure 8.5 represents the amplitude C of the first harmonic, according to the value of ε (we assume $A > 0$). The curves are indexed by $0 = A_0 < A_1 < A_2 < A_3$.

$$C = \frac{A}{\omega^2 - \omega_0^2} - \frac{3\varepsilon}{4} \frac{A^3}{(\omega^2 - \omega_0^2)^4}$$

or

$$A = (\omega^2 - \omega_0^2)C + \frac{3\varepsilon}{4} C^3$$

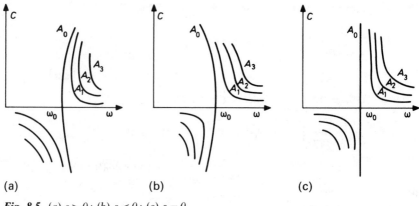

Fig. 8.5. (a) $\varepsilon > 0$; (b) $\varepsilon < 0$; (c) $\varepsilon = 0$.

Stability of the periodic solution

Since $\varphi(t, 0) \equiv 0$ it is reasonable to think that, under sufficiently regular conditions, $\varphi(t, \varepsilon)$ has the same stability properties as the solution $\varphi(t) \equiv 0$ of $dx/dt = A(t)x$.

To study this we return to the change of variable

$$y(t) = \exp(Bt)\,\Phi(0,\, t)x(t) = S(t)x(t) \tag{39}$$

where $\psi(t, \varepsilon) = S(t)\varphi(t, \varepsilon)$.

Through the transformation S, (39) becomes

$$\frac{dy}{dt} = By + S(t)g[t, S^{-1}(t)y, \varepsilon]$$

$$= By + h(t, y, \varepsilon) \tag{40}$$

Since $S(t)$ is bounded uniformly in t, and $S^{-1}(t)$ also, x and $S(t)x$ behave in the same way from the point of view of stability. Moreover h satisfies the condition (36) for a function $K(\varepsilon, r)$ so that, if we put $z = y - \psi$, z is a solution of the equation

$$\frac{dz}{dt} = Bz + h(t, y, \varepsilon) - h[t, \psi(t, \varepsilon), \varepsilon] \tag{41}$$

and the right-hand side satisfies

$$\|h(t, y, \varepsilon) - h[t, \psi(t, \varepsilon), \varepsilon]\| \leqslant K(\varepsilon, r)\|y - \psi(t, \varepsilon)\|$$

We can therefore deduce the following result from Theorem 1.2.4.

Theorem 2.2.5

Let

$$\frac{dx}{dt} = A(t)x + g(t, x, \varepsilon) \tag{42}$$

be a differential equation satisfying the assumptions of Theorem 2.2.2 (in particular, g satisfies (36)).

If the characteristic exponents of A have non-zero real parts (i.e. if the multipliers are not of modulus 1), the solution $\varphi(t, \varepsilon)$ constructed in Theorem 2.2.2 is of the same nature as the solution 0 of $dx/dt = A(t)x$.

Thus, if the real parts of the characteristic exponents are all strictly less than 0, $\varphi(t, \varepsilon)$ is asymptotically stable; if one of them is strictly greater than 0, it is unstable.

If the system $dz/dt = Bz$ has a saddle point, the system (41) has one also and presents stable and unstable sets which are tangential to those of $dz/dt = Bz$ if g is of class C^2.

See **Exercises 9 to 12** at the end of the chapter.

2.3. Critical systems; the bifurcation equation

We consider a periodic linear system

$$\frac{dx}{dt} = A(t) \tag{43}$$

where $A(t) \in \mathcal{L}(\mathbb{R}^d, \mathbb{R}^d)$ and is of period T. We assume that there exist non-zero solutions of (43) of period T. Consider a perturbation of (43), say

$$\frac{dx}{dt} = A(t)x + \varepsilon g(t, x, \varepsilon) \tag{44}$$

where g is of period T. We say that we have a critical system perturbation problem, since we are seeking, for $\varepsilon \leqslant \varepsilon_0$, solutions of (44) of period T.

Suppose, for simplicity, that all the solutions of (43) are of period T; then its resolvent Φ satisfies $\Phi(t, + T, 0) = \Phi(t, 0)$. Let $x(t) = \Phi(t, 0)y(t)$. Then x is a solution of (44) iff y is a solution of

$$\frac{dy}{dt} = \varepsilon \Phi(0, t)g[t, \Phi(t, 0)y, \varepsilon]$$

$$= \varepsilon h(t, y, \varepsilon) \tag{45}$$

where h is of period T. If y is a solution of period T of (45), x is a solution of period T of (44). Thus we reduce problems of the search for periodic solutions of critical systems to problems relative to equations of a particular type.

We shall see that we can proceed thus in many circumstances and that we can even use a variety of changes of variable to solve certain particular problems, especially for autonomous systems.

PERIODIC FUNCTION ASSOCIATED WITH A CRITICAL SYSTEM

Definition 2.3.1 (Hale—the standard form of a perturbed equation)
We shall say that a critical perturbation is in **standard form** if it is of the form

$$\frac{dx}{dt} = \varepsilon f(t, x, \varepsilon) \tag{46}$$

where $f(t, x, \varepsilon) \in \mathbb{R}^d$ is continuous and of period T.
This is a perturbation of the equation $dx/dt = 0$, whose solutions, constants, are functions of period T for every T.

We know already that if f does not depend on x (46) has a periodic solution only if

$$\int_0^T f(t, \varepsilon) \, dt = 0$$

To solve (46) we shall take again the steps which we have already considered: write the same formula as the formula of variation of constants for linear equations with non-zero right-hand side and iterate the operation. Unfortunately, as we have remarked, we shall not be able to find a periodic solution at each stage if the right-hand side is not of zero mean. We shall therefore, at each stage, transform the right-hand side to make it periodic.

Theorem 2.3.2
(a) Let

$$\frac{dx}{dt} = \varepsilon f(t, x, \varepsilon) \tag{46}$$

be a differential equation in $\mathbb{R}^d$ where f is a function of class C^1 for $t \in \mathbb{R}$, $0 \le \|x\| \le R$ and $0 \le |\varepsilon| \le \varepsilon_0$. Suppose that f has period T. Then $\forall \rho \le R$ there exists $\varepsilon_1 > 0$ such that $\forall \varepsilon$, $0 \le |\varepsilon| \le \varepsilon_1$, and $\forall a \in \mathbb{R}^d$, $0 \le |a| \le r < \rho$, there exists $Z(t, a, \varepsilon)$ of period T such that $Z(t, a, 0) \equiv a$, $|Z(t, x, \varepsilon)| \le \rho$ and

$$\int_0^T Z(s, a, \varepsilon) \, ds = 0$$

which satisfies

$$\frac{dZ}{dt}(t, a, \varepsilon) = \varepsilon f[t, Z(t, a, \varepsilon), \varepsilon] - \frac{\varepsilon}{T} \int_0^T f[u, Z(u, a, \varepsilon), \varepsilon] \, du$$

(b) $Z(t, a, \varepsilon)$ is continuously differentiable with respect to a and ε and can be constructed by successive approximation in the following way. Let $\varphi \in \mathscr{C}(T)$ and $H\varphi = a + $ the solution of period T and of mean zero of the equation

$$\frac{dg}{dt} = \varepsilon \left\{ f[t, \varphi(t), \varepsilon] - \frac{1}{T} \int_0^T f[u, \varphi(u), \varepsilon] \, du \right\} \tag{47}$$

The sequence $\varphi_0 = a$, $\varphi_1 = Ha, \ldots, \varphi_n = H\varphi_{n-1}$ converges, if $n \to \infty$, to $Z(t, a, \varepsilon)$ and the convergence is uniform. $Z(t, a, \varepsilon)$ is unique and is called the periodic function associated with (46).

Proof We note first that the right-hand side of (47) is of mean zero; thus there exists a solution of period T of (47) and if a constant is subtracted there exists one, and only one, solution of mean zero.

Let us denote by $\varepsilon \bar{f}(t)$ the right-hand side of (47); the solutions are of the form

$$g(t) = g(0) + \varepsilon \int_0^t \bar{f}(u)\, du$$

Let $g_i(t)$, $1 \leqslant i \leqslant d$, be the components of g, and $\bar{f}_i$ those of $\bar{f}$. If g is of mean zero

$$\int_0^T g_i(t)\, dt = 0$$

and there exists $0 < \vartheta_i < T$ such that $g_i(\vartheta_i) = 0$, i.e. such that

$$0 = g_i(0) + \varepsilon \int_0^{\vartheta_i} \bar{f}(u)\, du$$

Then

$$g_i(t) = \varepsilon \int_{\vartheta_i}^t \bar{f}(s)\, ds$$

and if ϑ is the vector with coordinates ϑ_i

$$g(t) = \varepsilon \int_\vartheta^t \bar{f}(s)\, ds$$

Thus

$$H\varphi(t) = a + \varepsilon \int_\vartheta^t \left\{ f[s, \varphi(s), \varepsilon] - \frac{1}{T} \int_0^T f[u, \varphi(u), \varepsilon]\, du \right\} ds$$

Suppose that $\|\varphi\|_T \leqslant \rho$. We put

$$M = \sup_{|\varepsilon| \leqslant \varepsilon_0} \int_0^T \|f(u, x, \varepsilon)\, du\|$$

for $\|x\| \leqslant \rho$; then $\|H\varphi\|_T \leqslant \|a\| + |\varepsilon|TM + |\varepsilon|M$. If $N = TM + M$ (independent of φ), $\|H\varphi\|_T \leqslant \|a\| + |\varepsilon|N$. Now let $\|a\| = r < \rho$. If $|\varepsilon| \leqslant (\rho - \|a\|)/N$, then $\|H\varphi\|_T \leqslant \rho$: $H\varphi \in B_\rho$ $\forall \varphi \in B_\rho$. Then if $\psi \in \mathscr{C}(T)$, $\|\psi\|_T \leqslant \rho$,

$$\|H\varphi - H\psi\|_T \leqslant |\varepsilon| \int_0^T \left\{ |f[s, \varphi(s), \varepsilon] - f[s, \psi(s), \varepsilon]| \right.$$

$$\left. + \frac{1}{T} \int_0^T \|f[u, \varphi(u), \varepsilon] - f[u, \psi(u), \varepsilon]\| \right\} ds$$

Since f is of class C^1 over $\|x\| \leqslant \rho$, $\|f(s, x, \varepsilon) - f(s, y, \varepsilon)\| \leqslant A\|x - y\|$ and there exists a finite number B such that

$$\|H\varphi - H\psi\| \leqslant |\varepsilon| B \|\varphi - \psi\|_T$$

If $\varepsilon_1 = 1/B$, then $\forall \varepsilon: |\varepsilon| \leqslant \varepsilon_1$, $\|H\varphi - H\psi\| \leqslant K\|\varphi - \psi\|_T$ where $K < 1$. Thus H, which depends on a and ε, is a contraction in B_ρ. In fact H is a contraction uniform in a and ε, i.e. there exists k, independent of a and ε, such that $\|H\varphi - H\psi\| \leqslant k\|\varphi - \psi\|_T$. For this it is sufficient to choose for A the supremum of the values of $(\partial f/\partial x)(s, x, \varepsilon)$ for $s \in \mathbb{R}$, $\|x\| \leqslant \rho$ and $|\varepsilon| \leqslant \varepsilon_0$, and then the supremum for B, and finally to put $\varepsilon_1 = 1/B$.

If it was found that f depended on a parameter λ, where $\|\lambda\| \leqslant \lambda_0$ in $\mathbb{R}^q$, we could then choose ε so that H would be a contraction uniform in $(a, \varepsilon, \lambda)$.

Thus there exists for all a and ε a fixed point $Z(t, a, \varepsilon)$ of H. We can prove immediately that if $\psi(t, a, \varepsilon) \in B_\rho$ and is of class C^1, $H\psi$ is also of class C^1; this is sufficient to justify differentiation under the integral sign. In those conditions, we can very easily verify that the fixed point $Z(t, a, \varepsilon)$ is of class C^1 from the fact of uniformity.

Thus we have found that $Z(t, a, \varepsilon)$ of class C^1 is periodic with period T and of mean zero. $Z(t, a, \varepsilon)$ is the limit in $\mathscr{C}(T)$ of the sequence $\varphi_0 \equiv a$, $\varphi_1 \equiv H\varphi_0, \ldots, \varphi_n \equiv H\varphi_{n-1}$; if $\varepsilon = 0$, $Ha \equiv a$ and therefore $\varphi_n \equiv a$ and $Z(t, a, 0) = a$. Naturally $Z \in B_\rho$. Finally

$$\frac{dZ}{dt}(t, a, \varepsilon) = \frac{d}{dt} H[Z(t, a, \varepsilon)]$$

$$= \varepsilon \left\{ f[t, Z(t, a, \varepsilon), \varepsilon] - \frac{1}{T} \int_0^T f[t, Z(t, a, \varepsilon), \varepsilon] \, dt \right\}$$

which completes the proof.

We recall that if f depends on a parameter λ $Z(t, a, \varepsilon, \lambda)$ is of class C^1.

Quality of the approximation

We have established that $\forall \varphi \in B_\rho$, $\psi \in B_\rho$: $\|H\varphi - H\psi\|_T \leqslant |\varepsilon| \|\varphi - \psi\|_T$. Applying this relation to φ_k and $Z(t, a, \varepsilon)$ we obtain the inequality

$$\|\varphi_{k+1}(t) - Z(t, a, \varepsilon)\|_T \leqslant |\varepsilon| B \|\varphi_k(t) - Z(t, a, \varepsilon)\|_T$$

and by iteration

$$\|\varphi_{k+1}(t) - Z(t, a, \varepsilon)\|_T \leqslant |\varepsilon|^{k+1} B^{k+1} \|a - Z(t, a, \varepsilon)\|_T$$

Since $\|a\| < \rho$ and $Z(t, a, \varepsilon) \in B_\rho$, φ_n approaches Z with the approximation

$$\|\varphi_n(t) - Z(t, a, \varepsilon)\| \leqslant |\varepsilon|^n B^n T 2\rho$$

Case where f is C^∞ relative to ε: practical method of approximation

In this case the approximations are at each stage C^∞ in ε. Consider φ_1:

$$\varphi_1(t) = a + \varepsilon \int_\vartheta^t \left[f(s, a, \varepsilon) - \frac{1}{T} \int_0^T f(u, a, \varepsilon) \, du \right] ds$$

The term in ε in the expansion of φ_1 is obtained by taking the constant term in the expansion of the integral. In the same way we obtain the term in ε^{k+1} in the expansion of φ_{k+1} by expanding the integral up to order ε^k and thus $f(s, \varphi_k, \varepsilon)$ up to order ε^k.

Thus, to obtain the expansion of $Z(t, a, \varepsilon)$ up to the term in ε^n we calculate $\varphi_0 \equiv a$, the expansion of φ_1 to the first order, starting from that the expansion of φ_2 up to $\varepsilon^2, \ldots$, starting from that the expansion of φ_{n-1} up to ε^{n-1}, that of φ_n up to ε^n.

Example Let

$$\frac{dx}{dt} = \varepsilon(1 + x + x^2 \cos t) = \varepsilon f(t, x)$$

Let $\varphi_0 \equiv a$; φ_1 is a solution of

$$\frac{dg}{dt} = \varepsilon \left[f(t, a) - \frac{1}{2\pi} \int_0^{2\pi} f(u, a)\, du \right]$$

Let $dg/dt = \varepsilon a^2 \cos t$; $g(t) = g(0) + \varepsilon a^2 \sin t$; g is of mean zero. If $g(0) = 0$, let $\varphi_1(t) = a + \varepsilon a^2 \sin t$.

φ_2 is a solution of

$$\frac{dg}{dt} = \varepsilon \left[f(t, \varphi_1) - \frac{1}{2\pi} \int_0^{2\pi} f(u, \varphi_1)\, du \right]$$

Let

$$\frac{dg}{dt} = \varepsilon \left\{ 1 + a + \varepsilon a^2 \sin t + (a + \varepsilon a^2 \sin t)^2 \cos t \right.$$

$$\left. - \frac{1}{2\pi} \int_0^{2\pi} [1 + a + \varepsilon a^2 \sin u + (a + \varepsilon a^2 \sin u)^2 \cos u]\, du \right\}$$

Restricting ourselves to the term in ε^2 we find

$$\frac{dg}{dt} = \varepsilon[1 + a + \varepsilon a^2 \sin t + a^2 \cos t + 2\varepsilon a^3 \sin t \cos t - 1 - a]$$

Let

$$g(t) = g(0) + \varepsilon a^2 \sin t - \frac{\varepsilon^2 a^2}{2} (2 \cos t + a \cos 2t)$$

and then since $\varphi_2 - a$ is periodic

$$\varphi_2(t) = a + \varepsilon a^2 \sin t - \frac{\varepsilon^2 a^2}{2} (2 \cos t + a \cos 2t)$$

DETERMINING EQUATION OR BIFURCATION EQUATION

By modifying the right-hand side of (46) we have been able to find a sequence of periodic functions, having a periodic limit $Z(t, a, \varepsilon)$, as solutions of

$$\frac{dZ}{dt}(t, a, \varepsilon) = \varepsilon f[t, Z(t, a, \varepsilon), \varepsilon] - \frac{\varepsilon}{T} \int_0^T f[u, Z(u, a, \varepsilon), \varepsilon] \, du$$

It is quite obvious that if the integral on the right-hand side is zero, Z is a solution of (46); what is remarkable is that every periodic solution of (46) can be obtained thus.

Theorem 2.3.3

Let us consider the conditions of Theorem 2.3.2. Let $Z(t, a, \varepsilon)$ be the periodic function associated with (46). If there exists a function $a(\varepsilon)$ defined for $0 \leqslant |\varepsilon| \leqslant \varepsilon_1$ satisfying

$$\int_0^T f\{t, Z[t, a(\varepsilon), \varepsilon]\} \, dt = 0 \tag{48}$$

then, in $\|x\| \leqslant \rho$, $Z[t, a(\varepsilon), \varepsilon]$ is a solution of (46) of period T.

Conversely, if there exists in $0 < |x| \leqslant \rho < R$ a solution $x(t, \varepsilon)$ of (46) of period T, defined and continuous for $0 < |\varepsilon| \leqslant \varepsilon_2 \leqslant \varepsilon_1$, then $x(t, \varepsilon) = Z(t, a(\varepsilon), \varepsilon)$. Equation (48) is called the **determining equation** or **bifurcation equation**.

Proof We have only to prove the converse. Let $x(t, \varepsilon)$ be the solution of period T, whose existence is assumed, and let

$$a(\varepsilon) = \frac{1}{T} \int_0^T x(t, \varepsilon) \, dt$$

$a(\varepsilon)$ is continuous and $\|a(\varepsilon)\| \leqslant \rho$ for $|\varepsilon| \leqslant \varepsilon_2$.

To this vector $a(\varepsilon)$ there corresponds for each ε a unique solution $z[t, a(\varepsilon), \varepsilon]$ of the system

$$\frac{dz}{dt}[t, a(\varepsilon), \varepsilon] = \varepsilon\left(f\{t, z[t, a(\varepsilon), \varepsilon], \varepsilon\} - \frac{1}{T} \int_0^T f\{u, z[u, a(\varepsilon), \varepsilon], \varepsilon\} \, du \right)$$

such that $z[t, a(0), 0] = a(0) = x(t, 0)$ since $x(t, 0)$ is a constant.

Now $x(t, \varepsilon)$ has exactly the properties of $z[t, a(\varepsilon), \varepsilon]$; in fact

$$\int_0^T \left\{ \frac{dx}{dt}(t, \varepsilon) - \varepsilon f[t, x(t, \varepsilon), \varepsilon] \right\} dt = 0$$

and since $x(T, \varepsilon) = x(0, \varepsilon)$ this relation implies that

$$\int_0^T f[t, x(t, \varepsilon), \varepsilon] \, dt = 0$$

and therefore that

$$\frac{d}{dt} x(t, \varepsilon) = \varepsilon\left\{ f[t, x(t, \varepsilon), \varepsilon] - \frac{1}{T} \int_0^T f[u, x(u, \varepsilon), \varepsilon]\, du \right\}$$

$$x(t, 0) = a(0)$$

Consequently $x(t, \varepsilon) \equiv z[t, a(\varepsilon), \varepsilon]$ and the relation

$$\int_0^T f[t, x(t, \varepsilon), \varepsilon]\, dt = 0$$

clearly implies

$$\int_0^T f\{t, z[t, a(\varepsilon), \varepsilon], \varepsilon\}\, dt = 0$$

and the bifurcation equation is indeed satisfied by the function $a(\varepsilon)$ that we have constructed.

Comments relative to the bifurcation equation

The bifurcation equation is a transcendental equation, in general very difficult to solve; in fact it assembles in a condensed form the set of necessary conditions which have to be introduced in the various methods of searching for periodic solutions.

Suppose for example that we wish to determine a periodic solution by means of its expansion in a Fourier series

$$x(t) = \sum_{n \in \mathbb{Z}} a_n \exp\left(\frac{2i\pi nt}{T}\right)$$

so that

$$\frac{dx}{dt} = \sum_{n \in \mathbb{Z}} \frac{2i\pi n}{T} a_n \exp\left(\frac{2i\pi nt}{T}\right)$$

We can try to expand in a Fourier series:

$$f[t, x(t), \varepsilon] = \sum_{n \in \mathbb{Z}} b_n \exp\left(\frac{2i\pi nt}{T}\right)$$

Naturally the coefficients b_n depend on the coefficients a_k and we must try to identify for each ε the coefficients b_n and $(2i\pi n/T)a_n$, which leads to an infinite number of equations; it is this set of equations which is represented by the bifurcation equation. We can establish, under suitable assumptions, that for sufficiently small ε and $a_0 \leqslant \rho$ we can determine $a_n, n \neq 0$, as a function of a_0. In fact, most methods of approximation resemble the one that we are using here and we shall look at an example in detail.

One advantage of the method that we have chosen to describe is that it is easy to establish the necessary and sufficient conditions of Theorem 2.3.3, as well as the convergence of the approximations towards the solution required. This is not always the case.

Another advantage is that it is possible to demonstrate a periodic solution and an approximation to it under more general conditions than those that are most often considered; we shall come back to this at the end of this section.

For the moment we shall indicate the practical method of solution.

Theorem 2.3.4 (Practical method)
Suppose that we have established the existence of a solution $Z[t, a(\varepsilon), \varepsilon]$; we can find an approximation to it up to order n as follows.

(a) Calculate, starting from a vector a, an approximation up to the term in ε^n of the associated periodic function $Z(t, a, \varepsilon)$ as we have explained. Let $Z_n(t, a, \varepsilon)$ be this approximation.

(b) Seek a function $a_n(\varepsilon)$ such that

$$\int_0^T f\{t, Z_n[t, a_n(\varepsilon), \varepsilon], \varepsilon\}\, dt$$

is of the form $\varepsilon^{n+1} h(a, \varepsilon)$ where h is continuous and $h(a, 0) = 0$. We do this by considering $\psi_n(a, \varepsilon)$ which is the expression

$$\int_0^T f\{t, Z_n[t, a_n(\varepsilon), \varepsilon], \varepsilon\}\, dt$$

in which we keep the terms only up to order n. If the Jacobian $\partial \psi_n / \partial a$ is non-zero for $|a| \leqslant \rho$ and $|\varepsilon| \leqslant \varepsilon_0$ the function $Z_n[t, a_n(\varepsilon), \varepsilon]$ is an approximation up to order n of $Z[t, a(\varepsilon), \varepsilon]$.

Proof Consider the bifurcation equation which determines $a(\varepsilon)$,

$$\Phi(a, \varepsilon) = \int_0^T f[t, Z(t, a, \varepsilon), \varepsilon] = 0$$

and the approximate equation which determines $a_n(\varepsilon)$,

$$\psi_n(a, \varepsilon) = \int_0^T f[t, Z_n(t, a, \varepsilon), \varepsilon]$$

$$= \varepsilon^{n+1} h(a, \varepsilon)$$

We know that

$$\forall a \quad |a| \leqslant \rho : \|Z_n(t, a, \varepsilon) - Z(t, a, \varepsilon)\| \leqslant A(n)\varepsilon^n$$

Therefore under the same conditions $\|\Phi(a, \varepsilon) - \psi_n(a, \varepsilon)\| \leqslant B(n)\varepsilon^n$. Suppose that

$$\left\| \frac{\partial \psi_n}{\partial a} \right\| \geqslant \alpha > 0$$

By hypothesis we assume that $a(\varepsilon)$ exists. Then

$$B(n)\varepsilon^n \geqslant \|\psi_n[a(\varepsilon), \varepsilon]\|$$

$$\geqslant \|\psi_n[a_n(\varepsilon), \varepsilon] - \psi_n[a(\varepsilon), \varepsilon]\| + \|\psi_n[a_n(\varepsilon), \varepsilon]\|$$

$$\geqslant |a_n(\varepsilon) - a(\varepsilon)| \left\|\frac{\partial\psi_n}{\partial a}\{a(\varepsilon) + \vartheta[a_n(\varepsilon) - a(\varepsilon)]\}\right\| + \varepsilon^{n+1}h(a, \varepsilon)$$

$$\geqslant |a_n(\varepsilon) - a(\varepsilon)|\alpha + B(n)\frac{\varepsilon^n}{2} \quad \text{for } \varepsilon \leqslant \varepsilon_1$$

Then

$$|a_n(\varepsilon) - a(\varepsilon)| \leqslant \frac{B(n)}{2}\varepsilon^n$$

and

$$\|Z_n[t, a_n(\varepsilon), \varepsilon] - Z_n[t, a(\varepsilon), \varepsilon]\| \leqslant \sup_{t\in[0,T]}\left\|\frac{\partial Z_n}{\partial a}[t, a(\varepsilon), a]\right\| |a_n(\varepsilon) - a(\varepsilon)|$$

$$\leqslant C(n)\varepsilon^n$$

Since

$$\|Z[t, a(\varepsilon), \varepsilon] - Z_n[t, a_n(\varepsilon), \varepsilon]\|$$

$$\leqslant \|Z[t, a(\varepsilon), \varepsilon] - Z_n[t, a(\varepsilon), \varepsilon]\| + \|Z_n[t, a(\varepsilon), \varepsilon] - Z_n[t, a_n(\varepsilon), \varepsilon]\|$$

$$\leqslant [A(n) + C(n)]\varepsilon^n$$

we obtain the result.

Approximation and ε^n-approximate solution

Theorem 2.3.5
The term of rank n in the sequence of approximation to $Z[t, a(\varepsilon), \varepsilon]$ is a function of period T; it is a $C\varepsilon^{n+1}$-approximate solution of (46) where C depends only on ρ, ε_0 and f.

Proof By definition

$$\frac{dZ_n}{dt}(t, a, \varepsilon) = \varepsilon\left\{f[t, Z_{n-1}(t, a, \varepsilon), \varepsilon] - \frac{1}{T}\int_0^T f[u, Z_{n-1}(u, a, \varepsilon), \varepsilon]\, du\right\}$$

Now we know that for $\varepsilon \leqslant \varepsilon_1$, $|a| \leqslant \rho$,

(a) $\exists M: |Z_{n-1}[t, a_n(\varepsilon), \varepsilon] - Z_n[t, a_n(\varepsilon), \varepsilon]| \leqslant M|\varepsilon|^n$

(b) $\exists N: \left|\int_0^T f\{u, Z_n[u, a_n(\varepsilon), \varepsilon], \varepsilon\}\, du\right| \leqslant N|\varepsilon|^{n+1}$

(c) f is Lipschitzian with respect to k uniformly in t and ε for $|x| \leqslant \rho$.

Then

$$\left|\frac{dZ_n}{dt} - \varepsilon f\{t, Z_n[t, a_n(\varepsilon), \varepsilon], \varepsilon\}\right|$$

$$\leqslant |\varepsilon|\,|f\{t, Z_{n-1}[t, a_n(\varepsilon), \varepsilon], \varepsilon\} - f\{t, Z_n[t, a_n(\varepsilon), \varepsilon], \varepsilon\}|$$

$$+ \frac{|\varepsilon|}{T}\int_0^T |f\{t, Z_{n-1}[t, a_n(\varepsilon), \varepsilon], \varepsilon\} - f\{t, Z_n[t, a_n(\varepsilon), \varepsilon], \varepsilon\}|\,dt$$

$$+ \frac{|\varepsilon|}{T}\int_0^T f\{u, Z_n[u, a_n(\varepsilon), \varepsilon], \varepsilon\}\,du$$

which is therefore bounded above by

$$kM|\varepsilon|^{n+1} + kM|\varepsilon|^{n+1} + \frac{|\varepsilon|}{T}N|\varepsilon|^{n+1} \leqslant C|\varepsilon|^{n+1}$$

Example

$$\frac{dx}{dt} = \varepsilon(1 + x + x^2\cos t) = \varepsilon f(t, x)$$

We have already calculated the following approximations to $Z(t, a, \varepsilon)$:

$$\varphi_0(t) \equiv a \qquad \varphi_1(t) = a + \varepsilon a^2 \sin t$$

$$\varphi_2(t) = a + \varepsilon a^2 \sin t - \frac{\varepsilon^2 a^2}{2}(2\cos t + a\cos 2t)$$

The corresponding bifurcation equations furnish the values of $a_0(\varepsilon)$, $a_1(\varepsilon)$ and $a_2(\varepsilon)$:

$$\frac{1}{2\pi}\int_0^{2\pi} f(t, a)\,dt = 1 + a$$

Therefore $a_0(\varepsilon) = -1$. Therefore $Z_0[t, a_0(\varepsilon), \varepsilon] = -1$.

$$\frac{1}{2\pi}\int_0^{2\pi} f[t, \varphi_1(t)]\,dt = 1 + a$$

Therefore $a_1(\varepsilon) = -1$. Therefore $Z_1[t, a_1(\varepsilon), \varepsilon] = -1 + \varepsilon \sin t$.

$$\frac{1}{2\pi}\int_0^{2\pi} f[t, \varphi_2(t)]\,dt = 1 + a - \varepsilon^2 a^3$$

Therefore $a_2(\varepsilon) = -(1 + \varepsilon^2)$. Therefore

$$Z_2[t, a_2(\varepsilon), \varepsilon] = -1 + \varepsilon \sin t - \varepsilon^2\left(1 + \cos t - \frac{1}{2}\cos 2t\right)$$

We can consider Z_0, Z_1, Z_2 as α-approximate solutions:

$$\frac{dZ_0}{dt} - \varepsilon f(t, Z_0) = \varepsilon \cos t$$

Therefore

$$\left\| \frac{dZ_0}{dt} - \varepsilon f(t, Z_0) \right\| \leqslant |\varepsilon|$$

$$\frac{dZ_1}{dt} - \varepsilon f(t, Z_n) = \varepsilon^2 (\sin t - \sin 2t) + \varepsilon^3 h_3(\varepsilon)$$

Therefore

$$\left\| \frac{dZ_1}{dt} - \varepsilon f(t, Z_1) \right\| \leqslant 2\varepsilon^2 [1 + k_3(\varepsilon)]$$

$$\frac{dZ_2}{dt} - \varepsilon f(t, Z_2) = \frac{3\varepsilon^3}{2} (\cos 2t + \sin 2t \cos t) + \varepsilon^4 h_4(\varepsilon)$$

Therefore

$$\left\| \frac{dZ_2}{dt} - \varepsilon f(t, Z_2) \right\| \leqslant \frac{3}{\sqrt{2}} \varepsilon^3 [1 + k_4(\varepsilon)]$$

Naturally

$$\lim_{\varepsilon \to 0} h_3(\varepsilon) = \lim_{\varepsilon \to 0} k_3(\varepsilon) = \lim_{\varepsilon \to 0} h_4(\varepsilon) = \lim_{\varepsilon \to 0} k_4(\varepsilon) = 0$$

TWO CASES OF SOLUTION OF THE BIFURCATION EQUATION

It is convenient now to give conditions for the existence of the function $a(\varepsilon)$ enabling us to construct $Z[t, a(\varepsilon), \varepsilon]$.

We shall first establish a simple theorem which is met, in one form or another, in all the methods.

Theorem 2.3.6 (Poincaré's theorem)
Let

$$\frac{dx}{dt} = \varepsilon f(t, x, \varepsilon) \tag{49}$$

be a differential equation in $\mathbb{R}^d$ where f is a function of period T, continuous and continuously differentiable with respect to ε and x for $t \in \mathbb{R}$, $0 \leqslant \|x\| \leqslant R$ and $0 \leqslant |\varepsilon| \leqslant \varepsilon_0$.

Suppose that there exists in $\mathbb{R}^d$ a vector a_0 such that

$$\int_0^T f(t, a_0, 0) \, dt = 0$$

and the Jacobian

$$\det \frac{\partial}{\partial a} \left[\int_0^T f(t, a_0, \varepsilon) \, dt \right]_{\varepsilon = 0} \neq 0$$

Then there exists in $\|x\| \leqslant R$ a solution $z(t, \varepsilon)$ of period T and such that $z(t, 0) = a_0$.

Proof The bifurcation equation relative to the approximation a_0 to $Z(t, a_0, \varepsilon)$ is

$$\int_0^T f(t, a_0, 0) \, dt = 0$$

It therefore has a solution a_0 and

$$\lim_{\varepsilon \to 0} Z(t, a_0, \varepsilon) = a_0$$

Therefore the second condition states that we can apply the theorem of implicit functions (cf. Chapter 1, Theorem 2.2.16) to the bifurcation equation

$$\int_0^T f\{t, Z[t, a(\varepsilon), \varepsilon], \varepsilon\} \, dt = 0$$

We shall indicate much more general conditions than those of Poincaré's theorem under which we can still affirm the existence of a periodic solution. It is an advantage of the method which we have presented that it lets us see these conditions appear; they arise frequently in practice as we shall see.

The following results are due essentially to Hale.

Systems invariant under certain transformations

If a system satisfies certain invariance conditions we can simplify the bifurcation equation and possibly find solutions even if the conditions of the preceding theorem are not met.

Definition 2.3.7 (Hale)

Let S be a matrix all of whose elements are zero except those on the diagonal which are $+1$ or -1 (S is $d \times d$, diagonal and real, and $S^2 = I$). If

$$\forall V \in \mathbb{R}^d \qquad Sf(-t, SV, \varepsilon) = -f(t, V, \varepsilon) \qquad (50)$$

we say that $f(t, V, \varepsilon)$ has the property (50) with respect to S.

If $f = (f_1, f_2, \ldots, f_d)$, (50) can be decomposed into d relations:

$$\text{if} \quad S_{ii} = 1 \qquad f_i(-t, SV, \varepsilon) = -f_i(t, V, \varepsilon)$$
$$\text{if} \quad S_{ii} = -1 \quad f_i(-t, SV, \varepsilon) = f_i(t, V, \varepsilon)$$

If $V = SV$, these relations indicate whether f_i is odd or even; if $S = I$, $f(t, V, \varepsilon)$ is odd for all V.

Theorem 2.3.8

Let $dx/dt = \varepsilon f(t, x, \varepsilon)$ be a critical system satisfying the conditions of Theorem 2.3.2 and having the property (50) relative to S. Let $a: \|a\| \leqslant r < \rho$ and let $Z(t, a, \varepsilon)$ be the periodic function associated with the system. Let

$$\Phi(a, \varepsilon) = \int_0^T f[t, Z(t, a, \varepsilon), \varepsilon]\, dt = 0$$

be the corresponding bifurcation equation and let $\Phi_1, \ldots, \Phi_d(a, \varepsilon)$ be the coordinates of $\Phi(a, \varepsilon)$.

For every a such that $Sa = a$ and every i such that $S_{ii} = 1$, $\Phi_i(a, \varepsilon) = 0$.

Proof Let $\varphi \in \mathscr{C}(T)$, $\|\varphi\|_T \leqslant r$ and such that $S\varphi(-t) = \varphi(t)$. Let H be the operator defined in Theorem 2.3.2 and a be such that $Sa = a$:

$$S[H\varphi](-t) = Sa + \varepsilon S \int_{-\vartheta}^{-t} \left\{ f[s, \varphi(s), \varepsilon] - \frac{1}{T} \int_{-T}^0 f[u, \varphi(u), \varepsilon]\, du \right\} ds$$

$$= a - \varepsilon \int_{\vartheta}^t \left\{ Sf[-s, \varphi(-s), \varepsilon] + \frac{1}{T} \int_0^T Sf[-u, \varphi(-u), \varepsilon]\, du \right\} ds$$

$$= a + \varepsilon \int_{\vartheta}^t \left\{ f[s, \varphi(s), \varepsilon] - \frac{1}{T} \int_0^T f[u, \varphi(u), \varepsilon]\, du \right\} ds$$

$$= H\varphi(t)$$

Then let

$$\varphi_0(t) \equiv a \qquad S\varphi_0(-t) = \varphi_0(t)$$

and therefore

$$S[H\varphi_0](-t) = H\varphi_0(t)$$

i.e.

$$S\varphi_1(-t) = \varphi_1(t)$$

and by recurrence

$$\forall n \in \mathbb{N} \qquad S\varphi_n(-t) = \varphi_n(t)$$

Consequently, in the limit when $n \to \infty$,

$$SZ(-t, a, \varepsilon) = Z(t, a, \varepsilon)$$

and therefore

$$f[-t, Z(-t, a, \varepsilon), \varepsilon] = -Sf[t, Z(t, a, \varepsilon), \varepsilon]$$

If $S_{ii}^2 = 1$, f_i is odd, of period T, and its mean is zero; therefore $\Phi_i(a, \varepsilon) = 0$.

This theorem enables us to reduce the number of scalar equations equivalent to the bifurcation equation, and it is this which permits us possibly to be released from the conditions of application of Poincaré's theorem. A simple example follows.

Theorem 2.3.9

If under the conditions of Theorem 2.3.8 $\forall V \in \mathbb{R}^d$ $f(-t, V, \varepsilon) = -f(t, V, \varepsilon)$, there exists $\forall a$: $\|a\| \leqslant r$ a periodic solution with initial value a. In dimension 2 there is therefore a centre.

See *Exercises 13 and 14* at the end of the chapter.

2.4. Examples

$$\frac{d^2x}{dt^2} + \omega_0{}^2 x = \varepsilon f\left(t, x, \frac{dx}{dt}\right) \qquad \frac{d^2x}{dt^2} + \omega_0{}^2 x = \varepsilon f\left(x, \frac{dx}{dt}\right)$$

All the examples which we shall study will be of one form or the other according as f depends on t or not. In the first case we shall assume that f is of period T and we shall describe as a free system the system for which $\varepsilon = 0$. This system has free oscillations of period $T_0 = 2\pi/\omega_0$; $\omega_0/2\pi$ is called the natural frequency.

Considering f as the input to the system and a solution as an output we shall call the latter a forced oscillation if it is a periodic solution of period T. In this case we say that there is synchronization; it may be that this is produced only when starting from a certain value which we shall call the synchronization threshold. We shall study also other phenomena (the presence of harmonics, etc.).

For an autonomous system we shall see whether there exist one or more solutions of unknown period T, possibly close to T_0. For this we shall adopt the change of variable which will allow us to put a critical system into standard form.

Changes of variable

In this type of problem we often use two particular changes of variable. The first is similar to the one that we have already indicated and is often called Van der Pol's change of variable. It consists in putting

$$x(t) = a(t) \cos \omega t + b(t) \sin \omega t$$

$$\frac{dx}{dt} = -a(t)\omega \sin \omega t + \omega b(t) \cos \omega t$$

The second, used by Bogolyubov, Krylov and Mitropolski (among others) is

$$x(t) = A(t) \cos[\omega t + \Phi(t)]$$

$$\frac{dx}{dt} = -\omega A(t) \sin[\omega t + \Phi(t)]$$

In both cases $T = 2\pi/\omega$ is the period of the solution sought. We shall study two examples in detail.

VAN DER POL'S EQUATION

This is the equation which we have already studied in Chapter 1. Let

$$\frac{d^2x}{dt^2} + \omega_0^2 x = \varepsilon(1 - x^2)\frac{dx}{dt} \tag{51}$$

Equation (51) is equivalent to the system

$$\left.\begin{array}{l} \dfrac{d}{dt}(x) = x' \\[2mm] \dfrac{d}{dt}(x') = -\omega_0^2 x + \varepsilon(1 - x^2)x' \end{array}\right\} \tag{52}$$

We seek a solution of period $T = 2\pi/\omega$ with $\omega^2 = \omega_0^2 + \lambda\varepsilon$. Using Van der Pol's change of variable

$$\left.\begin{array}{l} x(t) = a(t) \cos \omega t + b(t) \sin \omega t \\[2mm] x'(t) = -\omega a(t) \sin \omega t + \omega b(t) \cos \omega t \end{array}\right\} \tag{53}$$

and (52) becomes

$$\left.\begin{array}{l} \dfrac{da}{dt} \cos \omega t + \dfrac{db}{dt} \sin \omega t = 0 \\[3mm] \dfrac{da}{dt}(-\omega \sin \omega t) + \dfrac{db}{dt}(\omega \cos \omega t) + (\omega_0^2 - \omega^2)[a(t)\cos \omega t + b(t)\sin \omega t] \\[3mm] \hspace{6cm} = \varepsilon(1 - x^2)x' \end{array}\right\} \tag{54}$$

where we must express x and x' on the right-hand side of the second equation by using (53). We thus obtain the standard form of (51):

$$\left.\begin{array}{l} \dfrac{da}{dt} = -\dfrac{\varepsilon}{\omega}[\lambda x + (1 - x^2)x'] \sin \omega t = \varepsilon g_1[t, a(t), b(t)] \\[3mm] \dfrac{db}{dt} = \dfrac{\varepsilon}{\omega}[\lambda x + (1 - x^2)x'] \cos \omega t = \varepsilon g_2[t, a(t), b(t)] \end{array}\right\} \tag{55}$$

where x and x' must, as before, be replaced by the expression (53).

If $(a, b) \in \mathbb{R}^2$ is a constant vector, we shall find the first approximation to the bifurcation equation

$$\int_0^{2\pi/\omega} g(t, a, b) \, dt = 0$$

i.e.

$$\int_0^{2\pi/\omega} \lambda(a \cos \omega t + b \sin \omega t) \sin \omega t$$

$$+ \left[1 - (a \cos \omega t + b \sin \omega t)^2\right](-a\omega \sin \omega t + b\omega \cos \omega t) \sin \omega t = 0$$

$$\int_0^{2\pi/\omega} (a \cos \omega t + b \sin \omega t) \cos \omega t$$

$$+ \left[1 - (a \cos \omega t + b \sin \omega t)^2\right](-a\omega \sin \omega t + b\omega \cos \omega t) \cos \omega t = 0$$

Almost all the terms are zero and the first equation reduces to

$$0 = \int_0^{2\pi/\omega} \left[(\lambda b - a\omega) \sin^2 \omega t + (\omega a^3 - 2ab^2\omega) \sin^2 \omega t \cos^2 \omega t \right.$$

$$\left. + ab^2 \sin^4 \omega t\right] dt$$

i.e. $0 = \lambda b - a\omega + \frac{1}{4}(\omega a^3 - 2ab^2\omega) + \frac{3}{4}ab^2\omega$.
The second equation reduces to

$$0 = \lambda a + b\omega + \frac{1}{4}(-\omega b^3 + 2ba^2\omega) - \frac{3}{4}ba^2\omega$$

Thus we find

$$F_1(\lambda, a, b) = \lambda b - a\omega\left(1 - \frac{a^2 + b^2}{4}\right) = 0$$

$$F_2(\lambda, a, b) = \lambda a + b\omega\left(1 - \frac{a^2 + b^2}{4}\right) = 0$$

If we can find a, b, λ satisfying these two equations, and if

$$\frac{D(F_1, F_2)}{D(\lambda, a, b)} \neq 0$$

we can state by virtue of Poincaré's theorem that there exists a solution of period $2\pi/\omega$, where $\omega^2 = \omega_0^2 + \lambda\varepsilon$, whose limit when $\varepsilon \to 0$ is (a, b). Now

$$\frac{D(F_1, F_2)}{D(\lambda, a, b)} = \begin{vmatrix} b & \omega\left(-1 + \dfrac{a^2}{2} + \dfrac{a^2 + b^2}{4}\right) & \lambda + \dfrac{ab\omega}{2} \\ a & \lambda - \dfrac{ab\omega}{2} & \omega\left(1 - \dfrac{b^2}{2} - \dfrac{a^2 + b^2}{4}\right) \end{vmatrix}$$

$a = b = 0$ is clearly a solution ($x \equiv 0$ is a solution of (51)), but $\lambda = 0$, $a^2 + b^2 = 4$ is also a solution, for the Jacobian is then of rank 2.

Thus there exists a solution of period $2\pi/\omega_0$ up to the term in ε^2, with orbit $\omega^2 x^2 + x'^2 = 4\omega^2$ (for (51)) up to the term in ε.

DUFFING'S EQUATION

We are concerned with the equation

$$\frac{d^2 x}{dt^2} + \omega_0{}^2 x + \varepsilon\gamma x^3 = 0 \tag{56}$$

Let us make Van der Pol's change of variable with $\omega^2 = \omega_0{}^2 + \varepsilon\lambda$. Equation (56) becomes

$$\frac{da}{dt}\cos\omega t + \frac{db}{dt}\sin\omega t = 0$$

$$-\omega^2[a(t)\cos\omega t + b(t)\sin\omega t] + \omega\left(-\frac{da}{dt}\sin\omega t + \frac{db}{dt}\cos\omega t\right)$$

$$= -\omega_0{}^2[a(t)\cos\omega t + b(t)\sin\omega t]$$
$$- \varepsilon\gamma[a(t)\cos\omega t + b(t)\sin\omega t]^3$$

The standard form is then

$$\frac{da}{dt} = \varepsilon\left\{-\frac{\lambda}{\omega}[a(t)\cos\omega t + b(t)\sin\omega t]\right.$$

$$\left. +\frac{\gamma}{\omega}[a(t)\cos\omega t + b(t)\sin\omega t]^3\right\}\sin\omega t$$

$$= \varepsilon g_1[t, a(t), b(t)]$$

$$\frac{db}{dt} = \varepsilon\left\{\frac{\lambda}{\omega}[a(t)\cos\omega t + b(t)\sin\omega t]\right.$$

$$\left. -\frac{\gamma}{\omega}[a(t)\cos\omega t + b(t)\sin\omega t]^3\right\}\cos\omega t$$

$$= \varepsilon g_2[t, a(t), b(t)]$$

We can establish immediately that

$$g_1(-t, a, 0) = -g_1(t, a, 0)$$

$$g_2(-t, a, 0) = g_2(t, a, 0)$$

Let

$$S = \begin{pmatrix} 1 & 0 \\ 0 & -1 \end{pmatrix}: S\begin{pmatrix} a \\ 0 \end{pmatrix} = \begin{pmatrix} a \\ 0 \end{pmatrix} \quad\text{and}\quad Sg\left[-t, S\begin{pmatrix} a \\ 0 \end{pmatrix}\right] = -g\left[t, \begin{pmatrix} a \\ 0 \end{pmatrix}\right]$$

Let us take as a first approximation to the periodic function associated with the system the constant

$$\begin{pmatrix} a \\ 0 \end{pmatrix}$$

The bifurcation equation reduces to a scalar equation since

$$\int_0^{2\pi/\omega} g_1(t, a, 0) \, dt = 0$$

(cf. Theorem 2.3.8). The remaining equation is

$$a \int_0^{2\pi/\omega} (\lambda \cos^2 \omega t - \gamma a^2 \cos^4 \omega t) \, dt = 0$$

It is satisfied if $a = 0$ ($x \equiv 0$ is a solution of (56)) and if

$$a^2 = \frac{4 \lambda}{3 \gamma} \quad \text{or} \quad a^2 = \frac{4}{3} \frac{\omega^2 - \omega_0^2}{\varepsilon \gamma}$$

since the integral is equal to

$$\lambda \frac{\pi}{\omega} - \frac{3}{4} \gamma a^2 \frac{\pi}{\omega}$$

We note that from Theorem 2.3.8

$$\int_0^{2\pi/\omega} g_1[t, Z(t, a, 0), 0] \, dt = 0$$

exactly, and that

$$\int_0^{2\pi/\omega} g_2[t, Z(t, a, 0), 0] \, dt = 0$$

has a solution $a(\varepsilon)$ since

$$\frac{\partial}{\partial a} \left[\int_0^{2\pi/\omega} g_2(t, a, 0) \, dt \right]_{\varepsilon = 0} = -\frac{6}{4} \gamma \pi a \frac{1}{\omega_0} \neq 0$$

Thus this is a case in which we have not had recourse to Poincaré's theorem.

Therefore there exists an infinity of periodic solutions depending on the parameter a; the solution with initial value

$$\begin{pmatrix} a \\ 0 \end{pmatrix}$$

has as first approximation $x(t) = a \cos \omega t$, with $\omega^2 = \omega_0^2 + \frac{3}{4} \varepsilon \gamma a^2$ and with period

$$T = \frac{2\pi}{\omega_0} \left(1 - \frac{3}{8} \varepsilon \gamma \frac{a^2}{\omega_0^2} \right)$$

The solution with initial value

$$\begin{pmatrix} a \\ b \end{pmatrix}$$

has as first approximation

$$x(t) = a \cos \omega t + \frac{b}{\omega} \sin \omega t$$

$$\omega^2 = \omega_0^2 + \tfrac{3}{4}\varepsilon\gamma a^2$$

Thus all the solutions are periodic.

AUTONOMOUS PERTURBATION OF AN AUTONOMOUS SYSTEM HAVING A CENTRE

The preceding example is of an autonomous system

$$\frac{d^2x}{dt^2} + \omega_0^2 x = 0$$

perturbed by $\varepsilon\gamma x^3$, it is remarkable that in spite of the structural instability of the free system, the perturbed system is topologically equivalent since it too possesses a centre.

This arises from the conjunction of the following two facts.

(a) The autonomous system had a centre, the orbits not depending on the period.

(b) The perturbed system had an invariance property reducing the bifurcation equation to a single scalar equation; by choosing T as a function of the single parameter appearing in this equation we find for every initial value a solution whose period depends on this initial value.

Every time these conditions apply simultaneously, this phenomenon will be reproduced, as we shall explain in the following theorem.

Theorem 2.4.1 (Hale)
Let

$$\frac{d^2x}{dt^2} + \omega_0^2 x = \varepsilon f\left(x, \frac{dx}{dt}\right) \tag{57}$$

be an autonomous system where f is a function of class C^2.
Suppose that $f(0, 0) = 0$ and that one of the following properties is satisfied

(a) $f(u, v) = f(u, -v)$
(b) $f(-u, v) = -f(u, v)$

Then $\exists \varepsilon_1 > 0$ and $\rho > 0$: $\forall a \ \|a\| \leqslant \rho$ and $\forall \varepsilon \ |\varepsilon| \leqslant \varepsilon_1$ the solution with initial value $(a, 0)$ is periodic with period $T(a)$ and the system has a centre.

Proof The method consists in following the steps in the Duffing equation: we put $\omega^2 = \omega_0^2 + \varepsilon\lambda$, we make the Van der Pol change of variables and we establish that the bifurcation equation relative to the first approximation reduces to a single equation which allows us to determine λ, and therefore T, as a function of a.

THE NON-AUTONOMOUS DUFFING EQUATION; HARMONICS; SUB-HARMONICS

Suppose the system considered on page 505 is subjected to a periodic input. We study the equation

$$\frac{d^2x}{dt^2} + \omega_0^2 x + \varepsilon\gamma x^3 = A \cos \omega t \tag{58}$$

We have here the same equation as that on page 505 but this time we assume that $\omega_0^2 \sim k^2\omega^2$ where $k = p/q \in Q$, so that, having put the system into standard form, we shall be dealing with a critical system.

If $\varepsilon = 0$

$$x(t) = \frac{A}{\omega_0^2 - \omega^2} \cos \omega t$$

is a solution for $\omega \neq \omega_0$ with period $T = 2\pi/\omega$, whereas if $A = 0$ we know that there exists, for every initial value, a periodic solution of period

$$T = \frac{2\pi}{\omega_0}\left(1 - \frac{3}{8}\varepsilon\gamma\frac{a^2}{\omega_0^2}\right) \approx \frac{2\pi}{\omega_0}$$

We shall examine whether there are solutions in which are superposed terms of period $T = 2\pi/\omega$ and $T = 2\pi/k\omega \approx 2\pi/\omega_0$. For $k = p$ we say that we have a **harmonic** of order p; if $k = 1/n$ we have a **sub-harmonic** of order n. In general we can take $k = p/q \in Q$; the superposition of the two functions will be periodic.

We make the following change of variable after putting $k^2\omega^2 = \omega_0^2 + \lambda\varepsilon$:

$$x(t) = a(t)\cos k\omega t + b(t)\sin k\omega t + \frac{A}{\omega_0^2 - \omega^2}\cos \omega t$$

$$x'(t) = k\omega[-a(t)\sin k\omega t + b(t)\cos k\omega t] - \frac{A\omega}{\omega_0^2 - \omega^2}\sin \omega t$$

so that

$$\frac{da}{dt} = \frac{\varepsilon}{k\omega}\sin k\omega t\left\{\gamma\left[a(t)\cos k\omega t + b(t)\sin k\omega t + \frac{A}{\omega_0^2 - \omega^2}\cos \omega t\right]^3 \right.$$

$$\left. - \lambda[a(t)\cos k\omega t + b(t)\sin k\omega t]\right\}$$

$$\frac{db}{dt} = \frac{\varepsilon}{k\omega} \cos k\omega t \left\{ -\gamma \left[a(t) \cos k\omega t + b(t) \sin k\omega t + \frac{A}{{\omega_0}^2 - \omega^2} \cos \omega t \right]^3 \right.$$

$$\left. + \lambda [a(t) \cos k\omega t + b(t) \sin k\omega t] \right\}$$

a system which we shall denote by

$$\frac{da}{dt} = \varepsilon g_1[t, a(t), b(t)]$$

$$\frac{db}{dt} = \varepsilon g_2[t, a(t), b(t)]$$

We establish again that

$$g_1(-t, a, 0) = -g_1(t, a, 0)$$

$$g_2(-t, a, 0) = g_2(t, a, 0)$$

By considering the constant

$$\begin{pmatrix} a \\ 0 \end{pmatrix}$$

as a first approximation to the associated periodic function we know that the bifurcation equation reduces to

$$\int_0^{T(k)} g_2(t, a, 0) \, dt = 0$$

where $T(k)$ is the period of g, i.e. the lowest common multiple of $2\pi/\omega$ and $2\pi/k\omega$; if $k = p/q$ this number is $q2\pi/\omega$. By putting

$$B = \frac{A}{{\omega_0}^2 - \omega^2}$$

the determinant equation is therefore

$$\int_0^{2\pi q/\omega} \cos \frac{p}{q}\omega t \left[\lambda a \cos \frac{p}{q}\omega t - \gamma \left(a \cos \frac{p}{q}\omega t + B \cos \omega t \right)^3 \right] dt = 0$$

or

$$0 = \lambda a\pi - \gamma a^3 \frac{3}{4}\pi - 3\gamma \frac{a^2 B}{8} \int_0^{2\pi} [3 \cos(p+q)s + 3 \cos(p-q)s$$

$$+ \cos(3p+q)s + \cos(3p-q)s] \, ds$$

$$- 3\gamma a \frac{B^2}{8} \int_0^{2\pi} [2 + 2 \cos 2ps + 2 \cos 2qs$$

$$+ \cos 2(p+q)s + \cos 2(p-q)s] \, ds$$

$$-\gamma \frac{B^3}{8} \int_0^{2\pi} [3\cos(p+q)s + 3\cos(p-q)s$$

$$+ \cos(3q+p)s + \cos(3q-p)s]\,ds$$

We see from this expression that $p = q$, $p = 3q$ and $p = q/3$ are particular cases. Apart from these cases, the first approximation to the determinant equation is

$$0 = \lambda - \gamma a^2 \tfrac{3}{4} - \gamma 2B^2 \tfrac{3}{4}$$

or

$$k^2\omega^2 - \omega_0{}^2 = \tfrac{3}{4}\gamma\varepsilon(a^2 + 2B^2) \tag{59}$$

There exists therefore a solution

$$a \cos \omega k t + b \sin k\omega t + \frac{A}{\omega_0{}^2 - \omega^2} \cos \omega t \quad \text{where } k \neq 1, \tfrac{1}{3}, 3$$

if

$$\omega^2 \approx \frac{\omega_0{}^2}{k^2}\left[1 + \frac{3\gamma\varepsilon}{\omega_0{}^2}\left(a^2 + 2\frac{k^4 A^2}{\omega_0{}^4(k^2-1)^2}\right)\right]$$

The relation (59) gives the amplitude a as a function of ω; we can represent it as shown in Fig. 8.6 ($|\varepsilon_1| < |\varepsilon_2| < |\varepsilon_3|$).

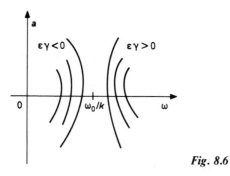

Fig. 8.6

If $k = 1$, $\tfrac{1}{3}$ or 3, we find the following bifurcation equations:

$$k = 1 \qquad \lambda a = \gamma \frac{3}{4}(a+B)^3 \qquad\qquad \text{with } B = -\frac{A}{\lambda\varepsilon}$$

$$k = \frac{1}{3} \qquad \lambda = \frac{3}{4}\gamma(a^2 + aB + 2B^2) \qquad \text{with } B = -\frac{9A}{8\omega^2}$$

$$k = 3 \qquad \lambda = \frac{\gamma}{4a}(3a^3 + B^3) \qquad\qquad \text{with } B = \frac{A}{8\omega^2}$$

If $k = \frac{1}{3}$

$$\omega^2 = 9\omega_0{}^2\left[1 + \frac{3\varepsilon\gamma}{4\omega_0{}^2}(a^2 + aB + 2B^2)\right]$$

The frequency response curve has the same appearance as in the previous case.
If $k = 3$

$$\omega^2 = \frac{\omega_0{}^2}{9}\left[1 + \frac{\gamma\varepsilon}{4a\omega_0{}^2}(3a^3 + B^3)\right]$$

We can therefore find, for small a, a harmonic of order 3 on condition that the amplitude of the forcing term is at least as small as $ka^{1/3}$. If $A = a^{1/3}A_0$

$$\omega^2 = \frac{\omega_0{}^2}{9}\left\{1 + \frac{\gamma\varepsilon}{4\omega_0{}^2}\left[3a^2 + \left(\frac{9A_0{}^2}{8\omega_0{}^2}\right)^3\right]\right\}$$

If A is very small, e.g. equal to $aA_1{}^{1/3}$,

$$\omega^2 = \frac{\omega_0{}^2}{9}\left[1 + \frac{\gamma\varepsilon}{4\omega_0{}^2}(3 + A_1)a^2\right]$$

If $k = 1$ and if A does not depend on ε, we see that we cannot find periodic solutions if ε tends to zero. (If $A = 0$, we find again the result of page 507.)

See *Exercises 15 to 18* at the end of the chapter.

2.5. The methods of Lindstedt and Poincaré

We quote these methods mainly because they are often met in the literature and because it is useful to compare them with the one which we have expounded. Let

$$\frac{\mathrm{d}x}{\mathrm{d}t} = A(t)x + f(t, x, \varepsilon) \tag{60}$$

be a differential equation where f and A are of period $T = 2\pi/\omega$. To study this it is natural to use Poincaré's theorem for expansion in powers of ε and of $y_0 - x_0$ of the solution $\varphi(t, 0, y_0)$ in the neighbourhood of the solution of $\mathrm{d}x/\mathrm{d}t = A(t)x$ with initial value x_0 (cf. Chapter 7, Theorem 4.1.1).

In Lindstedt's method we are looking for a solution of the form

$$x(t) = \varphi_0(t) + \varepsilon\varphi_1(t) + \varepsilon^2\varphi_2(t) + \dots$$

of period $2\pi/\omega$ where $\omega = \omega_0 + \varepsilon\omega_1 + \varepsilon^2\omega_2 + \dots$.

Formally this presents no problem; we put $x(t)$ in (60) and we find—in principle—φ_0, then $\varphi_1, \varphi_2, \dots$ by identifying the powers of ε.

Naturally we must stop at a certain point, and that is where the problems begin.

(a) How do we study the existence of a periodic solution, knowing only an expansion stopped at a point k?

(b) How do we interpret the fact that the solution constructed is the expansion of a periodic function?

Lindstedt does not really give an answer to the first question; as regards the second, he proposes to require that all the terms in the expansion be periodic. This condition, which is rather arbitrary, is not necessary; to convince oneself of this it is enough to observe that a periodic function can have in its series expansion, in powers of ε, terms which are not periodic, as for example

$$\cos(1 + \varepsilon)t = \cos t \cos \varepsilon t - \sin \varepsilon t$$

$$= \cos t - \varepsilon t \sin t - \frac{\varepsilon^2}{2} t^2 \cos t + \frac{\varepsilon^3}{3!} t^3 \sin t + \dots$$

The terms of the form $t \sin t$, $t^2 \cos t$, etc., are called **secular terms**.

By identifying the coefficients of the different powers of ε we write, in fact, the differential equations of which $\varphi_0, \varphi_1, \dots$ are solutions. In requiring that these functions be periodic, we are actually requiring that the conditions of the Fredholm alternative be satisfied at each stage.

These conditions determine the period and the initial value of a periodic solution. We sometimes call this operation 'suppression of secular terms'. It is nothing more than the application of the conditions of the Fredholm alternative. We shall see an example of this below.

First we must make it clear that this method can lead to divergent series not corresponding to any actual periodic solution. It nevertheless remains true that the method allows simple and rapid calculations whose results should be exploited with care, if in addition we have at our disposal other information on the existence of a periodic solution. In any case, we construct by this procedure a periodic function which is in general an ε-approximate solution; this is already a considerable result especially if we know also that there exists, in the vicinity of its orbit, a periodic solution.

Examples

(a) *Duffing's equation*: we are actually in the situation which we have just described, since we know that all the solutions are periodic. Let

$$\frac{d^2 x}{dt^2} + \omega_0{}^2 x + \varepsilon \gamma x^3 = 0 \tag{61}$$

To make the period ω obvious, we put $y(s) = x(s/\omega)$ and (61) becomes

$$\omega^2 \frac{d^2 y}{ds^2} + \omega_0{}^2 y + \varepsilon \gamma y^3 = 0 \tag{62}$$

We shall seek a solution of period 2π.

Let $y(s) = y_0(s) + \varepsilon y_1(s) + \ldots$, $\omega^2 = \omega_0{}^2 + \varepsilon \omega_1{}^2 + \ldots$. Substituting in (62) we find

$$(\omega_0{}^2 + \varepsilon \omega_1{}^2 + \ldots)\left(\frac{d^2 y_0}{ds^2} + \varepsilon \frac{d^2 y_1}{ds^2} + \ldots\right)$$

$$+ \omega_0{}^2(y_0 + \varepsilon y_1 + \ldots) + \varepsilon \gamma (y_0{}^3 + \ldots) = 0$$

This gives

$$\omega_0{}^2 \frac{d^2 y_0}{ds^2} + \omega_0{}^2 y_0 = 0$$

$$\omega_1{}^2 \frac{d^2 y_0}{ds^2} + \omega_0{}^2 \frac{d^2 y_1}{ds^2} + \omega_0{}^2 y_1 + \gamma y_0{}^3 = 0$$

... etc.

We can assume that

$$\frac{dy}{ds}(0) = 0$$

Then $y_0(s) = A \cos s$ and y_1 is a solution of

$$\frac{d^2 y_1}{ds^2} + y_1 = \frac{\omega_1{}^2}{\omega_0{}^2} A \cos s - \gamma \frac{A^3}{\omega_0{}^2} \cos^3 s \qquad (63)$$

or, in vector form

$$\frac{dY}{ds} = \begin{pmatrix} 0 & 1 \\ -1 & 0 \end{pmatrix} Y(s) + \begin{bmatrix} 0 \\ f(s) \end{bmatrix}$$

whose adjoint equation is

$$\frac{dZ}{ds} = -\begin{pmatrix} 0 & -1 \\ 1 & 0 \end{pmatrix} Z \quad \text{with solutions} \quad \begin{pmatrix} Z_1 \\ Z_2 \end{pmatrix}$$

The conditions of periodicity for solutions of (63) are

$$\int_0^{2\pi} f(s) Z_2(s) \, ds = 0$$

for every solution Z. Since Z_2 is a solution of

$$\frac{d^2 Z_2}{ds^2} + Z_2 = 0$$

the conditions are then

$$\int_0^{2\pi} (\omega_1{}^2 \cos s - \gamma A^2 \cos^3 s) \sin s \, ds$$

$$= 0 = \int_0^{2\pi} (\omega_1{}^2 \cos s - \gamma A^2 \cos^3 s) \cos s \, ds$$

The first relation is always satisfied and the second is $0 = \omega_1^2 \pi - \gamma A^2 \frac{3}{4}\pi$. Thus $\omega_1^2 = \frac{3}{4}\gamma A^2$ and (63) becomes

$$\frac{d^2 y_1}{ds^2} + y_1 = -\frac{\gamma A^3}{4\omega_0^2} \cos 3s$$

and

$$y_1(s) = B \cos s + C \sin s + \frac{\gamma A^3}{32\omega_0^2} \cos 3s$$

If we assume that $dy_1/ds = 0, C = 0$ and we put $B = 0$ since the term $\varepsilon B \cos s$ is negligible in comparison with $A \cos s$, we find as the approximation

$$x(t) = A \cos \omega t + \varepsilon\gamma \frac{A^3}{32\omega_0^2} \cos 3\omega t + \dots \quad \text{with } \omega^2 = \omega_0^2 + \frac{3}{4}\varepsilon\gamma A^2 + \dots$$

If we had wished to 'remove the secular terms', we should have solved Eqn (63) in the form

$$y_1(s) = a \cos s + \frac{A}{\omega_0^2} \int_0^s (\omega_1^2 \cos u - \gamma A^2 \cos^3 u) \sin(s - u) \, du$$

which would have made the secular term obvious:

$$\frac{A}{\omega_0^2} \sin s \left(\omega_1^2 \frac{s}{2} - \gamma A^2 \frac{3s}{8} \right)$$

Its removal implies that $\omega_1^2 = \gamma A^2 \frac{3}{4}$, which we have already found.

We note that, with reference to the result obtained in the second example of Section 2.4, we obtain a supplementary term in the approximation without excessive effort.

(b) *Van der Pol's equation*: this is

$$\frac{d^2 x}{dt^2} + x = \varepsilon(1 - x^2)\frac{dx}{dt}$$

We then put

$$y(s) = x\left(\frac{s}{\omega}\right): \quad \omega^2 \frac{d^2 y}{ds^2} + y = \varepsilon\omega(1 - x^2)\frac{dy}{ds}$$

If $y(s) = y_0(s) + \varepsilon y_1(s) + \dots$ and $\omega = 1 + \varepsilon\omega_1 + \dots$, by substitution

$$\frac{d^2 y_0}{ds^2} + y_0 = 0 \quad \text{and} \quad y_0(s) = A \cos s \quad (\text{if } y_0'(0) = 0)$$

$$\frac{d^2 y_1}{ds^2} + y_1 = -2\omega_1 \frac{d^2 y_0}{ds^2} + (1 - y_0^2)\frac{dy_0}{ds}$$

$$= 2\omega_1 A \cos s - A \sin s + A^3 \sin s \cos^2 s$$

The conditions for orthogonality are

$$\int_0^{2\pi} (2\omega_1 \cos s - \sin s + A^2 \sin s \cos^2 s) \cos s \, ds$$

$$= 0$$

$$= \int_0^{2\pi} (2\omega_1 \cos s - \sin s + A^2 \sin s \cos^2 s) \sin s \, ds$$

or $2\pi\omega_1 = 0$ and $-\pi + (A^2/8)2\pi = 0$, i.e. $\omega_1 = 0$ and $A^2 = 4$. We find again the approximation of the first example of Section 2.4:

$$x(t) = 2 \cos t + \varepsilon^2 \varphi_2(t) + \ldots \qquad \omega = 1 + \varepsilon^2 \omega_2 + \ldots$$

It was Poincaré who noted that Lindstedt's method could lead to series which did not correspond to actual solutions. He proposed another method, naturally correct, in which the approximations are not necessarily periodic.

With reference to the equation

$$\frac{d^2x}{dt^2} + x = \varepsilon f\left(x, \frac{dx}{dt}\right)$$

we seek a solution near $x_0(t) = A \cos t$ with initial value $A + \lambda(\varepsilon)$ and of period $2\pi + \varepsilon T_1$. In accordance with Poincaré's theorem we then put

$$\frac{d^2x}{dt^2} + x = \varepsilon f\left(x, \frac{dx}{dt}\right)$$

$$x(t) = A \cos t + \varepsilon \varphi_1(t) + \lambda \varphi_2(t) + \varepsilon \lambda \varphi_3(t) + \ldots$$

and we require that

$$\left.\begin{array}{l} x(2\pi + \varepsilon T_1, \varepsilon, \lambda) - x(0, \varepsilon, \lambda) = 0 \\[2mm] \dfrac{dx}{dt}(2\pi, \varepsilon T_1, \varepsilon, \lambda) - x(0, \varepsilon, \lambda) = 0 \end{array}\right\} \tag{64}$$

These conditions are met for $\varepsilon = 0$; if we can find $T_1(\varepsilon)$ and $\lambda(\varepsilon)$ removing the term in ε and such that the Jacobian is zero for $\varepsilon = 0$, we shall be able to state that there exists a periodic solution of which we have obtained the first term of the approximation.

In general Eqns (64) are of the same order of complexity as the bifurcation equations which strongly resemble them.

Part 3: Method of the mean

3.1. Generalities

Consider the equation

$$\frac{dx}{dt} = \varepsilon f(t, x) \tag{65}$$

where f has period T. We know that the bifurcation equation corresponding to a constant a is, as a first approximation,

$$\bar{f}(a) = \frac{1}{T} \int_0^T f(a, t) = 0$$

If a satisfies $\bar{f}(a) = 0$ and if $d\bar{f}(a)/dx$ is invertible, there exists, in a neighbourhood of a, a solution $\varphi(t, \varepsilon)$ of (65) of period T and such that $\varphi(t, 0) \equiv a$.

We can reformulate this result in the following way. If $f(t, x)$ is of period T, let us consider the two equations

$$\frac{dx}{dt} = \varepsilon f(t, x) \tag{66}$$

$$\frac{dy}{dt} = \frac{\varepsilon}{T} \int_0^T f(t, y) \, dt = \bar{f}(y) \tag{67}$$

If there exists a solution of (67), $y \equiv a$, such that $d\bar{f}(a)/dy$ is invertible, there exists a solution of (66), $\varphi(t, \varepsilon)$ of period T, such that

$$\lim_{\varepsilon \to 0} \varphi(t, \varepsilon) \equiv a$$

The solution to the 'averaged' system is thus an approximation to that of the initial system.

In its turn, the averaged system (67), which is an autonomous system, is approximated by the variational equation in the neighbourhood of a which is an equation with constant coefficients:

$$\frac{dz}{dt} = \varepsilon \frac{d\bar{f}}{dy}(a)z \tag{68}$$

Study of the eigenvalues of $d\bar{f}(a)/dy$ enables us to determine the stability of the solutions and the fact that $d\bar{f}(a)/dy$ is invertible allows us to establish the uniqueness of $\varphi(t, \varepsilon)$ in a neighbourhood of a. This is due to the following fact: all the eigenvalues of the resolvent of (68) are of the form $\exp(\varepsilon \lambda T)$ where λ is an eigenvalue of $d\bar{f}(a)/dy$ and is therefore non-zero, so that if ε is small enough $\varepsilon \lambda T \neq 2n\pi T \ \forall n \in \mathbb{N}$ and, as we have already remarked after Theorem 7.1.2, this implies that $z \equiv 0$ is the only periodic solution of (68).

To make clear this link between Eqn (66) and Eqn (67) we shall establish the following lemma.

Lemma 3.1.1 (Bogolyubov)
Let $\tilde{f}(t, \xi)$ be a periodic solution of

$$\frac{\partial \tilde{f}}{\partial t}(t, \xi) = f(t, \xi) - \bar{f}(\xi)$$

and let $x = \xi + \varepsilon \tilde{f}(t, \xi)$. x is a solution of (66) iff ξ is a solution of

$$\frac{d\xi}{dt} = \varepsilon \bar{f}(\xi) + \varepsilon^2 F(t, \xi, \varepsilon) \tag{69}$$

where F has period T, so that if ξ has period T, x also has period T.

Proof Since $\bar{f}$ is the mean of f, there exists a solution $\tilde{f}(t, \xi)$ of period T of the equation

$$\frac{\partial \tilde{f}}{\partial t}(t, \xi) = f(t, \xi) - \bar{f}(\xi)$$

For ε sufficiently small the change of variable $x \rightarrow \xi$ is close to the identity I and therefore bijective. We then calculate

$$\frac{dx}{dt} = \frac{d\xi}{dt} + \varepsilon \frac{\partial \tilde{f}}{\partial t}(t, \xi) + \varepsilon \frac{\partial \tilde{f}}{\partial \xi}(t, \xi) \frac{d\xi}{dt}$$

and x is a solution of (65) iff

$$\frac{dx}{dt} = \varepsilon f(t, x)$$

$$= \varepsilon f(t, x) - \varepsilon \bar{f}(\xi) + \varepsilon \bar{f}(\xi)$$

$$= \varepsilon \frac{\partial \tilde{f}}{\partial t}(t, \xi) + \left[I + \varepsilon \frac{\partial \tilde{f}}{\partial \xi}(t, \xi) \right] \frac{d\xi}{dt}$$

i.e.

$$\left[I + \varepsilon \frac{\partial \tilde{f}}{\partial \xi}(t, \xi) \right] \frac{d\xi}{dt} = \varepsilon \bar{f}(\xi) + \varepsilon \left[f(t, x) - \bar{f}(\xi) - \frac{\partial \tilde{f}}{\partial t}(t, \xi) \right]$$

$$= \varepsilon \bar{f}(\xi) + \varepsilon [f(t, x) - f(t, \xi)] + \varepsilon \left[f(t, \xi) - \bar{f}(\xi) - \frac{\partial \tilde{f}}{\partial t}(t, \xi) \right]$$

The last term is zero by construction,

$$f(t, x) - f(t, \xi) = \varepsilon \frac{\partial f}{\partial x}[t, \eta(\xi)] \tilde{f}(t, \xi)$$

Thus

$$\frac{d\xi}{dt} = \left[I + \varepsilon \frac{\partial \tilde{f}}{\partial \xi}(t, \xi) \right]^{-1} \left\{ \varepsilon \bar{f}(\xi) + \varepsilon^2 \frac{\partial f}{\partial x}[t, \eta(\xi)] \tilde{f}(t, \xi) \right\}$$

$$= \varepsilon \bar{f}(\xi) + \varepsilon^2 F(t, \xi, \varepsilon)$$

where F is certainly of period T.

Remark   If

$$f(t, x) = a(x) + b(x) \exp\left(2i\pi \, \frac{t}{T} \right)$$

$$\tilde{f}(t, x) = \frac{b(x)}{2i\pi} \, T \exp\left(2i\pi \, \frac{t}{T} \right)$$

If f is expanded in a Fourier series, we find a sum of such expressions.

In general we can prove the following theorem, which states clearly, in a wider framework, the reasoning we presented at the beginning of the section; the proof of the theorem uses Lemma 3.1.1.

Theorem 3.1.2

Let

$$f: \begin{cases} \mathbb{R} \times \mathbb{R}^n \times \mathbb{R}^+ \to \mathbb{R}^n \\ (t, x, \varepsilon) \rightsquigarrow f(t, x, \varepsilon) \end{cases}$$

be a continuous function having a partial derivative

$$\frac{\partial f}{\partial x} (t, x, \varepsilon)$$

and period T. Consider the equations

$$\frac{dx}{dt} = \varepsilon f(t, x, \varepsilon) \tag{70}$$

and

$$\frac{dx}{dt} = \varepsilon \, \frac{1}{T} \int_0^T f(t, x, 0) \, dt = \varepsilon \bar{f}(x) \tag{71}$$

Suppose there exists a solution $x \equiv a$ of (71) such that

$$\det \frac{d\bar{f}}{dx} (a) \neq 0$$

Then there exists $\varepsilon_0 > 0$ such that $\forall \varepsilon \leqslant \varepsilon_0$ there exists a solution of (70) $\varphi(t, \varepsilon)$, continuous over $(\mathbb{R} \times [0, \varepsilon_0])$, of period T and such that $\varphi(t, 0) \equiv a$.

The latter is unique within a neighbourhood of a. Moreover, if all the eigenvalues of $d\bar{f}(a)/dx$ have strictly negative real part, $\varphi(t, \varepsilon)$ is uniformly asymptotically stable $\forall \varepsilon \leqslant \varepsilon_0$. If, however, one of these eigenvalues has a strictly positive real part, $\varphi(t, \varepsilon)$ is unstable.

Comparison with Hale's method

(a) From the point of view of the search for periodic solutions, and approximations to them, the two methods are closely related.

If a is a constant chosen as an approximation of order zero to the periodic function associated with (65), the bifurcation equation of order zero is then

$$\int_0^T f(t, a)\, dt = 0$$

The approximation of order zero to the periodic solution of (65) is $\varphi_0(t) \equiv a$ where $\bar{f}(a) = 0$. If $d\bar{f}(a)/dx$ is invertible it corresponds to a unique periodic solution in the neighbourhood of a.

We note that if

$$\det \frac{d}{dx} \bar{f}(a) = 0$$

it may still be possible to apply Hale's method.

Let us now seek the approximation of order 1 to the periodic function associated with (65); this is $\varphi_1(t) = a + \psi(t, a)$ where $\psi(t, a)$ is a solution of

$$\frac{dg}{dt} = \varepsilon[f(t, a) - \bar{f}(a)] = \varepsilon \frac{\partial \tilde{f}}{\partial t}(t, a)$$

Choosing $\tilde{f}$ with zero mean, $\varphi_1(t) = a + \varepsilon \tilde{f}(t, a)$; this is what Bogolyubov calls an 'improved' approximation of the first order, a being such that $\bar{f}(a)$ is null.

The first-order approximation to the periodic solution is obtained starting from the bifurcation equation

$$0 = \int_0^T f[t, a(\varepsilon)]\, dt + \varepsilon \int_0^T \frac{\partial f}{\partial x}[t, a(\varepsilon)]\tilde{f}[t, a(\varepsilon)]\, dt \qquad (72)$$

Putting $a(\varepsilon) = a + \eta(\varepsilon)$ we find $\eta(\varepsilon) = b\varepsilon + O(\varepsilon)$; therefore

$$\varphi_1[t, a(\varepsilon)] = a + \varepsilon \tilde{f}(t, a) + \varepsilon^2 g(t, \varepsilon)$$

The first-order approximation is therefore the same.

To obtain a second-order approximation Bogolyubov proposes an equivalent of Lemma 3.1.1; we make the change of variable $x = \zeta + \tilde{f}(t, \zeta)$ and x is a solution of (65) if

$$\frac{d\zeta}{dt} = \varepsilon \bar{f}(\zeta) + \varepsilon^2 \frac{1}{T} \int_0^T \tilde{f}(t, \zeta) \frac{\partial f}{\partial x}(t, \zeta)\, dt + \varepsilon^3 G(t, \zeta, \varepsilon) \qquad (73)$$

If $a(\varepsilon)$ is a solution of the bifurcation equation (72), $\zeta = a(\varepsilon)$ is a solution of (73) up to order ε^3, so that to this accuracy Hale's first-order approximation is equal to Bogolyubov's second-order approximation. Beyond the second order it becomes extremely complicated to write the corresponding change of variable in Bogolyubov's method.

Thus the approximations are comparable and are rather simpler with Hale's method.

(b) Hale's method can obviously be generalized to equations in Banach spaces but it does not have an equivalent for the study of almost-periodic or integral solutions for which the method of the mean is particularly well suited. In fact the method of the mean has a more important field of application.

We shall return in the next section to the case of almost-periodic solutions.

Slow time, rapid-time

We shall give some indications, of a heuristic nature, of a way of interpreting the method of the mean; this interpretation is, in its turn, a basic idea for generalizations.

We take again

$$\frac{dx}{dt} = \varepsilon f(t, x) \tag{74}$$

where $f(t + T, x) = f(t, x) \; \forall t \in \mathbb{R}, \; \forall x \in \mathbb{R}^d$. Let $y_\varepsilon(s) = x(s/\varepsilon)$. Then

$$\frac{dy_\varepsilon}{ds}(s) = \frac{1}{\varepsilon}\frac{dx}{dt}\left(\frac{s}{\varepsilon}\right) = f\left[\frac{s}{\varepsilon}, x\left(\frac{s}{\varepsilon}\right)\right]$$

Then x is a solution of (74) iff y_ε is a solution of the equation

$$\frac{dy_\varepsilon}{ds}(s) = f\left[\frac{s}{\varepsilon}, y_\varepsilon(s)\right] \tag{75}$$

In this equation there occur two time scales, s/ε which for small ε varies 'rapidly', and s which varies 'relatively' slowly.

A solution of (75) satisfies

$$y_\varepsilon(s) = x(0) + \int_0^s f\left(\frac{u}{\varepsilon}, y_\varepsilon(u)\right) du$$

Now we can very easily prove that, if $a(t)$ is continuous and of period T, for every bounded continuous function φ

$$\int_0^s a\left(\frac{u}{\varepsilon}\right)\varphi(u)\, du \underset{\varepsilon \to 0}{\to} \int_0^s \varphi(u)\left[\frac{1}{T}\int_0^T a(v)\, dv\right] du$$

In fact

$$\int_0^s\left[a\left(\frac{u}{\varepsilon}\right) - \frac{1}{T}\int_0^T a(v)\, dv\right]\varphi(u)\, du = \varepsilon\int_0^T\left[a(u) - \frac{1}{T}\int_0^T a(v)\, dv\right]\varphi(u\varepsilon)\, du$$

$$+ \varepsilon\int_0^{2T}\left[a(u) - \frac{1}{T}\int_0^T a(v)\, dv\right]\varphi(u\varepsilon)\, du$$

$$\vdots$$

$$+ \varepsilon\int_{(k-1)T}^{kT}\left[a(u) - \frac{1}{T}\int_0^T a(v)\, dv\right]\varphi(u\varepsilon)\, du$$

$$+ \varepsilon\int_{kT}^{s/\varepsilon}\left[a(u) - \frac{1}{T}\int_0^T a(v)\, dv\right]\varphi(u\varepsilon)\, du$$

where $k = [s/\varepsilon T]$ is the integral part of $s/\varepsilon T$: $|\varphi(u\varepsilon)| \leqslant |\varphi(\varepsilon pT)| +$

$|\varphi(\varepsilon u) - \varphi(\varepsilon p T)|$ in each interval $\{\varepsilon p T \leqslant \varepsilon u \leqslant \varepsilon(p+1)T\}$: $\|\varphi(\varepsilon u) - \varphi(\varepsilon p T)\| \leqslant \eta$ if $\varepsilon \leqslant \alpha/T$. The integral relative to $\varphi(\varepsilon p T)$ is zero and each of the others is bounded above by

$$\eta\varepsilon \int_0^T \left| a(u) - \frac{1}{T} \int_0^T a(v)\,dv \right| du = A\eta\varepsilon$$

and the sum is bounded above by $s\eta A/T$, which establishes the result.

If in Eqn (74) $f(s, x) = a(s)x$ we obtain that

$$y_\varepsilon(s) \approx x(0) + \int_0^s y_\varepsilon(u)\left[\frac{1}{T} \int_0^T a(v)\,dv \right] du \quad \text{when } \varepsilon \to 0$$

$$= x(0) + \int_0^s x\left(\frac{u}{\varepsilon}\right)\left[\frac{1}{T} \int_0^T a(v)\,dv \right] du$$

$$= x(0) + \varepsilon \int_0^{s/\varepsilon} x(w)\left[\frac{1}{T} \int_0^T a(v)\,dv \right] dw$$

or

$$x(s) = x(0) + \varepsilon \int_0^s x(w)\bar{a}\,dw$$

Therefore

$$\frac{dx}{ds} = \varepsilon\bar{a}x$$

We must note that this reasoning can no longer be used for a function $f[s, x(s), \varepsilon]$. It nevertheless remains true that this way of considering two different time scales clarifies the result obtained by replacing Eqn (74) by the 'averaged' equation, and even suggests that, in the case where we have at the outset several time scales, we take the mean with respect to some of them, as we shall see in the following section.

Example: motion of a pendulum whose point of suspension oscillates. Let ϑ be the angle between the vertical and the pendulum measured from the lowest point and let $\omega/2\pi$ be the frequency of vibration of the point of suspension. The equation of motion is then

$$\frac{d^2\vartheta}{dt^2}(t) + \lambda\frac{d\vartheta}{dt}(t) + \left(\frac{g}{l} - a\frac{\omega^2}{l}\sin\omega t\right)\sin\vartheta(t) = 0 \tag{76}$$

where g denotes the acceleration due to gravity, l the length of the pendulum supposed of mass 1, a the amplitude of vibration of the support and λ a coefficient of friction.

We put

$$\varepsilon = \frac{a}{l} \qquad \varepsilon^2 k^2 = \frac{g}{l}\frac{1}{\omega^2} = \frac{\omega_0^2}{\omega^2} \qquad 2\varepsilon\rho = \frac{\lambda}{\omega} > 0$$

If $\Theta(\tau) = \vartheta(\tau/\varepsilon)$, (76) takes the form

$$\frac{d^2\Theta}{d\tau^2}(\tau) + 2\varepsilon\rho\,\frac{d\Theta}{d\tau}(\tau) + (\varepsilon^2 k^2 - \varepsilon\sin\tau)\sin\Theta(\tau) = 0 \qquad (77)$$

To obtain a system in standard form we put

$$\Theta(\tau) = \varphi(\tau) - \varepsilon\sin\tau\sin\varphi(\tau)$$

$$\frac{d\Theta}{d\tau}(\tau) = \varepsilon\Omega(\tau) - \varepsilon\cos\tau\sin\varphi(\tau)$$

Then

$$\frac{d\varphi}{d\tau} = \varepsilon\Omega(\tau)\big[1 - \varepsilon\sin\tau\cos\varphi(\tau)\big]^{-1}$$

$$= \varepsilon\Omega(\tau) + \varepsilon^2\Omega(\tau)\sin\tau\cos\varphi(\tau) + \ldots$$

$$= \varepsilon\Omega(\tau) + \varepsilon^2 G(\tau, \Omega, \varphi)$$

$$\frac{d\Omega}{d\tau} = \varepsilon\big[\Omega(\varepsilon)\cos\tau\cos\varphi(\tau) - 2\rho\Omega(\tau)\big] + 2\rho\cos\tau\sin\varphi(\tau)$$

$$\qquad - \big[k^2\sin\varphi(\tau) + \sin^2\tau\sin\varphi(\tau)\cos\varphi(\tau)\big]$$

The equation obtained by applying the method of the mean relative to τ varying over $[0, 2\pi]$ is then

$$\frac{d\psi}{d\tau} = \varepsilon\eta$$

$$\frac{d\eta}{d\tau} = \varepsilon(-2\rho\eta - k^2\sin^2\psi - \tfrac{1}{2}\sin\psi\cos\psi)$$

or

$$\frac{d^2\psi}{d\tau^2} + 2\rho\varepsilon\,\frac{d\psi}{d\tau} + \varepsilon^2\sin\psi(k^2 + \tfrac{1}{2}\cos\psi) = 0 \qquad (78)$$

$\psi = \pi$ is a solution.

We shall study the variational equation relative to this equilibrium assuming that $k^2 < \tfrac{1}{2}$. If $x = \psi - \pi$ and $y = dx/d\tau$, the system equivalent to (78) becomes

$$\frac{dx}{d\tau} = y$$

$$\frac{dy}{d\tau} = -2\rho\varepsilon y + \varepsilon^2\sin x(k^2 - \tfrac{1}{2}\cos x)$$

The linear variational equation in the neighbourhood of $x = 0$ is

$$\frac{dX}{dt} = \begin{pmatrix} 0 & 1 \\ \varepsilon^2(k^2 - \tfrac{1}{2}) & -2\rho\varepsilon \end{pmatrix} X$$

The eigenvalues of the matrix are $-\rho\varepsilon \pm \varepsilon[\rho^2 - (\tfrac{1}{2} - k^2)]^{1/2}$; since their real part is strictly negative, $x = 0$ is a stable solution.

Thus, if $2g/l = 2\omega_0^2 < (a^2/l^2)\omega^2$, the pendulum can describe a stable periodic motion of frequency $\omega/2\pi$ in the neighbourhood of $\vartheta = \pi$.

3.2. Case of almost-periodic equations and solutions

As we have indicated, we often need to consider functions in which several periods are superimposed, e.g.

$$\sum_{p \leqslant n} A_p(x) \cos \omega_p t$$

These functions are particular cases of almost-periodic functions, of which the following is the definition.

Definition 3.2.1
A continuous function

$$\mathbb{R} \times \mathbb{R}^n \to \mathbb{C}^p$$

$$(t, x) \rightsquigarrow f(t, x)$$

is called almost-periodic in t uniformly, for x in a compact set K if $\forall \eta > 0$ $\exists l(\eta) > 0$ such that in every interval of length $l(\eta)$ we can find $\tau > 0$: $\| f(t + \tau, x) - f(t, x) \| \leqslant \eta \ \forall t \in \mathbb{R}, \ \forall x \in K$.

The set of almost-periodic functions is a vector space closed for the topology of uniform convergence (for $x \in K$ fixed).

Example $\exp(ait) + \exp(bit)$ is an almost-periodic function. If a/b is rational it is actually periodic, but if this ratio is irrational it is almost-periodic but not periodic.

In general for every sequence

$$a_n: \sum_{n \in \mathbb{N}} |a_n| < \infty$$

and every sequence λ_n of real numbers

$$\sum_{n \leqslant N} a_n \exp(i\lambda_n t)$$

is almost-periodic.

We can show that the following proposition is true.

Proposition 3.2.2
Every almost-periodic function is bounded and uniformly continuous and

$$\lim_{T \to \infty} \frac{1}{T} \int_0^{a+T} f(u)\, du$$

exists for all a and is independent of a. We shall write

$$\bar{f}(x) = \lim_{T \to \infty} \int_0^T f(u, x)\, du$$

in the case where f, dependent on x, is almost-periodic uniformly in x. We shall say that $\bar{f}$ is the mean of f.

Theorem 3.1.2 rests essentially on Bogolyubov's lemma for change of variable. To establish this we must find $\tilde{f}(t, \xi)$, a periodic solution of

$$\frac{\partial \tilde{f}}{\partial t}(t, \xi) = f(t, \xi) - \bar{f}(\xi)$$

If instead of being periodic

$$f = \sum_{n \in \mathbb{N}} a_n(x) \exp(i\lambda_n t)$$

and if $\bar{f}(\xi)$ denotes the mean of f, the function

$$\sum_{n:\lambda_n \neq 0} a_n(x) \exp(i\lambda_n t) \frac{1}{i\lambda_n}$$

is a solution of this equation.

It is not surprising, then, that a comparable lemma exists for almost-periodic functions, to the extent that we can find $\tilde{f}(t, x, \varepsilon)$, almost-periodic, such that the difference

$$\frac{\partial \tilde{f}}{\partial t}(t, x, \varepsilon) - f(t, x) - \bar{f}(x)$$

is arbitrarily small.
This suffices to prove the following theorem.

Theorem 3.2.3 (Hale)
Let

$$\frac{dx}{dt} = \varepsilon f(t, x) + \varepsilon h(\varepsilon t, x) \tag{79}$$

be a differential equation where h, f, $\partial f(t, x)/\partial x$ and $\partial h(t, x)/\partial x$ are continuous, where f is almost-periodic uniformly in x over compact sets and where h is periodic with period T. Let

$$\bar{f}(\xi) = \lim_{T \to \infty} \frac{1}{T} \int_0^T f(t, \xi) \, d\xi$$

Let

$$\frac{dx}{dt} = \varepsilon \bar{f}(x) + \varepsilon h(\varepsilon t, x) = \varepsilon F(\varepsilon t, x) \qquad (80)$$

This is the system obtained by taking the mean relative to rapid time.

Suppose that there exists a periodic solution $\varphi_0(\varepsilon t)$ of period T/ε of (80). Finally let $\tau = \varepsilon t$ and let

$$\frac{dz}{d\tau} = \frac{\partial F}{\partial x} [\tau, \varphi_0(\tau)] z = Az \qquad (81)$$

be the variational equation of (80) in the neighbourhood of φ_0.

If the characteristic exponents of A have non-zero real parts, there exists ε_0 such that for all $\varepsilon \leqslant \varepsilon_0$ there exists $\varphi(t, \varepsilon)$, a solution of (79), almost-periodic and such that

$$\lim_{\varepsilon \to 0} \varphi(t, \varepsilon) - \varphi_0(\varepsilon t) = 0$$

and unique in the neighbourhood of φ_0. If the characteristic exponents have strictly negative real parts, $\varphi(t, \varepsilon)$ is uniformly asymptotically stable; if one of them is strictly positive $\varphi(t, \varepsilon)$ is unstable.

Example: pendulum with friction, periodic excitation and oscillating support   Denoting by ϑ the angle of the axis of the pendulum measured from the lowest point, by g the acceleration due to gravity, by l the length of the pendulum, by $a \sin(t/\varepsilon)$ the motion of the support, and by $k > 0$ the coefficient of friction, if F is the force of excitation the equation of motion is

$$\frac{d^2\vartheta}{dt^2} + k \frac{d\vartheta}{dt} + \left(\frac{g}{l} - \frac{a}{l\varepsilon^2} \sin \frac{t}{\varepsilon} \right) \sin \vartheta = F \cos \omega t$$

Suppose that $a/l = \varepsilon A$ and $g/l = 1$; then the equation becomes

$$\frac{d^2\vartheta}{dt^2} + k \frac{d\vartheta}{dt} + \left(1 - \frac{A}{\varepsilon} \sin \frac{t}{\varepsilon} \right) \sin \vartheta = F \cos \omega t \qquad (82)$$

equivalent to the system

$$\frac{dx}{dt} = y - A \cos \frac{t}{\varepsilon} \sin x$$

$$\frac{dy}{dt} = yA \cos \frac{t}{\varepsilon} \cos x - A^2 \left(\cos \frac{t}{\varepsilon} \right)^2 \sin x \cos x - \sin x$$

$$+ F \cos \omega t - k \left(y - A \cos \frac{t}{\varepsilon} \sin x \right) \qquad (83)$$

If

$$\frac{t}{\varepsilon} = \tau \qquad \begin{aligned} X(\tau) &= x(\varepsilon\tau) \\ Y(\tau) &= y(\varepsilon\tau) \end{aligned}$$

the system (83) becomes

$$\frac{dx}{dt}(\varepsilon\tau) = y(\varepsilon\tau) - A \cos \tau \sin x(\varepsilon\tau)$$

$$\begin{aligned} \frac{dy}{dt}(\varepsilon\tau) = {} & Ay(\varepsilon\tau) \cos \tau \cos \varepsilon\tau - A^2 \cos^2 \tau \sin x(\varepsilon\tau) \cos x(\varepsilon\tau) \\ & - \sin x(\varepsilon\tau) + F \cos \varepsilon\omega\tau \\ & - k[y(\varepsilon\tau) - A \cos \tau \sin x(\varepsilon\tau)] \end{aligned}$$

or

$$\frac{dX}{d\tau} = \varepsilon(Y - A \cos \tau \sin X)$$

$$\begin{aligned} \frac{dY}{d\tau} = {} & \varepsilon(AY \cos \tau \cos X - A^2 \cos^2 \tau \sin X \cos X - \sin X) \\ & + \varepsilon F \cos \varepsilon\omega\tau - k(Y - A \cos \tau \sin X)\varepsilon \end{aligned}$$

We have the conditions for application of Theorem 3.2.3. Let us take the mean relative to rapid time τ; we obtain the system

$$\frac{dX}{d\tau} = \varepsilon Y$$

$$\frac{dY}{d\tau} = \varepsilon\left(-\frac{A^2}{2}\sin X \cos X - \sin X\right) + \varepsilon F \cos \varepsilon\omega\tau - kY\varepsilon$$

and, returning to t,

$$\frac{dx}{dt} = y$$

$$\frac{dy}{dt} = -\sin x(1 + A^2 \cos x) + F \cos \omega t - ky$$

This system is equivalent to the equation

$$\frac{d^2\psi}{dt^2} + k\frac{d\psi}{dt} + \left(1 + \frac{A^2}{2}\cos \psi\right)\sin \psi = F \cos \omega t \tag{84}$$

We shall study the conditions under which Eqn (84) has a solution ψ_0 of period $2\pi/\omega$ such that the characteristic exponents of the linear variational equation relative to ψ_0 have negative real parts.

If F is zero $\psi(t) \equiv \pi$ is a solution; we therefore put $\eta = \psi - \pi$, and then η is a solution of

$$\frac{d^2\eta}{dt^2} + k\frac{d\eta}{dt} + \left(\frac{A^2}{2} - 1\right)\eta = F \cos \omega t + \left(\frac{A^2}{2} - 1\right)\eta + \sin \eta\left(1 - \frac{A^2}{2}\cos \eta\right)$$

$$= g(t, \eta, F) \tag{85}$$

We shall assume that F is small and $A^2 > 2$. Since $k \neq 0$ the system (85) is non-critical and we have the conditions for application of Theorem 2.2.2. There then exists, in a neighbourhood of $\eta = 0$, a periodic solution $\eta_0(t)$ and consequently a periodic solution $\psi_0(t)$ in a neighbourhood of π. For $F = 0$ the characteristic exponents have strictly negative real parts. Now (85) and the system with $F = 0$ are topologically equivalent if $F \leqslant \varepsilon_0$; their linearized systems are then topologically equivalent and the characteristic exponents of the linearized system of (85) are strictly negative for $\varepsilon \leqslant \varepsilon_0$ (cf. Chapter 2, Theorem 3.2.6 and following theorems).

Thus, there exists a solution of (82), almost-periodic and stable, which remains within a neighbourhood of $\vartheta = \pi$.

See **Exercises 19 and 20** at the end of the chapter.

Part 4: Autonomous systems in $\mathbb{R}^2$

Dynamic systems in $\mathbb{R}^2$ present the following very interesting peculiarity; every closed curve γ determines two distinct regions in $\mathbb{R}^2$, the interior and the exterior. If an equation $\mathrm{d}x/\mathrm{d}t = f(x)$ has a periodic solution with orbit γ, and if the Cauchy problem has at every point a unique solution, the solutions emerging from points interior to γ remain within γ; those starting from exterior points remain outside γ.

Every system topologically equivalent to the above one therefore has the same property.

Moreover we can prove the following theorem.

Theorem 4.1.1
Let $\mathrm{d}x/\mathrm{d}t = f(x)$ be a dynamic system in $\mathbb{R}^2$ where f is of class C^1. Suppose that there exists a periodic solution which is asymptotically orbitally stable. Let C be its orbit.

Let G be an open neighbourhood of C. Then $\exists \varepsilon > 0$: $\forall g$ satisfying $\|g - f\|_{1,G} < \varepsilon$ the system $\mathrm{d}y/\mathrm{d}t = g(y)$ has an orbitally stable periodic solution.

4.1. Two theorems concerning divergence

We shall denote an autonomous system in $\mathbb{R}^2$ by the form

$$\frac{\mathrm{d}x}{\mathrm{d}t} = P(x, y)$$

$$\frac{\mathrm{d}y}{\mathrm{d}t} = Q(x, y)$$

We shall assume that the vector V, of components P and Q, is of class C^1. We shall write div V for the quantity

$$\text{div } V = \frac{\partial P}{\partial x} + \frac{\partial Q}{\partial y}$$

The following theorem arises from Green's formula which in $\mathbb{R}^2$ has the following elementary form: if D is a bounded connected domain in $\mathbb{R}^2$ with boundary γ

$$\int_\gamma P \, dx - Q \, dy = \iint_D \text{div } V(x, y) \, dx \, dy$$

If γ is the orbit of a periodic solution, the first term is zero; therefore so also is the second, which establishes the following result.

Proposition 4.1.2
In every connected domain in which div V has fixed sign, there cannot be a periodic solution.

A second theorem peculiar to $\mathbb{R}^2$ results from the following fact: if we know one characteristic multiplier we automatically know the other since we know their product (cf. Theorem 1.1.3).

Theorem 4.1.3
Let

$$\frac{dx}{dt} = f(x) \tag{86}$$

be an equation in $\mathbb{R}^2$ where

$$f = \begin{pmatrix} P \\ Q \end{pmatrix}$$

is of class C^1. If $(x_0, y_0)(t)$ is a solution of (86) of period T, it is orbitally stable with asymptotic phase if

$$\int_0^T \left\{ \frac{\partial P}{\partial x} \left[x_0(t), y_0(t) \right] + \frac{\partial Q}{\partial y} \left[x_0(t), y_0(t) \right] \right\} dt < 0$$

It is unstable if this curvilinear integral along the orbit from (x_0, y_0) is strictly greater than zero.

Proof   The variational equation in the neighbourhood of (x_0, y_0) is

$$\left. \begin{aligned}
\frac{dx}{dt} &= \frac{\partial P}{\partial x} \left[x_0(t), y_0(t) \right] x + \frac{\partial P}{\partial y} \left[x_0(t), y_0(t) \right] y \\
\frac{dy}{dt} &= \frac{\partial Q}{\partial x} \left[x_0(t), y_0(t) \right] x + \frac{\partial Q}{\partial y} \left[x_0(t), y_0(t) \right] y
\end{aligned} \right\} \tag{87}$$

Since (x_0, y_0) is periodic one of the characteristic multipliers of (87) is 1, and the other is therefore equal to the product. Now we know (cf. Theorem 1.1.3) that the following holds:

$$\exp\left[\int_0^T \operatorname{tr} DF\binom{x_0}{y_0}(t)\,\mathrm{d}t\right] = \exp\left[\int_\gamma \operatorname{div} f(M)\,\mathrm{d}M\right]$$

It suffices then to apply Corollary 1.2.3.

See **Exercises 21 and 22** at the end of the chapter.

4.2. Poincaré–Bendixson theory

Let

$$\frac{\mathrm{d}x}{\mathrm{d}t} = f(x, y) \tag{88}$$

be an autonomous system in $\mathbb{R}^2$ defined over a domain at each point of which we assume that the Cauchy problem has a unique solution defined for $t \geqslant t_0$ $\forall t_0$.

We shall assume that f is continuous and shall define a **singular point** as a point at which $f(x, y) = 0$, and a **regular point** as a point at which this vector is not zero.

Definition 4.2.1

We define a **transverse segment** as a closed segment σ of which every point is regular and such that at each of these points $f(x, y)$ is not collinear with σ.

The following properties whose geometric interpretation is obvious can be proved very easily by using the uniqueness of the orbits passing through each point of the domain in which f is defined and continuous.

(a) Every regular point is an interior point of a transverse segment.

(b) At each point of a transverse segment there exists one orbit, and only one, which passes through it; all these orbits traverse the segment in the same direction. (In fact $f(x, y)$ which is continuous cannot change to the other side of a segment without vanishing or becoming collinear with that segment.)

(c) For every $\varepsilon > 0$, for every point interior to a transverse segment, there exists a disc γ_ε centred at P such that every solution emerging at the instant t_0 from a point within γ_ε leaves this disc before the instant $t_0 + \varepsilon$. (In fact $f(x, y)$ is minimized in the neighbourhood of every point interior to a transverse segment; therefore the 'velocity' is minimized on the orbits of solutions emerging from a point sufficiently close to P.)

(d) A closed orbit cuts a transverse segment in one point only.

(e) An arc of an orbit of finite length cuts a transverse segment in a finite number of points; the latter are in the same order as the parameters of the points of intersection (Fig. 8.7).

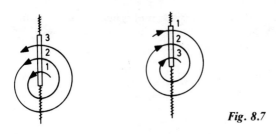

Fig. 8.7

It was the distribution of these points of intersection that we studied in Chapter 1 to establish the existence of a periodic solution of the Van der Pol equation.

Definition 4.2.2
We define a **positive semi-orbit** as a set of points C^+ associated with a solution φ: $C^+ = \{\varphi(t), t \geqslant t_0\}$. We define the **limit points of** C^+ as the set

$$L(C^+) = \{M \in \mathbb{R}^2 : \exists \{t_n\}_{n \in \mathbb{N}} \lim_{t_n \to \infty} \varphi(t_n) = M\}$$

Proposition 4.2.3
Let C^+ be a positive semi-orbit contained in a closed set. Let $L(C^+)$ be the set of limit points of C^+. If $L(C^+)$ contains a regular point P the orbit passing through P is defined $\forall t \in \mathbb{R}$.

Proof　There exists a sequence $\{P_n\}_{n \in \mathbb{N}}$ of points of C^+ such that

$$\lim_{n \to \infty} P_n = P$$

We can choose the parameter of C^+ so that $\varphi(t, P_n) = \varphi(t + t_n, P_0)$ (i.e. $P_n = \varphi(t_n, P_0)$); then

$$\lim_{n \to \infty} \varphi(t, P_n) = \varphi(t, P)$$

can be written

$$\lim_{n \to \infty} \varphi(t + t_n, P_0) = \varphi(t, P)$$

Now

$$\forall t \quad \lim_{n \to \infty} t + t_n = +\infty$$

Therefore $\varphi(t, P)$ is defined for all t.

Proposition 4.2.4

If C^+ and $L(C^+)$ have a common point P, C^+ is in fact a closed orbit corresponding to a periodic solution.

Proof Let t_0 be the parameter of P in a parametrization of C^+. Since P is in $L(C^+)$, there exists in every neighbourhood G of P a point of the orbit with parameter greater than $t_0 + n$ $\forall n$. If C^+ were not the orbit of a periodic solution there would then exist an infinity of points of intersection with every transverse segment contained in G, but the points of intersection following each other in the order of the parametrization could not have limit P.

The following theorem is a corollary of these propositions.

Theorem 4.2.5 (The Poincaré–Bendixson theorem)

Let C^+ be a positive semi-orbit contained in a closed bounded set. If we assume that all the points of $L(C^+)$ are regular, then $L(C^+)$ is a periodic orbit and either $C^+ = L(C^+)$ or $L(C^+)$ is a limit cycle.

We apply the Poincaré–Bendixson theorem mainly in the following way. We determine a positively invariant bounded closed set K; if all the points of K are regular, there exists a periodic orbit. A set is positively invariant if every positive semi-orbit having a point within K is entirely contained in K. Suppose for example that γ_1 and γ_2 are two curves delimiting an annular set and that f, at every point of γ_1, is an out-going vector and, at every point of γ_2, is an incoming vector (Fig. 8.8). Such an ensemble is clearly positively invariant.

Naturally there are numerous corollaries of Theorem 4.2.5. We quote one which is frequently applied.

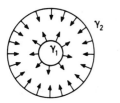

Fig. 8.8

Theorem 4.2.6

Let

$$\frac{d^2 x}{dt^2} + f(x)\frac{dx}{dt} + g(x) = 0 \quad \text{(Lienard's equation)} \tag{89}$$

We put

$$F(y) = \int_0^y f(u)\, du$$

and we assume that

(a) $\qquad f(x) = f(-x), \qquad f(0) < 0 \quad \text{and} \quad \lim_{|y| \to \infty} F(y) = +\infty$

Suppose that there exists y_0 such that $F(y) < 0$ if $0 < y < y_0$ and $F(y_0) = 0$. Moreover $F(y)$ is increasing for $y > y_0$.

(b) $\qquad xg(x) > 0 \quad \text{if } x \neq 0$

Then there exists a unique periodic solution and it is orbitally stable with asymptotic phase (cf. Exercise, Chapter 4, Section 3.2).
  This theorem is applied, for example, to Van der Pol's equation

$$\frac{d^2x}{dt^2} - \lambda^2(1 - x^2)\frac{dx}{dt} + x = 0 \quad \text{where } \lambda \neq 0$$

Theorem 4.2.6 can be proved by constructing, starting from a Lyapunov function, a positively invariant bounded closed set.

Exercises

1. Consider the following equation in $\mathbb{R}$:

$$\frac{d^2x}{dt^2} + \frac{dx}{dt} + \varepsilon x \sin t = 0 \tag{90}$$

Does there exist, for small enough ε, a solution of (90) of period 2π?

2. State explicitly, in dimension 2, the Dunford formula (from Definition 1.2.2) and study the eigenvalues and generalized proper spaces of $f(M)$.

3. Given the system

$$\left.\begin{array}{l} \dfrac{dx}{dt} = -y + x(x^2 + y^2)\sin\dfrac{\pi}{(x^2 + y^2)^{1/2}} \\[3mm] \dfrac{dy}{dt} = x + y(x^2 + y^2)\sin\dfrac{\pi}{(x^2 + y^2)^{1/2}} \end{array}\right\} \tag{91}$$

denote this system also by $dX/dt = F(X)$ where $X \in \mathbb{R}^2$.
  (a) Verify that $\varphi(t)$ defined by

$$x(t) = \tfrac{1}{2}\cos t$$
$$y(t) = \tfrac{1}{2}\sin t$$

  is a solution of (91).
  (b) Solve the variational equation $dz/dt = DF[\varphi(t)]z$ relative to φ. Hence

deduce that the resolvent of (91) satisfies

$$\Phi(t, 0) = \begin{pmatrix} \exp\left(-\dfrac{\pi t}{2}\right)\cos t & -\sin t \\[2ex] \exp\left(-\dfrac{\pi t}{2}\right)\sin t & \cos t \end{pmatrix}$$

(c) Verify that φ is orbitally stable with asymptotic phase.
(d) Study (91) by converting to polar coordinates.

4. Let $f(x, y)$ be a function of class C^1 and $k > 0$. Let

$$\frac{d^2 x}{dt^2} - k(1 - x^2)\frac{dx}{dt} + x = \varepsilon f\left(x, \frac{dx}{dt}\right) \tag{92}$$

Show that there exists $\varepsilon_0 > 0$ such that $\forall \varepsilon \leqslant \varepsilon_0$ (92) has a unique periodic solution in a neighbourhood of the unique periodic solution of Van der Pol's equation corresponding to $\varepsilon = 0$.

5. Study for small ε the stability of the system

$$\frac{d^2 x}{dt^2} + \frac{dx}{dt} + \varepsilon \sin tx = 0$$

6. If $\omega > 0$ and $\gamma > 0$ study, for small ε, the stability of the system

$$\frac{d^2 x}{dt^2} + (4\omega^2 + \gamma \varepsilon + \varepsilon \cos 2t)x = 0$$

7. Let f be the function of period 2π defined by $f(t) = \omega^2 + \varepsilon$ for $0 \leqslant t < \pi$ and $f(t) = \omega^2 - \varepsilon$ for $\pi \leqslant t < 2\pi$. Study the stability of the system

$$\frac{d^2 x}{dt^2} + f(t)x = 0$$

8. Study Mathieu's equation in the domain $a < 0$.

9. Prove Theorem 2.2.3 by putting, for every function f of $\mathscr{C}(T)$,

$$Hf(t) = [\exp(-\varepsilon BT) - I]\varepsilon \int_t^{t+T} \exp[\varepsilon B(t - u)]\{g[u, f(u), \varepsilon] + h(u)\}\, du$$

10. Let $dx/dt = Bx + g(\omega t, x, \varepsilon) + h(\omega t)$ where B, g and h satisfy the conditions of Theorem 2.2.3. Verify that there exists ε_1, ρ and ω_1, $0 < \rho < R$, $0 < \varepsilon_1 \leqslant \varepsilon_0$, such that $\forall \varepsilon < \varepsilon_1$, $\omega \geqslant \omega_1$, there exists in $\|x\| \leqslant \rho$ a unique solution $\varphi(t, \omega)$ of period T/ω such that

$$\lim_{\omega \to \infty} \varphi(t, \omega) = 0$$

11. Prove Theorem 2.2.4 by putting, for every function f of $\mathscr{C}(T)$,

$$Hf(t) = \left[\exp\left(-\frac{BT}{\varepsilon} \right) - I \right]^{-1} \frac{1}{\varepsilon} \int_t^{t+T} \exp\left[\frac{B(t-u)}{\varepsilon} \right] g[u, f(u), \varepsilon] \, du$$

(a) First assume that

$$B = \begin{pmatrix} \sigma_1^{\,2} & 0 \\ 0 & \sigma_2^{\,2} \end{pmatrix}$$

where $\sigma_1 \neq \sigma_2$.

(b) Pass to the general case.

12. Consider Duffing's equation with forced oscillation, studied above. Find the first coordinate $y_2(t)$ of the approximation $f_2(t) = Hf_1(t)$.

13. Let $A(t) \in \mathscr{L}(\mathbb{R}^d, \mathbb{R}^d)$ be continuous with period T and such that $A(t) = A(-t)$. By examining the properties of the equation $dx/dt = \varepsilon A(t)x$, show that all the characteristic multipliers of A are equal to 1.

14. (*Hale*) Consider the following equation in $\mathbb{R}$:

$$\frac{d^3 x}{dt^3} + \omega^2 \frac{dx}{dt} = \varepsilon f\left(x, \frac{dx}{dt}, \frac{d^2 x}{dt^2} \right)$$

We assume that

$$f\left(x, -\frac{dx}{dt}, \frac{d^2 x}{dt^2} \right) = -f\left(x, \frac{dx}{dt}, \frac{d^2 x}{dt^2} \right)$$

Let $dX/dt = A(\omega^2)X + \varepsilon f(t, X)$ be the corresponding system in $\mathbb{R}^3$. Putting

$$v^2 = \omega^2 + \varepsilon\lambda \quad \text{and} \quad X = \exp[A(v^2)t] \, Y$$

study the equation relating to Y, according to the values of the parameter λ.

15. Study the autonomous Duffing equation using the Bogolyubov change of variables.

16. Give in detail the proof of Theorem 2.4.1.

17. Study the Duffing equation with damping,

$$\frac{d^2 x}{dt^2} + \varepsilon\delta \frac{dx}{dt} + \omega_0^{\,2}(x + \varepsilon\gamma x^3) = \varepsilon A \cos \omega t$$

putting $\omega^2 = \omega_0^{\,2} + \lambda\varepsilon$.

18. Let

$$\frac{d^2 x}{dt^2} + x = \varepsilon(1 - x^2) \frac{dx}{dt} + \varepsilon A \cos \omega t$$

be an equation where $\varepsilon > 0$ and where $\omega^2 - 1 = \lambda\varepsilon$. Put the system into standard form and then write the first approximation to the bifurcation equation relative to (a, b), and to study this put $a = r \cos \vartheta$, $b = r \sin \vartheta$.

19. Let

$$\frac{d^2 f}{dt^2} + \varepsilon \beta \frac{df}{dt} + \frac{f}{9} + \varepsilon \gamma f^3 = A \cos \omega t \tag{93}$$

be a Duffing equation with periodic excitation and friction. By putting

$$f(t) = r \sin\left(\frac{\omega}{3} t + \varphi\right) - \frac{A}{\omega^2 - \frac{1}{9}} \cos \omega t$$

$$\frac{df}{dt} = r \frac{\omega}{3} \cos\left(\frac{\omega}{3} t + \varphi\right) + \frac{A\omega}{\omega^2 - \frac{1}{9}} \sin \omega t$$

transform (93) into a critical system

$$\frac{dr}{dt} = \varepsilon g_1(t, r, \varphi, \lambda)$$

$$\frac{d\varphi}{dt} = \varepsilon g_2(t, r, \varphi, \lambda)$$

where $\varepsilon = (\omega^2 - 1)/\lambda$. Then, using the method of the mean, study the harmonics of order 3 of (93).

20. Let

$$\frac{d^2 x}{dt^2} - \varepsilon(1 - x^2) \frac{dx}{dt} + x = 2 \sin \frac{t}{4} \tag{94}$$

be a Van der Pol equation with a non-zero right-hand side. By putting

$$x(t) = A(t) \cos t + B(t) \sin t + \frac{32}{15} \sin \frac{t}{4}$$

$$\frac{dx}{dt}(t) = -A(t) \sin t + B(t) \cos t + \frac{8}{15} \cos \frac{t}{4}$$

transform (94) into a critical system

$$\frac{dA}{dt} = \varepsilon f_1(t, A, B)$$

$$\frac{dB}{dt} = \varepsilon f_2(t, A, B)$$

Apply the method of the mean to study the existence of almost-periodic solutions, which for $\varepsilon = 0$ are of the form

$$\alpha \sin t + \beta \sin \frac{t}{4}$$

21. Let

$$\frac{d^2\vartheta}{dt^2} + k\frac{d\vartheta}{dt} + \lambda \sin \vartheta = 0$$

be the equation of a pendulum with friction ($k \neq 0$). Prove that no periodic solution exists.

22. Establish that the periodic solution of Van der Pol's equation

$$\frac{d^2x}{dt^2} - k^2(1 - x^2)\frac{dx}{dt} + x = 0$$

where $k \neq 0$, is orbitally asymptotically stable. For this the result of Theorem 3.2.12 in Chapter 2 concerning this equation can be used.

Bibliography

ARNOLD (1974): *Equations Différentielles Ordinaires*, Mir, Moscow. Supplementary chapters, Mir, Moscow, 1980.

BOGOLYUBOV and MIROPOLSKI (1962): *Méthodes Asymptotiques en Théorie des Oscillations Non-linéaires*, Gauthier-Villars, Paris.

CESARI (1971): *Asymptotic Behaviour and Stability Problems in Ordinary Differential Equations*, 3rd edn, Springer, Berlin.

CODDINGTON and LEVINSON (1955): *Theory of Ordinary Differential Equations*, McGraw-Hill, New York.

HALE (1969): *Ordinary Differential Equations*, Wiley-Interscience, New York.

HARTMANN (1964): *Ordinary Differential Equations*, Wiley-Interscience, New York.

HOCHSTADT (1970): *Integral Equations*, Wiley-Interscience, New York.

LEE and MARKUS (1967): *Foundations of Optimal Control Theory*, Wiley-Interscience, New York.

ROSEAU (1966): *Vibrations Non Linéaires*, Springer, Berlin.

ROUCHE, HABETS and LALOY (1977): *Stability Theory by Liapunov's Direct Method*, Springer, Berlin.

SCHWARTZ (1979): *Analyse Hilbertienne*, Hermann, Paris.

WASOV (1967): *Asymptotic Expansions for Ordinary Differential Equations*, Springer, Berlin.

YOSHIZAWA (1975): *Stability Theory and Existence of Periodic and Almost-periodic Solutions of Ordinary Differential Equations*, Springer, Berlin.

YOSIDA (1960): *Lectures on Differential and Integral Equations*, Wiley-Interscience, New York.

Index